Practical Data Science

A Guide to Building the Technology Stack for Turning Data Lakes into Business Assets

Andreas François Vermeulen

Apress®

Practical Data Science: A Guide to Building the Technology Stack for Turning Data Lakes into Business Assets

Andreas François Vermeulen
West Kilbride North Ayrshire, United Kingdom

ISBN-13 (pbk): 978-1-4842-3053-4
https://doi.org/10.1007/978-1-4842-3054-1

ISBN-13 (electronic): 978-1-4842-3054-1

Library of Congress Control Number: 2018934681

Managing Director, Apress Media LLC: Welmoed Spahr
Acquisitions Editor: Susan McDermott
Development Editor: Laura Berendson
Coordinating Editor: Rita Fernando

Cover designed by eStudioCalamar

Cover image designed by Freepik (www.freepik.com)

Distributed to the book trade worldwide by Springer Science+Business Media New York, 233 Spring Street, 6th Floor, New York, NY 10013. Phone 1-800-SPRINGER, fax (201) 348-4505, e-mail orders-ny@springer-sbm.com, or visit www.springeronline.com. Apress Media, LLC is a California LLC and the sole member (owner) is Springer Science+Business Media Finance Inc (SSBM Finance Inc). SSBM Finance Inc is a Delaware corporation.

For information on translations, please e-mail rights@apress.com, or visit www.apress.com/rights-permissions.

Apress titles may be purchased in bulk for academic, corporate, or promotional use. eBook versions and licenses are also available for most titles. For more information, reference our Print and eBook Bulk Sales web page at www.apress.com/bulk-sales.

Any source code or other supplementary material referenced by the author in this book is available to readers on GitHub via the book's product page, located at www.apress.com/9781484230534. For more detailed information, please visit www.apress.com/source-code.

Printed on acid-free paper

Table of Contents

About the Author ....**xv**

About the Technical Reviewer ....**xvii**

Acknowledgments ....**xix**

Introduction ....**xxi**

Chapter 1: Data Science Technology Stack **1**

Rapid Information Factory Ecosystem 1

Data Science Storage Tools.. 2

Schema-on-Write and Schema-on-Read 2

Data Lake 4

Data Vault... 4

Hubs 5

Links 5

Satellites.. 5

Data Warehouse Bus Matrix... 6

Data Science Processing Tools 6

Spark... 6

Spark Core... 7

Spark SQL.. 7

Spark Streaming.. 7

MLlib Machine Learning Library 7

GraphX... 8

Mesos.. 9

Akka 9

Cassandra 9

Kafka.. 10

    Kafka Core 10

    Kafka Streams 10

    Kafka Connect 10

Elastic Search 11

R.. 11

Scala 12

Python 12

MQTT (MQ Telemetry Transport).. 13

What's Next? 13

Chapter 2: Vermeulen-Krennwallner-Hillman-Clark 15

Windows 15

Linux 15

It's Now Time to Meet Your Customer 16

    Vermeulen PLC 16

    Krennwallner AG 17

    Hillman Ltd 18

    Clark Ltd 18

Processing Ecosystem 19

    Scala.. 20

    Apache Spark 20

    Apache Mesos 21

    Akka 21

    Apache Cassandra.. 21

    Kafka 22

    Message Queue Telemetry Transport 22

Example Ecosystem 22

    Python 23

    Is Python3 Ready? 23

R 26

    Development Environment 27

    R Packages 28

Sample Data 30

    IP Addresses Data Sets 32

    Customer Data Sets 35

    Logistics Data Sets 35

Summary 38

Chapter 3: Layered Framework 39

Definition of Data Science Framework 40

Cross-Industry Standard Process for Data Mining (CRISP-DM) 40

    Business Understanding 41

    Data Understanding 42

    Data Preparation 42

    Modeling 42

    Evaluation 42

    Deployment 43

Homogeneous Ontology for Recursive Uniform Schema 43

The Top Layers of a Layered Framework 44

    The Basics for Business Layer 45

    The Basics for Utility Layer 46

    The Basics for Operational Management Layer 47

    The Basics for Audit, Balance, and Control Layer 48

    The Basics for Functional Layer 49

Layered Framework for High-Level Data Science and Engineering 50

    Windows 51

    Linux 51

Summary 51

Chapter 4: Business Layer **53**

Business Layer 53

    The Functional Requirements 54

    The Nonfunctional Requirements 63

    Common Pitfalls with Requirements 78

Engineering a Practical Business Layer 81

    Requirements 81

    Requirements Registry 81

    Traceability Matrix 82

Summary 83

Chapter 5: Utility Layer **85**

Basic Utility Design 87

    Data Processing Utilities 89

    Maintenance Utilities 112

    Processing Utilities 114

Engineering a Practical Utility Layer 115

    Maintenance Utility 115

    Data Utility 116

    Processing Utility 116

Summary 117

Chapter 6: Three Management Layers **119**

Operational Management Layer 119

    Processing-Stream Definition and Management 119

    Parameters 120

    Scheduling 121

    Monitoring 123

    Communication 124

    Alerting 124

Audit, Balance, and Control Layer 125

    Audit 125

    Process Tracking 130

Data Provenance 130

Data Lineage... 130

Balance 131

Control.. 131

Yoke Solution 132

Producer 132

Consumer 133

Directed Acyclic Graph Scheduling... 134

Yoke Example 134

Cause-and-Effect Analysis System 140

Functional Layer... 141

Data Science Process 141

Start with a What-if Question 141

Take a Guess at a Potential Pattern... 141

Gather Observations and Use Them to Produce a Hypothesis........... 142

Use Real-World Evidence to Verify the Hypothesis....................... 142

Collaborate Promptly and Regularly with Customers and Subject Matter Experts As You Gain Insights 142

Summary.. 144

Chapter 7: Retrieve Superstep... 147

Data Lakes 148

Data Swamps... 149

1. Start with Concrete Business Questions 149

2. Data Governance 150

3. Data Quality 172

4. Audit and Version Management.. 172

Training the Trainer Model 172

Understanding the Business Dynamics of the Data Lake 173

R Retrieve Solution 173

Vermeulen PLC 173

Krennwallner AG 186

Hillman Ltd 188

Clark Ltd 194

Actionable Business Knowledge from Data Lakes 202

Engineering a Practical Retrieve Superstep 202

Vermeulen PLC 203

Krennwallner AG 209

Hillman Ltd 222

Clark Ltd 259

Connecting to Other Data Sources 261

SQLite 261

Date and Time 264

Other Databases 266

PostgreSQL 267

Microsoft SQL Server 267

MySQL 267

Oracle 268

Microsoft Excel 268

Apache Spark 269

Apache Cassandra 269

Apache Hive 270

Apache Hadoop 270

Amazon S3 Storage 271

Amazon Redshift 271

Amazon Web Services 272

Summary 273

Chapter 8: Assess Superstep 275

Assess Superstep 275

Errors 276

Accept the Error 276

Reject the Error 276

Correct the Error 277

Create a Default Value 277

Analysis of Data 277

Completeness 277

Uniqueness 278

Timeliness 278

Validity 278

Accuracy 278

Consistency 279

Practical Actions 279

Missing Values in Pandas 279

Engineering a Practical Assess Superstep 296

Vermeulen PLC 296

Krennwallner AG 339

Hillman Ltd 356

Clark Ltd 406

Summary 420

Chapter 9: Process Superstep **421**

Data Vault 422

Hubs 422

Links 422

Satellites 422

Reference Satellites 423

Time-Person-Object-Location-Event Data Vault 423

Time Section 424

Person Section 427

Object Section 430

Location Section 433

Event Section 436

Engineering a Practical Process Superstep 439

Time 439

Person 463

Object 476

Location 499

Event 507

Data Science Process 510

Roots of Data Science 510

Monte Carlo Simulation 515

Causal Loop Diagrams 515

Pareto Chart 517

Correlation Analysis 519

Forecasting 519

Data Science 524

Summary 525

Chapter 10: Transform Superstep 527

Transform Superstep 527

Dimension Consolidation 528

Sun Model 534

Building a Data Warehouse 542

Transforming with Data Science 547

Steps of Data Exploration and Preparation 547

Missing Value Treatment 547

Common Feature Extraction Techniques 556

Hypothesis Testing 562

T-Test 562

Chi-Square Test 565

Overfitting and Underfitting 566

Polynomial Features 567

Common Data-Fitting Issue 568

Precision-Recall 570

Precision-Recall Curve 570

Sensitivity and Specificity 571

F1-Measure 571

Receiver Operating Characteristic (ROC) Analysis Curves.................................... 577

Cross-Validation Test.. 579

Univariate Analysis.. 580

Bivariate Analysis.. 581

Multivariate Analysis... 581

Linear Regression 581

Simple Linear Regression.. 582

RANSAC Linear Regression 587

Hough Transform 589

Logistic Regression... 590

Simple Logistic Regression 590

Multinomial Logistic Regression 592

Ordinal Logistic Regression... 594

Clustering Techniques... 601

Hierarchical Clustering 601

Partitional Clustering 608

ANOVA 613

Principal Component Analysis (PCA) 615

Factor Analysis 615

Conjoint Analysis 619

Decision Trees... 624

Support Vector Machines, Networks, Clusters, and Grids 629

Support Vector Machines 629

Support Vector Networks.. 632

Support Vector Clustering.. 632

Data Mining... 633

Association Patterns.. 633

Classification Patterns... 635

Bayesian Classification.. 638

Sequence or Path Analysis 640

Forecasting 646

Pattern Recognition 646

Machine Learning 649

Supervised Learning 649

Unsupervised Learning 651

Reinforcement Learning 651

Bagging Data 652

Random Forests 655

Computer Vision (CV) 660

Natural Language Processing (NLP) 661

Text-Based 661

Speech-Based 662

Neural Networks 662

Gradient Descent 664

Regularization Strength 665

Simple Neural Network 665

TensorFlow 670

Basic TensorFlow 671

Real-World Uses of TensorFlow 676

The One-Arm Bandits (Contextual Version) 676

Processing the Digits on a Container 679

Summary 684

Chapter 11: Organize and Report Supersteps 685

Organize Superstep 685

Horizontal Style 686

Vertical Style 690

Island Style 693

Secure Vault Style 696

Association Rule Mining 699

Engineering a Practical Organize Superstep. 703

Report Superstep 718

    Summary of the Results 719

    Engineering a Practical Report Superstep... 722

Graphics 741

    Plot Options 742

    More Complex Graphs 752

    Summary 775

Pictures 776

    Channels of Images 776

    Cutting the Edge 778

    One Size Does Not Fit All 781

Showing the Difference.. 783

Summary.. 786

Closing Words 786

Index.. 787

About the Author

 **Andreas François Vermeulen** is a consulting manager for decision science, data science, data engineering, machine learning, robotics, artificial intelligence, computational analytics and business intelligence at Sopra-Steria and a doctoral researcher at the University of St Andrews, Scotland, on future concepts in massive distributed computing, mechatronics, big data, business intelligence, data science, data engineering, and deep learning. He owns and incubates the Rapid Information Factory data processing framework. He is active in developing next-generation processing frameworks and mechatronics engineering, with more than 37 years of international experience in data processing, software development, and system architecture. Andreas is a data scientist, doctoral trainer, corporate consultant, principal systems architect, and speaker/author/columnist on data science, distributed computing, big data, business intelligence, deep learning, and constraint programming. Andreas holds a bachelor's degree from North-West University, Potchefstroom, South Africa; a master of business administration degree from the University of Manchester, England; a master of business intelligence and data science degree from the University of Dundee, Scotland; and Ph.D. from the University of St Andrews.

About the Technical Reviewer

Chris Hillman is a principal data scientist working as part of an international team. With more than 20 years of experience in the analytics industry, Chris has works in various sectors, including life sciences, manufacturing, retail, and telecommunication. Using the latest technology, he specializes in producing actionable insights from large-scale analytical problems on parallel clusters. He has presented at conferences such as Strata, Hadoop world, and the IEEE big data streaming special interest group. Chris is currently studying for a Ph.D. in data science at the University of Dundee, applying big data analytics to the data produced from experimentation into the human proteome, and has published several research papers.

About the Technical Reviewer

Acknowledgments

To Denise: I am fortunate enough to have created a way of life I love . . . But you have given me the courage and determination to live it! Thanks for the time and patience to complete the book and numerous other mad projects.

To Laurence: Thank you for all the knowledge shared on accounting and finance.

To Chris: thank you. Your wisdom and insight made this great! Best of luck with your future.

To the staff at Apress: your skills transformed an idea into a book. Well done!

Introduction

People are talking about data lakes daily now. I consult on a regular basis with organizations on how to develop their data lake and data science strategy to serve their evolving and ever-changing business strategies. This requirement for agile and cost-effective information management is high on the priority list of senior managers worldwide.

It is a fact that many of the unknown insights are captured and stored in a massive pool of unprocessed data in the enterprise. These data lakes have major implications for the future of the business world. It is projected that combined data scientists worldwide will have to handle 40 zettabytes of data by 2020, an increase of 300 times since 2005.

There are numerous data sources that still must be converted into actionable business knowledge. This achievement will safeguard the future of the business that can achieve it.

The world's data producers are generating two-and-a-half quintillion bytes of new data every day. The addition of internet of things will cause this volume to be substantially higher. Data scientists and engineers are falling behind on an immense responsibility.

By reading this introduction, you are already an innovative person who wants to understand this advanced data structure that one and all now desire to tame.

To tame your data lake, you will require practical data science.

I propose to teach you how to tame this beast. I am familiar with the skills it takes to achieve this goal. I will guide you with the sole purpose of you learning and expanding while mastering the practical guidance in this book.

I will chaper one you from the data lake to the final visualization and storytelling.

You will understand what is in your business's data lake and how to apply data science to it.

Think of the process as comparable to a natural lake. It is vital to establish a sequence of proficient techniques with the lake, to obtain pure water in your glass.

Do not stress, as by the end of this book, you will have shared in more than 37 years of working experience with data and extracting actionable business knowledge. I will share with you the experience I gained in working with data on an international scale.

You will be offered a processing framework that I use on a regular basis to tame data lakes and the collection of monsters that live in and around those lakes.

I have included examples at the end of each chapter, along with code, that more serious data scientists can use as you progress through the book. Note, however, that it is not required for you to complete the examples in order to understand the concepts in each chapter.

So, welcome to a walk-through of a characteristic data lake project, using practical data science techniques and insights.

The objective of the rest of this introduction is to explain the fundamentals of data science.

Data Science

In 1960, Peter Naur started using the term *data science* as a substitute for *computer science*. He stated that to work with data, you require more than just computer science. I agree with his declaration.

Data science is an interdisciplinary science that incorporates practices and methods with actionable knowledge and insights from data in heterogeneous schemas (structured, semi-structured, or unstructured). It amalgamates the scientific fields of data exploration with thought-provoking research fields such as data engineering, information science, computer science, statistics, artificial intelligence, machine learning, data mining, and predictive analytics.

For my part, as I enthusiastically research the future use of data science, by translating multiple data lakes, I have gained several valuable insights. I will explain these with end-to-end examples and share my insights on data lakes. This book explains vital elements from these sciences that you will use to process your data lake into actionable knowledge. I will guide you through a series of recognized science procedures for data lakes. These core skills are a key set of assets to perfect as you begin your encounters with data science.

Data Analytics

Data analytics is the science of fact-finding analysis of raw data, with the goal of drawing conclusions from the data lake. Data analytics is driven by certified algorithms to statistically define associations between data that produce insights.

The perception of certified algorithms is exceptionally significant when you want to convince other business people of the importance of the data insights you have gleaned.

Note You should not be surprised if you are regularly asked the following: Substantiate it! How do you know it is correct?

The best answer is to point to a certified and recognized algorithm that you have used. Associate the algorithm to your business terminology to achieve success with your projects.

Machine Learning

The business world is buzzing with activities and ideas about machine learning and its application to numerous business environments. Machine learning is the capability of systems to learn without explicit software development. It evolved from the study of pattern recognition and computational learning theory.

The impact is that, with the appropriate processing and skills, you can augment your own data capabilities. Training enables a processing environment to complete several magnitudes of discoveries in the time it takes to have a cup of coffee.

Note Work smarter, not harder! Offload your data science to machines. They are faster and more consistent in processing your data lakes.

This skill is an essential part of achieving major gains in shortening the data-to-knowledge cycle. This book will cover the essential practical ground rules in later chapters.

Data Mining

Data mining is processing data to isolate patterns and establish relationships between data entities within the data lake. For data mining to be successful, there is a small number of critical data-mining theories that you must know about data patterns.

In later chapters, I will expand on how you can mine your data for insights. This will help you to discover new actionable knowledge.

Statistics

Statistics is the study of the collection, analysis, interpretation, presentation, and organization of data. Statistics deals with all aspects of data, including the planning of data collection, in terms of the design of surveys and experiments.

Data science and statistics are closely related. I will show you how to run through series of statistics models covering data collection, population, and samples to enhance your data science deliveries.

This book devotes later chapters to how you amalgamate these into an effective and efficient process.

Algorithms

An algorithm is a self-contained step-by-step set of processes to achieve a specific outcome. Algorithms execute calculations, data processing, or automated reasoning tasks with repeatable outcomes.

Algorithms are the backbone of the data science process. You should assemble a series of methods and procedures that will ease the complexity and processing of your specific data lake.

I will discuss numerous algorithms and good practices for performing practical data science throughout the book.

Data Visualization

Data visualization is your key communication channel with the business. It consists of the creation and study of the visual representation of business insights. Data science's principal deliverable is visualization. You will have to take your highly technical results and transform them into a format that you can show to non-specialists.

The successful transformation of data results to actionable knowledge is a skill set I will cover in detail in later chapters. If you master the visualization skill, you will be most successful in data science.

Storytelling

Data storytelling is the process of translating data analyses into layperson's terms, in order to influence a business decision or action. You can have the finest data science, but without the business story to translate your findings into business-relevant actions, you will not succeed.

I will provide details and practical insights into what to check for to ensure that you have the proper story and actions.

What Next?

I will demonstrate, using the core knowledge of the underlining science, how you can make a competent start to handle the transformation process of your data lake into actionable knowledge. The sole requirement is to understand the data science of your own data lake. Start rapidly to discover what data science reveals about your business. You are the master of your own data lake.

You will have to build familiarity with the data lake and what is flowing into the structure. My advice is to apply the data science on smaller scale activities, for insights from the data lake.

Note   Experiment—push the boundaries of your insights.

CHAPTER 1

Data Science Technology Stack

The Data Science Technology Stack covers the data processing requirements in the Rapid Information Factory ecosystem. Throughout the book, I will discuss the stack as the guiding pattern.

In this chapter, I will help you to recognize the basics of data science tools and their influence on modern data lake development. You will discover the techniques for transforming a data vault into a data warehouse bus matrix. I will explain the use of Spark, Mesos, Akka, Cassandra, and Kafka, to tame your data science requirements.

I will guide you in the use of elastic search and MQTT (MQ Telemetry Transport), to enhance your data science solutions. I will help you to recognize the influence of R as a creative visualization solution. I will also introduce the impact and influence on the data science ecosystem of such programming languages as R, Python, and Scala.

Rapid Information Factory Ecosystem

The Rapid Information Factory ecosystem is a convention of techniques I use for my individual processing developments. The processing route of the book will be formulated on this basis, but you are not bound to use it exclusively. The tools I discuss in this chapter are available to you without constraint. The tools can be used in any configuration or permutation that is suitable to your specific ecosystem.

I recommend that you begin to formulate an ecosystem of your own or simply adopt mine. As a prerequisite, you must become accustomed to a set of tools you know well and can deploy proficiently.

1

© Andreas François Vermeulen 2018
A. F. Vermeulen, *Practical Data Science*, https://doi.org/10.1007/978-1-4842-3054-1_1

Note   Remember: Your data lake will have its own properties and features, so adopt your tools to those particular characteristics.

Data Science Storage Tools

This data science ecosystem has a series of tools that you use to build your solutions. This environment is undergoing a rapid advancement in capabilities, and new developments are occurring every day.

I will explain the tools I use in my daily work to perform practical data science. Next, I will discuss the following basic data methodologies.

Schema-on-Write and Schema-on-Read

There are two basic methodologies that are supported by the data processing tools. Following is a brief outline of each methodology and its advantages and drawbacks.

Schema-on-Write Ecosystems

A traditional relational database management system (RDBMS) requires a schema before you can load the data. To retrieve data from my structured data schemas, you may have been running standard SQL queries for a number of years.

Benefits include the following:

- In traditional data ecosystems, tools assume schemas and can only work once the schema is described, so there is only one view on the data.

- The approach is extremely valuable in articulating relationships between data points, so there are already relationships configured.

- It is an efficient way to store "dense" data.

- All the data is in the same data store.

On the other hand, schema-on-write isn't the answer to every data science problem. Among the downsides of this approach are that

- Its schemas are typically purpose-built, which makes them hard to change and maintain.

- It generally loses the raw/atomic data as a source for future analysis.

- It requires considerable modeling/implementation effort before being able to work with the data.

- If a specific type of data can't be stored in the schema, you can't effectively process it from the schema.

At present, schema-on-write is a widely adopted methodology to store data.

Schema-on-Read Ecosystems

This alternative data storage methodology does not require a schema before you can load the data. Fundamentally, you store the data with minimum structure. The essential schema is applied during the query phase.

Benefits include the following:

- It provides flexibility to store unstructured, semi-structured, and disorganized data.

- It allows for unlimited flexibility when querying data from the structure.

- Leaf-level data is kept intact and untransformed for reference and use for the future.

- The methodology encourages experimentation and exploration.

- It increases the speed of generating fresh actionable knowledge.

- It reduces the cycle time between data generation to availability of actionable knowledge.

Schema-on-read methodology is expanded on in Chapter 6.

I recommend a hybrid between schema-on-read and schema-on-write ecosystems for effective data science and engineering. I will discuss in detail why this specific ecosystem is the optimal solution when I cover the functional layer's purpose in data science processing.

Data Lake

A data lake is a storage repository for a massive amount of raw data. It stores data in native format, in anticipation of future requirements. You will acquire insights from this book on why this is extremely important for practical data science and engineering solutions. While a schema-on-write data warehouse stores data in predefined databases, tables, and records structures, a data lake uses a less restricted schema-on-read-based architecture to store data. Each data element in the data lake is assigned a distinctive identifier and tagged with a set of comprehensive metadata tags.

A data lake is typically deployed using distributed data object storage, to enable the schema-on-read structure. This means that business analytics and data mining tools access the data without a complex schema. Using a schema-on-read methodology enables you to load your data as is and start to get value from it instantaneously.

I will discuss and provide more details on the reasons for using a schema-on-read storage methodology in Chapters 6–11.

For deployment onto the cloud, it is a cost-effective solution to use Amazon's Simple Storage Service (Amazon S3) to store the base data for the data lake. I will demonstrate the feasibility of using cloud technologies to provision your data science work. It is, however, not necessary to access the cloud to follow the examples in this book, as they can easily be processed using a laptop.

Data Vault

Data vault modeling, designed by Dan Linstedt, is a database modeling method that is intentionally structured to be in control of long-term historical storage of data from multiple operational systems. The data vaulting processes transform the schema-on-read data lake into a schema-on-write data vault. The data vault is designed into the schema-on-read query request and then executed against the data lake.

I have also seen the results stored in a schema-on-write format, to persist the results for future queries. The techniques for both methods are discussed in Chapter 9. At this point, I expect you to understand only the rudimentary structures required to formulate a data vault.

The structure is built from three basic data structures: hubs, inks, and satellites. Let's examine the specific data structures, to clarify why they are compulsory.

Hubs

Hubs contain a list of unique business keys with low propensity to change. They contain a surrogate key for each hub item and metadata classification of the origin of the business key.

The hub is the core backbone of your data vault, and in Chapter 9, I will discuss in more detail how and why you use this structure.

Links

Associations or transactions between business keys are modeled using link tables. These tables are essentially many-to-many join tables, with specific additional metadata.

The link is a singular relationship between hubs to ensure the business relationships are accurately recorded to complete the data model for the real-life business. In Chapter 9, I will explain how and why you would require specific relationships.

Satellites

Hubs and links form the structure of the model but store no chronological characteristics or descriptive characteristics of the data. These characteristics are stored in appropriated tables identified as satellites.

Satellites are the structures that store comprehensive levels of the information on business characteristics and are normally the largest volume of the complete data vault data structure. In Chapter 9, I will explain how and why these structures work so well to model real-life business characteristics.

The appropriate combination of hubs, links, and satellites helps the data scientist to construct and store prerequisite business relationships. This is a highly in-demand skill for a data modeler.

The transformation to this schema-on-write data structure is discussed in detail in Chapter 9, to point out why a particular structure supports the processing methodology. I will explain in that chapter why you require particular hubs, links, and satellites.

Data Warehouse Bus Matrix

The Enterprise Bus Matrix is a data warehouse planning tool and model created by Ralph Kimball and used by numerous people worldwide over the last 40+ years. The bus matrix and architecture builds upon the concept of conformed dimensions that are interlinked by facts.

The data warehouse is a major component of the solution required to transform data into actionable knowledge. This schema-on-write methodology supports business intelligence against the actionable knowledge. In Chapter 10, I provide more details on this data tool and give guidance on its use.

Data Science Processing Tools

Now that I have introduced data storage, the next step involves processing tools to transform your data lakes into data vaults and then into data warehouses. These tools are the workhorses of the data science and engineering ecosystem. Following are the recommended foundations for the data tools I use.

Spark

Apache Spark is an open source cluster computing framework. Originally developed at the AMP Lab of the University of California, Berkeley, the Spark code base was donated to the Apache Software Foundation, which now maintains it as an open source project. This tool is evolving at an incredible rate.

IBM is committing more than 3,500 developers and researchers to work on Spark-related projects and formed a dedicated Spark technology center in San Francisco to pursue Spark-based innovations.

SAP, Tableau, and Talend now support Spark as part of their core software stack. Cloudera, Hortonworks, and MapR distributions support Spark as a native interface.

Spark offers an interface for programming distributed clusters with implicit data parallelism and fault-tolerance. Spark is a technology that is becoming a de-facto standard for numerous enterprise-scale processing applications.

I discovered the following modules using this tool as part of my technology toolkit.

Spark Core

Spark Core is the foundation of the overall development. It provides distributed task dispatching, scheduling, and basic I/O functionalities.

This enables you to offload the comprehensive and complex running environment to the Spark Core. This safeguards that the tasks you submit are accomplished as anticipated. The distributed nature of the Spark ecosystem enables you to use the same processing request on a small Spark cluster, then on a cluster of thousands of nodes, without any code changes. In Chapter 10, I will discuss how you accomplish this.

Spark SQL

Spark SQL is a component on top of the Spark Core that presents a data abstraction called Data Frames. Spark SQL makes accessible a domain-specific language (DSL) to manipulate data frames. This feature of Spark enables ease of transition from your traditional SQL environments into the Spark environment. I have recognized its advantage when you want to enable legacy applications to offload the data from their traditional relational-only data storage to the data lake ecosystem.

Spark Streaming

Spark Streaming leverages Spark Core's fast scheduling capability to perform streaming analytics. Spark Streaming has built-in support to consume from Kafka, Flume, Twitter, ZeroMQ, Kinesis, and TCP/IP sockets. The process of streaming is the primary technique for importing data from the data source to the data lake.

Streaming is becoming the leading technique to load from multiple data sources. I have found that there are connectors available for many data sources. There is a major drive to build even more improvements on connectors, and this will improve the ecosystem even further in the future.

In Chapters 7 and 11, I will discuss the use of streaming technology to move data through the processing layers.

MLlib Machine Learning Library

Spark MLlib is a distributed machine learning framework used on top of the Spark Core by means of the distributed memory-based Spark architecture.

In Spark 2.0, a new library, `spark.mk`, was introduced to replace the RDD-based data processing with a DataFrame-based model. It is planned that by the introduction of Spark 3.0, only DataFrame-based models will exist.

Common machine learning and statistical algorithms have been implemented and are shipped with MLlib, which simplifies large-scale machine learning pipelines, including

- Dimensionality reduction techniques, such as singular value decomposition (SVD) and principal component analysis (PCA)

- Summary statistics, correlations, stratified sampling, hypothesis testing, and random data generation

- Collaborative filtering techniques, including alternating least squares (ALS)

- Classification and regression: support vector machines, logistic regression, linear regression, decision trees, and naive Bayes classification

- Cluster analysis methods, including k-means and latent Dirichlet allocation (LDA)

- Optimization algorithms, such as stochastic gradient descent and limited-memory BFGS (L-BFGS)

- Feature extraction and transformation functions

In Chapter 10, I will discuss the use of machine learning proficiency to support the automatic processing through the layers.

GraphX

GraphX is a powerful graph-processing application programming interface (API) for the Apache Spark analytics engine that can draw insights from large data sets. GraphX provides outstanding speed and capacity for running massively parallel and machine-learning algorithms.

The introduction of the graph-processing capability enables the processing of relationships between data entries with ease. In Chapters 9 and 10, I will discuss the use of a graph database to support the interactions of the processing through the layers.

Mesos

Apache Mesos is an open source cluster manager that was developed at the University of California, Berkeley. It delivers efficient resource isolation and sharing across distributed applications. The software enables resource sharing in a fine-grained manner, improving cluster utilization.

The Enterprise version of Mesos is Mesosphere Enterprise DC/OS. This runs containers elastically, and data services support Kafka, Cassandra, Spark, and Akka.

In microservices architecture, I aim to construct a service that spawns granularity, processing units and lightweight protocols through the layers. In Chapter 6, I will discuss the use of fine-grained microservices know-how to support data processing through the framework.

Akka

The toolkit and runtime methods shorten development of large-scale data-centric applications for processing. Akka is an actor-based message-driven runtime for running concurrency, elasticity, and resilience processes. The use of high-level abstractions such as actors, streams, and futures facilitates the data science and engineering granularity processing units.

The use of actors enables the data scientist to spawn a series of concurrent processes by using a simple processing model that employs a messaging technique and specific predefined actions/behaviors for each actor. This way, the actor can be controlled and limited to perform the intended tasks only. In Chapter 7-11, I will discuss the use of different fine-grained granularity processes to support data processing throughout the framework.

Cassandra

Apache Cassandra is a large-scale distributed database supporting multi–data center replication for availability, durability, and performance.

I use DataStax Enterprise (DSE) mainly to accelerate my own ability to deliver real-time value at epic scale, by providing a comprehensive and operationally simple data management layer with a unique always-on architecture built in Apache Cassandra. The standard Apache Cassandra open source version works just as well, minus some extra

9

management it does not offer as standard. I will just note that, for graph databases, as an alternative to GraphX, I am currently also using DataStax Enterprise Graph. In Chapter 7-11, I will discuss, the use of these large-scale distributed database models to process data through data science structures.

Kafka

This is a high-scale messaging backbone that enables communication between data processing entities. The Apache Kafka streaming platform, consisting of Kafka Core, Kafka Streams, and Kafka Connect, is the foundation of the Confluent Platform.

The Confluent Platform is the main commercial supporter for Kafka (see `www.confluent.io/`). Most of the Kafka projects I am involved with now use this platform. Kafka components empower the capture, transfer, processing, and storage of data streams in a distributed, fault-tolerant manner throughout an organization in real time.

Kafka Core

At the core of the Confluent Platform is Apache Kafka. Confluent extends that core to make configuring, deploying, and managing Kafka less complex.

Kafka Streams

Kafka Streams is an open source solution that you can integrate into your application to build and execute powerful stream-processing functions.

Kafka Connect

This ensures Confluent-tested and secure connectors for numerous standard data systems. Connectors make it quick and stress-free to start setting up consistent data pipelines. These connectors are completely integrated with the platform, via the schema registry.

Kafka Connect enables the data processing capabilities that accomplish the movement of data into the core of the data solution from the edge of the business ecosystem. In Chapter 7-11, I will discuss the use of this messaging pipeline to stream data through the configuration.

Elastic Search

Elastic search is a distributed, open source search and analytics engine designed for horizontal scalability, reliability, and stress-free management. It combines the speed of search with the power of analytics, via a sophisticated, developer-friendly query language covering structured, unstructured, and time-series data. In Chapter 11, I will discuss, the use of this elastic search to categorize data within the framework.

R

R is a programming language and software environment for statistical computing and graphics. The R language is widely used by data scientists, statisticians, data miners, and data engineers for developing statistical software and performing data analysis.

The capabilities of R are extended through user-created packages using specialized statistical techniques and graphical procedures. A core set of packages is contained within the core installation of R, with additional packages accessible from the Comprehensive R Archive Network (CRAN).

Knowledge of the following packages is a must:

- `sqldf` (data frames using SQL): This function reads a file into R while filtering data with an sql statement. Only the filtered part is processed by R, so files larger than those R can natively import can be used as data sources.

- `forecast` (forecasting of time series): This package provides forecasting functions for time series and linear models.

- `dplyr` (data aggregation): Tools for splitting, applying, and combining data within R

- `stringr` (string manipulation): Simple, consistent wrappers for common string operations

- RODBC, RSQLite, and RCassandra database connection packages: These are used to connect to databases, manipulate data outside R, and enable interaction with the source system.

- `lubridate` (time and date manipulation): Makes dealing with dates easier within R

11

- `ggplot2` (data visualization): Creates elegant data visualizations, using the grammar of graphics. This is a super-visualization capability.

- `reshape2` (data restructuring): Flexibly restructures and aggregates data, using just two functions: `melt` and `dcast` (or `acast`).

- `randomForest` (random forest predictive models): Leo Breiman and Adele Cutler's random forests for classification and regression

- `gbm` (generalized boosted regression models): Yoav Freund and Robert Schapire's AdaBoost algorithm and Jerome Friedman's gradient boosting machine

I will discuss each of these packages as I guide you through the book. In Chapter 6, I will discuss, the use of R to process the sample data within the sample framework. I will provide examples that demonstrate the basic ideas and engineering behind the framework and the tools.

Please note that there are many other packages in CRAN, which is growing on a daily basis. Investigating the different packages to improve your capabilities in the R environment is time well spent.

Scala

Scala is a general-purpose programming language. Scala supports functional programming and a strong static type system. Many high-performance data science frameworks are constructed using Scala, because of its amazing concurrency capabilities. Parallelizing masses of processing is a key requirement for large data sets from a data lake. Scala is emerging as the de-facto programming language used by data-processing tools. I provide guidance on how to use it, in the course of this book. Scala is also the native language for Spark, and it is useful to master this language.

Python

Python is a high-level, general-purpose programming language created by Guido van Rossum and released in 1991. It is important to note that it is an interpreted language: Python has a design philosophy that emphasizes code readability. Python uses a dynamic type system and automatic memory management and supports multiple programming paradigms (object-oriented, imperative, functional programming, and procedural).

Thanks to its worldwide success, it has a large and comprehensive standard library. The Python Package Index (PyPI) (`https://pypi.python.org/pypi`) supplies thousands of third-party modules ready for use for your data science projects. I provide guidance on how to use it, in the course of this book.

I suggest that you also install Anaconda. It is an open source distribution of Python that simplifies package management and deployment of features (see `www.continuum.io/downloads`).

MQTT (MQ Telemetry Transport)

MQTT stands for *MQ Telemetry Transport*. The protocol uses publish and subscribe, extremely simple and lightweight messaging protocols. It was intended for constrained devices and low-bandwidth, high-latency, or unreliable networks. This protocol is perfect for machine-to-machine- (M2M) or Internet-of-things-connected devices.

MQTT-enabled devices include handheld scanners, advertising boards, footfall counters, and other machines. In Chapter 7, I will discuss how and where you can use MQTT technology and how to make use of the essential benefits it generates. The apt use of this protocol is critical in the present and future data science environments. In Chapter 11, will discuss the use of MQTT for data collection and distribution back to the business.

What's Next?

As things change daily in the current ecosphere of ever-evolving and -increasing collections of tools and technological improvements to support data scientists, feel free to investigate technologies not included in the preceding lists. I have acquainted you with my toolbox. This Data Science Technology Stack has served me well, and I will show you how to use it to achieve success.

Note   My hard-earned advice is to practice with your tools. Make them your own! Spend time with them, cultivate your expertise.

## CHAPTER 2

Vermeulen-Krennwallner-Hillman-Clark

Let's begin by constructing a customer. I have created a fictional company for which you will perform the practical data science as your progress through this book. You can execute your examples in either a Windows or Linux environment. You only have to download the desired example set.

Any source code or other supplementary material referenced in this book is available to readers on GitHub, via this book's product page, located at `www.apress.com/9781484230534`.

Windows

I suggest that you create a directory called `c:\VKHCG` to process all the examples in this book. Next, from GitHub, download and unzip the `DS_VKHCG_Windows.zip` file into this directory.

Linux

I also suggest that you create a directory called `./VKHCG`, to process all the examples in this book. Then, from GitHub, download and untar the `DS_VKHCG_Linux.tar.gz` file into this directory.

Warning   If you change this directory to a new location, you will be required to change everything in the sample scripts to this new location, to get maximum benefit from the samples.

© Andreas François Vermeulen 2018
A. F. Vermeulen, *Practical Data Science*, https://doi.org/10.1007/978-1-4842-3054-1_2

These files are used to create the sample company's script and data directory, which I will use to guide you through the processes and examples in the rest of the book.

It's Now Time to Meet Your Customer

Vermeulen-Krennwallner-Hillman-Clark Group (VKHCG) is a hypothetical medium-size international company. It consists of four subcompanies: Vermeulen PLC, Krennwallner AG, Hillman Ltd, and Clark Ltd.

Vermeulen PLC

Vermeulen PLC is a data processing company that processes all the data within the group companies, as part of their responsibility to the group. The company handles all the information technology aspects of the business.

This is the company for which you have just been hired to be the data scientist. Best of luck with your future.

The company supplies

- Data science

- Networks, servers, and communication systems

- Internal and external web sites

- Data analysis business activities

- Decision science

- Process automation

- Management reporting

For the purposes of this book, I will explain what other technologies you need to investigate at every section of the framework, but the examples will concentrate only on specific concepts under discussion, as the overall data science field is more comprehensive than the few selected examples.

By way of examples, I will assist you in building a basic Data Science Technology Stack and then advise you further with additional discussions on how to get the stack to work at scale.

The examples will show you how to process the following business data:

- Customers

- Products

- Location

- Business processes

- A number of handy data science algorithms

I will explain how to

- Create a network routing diagram using geospatial analysis

- Build a directed acyclic graph (DAG) for the schedule of jobs, using graph theory

If you want to have a more detailed view of the company's data, take a browse at these data sets in the company's sample directory (./VKHCG/01-Vermeulen/00-RawData).

Later in this chapter, I will give you a more detailed walk-through of each data set.

Krennwallner AG

Krennwallner AG is an advertising and media company that prepares advertising and media content for the customers of the group.

It supplies

- Advertising on billboards

- Advertising and content management for online delivery

- Event management for key customers

Via a number of technologies, it records who watches what media streams. The specific requirement we will elaborate is how to identify the groups of customers who will have to see explicit media content. I will explain how to

- Pick content for specific billboards

- Understand online web site visitors' data per country

- Plan an event for top-10 customers at Neuschwanstein Castle

If you want to have a more in-depth view of the company's data, have a glance at the sample data sets in the company's sample directory (`./VKHCG/02-Krennwallner/00-RawData`).

Hillman Ltd

The Hillman company is a supply chain and logistics company. It provisions a worldwide supply chain solution to the businesses, including

- Third-party warehousing

- International shipping

- Door-to-door logistics

The principal requirement that I will expand on through examples is how you design the distribution of a customer's products purchased online.

Through the examples, I will follow the product from factory to warehouse and warehouse to customer's door.

I will explain how to

- Plan the locations of the warehouses within the United Kingdom

- Plan shipping rules for best-fit international logistics

- Choose what the best packing option is for shipping containers for a given set of products

- Create an optimal delivery route for a set of customers in Scotland

If you want to have a more detailed view of the company's data, browse the data sets in the company's sample directory (`./VKHCG/ 03-Hillman/00-RawData`).

Clark Ltd

The Clark company is a venture capitalist and accounting company that processes the following financial responsibilities of the group:

- Financial insights

- Venture capital management

- Investments planning

- Forex (foreign exchange) trading

I will use financial aspects of the group companies to explain how you apply practical data science and data engineering to common problems for the hypothetical financial data.

I will explain to you how to prepare

- A simple forex trading planner
- Accounting ratios
  - Profitability
  - Gross profit for sales
  - Gross profit after tax for sales
  - Return on capital employed (ROCE)
  - Asset turnover
  - Inventory turnover
  - Accounts receivable days
  - Accounts payable days

Processing Ecosystem

Five years ago, VKHCG consolidated its processing capability by transferring the concentrated processing requirements to Vermeulen PLC to perform data science as a group service. This resulted in the other group companies sustaining 20% of the group business activities; however, 90% of the data processing of the combined group's business activities was reassigned to the core team. Vermeulen has since consolidated Spark, Python, Mesos, Akka, Cassandra, Kafka, elastic search, and MQTT (MQ Telemetry Transport) processing into a group service provider and processing entity.

I will use R or Python for the data processing in the examples. I will also discuss the complementary technologies and advise you on what to consider and request for your own environment.

Note   The complementary technologies are used regularly in the data science environment. Although I cover them briefly, that does not make them any less significant.

VKHCG uses the R processing engine to perform data processing in 80% of the company business activities, and the other 20% is done by Python. Therefore, we will prepare an R and a Python environment to perform the examples. I will quickly advise you on how to obtain these additional environments, if you require them for your own specific business requirements.

I will cover briefly the technologies that we are **not** using in the examples but that are known to be beneficial.

Scala

Scala is popular in the data science community, as it supports massive parallel processing in an at-scale manner. You can install the language from the following core site: `www.scala-lang.org/download/`. Cheat sheets and references are available to guide you to resources to help you master this programming language.

Note   Many of my larger clients are using Scala as their strategical development language.

Apache Spark

Apache Spark is a fast and general engine for large-scale data processing that is at present the fastest-growing processing engine for large-scale data science projects. You can install the engine from the following core site: `http://spark.apache.org/`. For large-scale projects, I use the Spark environment within DataStax Enterprise (`www.datastax.com`), Hortonworks (`https://hortonworks.com/`), Cloudera (`www.cloudera.com/`), and MapR (`https://mapr.com/`).

Note   Spark is now the most sought-after common processing engine for at-scale data processing, with support increasing by the day. I recommend that you master this engine, if you want to advance your career in data science at-scale.

Apache Mesos

Apache Mesos abstracts CPU, memory, storage, and additional computation resources away from machines (physical or virtual), enabling fault-tolerant and elastic distributed systems to effortlessly build and run processing solutions effectively. It is industry proven to scale to 10,000s of nodes. This empowers the data scientist to run massive parallel analysis and processing in an efficient manner. The processing environment is available from the following core site: `http://mesos.apache.org/`. I want to give Mesosphere Enterprise DC/OS an honorable mention, as I use it for many projects. See `https://mesosphere.com`, for more details.

Note   Mesos is a cost-effective processing approach supporting growing dynamic processing requirements in an at-scale processing environment.

Akka

Akka supports building powerful concurrent and distributed applications to perform massive parallel processing, while sharing the common processing platform at-scale. You can install the engine from the following core site: `http://akka.io/`. I use Akka processing within the Mesosphere Enterprise DC/OS environment.

Apache Cassandra

Apache Cassandra database offers support with scalability and high availability, without compromising performance. It has linear scalability and a reputable fault-tolerance, as it is widely used by numerous big companies. You can install the engine from the following core site: `http://cassandra.apache.org/`.

I use Cassandra processing within the Mesosphere Enterprise DC/OS environment and DataStax Enterprise for my Cassandra installations.

Note   I recommend that you consider Cassandra as an at-scale database, as it supports the data science environment with stable data processing capability.

Kafka

Kafka is used for building real-time data pipelines and streaming apps. It is horizontally scalable, fault-tolerant, and impressively fast. You can install the engine from the following core site: `http://kafka.apache.org/`. I use Kafka processing within the Mesosphere Enterprise DC/OS environment, to handle the ingress of data into my data science environments.

Note   I advise that you look at Kafka as a data transport, as it supports the data science environment with robust data collection facility.

Message Queue Telemetry Transport

Message Queue Telemetry Transport (MQTT) is a machine-to-machine (M2M) and Internet of things connectivity protocol. It is an especially lightweight publish/subscribe messaging transport. It enables connections to locations where a small code footprint is essential, and lack of network bandwidth is a barrier to communication. See `http://mqtt.org/` for details.

Note   This protocol is common in sensor environments, as it provisions the smaller code footprint and lower bandwidths that sensors demand.

Now that I have covered the items you should know about but are not going to use in the examples, let's look at what you will use.

Example Ecosystem

The examples require the following environment. The two setups required within VKHCG's environment are Python and R.

Python

Python is a high-level programming language created by Guido van Rossum and first released in 1991. Its reputation is growing, as today, various training institutes are covering the language as part of their data science prospectus.

I suggest you install Anaconda, to enhance your Python development. It is an open source distribution of Python that simplifies package management and deployment of features (see `www.continuum.io/downloads`).

Ubuntu

I use an Ubuntu desktop and server installation to perform my data science (see `www.ubuntu.com/`), as follows:

```
sudo apt-get install python3 python3-pip python3-setuptools
```

CentOS/RHEL

If you want to use CentOS/RHEL, I suggest you employ the following install process:

```
sudo yum install python3 python3-pip python3-setuptools
```

Windows

If you want to use Windows, I suggest you employ the following install process.
Download the software from `www.python.org/downloads/windows/`.

Is Python3 Ready?

Once installation is completed, you must test your environment as follows:

```
Python3 --version
```

On success, you should see a response like this

```
Python 3.4.3+
```

Congratulations, Python is now ready.

Python Libraries

One of the most important features of Python is its libraries, which are extensively available and make it stress-free to include verified data science processes into your environment.

To investigate extra packages, I suggest you review the PyPI—Python Package Index (`https://pypi.python.org/`).

You have to set up a limited set of Python libraries to enable you to complete the examples.

Warning   Please ensure that you have verified all the packages you use. Remember: Open source is just that—open. Be vigilant!

Pandas

This provides a high-performance set of data structures and data-analysis tools for use in your data science.

Ubuntu

Install this by using

```
sudo apt-get install python-pandas
```

Centos/RHEL

Install this by using

```
yum install python-pandas
```

PIP

Install this by using

```
pip install pandas
```

More information on Pandas development is available at `http://pandas.pydata.org/`. I suggest following the cheat sheet (`https://github.com/pandas-dev/pandas/blob/master/doc/cheatsheet/Pandas_Cheat_Sheet.pdf`), to guide you through the basics of using Pandas.

I will explain, via examples, how to use these Pandas tools.

Note   I suggest that you master this package, as it will support many of your data loading and storing processes, enabling overall data science processing.

Matplotlib

Matplotlib is a Python 2D and 3D plotting library that can produce various plots, histograms, power spectra, bar charts, error charts, scatterplots, and limitless advance visualizations of your data science results.

Ubuntu

Install this by using

```
sudo apt-get install python-matplotlib
```

CentOS/RHEL

Install this by using

```
Sudo yum install python-matplotlib
```

PIP

Install this by using:

```
pip install matplotlib
```

Explore `http://matplotlib.org/` for more details on the visualizations that you can accomplish with exercises included in these packages.

Note   I recommend that you spend time mastering your visualization skills. Without these skills, it is nearly impossible to communicate your data science results.

NumPy

NumPy is the fundamental package for scientific computing, based on a general homogeneous multidimensional array structure with processing tools. Explore www.numpy.org/ for further details.

I will use some of the tools in the examples but suggest you practice with the general tools, to assist you with your future in data science.

SymPy

SymPy is a Python library for symbolic mathematics. It assists you in simplifying complex algebra formulas before including them in your code. Explore www.sympy.org for details on this package's capabilities.

Scikit-Learn

Scikit-Learn is an efficient set of tools for data mining and data analysis packages. It provides support for data science classification, regression, clustering, dimensionality reduction, and preprocessing for feature extraction and normalization. This tool supports both supervised learning and unsupervised learning processes.

I will use many of the processes from this package in the examples. Explore http://scikit-learn.org for more details on this wide-ranging package.

Congratulations. You are now ready to execute the Python examples. Now, I will guide you through the second setup for the R environment.

R

R is the core processing engine for statistical computing and graphics. Download the software from www.r-project.org/ and follow the installation guidance for the specific R installation you require.

Ubuntu

Install this by using

```
sudo apt-get install r-base
```

CentOS/RHEL

Install this by using

```
sudo yum install R
```

Windows

From `https://cran.r-project.org/bin/windows/base/`, install the software that matches your environment.

Development Environment

VKHCG uses the RStudio development environment for its data science and engineering within the group.

R Studio

RStudio produces a stress-free R ecosystem containing a code editor, debugging, and a visualization toolset. Download the relevant software from `www.rstudio.com/` and follow the installation guidance for the specific installation you require.

Ubuntu

Install this by using

```
wget https://download1.rstudio.org/rstudio-1.0.143-amd64.deb
sudo dpkg -i *.deb
rm *.deb
```

CentOS/RHEL

Install this by using

```
wget https://download1.rstudio.org/rstudio-1.0.143-x86_64.rpm
sudo yum install --nogpgcheck rstudio-1.0.143-x86_64.rpm
```

Windows

Install `https://download1.rstudio.org/RStudio-1.0.143.exe`.

R Packages

I suggest the following additional R packages to enhance the default R environment.

Data.Table Package

Data.Table enables you to work with data files more effectively. I suggest that you practice using Data.Table processing, to enable you to process data quickly in the R environment and empower you to handle data sets that are up to 100GB in size.

The documentation is available at `https://cran.r-project.org/web/packages/data.table/data.table.pdf`. See `https://CRAN.R-project.org/package=data.table` for up-to-date information on the package.

To install the package, I suggest that you open your RStudio IDE and use the following command:

```
install.packages ("data.table")
```

ReadR Package

The ReadR package enables the quick loading of text data into the R environment. The documentation is available at `https://cran.r-project.org/web/packages/readr/readr.pdf`. See `https://CRAN.R-project.org/package=readr` for up-to-date information on the package.

To install the package, I advise you to open your RStudio IDE and use the following command:

```
install.packages("readr")
```

I suggest that you practice by importing and exporting different formats of files, to understand the workings of this package and master the process. I also suggest that you investigate the following functions in depth in the ReadR package:

- `Spec_delim()`: Supports getting the specifications of the file without reading it into memory

- `read_delim()`: Supports reading of delimited files into the R environment

- `write_delim()`: Exports data from an R environment to a file on disk

JSONLite Package

This package enables you to process JSON files easily, as it is an optimized JSON parser and generator specifically for statistical data.

The documentation is at https://cran.r-project.org/web/packages/jsonlite/ jsonlite.pdf. See https://CRAN.R-project.org/package=jsonlite for up-to-date information on the package. To install the package, I suggest that you open your RStudio IDE and use the following command:

```
install.packages ("jsonlite")
```

I also suggest that you investigate the following functions in the package:

- fromJSON(): This enables you to import directly into the R environment from a JSON data source.

- prettify(): This improves the human readability by formatting the JSON, so that a human can read it easier.

- minify(): Removes all the JSON indentation/whitespace to make the JSON machine readable and optimized

- toJSON(): Converts R data into JSON formatted data

- read_json(): Reads JSON from a disk file

- write_json(): Writes JSON to a disk file

Ggplot2 Package

Visualization of data is a significant skill for the data scientist. This package supports you with an environment in which to build a complex graphic format for your data. It is so successful at the task of creating detailed graphics that it is called "The Grammar of Graphics." The documentation is located at https://cran.r-project.org/web/ packages/ ggplot2/ ggplot2.pdf. See https://CRAN.R-project.org/package= ggplot2 for up-to-date information on the package.

To install the package, I suggest that you to open your RStudio IDE and use the following command:

```
install.packages("ggplot2")
```

I recommend that you master this package to empower you to transform your data into a graphic you can use to demonstrate to your business the value of the results.

The packages we now have installed will support the examples.

Amalgamation of R with Spark

I want to discuss an additional package because I see its mastery as a major skill you will require to work with current and future data science. This package is interfacing the R environment with the distributed Spark environment and supplies an interface to Spark's built-in machine-learning algorithms. A number of my customers are using Spark as the standard interface to their data environments.

Understanding this collaboration empowers you to support the processing of at-scale environments, without major alterations in the R processing code. The documentation is at `https://cran.r-project.org/web/packages/sparklyr/` sparklyr.pdf. See `https://CRAN.R-project.org/package=sparklyr` for up-to-date information on the package.

To install the package, I suggest that you open your RStudio IDE and use the following command:

```
install.packages("sparklyr")
```

`sparklyr` is a direct R interface for Apache Spark to provide a complete `dplyr` back end. Once the filtering and aggregate of Spark data sets is completed downstream in the at-scale environment, the package imports the data into the R environment for analysis and visualization.

Sample Data

This book uses data for several examples. In the following section, I will explain how to use the VKHCG environment you installed to create the data sets that I will use in these examples.

Note: The processing of this sample data is spread out over the book. I am only giving you a quick introduction to the data. I will discuss each of the data sets in more detail once we start processing the data in later chapters. At this point, simply take note of the data locations and general formats.

This is the minimum data you will need to complete the examples.

Note   Please select a home directory for your examples:

If on Windows, I suggest C:/VKHCG.

If on Linux, I suggest $home/VKHCG.

Set up basic data sets processing in R by executing the following in your R environment:

```
###########################################################
rm(list=ls()) #will remove ALL objects
###########################################################
MY_INSTALL_DIR = "<selected home directory>"
###########################################################

if (file.exists (MY_INSTALL_DIR)==0) dir.create(MY_INSTALL_DIR)
subdirname = paste0(MY_INSTALL_DIR,  "/Vermeulen")
if (file.exists(subdirname)==0) dir.create(subdirname)
###########################################################
setwd(MY_INSTALL_DIR)
###########################################################
if (length(sessionInfo()$otherPkgs) > 0)
lapply(paste('package:',names(sessionInfo()$otherPkgs),sep=""),

detach,character.only=TRUE,unload=TRUE)
###########################################################
install.packages("readr")
###########################################################
install.packages("data.table")
###########################################################
```

Now, I will discuss the data sets you will use throughout the book.

Note   I am discussing only the descriptions of the data sources. It is not required that you load the data into R now. There will be sufficient time while processing the examples to load and process the data.

IP Addresses Data Sets

The network in VKHCG uses IP version 4 network addresses. The IPv4 protocol uses a structured addressing format of the following structure:

IP Address = w.x.y.z

The four sections can each hold the values 0 to 255. There are 2 to the power 32 IP addresses in the IPv4 protocol, so in universal terms, over 4 billion addresses are possible.

The following are the agreed formulas when dealing with IP4 addresses. Given an IP Address = w.x.y.z, the IP Number = $16777216 \cdot w + 65536 \cdot x + 256 \cdot y + z$. Given an IP Number, then:

- w = int (IP Number / 16777216) % 256

- x = int (IP Number / 65536) % 256

- y = int (IP Number / 256) % 256

- z = int (IP Number) % 256

That generates IP Address = w.x.y.z.

Addresses are classified as being of Class A, B, C, or D.

Class	1st Octet Decimal Range (w)
A	1–126*
B	128–191
C	192–223
D	224–239
E	240–254

Class A addresses 127.0.0.0 to 127.255.255.255 are reserved for loopback and diagnostic functions.

The following have been assigned as private addresses.

Class	Private Networks	Subnet Mask	Address Range
A	10.0.0.0	255.0.0.0	10.0.0.0–10.255.255.255
B	172.16.0.0–172.31.0.0	255.240.0.0	172.16.0.0–172.31.255.255
C	192.168.0.0	255.255.0.0	192.168.0.0–192.168.255.255

These addresses can be used by any company network within their internal network.

I have generated a series of IP addresses using the Class C address (192.168.0.1–192.168.0.255), i.e., 255 addresses that you will require for the examples.

The following data is for the examples:

In VKHCG\01-Vermeulen\00-RawData:

A Class C address block for internal network usage:

Data file: IP_DATA_C_VKHCG.csv

Type of file: Comma-separated values (CSV)

Amount of record: 255

Columns in data:

Column	Description
w	Integer between 0 and 255
x	Integer between 0 and 255
y	Integer between 0 and 255
z	Integer between 0 and 255
IP Address	text of format w.x.y.z
IP Number	Integer

Let's investigate the next data set.

In VKHCG\01-Vermeulen\00-RawData (this data set holds guidelines for which IP number is allocated to which location within the company's customer network):

Type of file: Comma-separated values (CSV)

Data file: IP_DATA_ALL.csv

Amount of records: 1,247,502

Columns in data:

Column	Description
ID	Text ID
Country	Country code
PlaceName	Name of location
PostCode	Post code of location
Latitude	Latitude of location
Longitude	Longitude of location
FirstIPNumber	First IP number in location
LastIPNumber	Last IP number in location

Let's investigate the next data set.

In VKHCG\01-Vermeulen\00-RawData (this data set holds which IP Number is assigned to which location within the company's own outside network):

Type of file: Comma-separated values (CSV)

Data file: IP_DATA_CORE.csv

Amount of records: 3,562

Columns in data:

Column	Description
ID	Text ID
Country	Country code
PlaceName	Name of location
PostCode	Post code of location
Latitude	Latitude of location
Longitude	Longitude of location
FirstIPNumber	First IP number in location
LastIPNumber	Last IP number in location

Customer Data Sets

VKHCG groups its customers onto billboards that it pays for on a per-billboard pricing model.

In VKHCG\ 02-Krennwallner\00-RawData (this data set holds the location of all the customer billboards):

> Type of File: comma-separated values (CSV)
>
> Data file: DE_Billboard_Locations.csv
>
> Amount of Records: 8,873
>
> Columns in Data:

Column	Description
ID	Text ID of billboard
Country	Country code of billboard
PlaceName	Name of billboard location
Latitude	Latitude of billboard location
Longitude	Longitude of billboard location

Logistics Data Sets

VKHCG has several warehouses and shops. I have grouped the locations of these buildings in three data sets.

Post Codes

In VKHCG\03-Hillman\00-RawData (data set one holds a complete United Kingdom post code list):

> Type of File: comma-separated values (CSV)
>
> Data file: GB_Postcode_Full.csv
>
> Amount of Records: 1,714,591

Columns in Data:

Column	Description
ID	Text ID for UK post code
Country	Country code
PostCode	UK post code
PlaceName	UK location name
Region	UK region
RegionCode	UK region code
AreaName	UK area name

Warehouse Data Set

In VKHCG\03-Hillman\00-RawData (data set two holds complete United Kingdom warehouse locations):

Type of file: comma-separated values (CSV)

Data file: GB_Postcode_Warehouse.csv

Amount of records: 3,005

Columns in data:

Column	Description
id	Text location ID
postcode	Post code for warehouse
latitude	Latitude for warehouse
longitude	Longitude for warehouse

Shop Data Set

In VKHCG\03-Hillman\00-RawData (data set three holds complete United Kingdom shop locations):

Type of file: Comma-separated values (CSV)

Data file: GB_Postcodes_Shops.csv

Amount of records: 1,048,575

Columns in data:

Column	Description
id	Text location ID
postcode	Post code for shop
latitude	Latitude for shop
longitude	Longitude for shop
FirstCode	Post code part 1 for shop
SecondCode	Post code part 2 for shop

Exchange Rate Data Set

In VKHCG\04-Clark\00-RawData (data set one holds exchange rates against the euro for a period 4,697 days):

Type of File: Comma-separated values (CSV) Pivot Table

Data file: Euro_ExchangeRates.csv

Amount of records: 4,697

Columns in data:

Column	Description
Date	Exchange rate date row
Exchange rate codes	41 columns with Exchange rate codes
Exchange rates	Grid of values of specific rates at the cross-section of the row and columns

Profit-and-Loss Statement Data Set

In VKHCG\04-Clark\00-RawData (data set two holds profit-and-loss statement results):

Type of file: Comma-separated values (CSV)

Data file: Profit_And_Loss.csv

Amount of records: 2,442

Columns in data:

Column	Description
QTR	Year/Quarter
TypeOfEntry	Profit and loss entry type
ProductClass1	Class entry one
ProductClass2	Class entry two
ProductClass3	Class entry three
Amount	Amount price
QTY	Quantity of items

Summary

I have now introduced you to the company, to enable you to complete the examples in the later chapters. Next, I will cover the layered framework, to introduce you to the basic framework for Practical Data Science.

CHAPTER 3

Layered Framework

In this chapter, I will introduce you to new concepts that enable us to share insights on a common understanding and terminology. I will define the Data Science Framework in detail, while introducing the Homogeneous Ontology for Recursive Uniform Schema (HORUS). I will take you on a high-level tour of the top layers of the framework, by explaining the fundamentals of the business layer, utility layer, operational management layer, plus audit, balance, and control layers.

I will discuss how to engineer a layered framework for improving the quality of data science when you are working in a large team in parallel with common business requirements.

Warning   If you are doing small-scale data science, you may be tempted to perform a quick analysis or pilot of the data at hand, without following the suggested framework. Please proceed with caution. I have spent countless hours restructuring small-scale data science pilots that cause problems when they are transferred to large-scale production ecosystems.

I am not suggesting that you build the complete set of layers only for smaller projects, only that you ensure that your project will effortlessly fit into the big processing world of production-size data science. You must understand that successful data science is based on the capability to go from pilot project to production at rapid speed.

Note   Do not be surprised, after you demonstrate your state-of-the-art data science discovery that saves the business millions, to be faced with demands to deploy within hours. The ability to perform this feat is a highly valued skill. The framework recommended will assist in accomplishing this with ease!

© Andreas François Vermeulen 2018
A. F. Vermeulen, *Practical Data Science*, https://doi.org/10.1007/978-1-4842-3054-1_3

Definition of Data Science Framework

Data science is a series of discoveries. You work toward an overall business strategy of converting raw unstructured data from your data lake into actionable business data. This process is a cycle of discovering and evolving your understanding of the data you are working with to supply you with metadata that you need.

My suggestion is that you build a basic framework that you use for your data processing. This will enable you to construct a data science solution and then easily transfer it to your data engineering environments.

I use the following framework for my projects. It works for projects from small departmental to at-scale internationally distributed deployments, as the framework has a series of layers that enables you to follow a logical building process and then use your data processing and discoveries across many projects.

Cross-Industry Standard Process for Data Mining (CRISP-DM)

CRISP-DM was generated in 1996, and by 1997, it was extended via a European Union project, under the ESPRIT funding initiative. It was the majority support base for data scientists until mid-2015. The web site that was driving the Special Interest Group disappeared on June 30, 2015, and has since reopened. Since then, however, it started losing ground against other custom modeling methodologies. The basic concept behind the process is still valid, but you will find that most companies do not use it as is, in any projects, and have some form of modification that they employ as an internal standard.

An overview of the basics is provided in Figure 3-1.

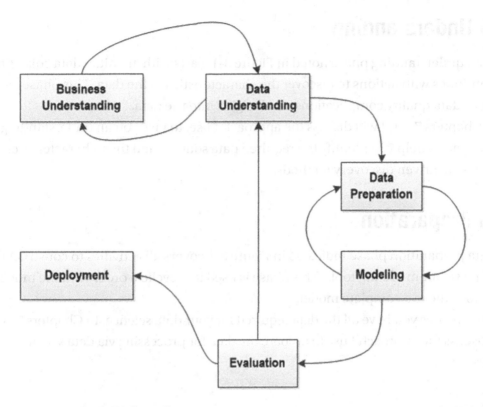

Figure 3-1. *CRISP-DM flowchart*

Business Understanding

This initial phase indicated in Figure 3-1 concentrates on discovery of the data science goals and requests from a business perspective. I will discuss the approach I use in Chapter 4.

Many businesses use a decision model or process mapping tool that is based on the Decision Model and Notation (DMN) standard. DMN is a standard published by the Object Management Group (see www.omg.org/spec/DMN/). The DMN standard offers businesses a modeling notation for describing decision management flows and business rules. I will not venture deeper into this notation in this book, but take note of it, as I have noted its use by my clients.

Data Understanding

The data understanding phase noted in Figure 3-1 starts with an initial data collection and continues with actions to discover the characteristics of the data. This phase identifies data quality complications and insights into the data.

In Chapters 7–11, I will discuss the approach I use, to give you an understanding of the data and to help you identify the required data sources and their characteristics via data science–driven discovery methods.

Data Preparation

The data preparation phase indicated in Figure 3-1 covers all activities to construct the final data set for modeling tools. This phase is used in a cyclical order with the modeling phase, to achieve a complete model.

This ensures you have all the data required for your data science. In Chapters 7 and 8, I will discuss the approach I use to prepare the data for processing via data science models.

Modeling

In this phase noted in Figure 3-1, different data science modeling techniques are nominated and evaluated for accomplishing the prerequisite outcomes, as per the business requirements. It returns to the data preparation phase in a cyclical order until the processes achieve success. In Chapters 9–11, I will discuss the approach I use to formulate and build the different models that apply to the data sets.

Evaluation

At this stage shown in Figure 3-1, the process should deliver high-quality data science. Before proceeding to final deployment of the data science, validation that the proposed data science solution achieves the business objectives is required. If this fails, the process returns to the data understanding phase, to improve the delivery.

Deployment

Creation of the data science is generally not the end of the project. Once the data science is past the development and pilot phases, it has to go to production. In the rest of the book, I will discuss how you successfully reach the end goal of delivering a practical data science solution to production. At this point, you should have a basic understanding of CRISP-DM, and we can now proceed to look at the process I have been using with success to create my data science projects. I will now proceed to the core framework I use.

Homogeneous Ontology for Recursive Uniform Schema

The Homogeneous Ontology for Recursive Uniform Schema (HORUS) is used as an internal data format structure that enables the framework to reduce the permutations of transformations required by the framework. The use of HORUS methodology results in a hub-and-spoke data transformation approach. External data formats are converted to HORUS format, and then a HORUS format is transformed into any other external format.

The basic concept is to take native raw data and then transform it first to a single format. That means that there is only one format for text files, one format for JSON or XML, one format for images and video. Therefore, to achieve any-to-any transformation coverage, the framework's only requirements are a data-format-to-HORUS and HURUS-to-data-format converter.

Table 3-1 demonstrates how by using of the hub-and-spoke methodology on text files to reduce the amount of convertor scripts you must achieve an effective data science solution. You can see that at only 100 text data sets, you can save 98% on a non-hub-and-spoke framework. I advise that you invest time in getting the hub-and-spoke framework to work in your environment, as the savings will reward anytime you invest. Similar deductions are feasible by streamlining other data types.

Table 3-1. *Advantages of Using Hub-and-Spoke Data Conversions*

Data Formats	Normal Convertors Required	HORUS Convertors Required (hub-and-spoke)	% Saved
2	4	4	0.00%
3	9	6	33.33%
4	16	8	50.00%
5	25	10	60.00%
6	36	12	66.67%
7	49	14	71.43%
8	64	16	75.00%
9	81	18	77.78%
10	100	20	80.00%
25	625	50	92.00%
50	2500	100	96.00%
100	10000	200	98.00%

Details of the requirements of HORUS will be covered in later chapters. At this point, simply take note that you will need some form of common data format internal to your data science process, to pass data from one format to another. I use HORUS principally for data processing and have had success with it for years.

The Top Layers of a Layered Framework

The top layers are to support a long-term strategy of creating a Center of Excellence for your data science work. The layers shown in Figure 3-1 enable you to keep track of all the processing and findings you achieve.

The use of a Center of Excellence framework is required, if you are working in a team environment. Using the framework will prepare you to work in an environment, with multiple data scientists working on a common ecosystem.

If you are just data wrangling a few data files, and your data science skills must yield quick insights, I would advise that you single out what you require from the overall framework. Remember: All big systems were a small pilot at some point. The framework will enable you to turn your small project into a big success, without having to do major restructuring along the route to production.

Let's discuss the framework (Figure 3-2).

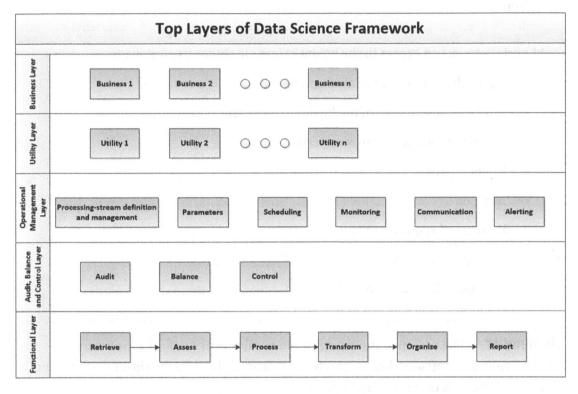

Figure 3-2. *Top layers of the data science framework*

So, now that you've seen them, let's look at these layers in further detail.

The Basics for Business Layer

The business layer indicated in Figure 3-2 is the principal source of the requirements and business information needed by data scientists and engineers for processing.

Tip   You can never have too much information that explains the detailed characteristics and requirements of your business.

Data science has matured. The years of providing a quick half-baked analysis, then rushing off to the next analysis are, to a great extent, over. You are now required to ensure the same superior quality and consistency to analysis as you would for any other business aspect.

Remember   You are employed by the business. Look after it, and it will look after you.

Material you would expect in the business layer includes the following:

- Up-to-date organizational structure chart
- Business description of the business processes you are investigating
- List of your subject matter experts
- Project plans
- Budgets
- Functional requirements
- Nonfunctional requirements
- Standards for data items

Remember   Data science should apply identical sets of rules, as in software engineering and statistics data research. I will discuss the business layer in detail and share my experiences with it in Chapter 4.

The Basics for Utility Layer

I define a utility as a process that is used by many data science projects to achieve common processing methodologies. The utility layer, as shown in Figure 3-2, is a common area in which you store all your utilities. You will be faced regularly with immediate requirements demanding that you fix something in the data or run your "magical" algorithm. You may have spent hours preparing for this moment, only to find that you cannot locate that awesome utility, or you have a version that you cannot be sure of working properly. Been there—and have got the scars to prove it.

Tip   Collect your utilities (including source code) in one central location. Keep detailed records of every version.

If you have data algorithms that you use regularly, I strongly suggest you keep any proof and credentials that show the process to be a high-quality and industry-accepted algorithm. Painful experience has proven to me that you will get tested, and it is essential to back up your data science 100%. The additional value created is the capability to get multiple teams to work on parallel projects and know that each data scientist or engineer is working with clear standards.

In Chapter 5, I will discuss the interesting and useful utilities that I believe every single data scientist has a responsibility to know.

The Basics for Operational Management Layer

As referred to in Figure 3-2, operations management is an area of the ecosystem concerned with designing and controlling the process chains of a production environment and redesigning business procedures. This layer stores what you intend to process. It is where you plan your data science processing pipelines.

The operations management layer is where you record

- Processing-stream definition and management
- Parameters
- Scheduling
- Monitoring
- Communication
- Alerting

The operations management layer is the common location where you store any of the processing chains you have created for your data science.

Note   Operations management is how you turn data science experiments into long-term business gains.

I will discuss the particulars of a number of subcomponents later in Chapter 6, covering operations management.

The Basics for Audit, Balance, and Control Layer

The audit, balance, and control layer, represented in Figure 3-2, is the area from which you can observe what is currently running within your data science environment. It records

- Process-execution statistics

- Balancing and controls

- Rejects and error-handling

- Codes management

The three subareas are utilized in following manner.

Audit

The audit sublayer, indicated in Figure 3-2, records any process that runs within the environment. This information is used by data scientists and engineers to understand and plan improvements to the processing.

Tip   Make sure your algorithms and processing generate a good and complete audit trail.

My experiences have taught me that a good audit trail is essential. The use of the native audit capability of the Data Science Technology Stack will provide you with a quick and effective base for your auditing. The audit information I find essential and useful is discussed in detail in Chapter 6. Following is a quick introduction to what you could expect in this topic.

Balance

The balance sublayer in Figure 3-2 ensures that the ecosystem is balanced across the available processing capability or has the capability to top-up capability during periods of extreme processing. The processing on demand capability of a cloud ecosystem is highly desirable for this purpose.

Tip   Plan your capability as a combination of always-on and top-up processing. That way, you get maximum productivity from your processing cycle.

Using the audit trail, it is possible to adapt to changing requirements and forecast what you will require to complete the schedule of work you submitted to the ecosystem.

In the always-on and top-up ecosystem you can build, you can balance your processing requirements by removing or adding resources dynamically as you move through the processing pipe. I suggest that you enable your code to balance its ecosystem as regularly as possible. An example would be during end-of-month processing, you increase your processing capacity to sixty nodes, to handle the extra demand of the end-of-month run. The rest of the month, you run at twenty nodes during business hours. During weekends and other slow times, you only run with five nodes. Massive savings can be generated in this manner.

Control

The control sublayer, indicated in Figure 3-2, controls the execution of the current active data science processes in a production ecosystem. The control elements are a combination of the control element within the Data Science Technology Stack's individual tools plus a custom interface to control the primary workflow. The control also ensures that when processing experiences an error, it can attempt a recovery, as per your requirements, or schedule a clean-up utility to undo the error. Chapter 6 will discuss in more detail how, what, and why you need specific interfaces and utilities.

Most of my bigger customers have separate discovery environments set up, in which you can run jobs without any limiting control mechanisms. This enables data scientists to concentrate on the models and processing and not on complying with the more controlled production requirements.

The Basics for Functional Layer

The functional layer of the data science ecosystem shown in Figure 3-2 is the main layer of programming required. The functional layer is the part of the ecosystem that executes the comprehensive data science. It consists of several structures.

- Data models
- Processing algorithms
- Provisioning of infrastructure

My processing algorithm is spread across six supersteps of processing, as follows:

1. *Retrieve*: This super step contains all the processing chains for retrieving data from the raw data lake via a more structured format.

2. *Assess*: This superstep contains all the processing chains for quality assurance and additional data enhancements.

3. *Process*: This superstep contains all the processing chains for building the data vault.

4. *Transform*: This superstep contains all the processing chains for building the data warehouse.

5. *Organize*: This superstep contains all the processing chains for building the data marts.

6. *Report*: This superstep contains all the processing chains for building virtualization and reporting the actionable knowledge.

These supersteps are discussed in detail in chapters specifically devoted to them (Chapters 6–11), to enable you to master the individual supersteps and the tools from the Data Science Technology Stack that are relevant.

Layered Framework for High-Level Data Science and Engineering

Any source code or other supplementary material referenced by me in this book is available to readers on GitHub, via this book's product page, located at `www.apress.com/9781484230534`.

The layered framework is engineered with the following structures. The following scripts will create for you a complete framework in the `C:\VKHCG\05-DS\00-Framework` directory, if you are on Windows, or `./VKHCG/05-DS/00-Framework`, if you are using a Linux environment.

Windows

If you prefer Windows and Python, in the source code for Chapter 2, you will find the relevant Python script, identified as `Build_Win_Framework.py`.

If you prefer Windows and R, in the source code for Chapter 2, you will find the relevant Python script, identified as `Build_Win_Framework.r`.

Linux

If you prefer Linux and Python, in the source code for Chapter 2, you will find the relevant Python script, identified as `Build_ Linux _Framework.py`.

If you prefer Linux and R, in the source code for Chapter 2, you will find the relevant Python script, identified as `Build_ Linux_Framework.r`.

Summary

I have explained the basic layered framework that the data science solution requires. You should have a general high-level understanding of how your solution fits together, to achieve success in your own data science solution.

In the next chapters, I will provide details for individual components, with practical examples.

CHAPTER 4

Business Layer

In this chapter, I define the business layer in detail, clarifying why, where, and what functional and nonfunctional requirements are presented in the data science solution. With the aid of examples, I will help you to engineer a practical business layer and advise you, as I explain the layer in detail and discuss methods to assist you in performing good data science.

The business layer is the transition point between the nontechnical business requirements and desires and the practical data science, where, I suspect, most readers of this book will have a tendency to want to spend their careers, doing the perceived more interesting data science. The business layer does not belong to the data scientist 100%, and normally, its success represents a joint effort among such professionals as business subject matter experts, business analysts, hardware architects, and data scientists.

Business Layer

The business layer is where we record the interactions with the business. This is where we convert business requirements into data science requirements.

If you want to process data and wrangle with your impressive data science skills, this chapter may not be the start of a book about practical data science that you would expect. I suggest, however, that you read this chapter, if you want to work in a successful data science group. As a data scientist, you are not in control of all aspects of a business, but you have a responsibility to ensure that you identify the true requirements.

Warning   I have seen too many data scientists blamed for bad science when the issue was bad requirement gathering.

© Andreas François Vermeulen 2018
A. F. Vermeulen, *Practical Data Science*, https://doi.org/10.1007/978-1-4842-3054-1_4

The Functional Requirements

Functional requirements record the detailed criteria that must be followed to realize the business's aspirations from its real-world environment when interacting with the data science ecosystem. These requirements are the business's view of the system, which can also be described as the "Will of the Business."

Tip   Record all the business's aspirations. Make everyone supply their input. You do not want to miss anything, as later additions are expensive and painful for all involved.

I use the MoSCoW method (Table 4-1) as a prioritization technique, to indicate how important each requirement is to the business. I revisit all outstanding requirements before each development cycle, to ensure that I concentrate on the requirements that are of maximum impact to the business at present, as businesses evolve, and you must be aware of their true requirements.

Table 4-1. *MoSCoW Options*

Must have	Requirements with the priority "must have" are critical to the current delivery cycle.
Should have	Requirements with the priority "should have" are important but not necessary to the current delivery cycle.
Could have	Requirements prioritized as "could have" are those that are desirable but not necessary, that is, nice to have to improve user experience for the current delivery cycle.
Won't have	Requirements with a "won't have" priority are thoseidentified by stakeholders as the least critical, lowest payback requests, or just not appropriate at that time in the delivery cycle.

General Functional Requirements

As a [user role] I want [goal] so that [business value] is achieved.

Specific Functional Requirements

The following requirements specific to data science environments will assist you in creating requirements that enable you to transform a business's aspirations into technical descriptive requirements. I have found these techniques highly productive in aligning requirements with my business customers, while I can easily convert or extend them for highly technical development requirements.

Data Mapping Matrix

The data mapping matrix is one of the core functional requirement recording techniques used in data science. It tracks every data item that is available in the data sources. I advise that you keep this useful matrix up to date as you progress through the processing layers.

Sun Models

The base sun models process was developed by Professor Mark Whitehorn (`www.penguinsoft.co.uk/mark.html`) and was taught to me in a master's degree course. (Thanks, Mark.)

I have added changes and evolved the process for my own consultancy process, mixing it with Dr. Ralph Kimball's wisdoms, Bill Inmon's instructions, and the great insights of several other people I have met in the course of my career.

These techniques will continue to evolve with fluctuations in data science, and we will have to adapt. So, here is version 2.0 sun models.

Tip   Learn to evolve continually to survive!

The sun models is a requirement mapping technique that assists you in recording requirements at a level that allows your nontechnical users to understand the intent of your analysis, while providing you with an easy transition to the detailed technical modeling of your data scientist and data engineer.

Note   Over the next few pages, I will introduce several new concepts. Please read on, as the section will help in explaining the complete process.

I will draft a sun model set as part of an example and explain how it guides your development and knowledge insights. First, I must guide you through the following basic knowledge, to ensure that you have the background you require to navigate subsequent chapters.

This sun model is for fact LivesAt and supports three dimensions: person, location, and date. Figure 4-1 shows a typical sun model.

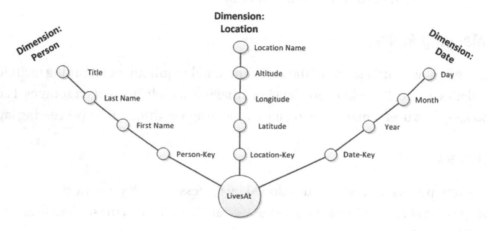

Figure 4-1. *Typical sun model*

This can be used to store the fact that a specific person lived at a specific location since a specific date. So, if you want to record that Dr Jacob Roggeveen lived on Rapa Nui, Easter Island, since April 5, 1722, your data would fit into the sun model, as seen in Figure 4-2.

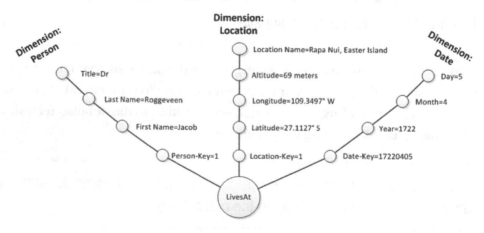

Figure 4-2. *Dr Jacob Roggeveen record in sun model*

See if you can record your own information in the same sun model. Next, I'll discuss the components of the sun model.

Dimensions

A dimension is a structure that categorizes facts and measures, to enable you to respond to business questions. A slowly changing dimension is a data structure that stores the complete history of the data loads in the dimension structure over the life cycle of the data lake. There are several types of Slowly Changing Dimensions (SCDs) in the data warehousing design toolkit that enable different recording rules of the history of the dimension.

SCD Type 1—Only Update

This dimension contains only the latest value for specific business information.

Example: Dr Jacob Roggeveen lived on Rapa Nui, Easter Island, but now lives in Middelburg, Netherlands.

Load	First Name	Last Name	Address
1	Jacob	Roggeveen	Rapa Nui, Easter Island
2	Jacob	Roggeveen	Middelburg, Netherlands

The SCD Type 1 records only the first place Dr Jacob Roggeveen lived and ignores the second load's changes.

Person		Location
First Name	Last Name	Address
Jacob	Roggeveen	Rapa Nui, Easter Island

I find this useful for storing the first address you register for a user, or the first product he or she has bought.

SCD Type 2—Keeps Complete History

This dimension contains the complete history of the specific business information. There are three ways to construct a SCD Type 2.

1. *Versioning*: The SCD keeps a log of the version number, allowing you to read the data back in the order in which the changes were made.

2. *Flagging*: The SCD flags the latest version of the business data row as a 1 and the rest as 0. This way, you can always get the latest row.

3. *Effective Date*: The SCD keeps records of the period in which the specific value was valid for the business data row.

Hint   The best SCD for data science is SCD Type 2 with Effective Date.

Example: Dr Jacob Roggeveen lived in Middelburg, Netherlands, and then in Rapa Nui, Easter Island, but now lives in Middelburg, Netherlands.

Load	First Name	Last Name	Address
1	Jacob	Roggeveen	Middelburg, Netherlands
2	Jacob	Roggeveen	Rapa Nui, Easter Island
3	Jacob	Roggeveen	Middelburg, Netherlands

SCD Type 2 records only changes in the location Dr Jacob Roggeveen lived, as it loads the three data loads, using versioning.

Person			Location	
First Name	Last Name	Version	Address	Version
Jacob	Roggeveen	1	Middelburg, Netherlands	1
			Rapa Nui, Easter Island	2
			Middelburg, Netherlands	3

I find this useful for storing all the addresses at which your customer has lived, without knowing from what date to what date he or she was at each address.

SCD Type 2 records only changes to the places Dr Jacob Roggeveen has lived, each time it loads the three data loads, using flagging.

Person			Location	
First Name	Last Name	Flag	Address	Flag
Jacob	Roggeveen	1	Middelburg, Netherlands	0
			Rapa Nui, Easter Island	0
			Middelburg, Netherlands	1

I find this useful when I want to store all the addresses at which a customer has lived, without knowing from what date till what date he or she was at each address, or the order he or she lived at each address. The flag only shows what the latest address is.

SCD Type 2 records only changes in the location Dr Jacob Roggeveen lived, as it loads the three data loads, using effective date.

Person			Location	
First Name	Last Name	Effective Date	Address	Effective Date
Jacob	Roggeveen	5 April 1722	Middelburg, Netherlands	1 February 1659
			Rapa Nui, Easter Island	5 April 1722
			Middelburg, Netherlands	31 January 1729

I find this useful when I want to store all the addresses at which a customer lived, in the order they lived at those locations, plus when they moved to that address.

SCD Type 3—Transition Dimension

This dimension records only the transition for the specific business information.

Example: Dr Jacob Roggeveen lived in Middelburg, Netherlands, then in Rapa Nui, Easter Island, but now lives in Middelburg, Netherlands.

Load	First Name	Last Name	Address
1	Jacob	Roggeveen	Middelburg, Netherlands
2	Jacob	Roggeveen	Rapa Nui, Easter Island
3	Jacob	Roggeveen	Middelburg, Netherlands

The transition dimension records only the last address change, by keeping one record with the last change in value and the current value. The three load steps are as follows:

Step 1: Lives in Middelburg, Netherlands

Location	
AddressPrevious	Address
	Middelburg, Netherlands

Step 2: Moves from Middelburg, Netherlands, to Rapa Nui, Easter Island

Location	
AddressPrevious	Address
Middelburg, Netherlands	Rapa Nui, Easter Island

Step 3: Moves from Rapa Nui, Easter Island, to Middelburg, Netherlands

Location	
AddressPrevious	Address
Rapa Nui, Easter Island	Middelburg, Netherlands

I use SCD Type 3 to record the transitions to the current values of the customer.

SCD Type 4—Fast-Growing Dimension.
This dimension handles specific business information with a high change rate. This enables the data to track the fast changes, without overwhelming the storage requirements.

In Chapter 9, I will show you how to design and deploy these dimensions for your data science.

Facts

A fact is a measurement that symbolizes a fact about the managed entity in the real world. In Chapter 9, I will show you how to design and deploy facts into the data science ecosystem.

Intra-Sun Model Consolidation Matrix

The intra-sun model consolidation matrix is a tool that helps you to identify common dimensions between sun models. Let's assume our complete set of sun models is only three models.

Tip   When drafting sun models, it is better to draft too many than too few. Use them to record your requirements in detail.

Sun Model One

The first sun model (Figure 4-3) records the relationship for Birth, with its direct dependence on the selectors Event, Date, and Time.

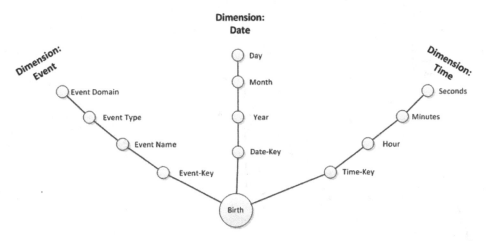

Figure 4-3. *Sun Model One*

Sun Model Two

The second sun model (Figure 4-4) records the relationship for LivesAt, with its direct dependence on the selectors Person, Date, Time, and Location.

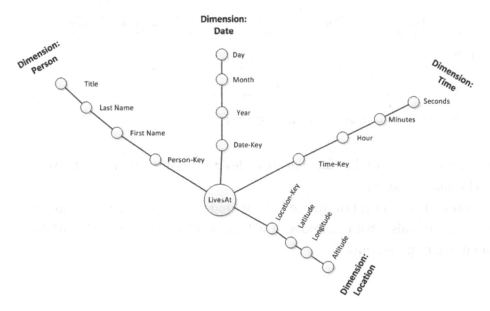

Figure 4-4. *Sun Model Two*

Sun Model Three

The third sun model (Figure 4-5) records the relationship for Owns, with its direct dependence on the selectors Object, Person, Date, and Time.

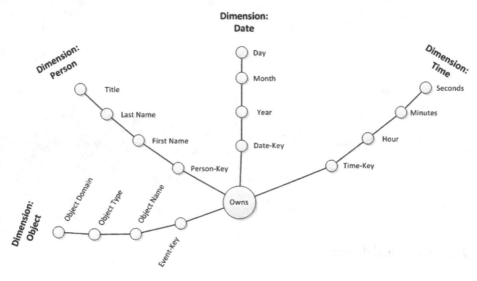

Figure 4-5. *Sun Model Three*

I then consolidate the different dimensions and facts onto a single matrix (Figure 4-6).

	Date (SCD 0)	Time (SCD 0)	Person (SCD 2)	Object (SCD 2)	Location (SCD 2)	Event (SCD 2)
Birth	X	X				X
LivesAt	X	X	X		X	
Owns	X	X	X	X		

Figure 4-6. *Intra-Sun model Linkage matrix*

The matrix will be used in Chapter 10, to design and implement the transform process for the data science.

So, now that you know how to use the sun models to capture the functional requirements, it is also that you understand how to achieve the collection of the nonfunctional requirements.

The Nonfunctional Requirements

Nonfunctional requirements record the precise criteria that must be used to appraise the operation of a data science ecosystem.

Accessibility Requirements

Accessibility can be viewed as the "ability to access" and benefit from some system or entity. The concept focuses on enabling access for people with disabilities, or special needs, or enabling access through assistive technology.

Assistive technology covers the following:

- *Levels of blindness support*: Must be able to increase font sizes or types to assist with reading for affected people

- *Levels of color-blindness support*: Must be able to change a color palette to match individual requirements

- *Use of voice-activated commands to assist disabled people*: Must be able to use voice commands for individuals that cannot type commands or use a mouse in normal manner

Audit and Control Requirements

Audit is the ability to investigate the use of the system and report any violations of the system's data and processing rules. Control is making sure the system is used in the manner and by whom it is pre-approved to be used.

An approach called role-based access control (RBAC) is the most commonly used approach to restricting system access to authorized users of your system. RBAC is an access-control mechanism formulated around roles and privileges. The components of RBAC are role-permissions—user-role and role-role relationships that together describe the system's access policy.

These audit and control requirements are also compulsory, by regulations on privacy and processing. Please check with your local information officer which precise rules apply. I will discuss some of the laws and regulations I have seen over the last years in Chapter 6, which will cover these in detail.

Availability Requirements

Availability is as a ratio of the expected uptime of a system to the aggregate of the downtime of the system. For example, if your business hours are between 9h00 and 17h00, and you cannot have more than 1 hour of downtime during your business hours, you require 87.5% availability.

Take note that you specify precisely at what point you expect the availability. If you are measuring at the edge of the data lake, it is highly possible that you will sustain 99.99999% availability with ease. The distributed and fault-tolerant nature of the data lake technology would ensure a highly available data lake. But if you measure at critical points in the business, you will find that at these critical business points, the requirements are more specific for availability.

Record your requirements in the following format:

Component C will be entirely operational for P% of the time over an uninterrupted measured period of D days.

Your customers will understand this better than the general "24/7" or "business hours" terminology that I have seen used by some of my previous customers. No system can achieve these general requirement statements.

The business will also have periods of high availability at specific periods during the day, week, month, or year. An example would be every Monday morning the data science results for the weekly meeting has to be available. This could be recorded as the following:

Weekly reports must be entirely operational for 100% of the time between 06h00 and 10h00 every Monday for each office.

Note Think what this means to a customer that has worldwide offices over several time zones. Be sure to understand every requirement fully!

The correct requirements are

- London's weekly reports must be entirely operational for 100% of the time between 06h00 and 10h00 (Greenwich Mean Time or British Daylight Time) every Monday.

- New York's weekly reports must be entirely operational for 100% of the time between 06h00 and 10h00 (Eastern Standard Time or Eastern Daylight Time) every Monday.

Note   You can clearly see that these requirements are now more precise than the simple general requirement.

Identify single points of failure (SPOFs) in the data science solution. Ensure that you record this clearly, as SPOFs can impact many of your availability requirements indirectly.

Highlight that those dependencies between components that may not be available at the same time must be recorded and requirements specified, to reflect this availability requirement fully.

Note   Be aware that the different availability requirements for different components in the same solution are the optimum requirement recording option.

Backup Requirements

A backup, or the process of backing up, refers to the archiving of the data lake and all the data science programming code, programming libraries, algorithms, and data models, with the sole purpose of restoring these to a known good state of the system, after a data loss or corruption event.

Remember: Even with the best distribution and self-healing capability of the data lake, you have to ensure that you have a regular and appropriate backup to restore. Remember a backup is only valid if you can restore it.

The merit of any system is its ability to return to a good state. This is a critical requirement. For example, suppose that your data scientist modifies the system with a new algorithm that erroneously updates an unknown amount of the data in the data lake. Oh, yes, that silent moment before every alarm in your business goes mad! You want to be able at all times to return to a known good state via a backup.

Warning   Please ensure that you can restore your backups in an effective and efficient manner. The process is backup-and-restore. Just generating backups does not ensure survival. Understand the impact it has to the business if it goes back two hours or what happens while you restore.

Capacity, Current, and Forecast

Capacity is the ability to load, process, and store a specific quantity of data by the data science processing solution. You must track the current and forecast the future requirements, because as a data scientist, you will design and deploy many complex models that will require additional capacity to complete the processing pipelines you create during your processing cycles.

Warning   I have inadvertently created models that generate several terabytes of workspace requirements, simply by setting the parameters marginally too in-depth than optimal. Suddenly, my model was demanding disk space at an alarming rate!

Capacity

Capacity is measured per the component's ability to consistently maintain specific levels of performance as data load demands vary in the solution. The correct way to record the requirement is

Component C will provide P% capacity for U users, each with M MB of data during a time frame of T seconds.

Example:

The data hard drive will provide 95% capacity for 1000 users, each with 10MB of data during a time frame of 10 minutes.

Warning   Investigate the capacity required to perform a full rebuild in one process. I advise researching new cloud on-demand capacity, for disaster recovery or capacity top-ups. I have been consulted after major incidents that crippled a company for weeks, owing to a lack of proper capacity top-up plans.

Concurrency

Concurrency is the measure of a component to maintain a specific level of performance when under multiple simultaneous loads conditions.

The correct way to record the requirement is

Component C will support a concurrent group of U users running predefined acceptance script S simultaneously.

Example:

The memory will support a concurrent group of 100 users running a sort algorithm of 1000 records simultaneously.

Note   Concurrency is the ability to handle a subset of the total user base effectively. I have found that numerous solutions can handle substantial volumes of users with as little as 10% of the users' running concurrently.

Concurrency is an important requirement to ensure an effective solution at the start. Capacity can be increased by adding extra processing resources, while concurrency normally involves complete replacements of components.

Design Tip   If on average you have short-running data science algorithms, you can support high concurrency to maximum capacity ratio. But if your average running time is higher, your concurrency must be higher too. This way, you will maintain an effective throughput performance:

Throughput Capacity

This is how many transactions at peak time the system requires to handle specific conditions.

Storage (Memory)

This is the volume of data the system will persist in memory at runtime to sustain an effective processing solution.

Tip   Remember: You can never have too much or too slow memory.

Storage (Disk)

This is the volume of data the system stores on disk to sustain an effective processing solution.

Tip   Make sure that you have a proper mix of disks, to ensure that your solutions are effective.

You will need short-term storage on fast solid-state drives to handle the while-processing capacity requirements.

Warning   There are data science algorithms that produce larger data volumes during data processing than the input or output data.

The next requirement is your long-term storage. The basic rule is to plan for bigger but slower storage. Investigate using clustered storage, whereby two or more storage servers work together to increase performance, capacity, and reliability. Clustering distributes workloads to each server and manages the transfer of workloads between servers, while ensuring availability.

The use of clustered storage will benefit you in the long term, during periods of higher demand, to scale out vertically with extra equipment.

Tip   Ensure that the server network is more than capable of handling any data science load. Remember: The typical data science algorithm requires massive data sets to work effectively.

The big data evolution is now bringing massive amounts of data into the processing ecosystem. So, make sure you have enough space to store any data you need.

Warning   If you have a choice, do not share disk storage or networks with a transactional system. The data science will consume any spare capacity on the shared resources. It is better to have a lower performance dedicated set of resources than to share a volatile process.

Storage (GPU)

This is the volume of data the system will persist in GPU memory at runtime to sustain an effective parallel processing solution, using the graphical processing capacity of the solution. A CPU consists of a limited amount of cores that are optimized for sequential serial processing, while a GPU has a massively parallel architecture consisting of thousands of smaller, more efficient cores intended for handling massive amounts of multiple tasks simultaneously.

The big advantage is to connect an effective quantity of very high-speed memory as closely as possible to these thousands of processing units, to use this increased capacity. I am currently deploying systems such as Kinetic DB and MapD, which are GPU-based data base engines. This improves the processing of my solutions by factors of a hundred in speed. I suspect that we will see key enhancements in the capacity of these systems over the next years.

Tip   Investigate a GPU processing grid for your high-performance processing. It is an effective solution with the latest technology.

Year-on-Year Growth Requirements

The biggest growth in capacity will be for long-term storage. These requirements are specified as how much capacity increases over a period.

The correct way to record the requirement is

Component C will be responsible for necessary growth capacity to handle additional M MB of data within a period of T.

Configuration Management

Configuration management (CM) is a systems engineering process for establishing and maintaining consistency of a product's performance, functional, and physical attributes against requirements, design, and operational information throughout its life.

Deployment

A methodical procedure of introducing data science to all areas of an organization is required. Investigate how to achieve a practical continuous deployment of the data science models. These skills are much in demand, as the processes model changes more frequently as the business adopts new processing techniques.

Documentation

Data science requires a set of documentation to support the story behind the algorithms. I will explain the documentation required at each stage of the processing pipe, as I guide you through Chapters 5–11.

Disaster Recovery

Disaster recovery (DR) involves a set of policies and procedures to enable the recovery or continuation of vital technology infrastructure and systems following a natural or human-induced disaster.

Efficiency (Resource Consumption for Given Load)

Efficiency is the ability to accomplish a job with a minimum expenditure of time and effort. As a data scientist, you are required to understand the efficiency curve of each of your modeling techniques and algorithms. As I suggested before, you must practice with your tools at different scales.

Tip   If it works at a sample 100,000 data points, try 200,000 data points or 500,000 data points. Make sure you understand the scaling dynamics of your tools.

Effectiveness (Resulting Performance in Relation to Effort)

Effectiveness is the ability to accomplish a purpose; producing the precise intended or expected result from the ecosystem. As a data scientist, you are required to understand the effectiveness curve of each of your modeling techniques and algorithms. You must ensure that the process is performing only the desired processing and has no negative side effects.

Extensibility

The ability to add extra features and carry forward customizations at next-version upgrades within the data science ecosystem. The data science must always be capable of being extended to support new requirements.

Failure Management

Failure management is the ability to identify the root cause of a failure and then successfully record all the relevant details for future analysis and reporting. I have found that most of the tools I would include in my ecosystem have adequate fault management and reporting capability already built in to their native internal processing.

Tip   I found it takes a simple but well-structured set of data science processes to wrap the individual failure logs into a proper failure-management system. Apply normal data science to it, as if it is just one more data source.

I always stipulate the precise expected process steps required when a failure of any component of the ecosystem is experienced during data science processing. Acceptance script S completes and reports every one of the X faults it generates. As a data scientist, you are required to log any failures of the system, to ensure that no unwanted side effects are generated that may cause a detrimental impact to your customers.

Fault Tolerance

Fault tolerance is the ability of the data science ecosystem to handle faults in the system's processing. In simple terms, no single event must be able to stop the ecosystem from continuing the data science processing.

Here, I normally stipulate the precise operational system monitoring, measuring, and management requirements within the ecosystem, when faults are recorded. Acceptance script S withstands the X faults it generates. As a data scientist, you are required to ensure that your data science algorithms can handle faults and recover from them in an orderly manner.

Latency

Latency is the time it takes to get the data from one part of the system to another. This is highly relevant in the distributed environment of the data science ecosystems. Acceptance script S completes within T seconds on an unloaded system and within T2 seconds on a system running at maximum capacity, as defined in the concurrency requirement.

Tip   Remember: There is also internal latency between components that make up the ecosystem that is not directly accessible to users. Make sure you also note these in your requirements.

Interoperability

Insist on a precise ability to share data between different computer systems under this section. Explain in detail what system must interact with what other systems. I normally investigate areas such as communication protocols, locations of servers, operating systems for different subcomponents, and the now-important end user's Internet access criteria.

Warning   Be precise with requirements, as open-ended interoperability can cause unexpected complications later in the development cycle.

Maintainability

Insist on a precise period during which a specific component is kept in a specified state. Describe precisely how changes to functionalities, repairs, and enhancements are applied while keeping the ecosystem in a known good state.

Modifiability

Stipulate the exact amount of change the ecosystem must support for each layer of the solution.

Tip   State what the limits are for specific layers of the solution. If the database can only support 2024 fields to a table, share that information!

Network Topology

Stipulate and describe the detailed network communication requirements within the ecosystem for processing. Also, state the expected communication to the outside world, to drive successful data science.

Note   Owing to the high impact on network traffic from several distributed data science algorithms processing, it is required that you understand and record the network necessary for the ecosystem to operate at an acceptable level.

Privacy

I suggest listing the exact privacy laws and regulations that apply to this ecosystem. Make sure you record the specific laws and regulations that apply. Seek legal advice if you are unsure.

This is a hot topic worldwide, as you will process and store other people's data and execute algorithms against this data. As a data scientist, you are responsible for your actions.

Warning   Remember: A privacy violation will result in a fine!

Tip   I hold liability insurance against legal responsibility claims for the data I process.

Quality

Specify the rigorous faults discovered, faults delivered, and fault removal efficiency at all levels of the ecosystem. Remember: Data quality is a functional requirement. This is a nonfunctional requirement that states the quality of the ecosystem, not the data flowing through it.

Recovery/Recoverability

The ecosystem must have a clear-cut mean time to recovery (MTTR) specified. The MTTR for specific layers and components in the ecosystem must be separately specified.

I typically measure in hours, but for other extra-complex systems, I measure in minutes or even seconds.

Reliability

The ecosystem must have a precise mean time between failures (MTBF). This measurement of availability is specified in a pre-agreed unit of time. I normally measure in hours, but there are extra sensitive systems that are best measured in years.

Resilience

Resilience is the capability to deliver and preserve a tolerable level of service when faults and issues to normal operations generate complications for the processing. The ecosystem must have a defined ability to return to the original form and position in time, regardless of the issues it has to deal with during processing.

Resource Constraints

Resource constraints are the physical requirements of all the components of the ecosystem. The areas of interest are processor speed, memory, disk space, and network bandwidth, plus, normally, several other factors specified by the tools that you deploy into the ecosystem.

Tip   Discuss these requirements with your system's engineers. This is not normally the area in which data scientists work.

Reusability

Reusability is the use of pre-built processing solutions in the data science ecosystem development process. The reuse of preapproved processing modules and algorithms is highly advised in the general processing of data for the data scientists. The requirement here is that you use approved and accepted standards to validate your own results.

Warning   I always advise that you use methodologies and algorithms that have proven lineage. An approved algorithm will guarantee acceptance by the business. Do not use unproven ideas!

Scalability

Scalability is how you get the data science ecosystem to adapt to your requirements. I use three scalability models in my ecosystem: horizontal, vertical, and dynamic (on-demand).

Horizontal scalability increases capacity in the data science ecosystem through more separate resources, to improve performance and provide high availability (HA). The ecosystem grows by scale out, by adding more servers to the data science cluster of resources.

Tip   Horizontal scalability is the proven way to handle full-scale data science ecosystems.

Warning   Not all models and algorithms can scale horizontally. Test them first. I would counsel against making assumptions.

Vertical scalability increases capacity by adding more resources (more memory or an additional CPU) to an individual machine.

Warning   Make sure that you size your data science building blocks correctly at the start, as vertical scaling of clusters can get expensive and complex to swap at later stages.

Dynamic (on-demand) scalability increases capacity by adding more resources, using either public or private cloud capability, which can be increased and decreased on a pay-as-you-go model. This is a hybrid model using a core set of resources that is the minimum footprint of the system, with additional burst agreements to cover any planned or even unplanned extra scalability increases in capacity that the system requires.

I'd like to discuss scalability for your power users. Traditionally, I would have suggested high-specification workstations, but I have found that you will serve them better by providing them access to a flexible horizontal scalability on-demand environment. This way, they use what they need during peak periods of processing but share the capacity with others when they do not require the extra processing power.

Security

One of the most important nonfunctional requirements is security. I specify security requirements at three levels.

Privacy

I would specifically note requirements that specify protection for sensitive information within the ecosystem. Types of privacy requirements to note include data encryption for database tables and policies for the transmission of data to third parties.

Tip   Sources for privacy requirements are legislative or corporate. Please consult your legal experts.

Physical

I would specifically note requirements for the physical protection of the system. Include physical requirements such as power, elevated floors, extra server cooling, fire prevention systems, and cabinet locks.

Warning   Some of the high-performance workstations required to process data science have stringent power requirements, so ensure that your data scientists are in a preapproved environment, to avoid overloading the power grid.

Access

I purposely specify detailed access requirements with defined account types/groups and their precise access rights.

Tip   I use role-based access control (RBAC) to regulate access to data science resources, based on the roles of individual users within the ecosystem and not by their separate names. This way, I simply move the role to a new person, without any changes to the security profile.

Testability

International standard IEEE 1233-1998 states that *testability* is the "degree to which a requirement is stated in terms that permit establishment of test criteria and performance of tests to determine whether those criteria have been met." In simple terms, if your requirements are not testable, do not accept them.

Remember   A lower degree of testability results in increased test effort. I have spent too many nights creating tests for requirements that are unclear.

Following is a series of suggestions, based on my experience.

Controllability

Knowing the precise degree to which I can control the state of the code under test, as required for testing, is essential. The algorithms used by data science are not always controllable, as they include random start points to speed the process. Running distributed algorithms is not easy to deal with, as the distribution of the workload is not under your control.

Isolate Ability

The specific degree to which I can isolate the code under test will drive most of the possible testing. A process such as deep learning includes non-isolation, so do not accept requirements that you cannot test, owing to not being able to isolate them.

Understandability

I have found that most algorithms have undocumented "extra features" or, in simple terms, "got-you" states. The degree to which the algorithms under test are documented directly impacts the testability of requirements.

Automatability

I have found the degree to which I can automate testing of the code directly impacts the effective and efficient testing of the algorithms in the ecosystem. I am an enthusiast of known result inline testing. I add code to my algorithms that test specific sub-sessions, to ensure that the new code has not altered the previously verified code.

Common Pitfalls with Requirements

I just want to list a sample of common pitfalls I have noted while performing data science for my customer base. If you are already aware of these pitfalls, well done!

Many seem obvious; however I regularly work on projects in which these pitfalls have cost my clients millions of dollars before I was hired. So, let's look at some of the more common pitfalls I encounter regularly.

Weak Words

Weak words are subjective or lack a common or precise definition. The following are examples in which weak words are included and identified:

- Users must *easily* access the system.

    What is "easily"?

- Use *reliable* technology.

    What is "reliable"?

- *State-of-the-art* equipment

    What is "state-of-the-art"?

- Reports must run *frequently*.

    What is "frequently"?

- *User-friendly* report layouts

  What is "user-friendly"?

- *Secure* access to systems.

  What is "secure?

- *All* data must be *immediately* reported.

  What is "all"? What is "immediately"?

I could add many more examples. But you get the common theme. Make sure the wording of your requirements is precise and specific. I have lamentably come to understand that various projects end in disaster, owing to weak words.

Unbounded Lists

An unbounded list is an incomplete list of items. Examples include the following:

- Accessible at least from London and New York.

  Do I connect only London to New York? What about the other 20 branches?

- Including, but not limited to, London and New York offices must have access.

  So, is the New Delhi office not part of the solution?

Make sure your lists are complete and precise. This prevents later issues caused by requirements being misunderstood.

Implicit Collections

When collections of objects within requirements are not explicitly defined, you or your team will assume an incorrect meaning. See the following example:

The solution must support TCP/IP and other network protocols supported by existing users with Linux.

- What is meant by "existing user"?

- What belongs to the collection of "other network protocols"?

- What specific protocols of TCP/IP are included?

- "Linux" is a collection of operation systems from a number of vendors, with many versions and even revisions. Do you support all the different versions or only one of them?

Make sure your collections are explicit and precise. This prevents later issues from requirements being misunderstood.

Ambiguity

Ambiguity occurs when a word within the requirement has multiple meanings. Examples are listed following.

Vagueness

The system must pass between 96–100% of the test cases using current standards for data science.

What are the "current standards"? This is an example of an unclear requirement!

Subjectivity

The report must easily and seamlessly integrate with the web sites.

"Easily" and "seamlessly" are highly subjective terms where testing is concerned.

Optionality:

The solution should be tested under as many hardware conditions as possible.

"As possible" makes this requirement optional. What if it fails testing on every hardware setup? Is that okay with your customer?

Under-specification

The solution must support Hive 2.1 and other database versions.

Do other database versions only include other Hive databases, or also others such as HBase version 1.0 and Oracle version 10i?

Under-reference

Users must be able to complete all previously defined reports in less than three minutes 90% of the day.

What are these "previously defined" reports? This is an example of an unclear requirement.

Engineering a Practical Business Layer

Any source code or other supplementary material referenced by me in this book is available to readers on GitHub, via this book's product page, located at `www.apress.com/9781484230534`. The business layer follows general business analysis and project management principals. I suggest a practical business layer consist of a minimum of three primary structures.

Note See the following source code from Chapter 2: `./VKHCG/05-DS/5000-BL/`.

For the business layer, I suggest using a directory structure, such as `./VKHCG/05-DS/5000-BL`. This enables you to keep your solutions clean and tidy for a successful interaction with a standard version-control system.

Requirements

Note See the following source code from Chapter 2: `./VKHCG/05-DS/5000-BL/0300-Requirements`.

Every requirement must be recorded with full version control, in a requirement-per-file manner. I suggest a numbering scheme of 000000-00, which supports up to a million requirements with up to a hundred versions of each requirement.

Requirements Registry

Note See the following source code from Chapter 2: `./VKHCG/05-DS/5000-BL/0100-Requirements-Registry`.

Keep a summary registry of all requirements in one single file, to assist with searching for specific requirements. I suggest you have a column with the requirement number, MoSCoW, a short description, date created, date last version, and status. I normally use the following status values:

- In-Development

- In-Production

- Retired

The register acts as a control for the data science environment's requirements. An example of a template is shown in Figure 4-7.

Practical Data Science						
Requirement Number	MoSCoW	Requirement Description	Date Created	Date Last Version	Status	Notes
R-000001-00						
R-000001-01						
R-000001-02						
R-000001-03						
R-000001-04						
R-000001-05						

Figure 4-7. *Requirements registry template*

Traceability Matrix

Note   See the following source code from Chapter 2: `./VKHCG/05-DS/5000-BL/0200-Traceability-Matrix`.

Create a traceability matrix against each requirement and the data science process you developed, to ensure that you know what data science process supports which requirement. This ensures that you have complete control of the environment. Changes are easy if you know how everything interconnects. An example of a template is shown in Figure 4-8.

Practical Data Science										
Trace Number	Input Source	Input Field	Process Rule	Output Source	Output Field	RAPTOR Step	Date Created	Date Last Version	Status	Notes
T-000001-00										
T-000001-01										
T-000001-02										
T-000001-03										
T-000001-04										
T-000001-05										

Figure 4-8. *Traceability matrix template*

Summary

Well done! You now have a business layer, and you know in detail what is expected of your data science environment. Remember: The business layer must support the comprehensive collection of entire sets of requirements, to be used successfully by the data scientists.

In the next chapter, on the utility layer, I will continue to steer you on the path to practical data science, to enable your data science environment to utilize a reusable processing ecosystem.

CHAPTER 5

Utility Layer

The utility layer is used to store repeatable practical methods of data science. The objective of this chapter is to define how the utility layer is used in the ecosystem.

Utilities are the common and verified workhorses of the data science ecosystem. The utility layer is a central storehouse for keeping all one's solutions utilities in one place. Having a central store for all utilities ensures that you do not use out-of-date or duplicate algorithms in your solutions. The most important benefit is that you can use stable algorithms across your solutions.

Tip Collect all your utilities (including source code) in one central place. Keep records on all versions for future reference.

If you use algorithms, I suggest that you keep any proof and credentials that show that the process is a high-quality, industry-accepted algorithm. Hard experience has taught me that you are likely to be tested, making it essential to prove that your science is 100% valid.

The additional value is the capability of larger teams to work on a similar project and know that each data scientist or engineer is working to the identical standards. In several industries, it is a regulated requirement to use only sanctioned algorithms.

On May 25, 2018, a new European Union General Data Protection Regulation (GDPR) goes into effect. The GDPR has the following rules:

- You must have valid consent as a legal basis for processing.

  For any utilities you use, it is crucial to test for consent. In Chapters 7–11, I will discuss how to test for consent at each step of the process.

© Andreas François Vermeulen 2018
A. F. Vermeulen, *Practical Data Science*, https://doi.org/10.1007/978-1-4842-3054-1_5

- ·You must assure transparency, with clear information about what data is collected and how it is processed.

  Utilities must generate complete audit trails of all their activities. I will cover this in detail in Chapter 6, about the requirement for an audit trail.

- You must support the right to accurate personal data.

  Utilities must use only the latest accurate data. I will discuss techniques on assessing data processing in Chapter 8, covering the assess superstep.

- You must support the right to have personal data erased.

  Utilities must support the removal of all information on a specific person. I will also discuss what happens if the "right to be forgotten" is requested.

Warning   The "right to be forgotten" is a request that demands that you remove a person(s) from all systems immediately. Noncompliance with such a request will result in a fine.

This sounds easy at first, but take warning from my experiences and ensure that this request is implemented with care.

In Chapters 7–11, I will discuss how to process this complex request at each step of the way.

- You must have approval to move data between service providers.

  I advise you to make sure you have 100% approval to move data between data providers. If you move the data from your customer's systems to your own systems without clear approval, both you and your customer may be in trouble with the law.

- You must support the right not to be subject to a decision based solely on automated processing.

    This item is the subject of debate in many meetings that I attend. By the nature of what we as data scientists perform, we are conducting, more or less, a form of profiling. The actions of our utilities support decisions from our customers. The use of approved algorithms in our utilities makes compliance easier.

Warning   Noncompliance with GDPR might incur fines of 4% of global turnover.

Demonstrate compliance by maintaining a record of all data processing activities. This will be discussed in detail in Chapter 6, in the section "Audit."

In France, you must use only approved health processing rules from the National Health Data Institute. Processing of any personal health data is prohibited without the provision of an individual's explicit consent. In the United States, the Health Insurance Portability and Accountability Act of 1996 (HIPAA) guides any processing of health data.

Warning   Noncompliance with HIPAA could incur fines of up to $50,000 per violation.

I suggest you investigate the rules and conditions for processing any data you handle. In addition, I advise you to get your utilities certified, to show compliance. Discuss with your chief data officer what procedures are used and which prohibited procedures require checking.

Basic Utility Design

The basic utility must have a common layout to enable future reuse and enhancements. This standard makes the utilities more flexible and effective to deploy in a large-scale ecosystem.

I use a basic design (Figure 5-1) for a processing utility, by building it a three-stage process.

1. Load data as per input agreement.

2. Apply processing rules of utility.

3. Save data as per output agreement.

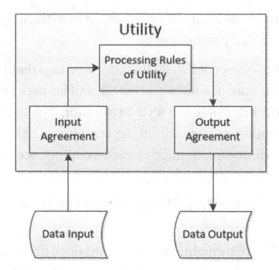

Figure 5-1. *Basic utility design*

The main advantage of this methodology in the data science ecosystem is that you can build a rich set of utilities that all your data science algorithms require. That way, you have a basic pre-validated set of tools to use to perform the common processing and then spend time only on the custom portions of the project.

You can also enhance the processing capability of your entire project collection with one single new utility update.

Note   I spend more than 80% of my non-project work time designing new utilities and algorithms to improve my delivery capability.

I suggest that you start your utility layer with a small utility set that you know works well and build the set out as you go along.

In this chapter, I will guide you through utilities I have found to be useful over my years of performing data science. I have split the utilities across various layers in the ecosystem, to assist you in connecting the specific utility to specific parts of the other chapters.

There are three types of utilities

- Data processing utilities

- Maintenance utilities

- Processing utilities

I will discuss these types in detail over the next few pages.

Data Processing Utilities

Data processing utilities are grouped for the reason that they perform some form of data transformation within the solutions.

Retrieve Utilities

Utilities for this superstep contain the processing chains for retrieving data out of the raw data lake into a new structured format. I will give specific implementations of these utilities in Chapter 7.

I suggest that you build all your retrieve utilities to transform the external raw data lake format into the Homogeneous Ontology for Recursive Uniform Schema (HORUS) data format that I have been using in my projects.

HORUS is my core data format. It is used by my data science framework, to enable the reduction of development work required to achieve a complete solution that handles all data formats.

I will provide details on format in the remaining chapters. At this point, just take note that HORUS exists.

Note   HORUS is my foundation and the solution to my core format requirements. If you prefer, create your own format, but feel free to use mine.

For demonstration purposes, I have selected the HORUS format to be CSV-based. I would normally use a JSON-based Hadoop ecosystem, or a distributed database such as Cassandra, to hold the core HORUS format.

Tip   Check the sample directory `C:\VKHCG\05-DS\9999-Data\` for the subsequent code and data.

I recommend the following retrieve utilities as a good start.

Text-Delimited to HORUS

These utilities enable your solution to import text-based data from your raw data sources. I will demonstrate this utility in Chapter 7, with sample data.

Example: This utility imports a list of countries in CSV file format into HORUS format.

```
# Utility Start CSV to HORUS ===================================
# Standard Tools
#=============================================================
import pandas as pd
# Input Agreement ============================================
sInputFileName='C:/VKHCG/05-DS/9999-Data/Country_Code.csv'
InputData=pd.read_csv(sInputFileName,encoding="latin-1")
print('Input Data Values ===================================')
print(InputData)
print('=====================================================')
# Processing Rules ===========================================
ProcessData=InputData

# Remove columns ISO-2-Code and ISO-3-CODE

ProcessData.drop('ISO-2-CODE', axis=1,inplace=True)
ProcessData.drop('ISO-3-Code', axis=1,inplace=True)

# Rename Country and ISO-M49
ProcessData.rename(columns={'Country': 'CountryName'}, inplace=True)
ProcessData.rename(columns={'ISO-M49': 'CountryNumber'}, inplace=True)

# Set new Index
ProcessData.set_index('CountryNumber', inplace=True)

# Sort data by CurrencyNumber
ProcessData.sort_values('CountryName', axis=0, ascending=False,
inplace=True)

print('Process Data Values ===================================')
print(ProcessData)
print('=====================================================')
# Output Agreement ===========================================
OutputData=ProcessData
```

```
sOutputFileName='C:/VKHCG/05-DS/9999-Data/HORUS-CSV-Country.csv'
OutputData.to_csv(sOutputFileName, index = False)

print('CSV to HORUS - Done')
# Utility done ================================================
```

XML to HORUS

These utilities enable your solution to import XML-based data from your raw data sources. I will demonstrate this utility in Chapter 7, with sample data.

Example: This utility imports a list of countries in XML file format into HORUS format.

```
# Utility Start XML to HORUS ===================================
# Standard Tools
#==============================================================
import pandas as pd
import xml.etree.ElementTree as ET
#==============================================================
def df2xml(data):
    header = data.columns
    root = ET.Element('root')
    for row in range(data.shape[0]):
        entry = ET.SubElement(root,'entry')
        for index in range(data.shape[1]):
            schild=str(header[index])
            child = ET.SubElement(entry, schild)
            if str(data[schild][row]) != 'nan':
                child.text = str(data[schild][row])
            else:
                child.text = 'n/a'
            entry.append(child)
    result = ET.tostring(root)
    return result
#==============================================================
```

```python
def xml2df(xml:data):
    root = ET.XML(xml:data)
    all_records = []
    for i, child in enumerate(root):
        record = {}
        for subchild in child:
            record[subchild.tag] = subchild.text
        all_records.append(record)
    return pd.DataFrame(all_records)

#===============================================================
# Input Agreement ==============================================
#===============================================================
sInputFileName='C:/VKHCG/05-DS/9999-Data/Country_Code.xml'

InputData = open(sInputFileName).read()

print('=======================================================')
print('Input Data Values =====================================')
print('=======================================================')
print(InputData)
print('=======================================================')

#===============================================================
# Processing Rules =============================================
#===============================================================
ProcessDataXML=InputData

# XML to Data Frame
ProcessData=xml2df(ProcessDataXML)

# Remove columns ISO-2-Code and ISO-3-CODE

ProcessData.drop('ISO-2-CODE', axis=1,inplace=True)
ProcessData.drop('ISO-3-Code', axis=1,inplace=True)

# Rename Country and ISO-M49
ProcessData.rename(columns={'Country': 'CountryName'}, inplace=True)
ProcessData.rename(columns={'ISO-M49': 'CountryNumber'}, inplace=True)
```

```
# Set new Index
ProcessData.set_index('CountryNumber', inplace=True)

# Sort data by CurrencyNumber
ProcessData.sort_values('CountryName', axis=0, ascending=False,
inplace=True)

print('=====================================================')
print('Process Data Values ================================')
print('=====================================================')
print(ProcessData)
print('=====================================================')
#=====================================================================
# Output Agreement ================================================
#=====================================================================
OutputData=ProcessData

sOutputFileName='C:/VKHCG/05-DS/9999-Data/HORUS-XML-Country.csv'
OutputData.to_csv(sOutputFileName, index = False)

print('=====================================================')
print('XML to HORUS - Done')
print('=====================================================')
# Utility done =====================================================
```

JSON to HORUS

These utilities enable your solution to import XML-based data from your raw data sources. I will demonstrate this utility in Chapter 7, with sample data.

Example: This utility imports a list of countries in JSON file format into HORUS format.

```
# Utility Start JSON to HORUS =====================================
# Standard Tools
#=====================================================================
import pandas as pd
```

```
# Input Agreement ==========================================
sInputFileName='C:/VKHCG/05-DS/9999-Data/Country_Code.json'
InputData=pd.read_json(sInputFileName,
                       orient='index',
                       encoding="latin-1")
print('Input Data Values ===================================')
print(InputData)
print('=====================================================')
# Processing Rules ==========================================
ProcessData=InputData

# Remove columns ISO-2-Code and ISO-3-CODE

ProcessData.drop('ISO-2-CODE', axis=1,inplace=True)
ProcessData.drop('ISO-3-Code', axis=1,inplace=True)

# Rename Country and ISO-M49
ProcessData.rename(columns={'Country': 'CountryName'}, inplace=True)
ProcessData.rename(columns={'ISO-M49': 'CountryNumber'}, inplace=True)

# Set new Index
ProcessData.set_index('CountryNumber', inplace=True)

# Sort data by CurrencyNumber
ProcessData.sort_values('CountryName', axis=0, ascending=False,
inplace=True)

print('Process Data Values ===================================')
print(ProcessData)
print('=====================================================')
# Output Agreement ==========================================
OutputData=ProcessData

sOutputFileName='C:/VKHCG/05-DS/9999-Data/HORUS-JSON-Country.csv'
OutputData.to_csv(sOutputFileName, index = False)

print('JSON to HORUS - Done')
# Utility done ==============================================
```

Database to HORUS

These utilities enable your solution to import data from existing database sources. I will demonstrate this utility in Chapter 7, with sample data.

Example: This utility imports a list of countries in SQLite data format into HORUS format.

```
# Utility Start Database to HORUS ===================================
# Standard Tools
#==================================================================
import pandas as pd
import sqlite3 as sq
# Input Agreement ==================================================
sInputFileName='C:/VKHCG/05-DS/9999-Data/utility.db'
sInputTable='Country_Code'
conn = sq.connect(sInputFileName)
sSQL='select * FROM ' + sInputTable + ';'
InputData=pd.read_sql_query(sSQL, conn)

print('Input Data Values ===================================')
print(InputData)
print('=====================================================')
# Processing Rules =================================================
ProcessData=InputData

# Remove columns ISO-2-Code and ISO-3-CODE

ProcessData.drop('ISO-2-CODE', axis=1,inplace=True)
ProcessData.drop('ISO-3-Code', axis=1,inplace=True)

# Rename Country and ISO-M49
ProcessData.rename(columns={'Country': 'CountryName'}, inplace=True)
ProcessData.rename(columns={'ISO-M49': 'CountryNumber'}, inplace=True)

# Set new Index
ProcessData.set_index('CountryNumber', inplace=True)

# Sort data by CurrencyNumber
ProcessData.sort_values('CountryName', axis=0, ascending=False, inplace=True)
```

```
print('Process Data Values ==================================')
print(ProcessData)
print('===================================================')
# Output Agreement ==========================================
OutputData=ProcessData

sOutputFileName='C:/VKHCG/05-DS/9999-Data/HORUS-CSV-Country.csv'
OutputData.to_csv(sOutputFileName, index = False)

print('Database to HORUS - Done')
# Utility done ==============================================
```

There are also additional expert utilities that you may want.

Picture to HORUS

These expert utilities enable your solution to convert a picture into extra data. These utilities identify objects in the picture, such as people, types of objects, locations, and many more complex data features. I will discuss this in more detail in Chapter 7.

Example: This utility imports a picture of a dog called Angus in JPG format into HORUS format.

```
# Utility Start Picture to HORUS ===================================
# Standard Tools
#==================================================================
from scipy.misc import imread
import pandas as pd
import matplotlib.pyplot as plt
import numpy as np
# Input Agreement ==========================================
sInputFileName='C:/VKHCG/05-DS/9999-Data/Angus.jpg'
InputData = imread(sInputFileName, flatten=False, mode='RGBA')

print('Input Data Values ==================================')
print('X: ',InputData.shape[0])
print('Y: ',InputData.shape[1])
print('RGBA: ', InputData.shape[2])
print('===================================================')
```

```
# Processing Rules ==========================================
ProcessRawData=InputData.flatten()
y=InputData.shape[2] + 2
x=int(ProcessRawData.shape[0]/y)
ProcessData=pd.DataFrame(np.reshape(ProcessRawData, (x, y)))
sColumns= ['XAxis','YAxis','Red', 'Green', 'Blue','Alpha']
ProcessData.columns=sColumns
ProcessData.index.names =['ID']
print('Rows: ',ProcessData.shape[0])
print('Columns :',ProcessData.shape[1])
print('=======================================================')
print('Process Data Values ==================================')
print('=======================================================')
plt.imshow(InputData)
plt.show()
print('=======================================================')
# Output Agreement ==========================================
OutputData=ProcessData
print('Storing File')
sOutputFileName='C:/VKHCG/05-DS/9999-Data/HORUS-Picture.csv'
OutputData.to_csv(sOutputFileName, index = False)
print('=======================================================')
print('Picture to HORUS - Done')
print('=======================================================')
# Utility done ==============================================
```

Video to HORUS

These expert utilities enable your solution to convert a video into extra data. These utilities identify objects in the video frames, such as people, types of objects, locations, and many more complex data features. I will provide more detail on these in Chapter 7.

Example: This utility imports a movie in MP4 format into HORUS format. The process is performed in two stages.

Movie to Frames

```
# Utility Start Movie to HORUS (Part 1) =======================
# Standard Tools
#===============================================================
import os
import shutil
import cv2
#===============================================================
sInputFileName='C:/VKHCG/05-DS/9999-Data/dog.mp4'
sDataBaseDir='C:/VKHCG/05-DS/9999-Data/temp'
if os.path.exists(sDataBaseDir):
    shutil.rmtree(sDataBaseDir)
if not os.path.exists(sDataBaseDir):
    os.makedirs(sDataBaseDir)
print('======================================================')
print('Start Movie to Frames')
print('======================================================')
vidcap = cv2.VideoCapture(sInputFileName)
success,image = vidcap.read()
count = 0
while success:
    success,image = vidcap.read()
    sFrame=sDataBaseDir + str('/dog-frame-' + str(format(count, '04d'))
    + '.jpg')
    print('Extracted: ', sFrame)
    cv2.imwrite(sFrame, image)
    if os.path.getsize(sFrame) == 0:
        count += -1
        os.remove(sFrame)
        print('Removed: ', sFrame)
    if cv2.waitKey(10) == 27: # exit if Escape is hit
        break
    count += 1
```

```
print('=======================================================')
print('Generated : ', count, ' Frames')
print('=======================================================')
print('Movie to Frames HORUS - Done')
print('=======================================================')
# Utility done ===============================================
```

I have now created frames and need to load them into HORUS.

Frames to Horus

```
# Utility Start Movie to HORUS (Part 2) =====================
# Standard Tools
#===========================================================
from scipy.misc import imread
import pandas as pd
import matplotlib.pyplot as plt
import numpy as np
import os
# Input Agreement ===========================================
sDataBaseDir='C:/VKHCG/05-DS/9999-Data/temp'
f=0
for file in os.listdir(sDataBaseDir):
    if file.endswith(".jpg"):
        f += 1
        sInputFileName=os.path.join(sDataBaseDir, file)
        print('Process : ', sInputFileName)

        InputData = imread(sInputFileName, flatten=False, mode='RGBA')

        print('Input Data Values ===================================')
        print('X: ',InputData.shape[0])
        print('Y: ',InputData.shape[1])
        print('RGBA: ', InputData.shape[2])
        print('=====================================================')
```

```
        # Processing Rules ============================================
        ProcessRawData=InputData.flatten()
        y=InputData.shape[2] + 2
        x=int(ProcessRawData.shape[0]/y)
        ProcessFrameData=pd.DataFrame(np.reshape(ProcessRawData, (x, y)))
        ProcessFrameData['Frame']=file
        print('=====================================================')
        print('Process Data Values ===================================')
        print('=====================================================')
        plt.imshow(InputData)
        plt.show()

        if f == 1:
            ProcessData=ProcessFrameData
        else:
            ProcessData=ProcessData.append(ProcessFrameData)
if f > 0:
    sColumns= ['XAxis','YAxis','Red', 'Green', 'Blue','Alpha','FrameName']
    ProcessData.columns=sColumns
    print('=====================================================')
    ProcessFrameData.index.names =['ID']
    print('Rows: ',ProcessData.shape[0])
    print('Columns :',ProcessData.shape[1])
    print('=====================================================')
    # Output Agreement ===========================================
    OutputData=ProcessData
    print('Storing File')
    sOutputFileName='C:/VKHCG/05-DS/9999-Data/HORUS-Movie-Frame.csv'
    OutputData.to_csv(sOutputFileName, index = False)
print('=====================================================')
print('Processed ; ', f,' frames')
print('=====================================================')
print('Movie to HORUS - Done')
print('=====================================================')
# Utility done ===============================================
```

Audio to HORUS

These expert utilities enable your solution to convert an audio into extra data. These utilities identify objects in the video frames, such as people, types of objects, locations, and many more complex data features. I will discuss these in more detail in Chapter 7.

Example: This utility imports a set of audio files in WAV format into HORUS format.

```python
# Utility Start Audio to HORUS ================================
# Standard Tools
#=============================================================
from scipy.io import wavfile
import pandas as pd
import matplotlib.pyplot as plt
import numpy as np
#=============================================================
def show_info(aname, a,r):
    print ('----------------')
    print ("Audio:", aname)
    print ('----------------')
    print ("Rate:", r)
    print ('----------------')
    print ("shape:", a.shape)
    print ("dtype:", a.dtype)
    print ("min, max:", a.min(), a.max())
    print ('----------------')
    plot_info(aname, a,r)
#=============================================================
def plot_info(aname, a,r):
    sTitle= 'Signal Wave - '+ aname + ' at ' + str(r) + 'hz'
    plt.title(sTitle)
    sLegend=[]
    for c in range(a.shape[1]):
        sLabel = 'Ch' + str(c+1)
        sLegend=sLegend+[str(c+1)]
        plt.plot(a[:,c], label=sLabel)
    plt.legend(sLegend)
    plt.show()
```

```
#=================================================================
sInputFileName='C:/VKHCG/05-DS/9999-Data/2ch-sound.wav'
print('====================================================')
print('Processing : ', sInputFileName)
print('====================================================')
InputRate, InputData = wavfile.read(sInputFileName)
show_info("2 channel", InputData,InputRate)
ProcessData=pd.DataFrame(InputData)
sColumns= ['Ch1','Ch2']
ProcessData.columns=sColumns
OutputData=ProcessData
sOutputFileName='C:/VKHCG/05-DS/9999-Data/HORUS-Audio-2ch.csv'
OutputData.to_csv(sOutputFileName, index = False)
#=================================================================
sInputFileName='C:/VKHCG/05-DS/9999-Data/4ch-sound.wav'
print('====================================================')
print('Processing : ', sInputFileName)
print('====================================================')
InputRate, InputData = wavfile.read(sInputFileName)
show_info("4 channel", InputData,InputRate)
ProcessData=pd.DataFrame(InputData)
sColumns= ['Ch1','Ch2','Ch3', 'Ch4']
ProcessData.columns=sColumns
OutputData=ProcessData
sOutputFileName='C:/VKHCG/05-DS/9999-Data/HORUS-Audio-4ch.csv'
OutputData.to_csv(sOutputFileName, index = False)
#=================================================================
sInputFileName='C:/VKHCG/05-DS/9999-Data/6ch-sound.wav'
print('====================================================')
print('Processing : ', sInputFileName)
print('====================================================')
InputRate, InputData = wavfile.read(sInputFileName)
show_info("6 channel", InputData,InputRate)
ProcessData=pd.DataFrame(InputData)
sColumns= ['Ch1','Ch2','Ch3', 'Ch4', 'Ch5','Ch6']
ProcessData.columns=sColumns
```

```
OutputData=ProcessData
sOutputFileName='C:/VKHCG/05-DS/9999-Data/HORUS-Audio-6ch.csv'
OutputData.to_csv(sOutputFileName, index = False)
#===============================================================
sInputFileName='C:/VKHCG/05-DS/9999-Data/8ch-sound.wav'
print('=======================================================')
print('Processing : ', sInputFileName)
print('=======================================================')
InputRate, InputData = wavfile.read(sInputFileName)
show_info("8 channel", InputData,InputRate)
ProcessData=pd.DataFrame(InputData)
sColumns= ['Ch1','Ch2','Ch3', 'Ch4', 'Ch5','Ch6','Ch7','Ch8']
ProcessData.columns=sColumns
OutputData=ProcessData
sOutputFileName='C:/VKHCG/05-DS/9999-Data/HORUS-Audio-8ch.csv'
OutputData.to_csv(sOutputFileName, index = False)
print('=======================================================')
print('Audio to HORUS - Done')
print('=======================================================')
#===============================================================
# Utility done ================================================
#===============================================================
```

Data Stream to HORUS

These expert utilities enable your solution to handle data streams. Data streams are evolving as the fastest-growing data collecting interface at the edge of the data lake. I will offer extended discussions and advice later in the book on the use of data streaming in the data science ecosystem.

Tip   I use a package called python-confluent-kafka for my Kafka streaming requirements. I have also used PyKafka with success.

In the Retrieve superstep of the functional layer (Chapter 7), I dedicate more text to clarifying how to use and generate full processing chains to retrieve data from your data lake, using optimum techniques.

Assess Utilities

Utilities for this superstep contain all the processing chains for quality assurance and additional data enhancements. I will provide specific implementations for these utilities in Chapter 8.

The assess utilities ensure that the data imported via the Retrieve superstep are of a good quality, to ensure it conforms to the prerequisite standards of your solution. I perform feature engineering at this level, to improve the data for better processing success in the later stages of the data processing.

There are two types of assess utilities:

Feature Engineering

Feature engineering is the process by which you enhance or extract data sources, to enable better extraction of characteristics you are investigating in the data sets. Following is a small subset of the utilities you may use. I will cover many of these in Chapter 8.

Fixers Utilities

Fixers enable your solution to take your existing data and fix a specific quality issue.
Examples include

- Removing leading or lagging spaces from a data entry

  Example in Python:

  ```python
  baddata = "    Data Science with too many spaces is
  bad!!!      "
  print('>',baddata,'<')
  cleandata=baddata.strip()
  print('>',cleandata,'<')
  ```

- Removing nonprintable characters from a data entry

  Example in Python:

  ```python
  import string
  printable = set(string.printable)
  ```

```
baddata = "Data\x00Science with\x02 funny characters is
\x10bad!!!"
cleandata=''.join(filter(lambda x: x in string.printable,
baddata))
print(cleandata)
```

- Reformatting data entry to match specific formatting criteria. Convert 2017/01/31 to 31 January 2017

Example in Python:

```
import datetime as dt
baddate = dt.date(2017, 1, 31)
baddata=format(baddate,'%Y-%m-%d')
print(baddata)
gooddate = dt.datetime.strptime(baddata,'%Y-%m-%d')
gooddata=format(gooddate,'%d %B %Y')
print(gooddata)
```

In Chapter 8, I will consider different complementary utilities.

Adders Utilities

Adders use existing data entries and then add additional data entries to enhance your data. Examples include

- Utilities that look up extra data against existing data entries in your solution. A utility can use the United Nations' ISO M49 for the countries list, to look up 826, to set the country name to United Kingdom. Another utility uses ISO alpha-2 lookup to GB to return the country name back to United Kingdom.

- Zoning data that is added by extra data entries based on a test. The utility can indicate that the data entry is valid, i.e., you found the code in the lookup. A utility can indicate that your data entry for bank balance is either in the black or the red.

I will discuss many of these utilities in Chapter 8.

Process Utilities

Utilities for this superstep contain all the processing chains for building the data vault. In Chapter 9, I will provide specific implementations for these utilities.

I will discuss the data vault's (Time, Person, Object, Location, Event) design, model, and inner workings in detail during the Process supersstep of the functional layer (Chapter 9).

For the purposes of this chapter, I will at this point introduce the data vault as a data structure that uses well-structured design to store data with full history. The basic elements of the data vault are hubs, satellites, and links. Full details on the structure and how to build the data vault is explained in Chapter 9, covering the Process superstep.

In this chapter, I will note only some basic concepts. There are three basic process utilities.

Data Vault Utilities

The data vault is a highly specialist data storage technique that was designed by Dan Linstedt. The data vault is a detail-oriented, historical-tracking, and uniquely linked set of normalized tables that support one or more functional areas of business. It is a hybrid approach encompassing the best of breed between 3rd normal form (3NF) and star schema. I will discuss the use of this configuration in detail in Chapter 9. A basic example is shown in Figure 5-2.

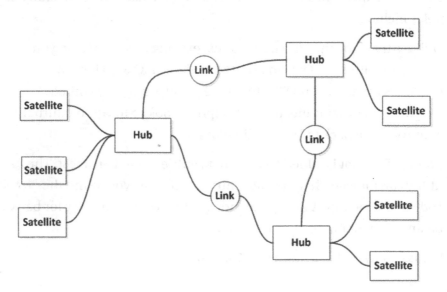

Figure 5-2. *Simple Data Vault*

Hub Utilities

Hub utilities ensure that the integrity of the data vault's (Time, Person, Object, Location, Event) hubs is 100% correct, to verify that the vault is working as designed.

Satellite Utilities

Satellite utilities ensure the integrity of the specific satellite and its associated hub.

Link Utilities

Link utilities ensure the integrity of the specific link and its associated hubs. As the data vault is a highly structured data model, the utilities in the Process superstep of the functional layer (see Chapter 9) will assist you in building your own solution.

Transform Utilities

Utilities for this superstep contain all the processing chains for building the data warehouse from the results of your practical data science. In Chapter 10, I will provide specific implementations for these utilities.

I will discuss the data science and data warehouse design, model, and inner workings in detail in Chapter 10. I will, therefore, only note some basic data warehouse concepts in this chapter.

In the Transform superstep, the system builds dimensions and facts to prepare a data warehouse, via a structured data configuration, for the algorithms in data science to use to produce data science discoveries. There are two basic transform utilities.

Dimensions Utilities

The dimensions use several utilities to ensure the integrity of the dimension structure. Concepts such as conformed dimension, degenerate dimension, role-playing dimension, mini-dimension, outrigger dimension, slowly changing dimension, late-arriving dimension, and dimension types (0, 1, 2, 3) will be discussed in detail in Chapter 10. They all require specific utilities to ensure 100% integrity of the dimension structure.

Pure data science algorithms are the most used at this point in your solution. I will discuss extensively in Chapter 10 what data science algorithms are required to perform practical data science. I will ratify that the most advanced of these are standard algorithms, which will result in common utilities.

Facts Utilities

These consist of a number of utilities that ensure the integrity of the dimensions structure and the facts. There are various statistical and data science algorithms that can be applied to the facts that will result in additional utilities.

Note   The most important utilities for your data science will be transform utilities, as they hold the accredited data science you need for your solution to be successful.

Data Science Utilities

There are several data science–specific utilities that are required for you to achieve success in the data processing ecosystem.

Data Binning or Bucketing

Binning is a data preprocessing technique used to reduce the effects of minor observation errors. Statistical data binning is a way to group a number of more or less continuous values into a smaller number of "bins."

Example: Open your Python editor and create a file called DU-Histogram.py in the directory C:\VKHCG\05-DS\4000-UL\0200-DU. Now copy this code into the file, as follows:

```
import numpy as np
import matplotlib.mlab as mlab
import matplotlib.pyplot as plt

np.random.seed(0)

# example data
mu = 90  # mean of distribution
sigma = 25  # standard deviation of distribution
x = mu + sigma * np.random.randn(5000)

num_bins = 25

fig, ax = plt.subplots()
```

```
# the histogram of the data
n, bins, patches = ax.hist(x, num_bins, normed=1)

# add a 'best fit' line
y = mlab.normpdf(bins, mu, sigma)
ax.plot(bins, y, '--')
ax.set_xlabel('Example Data')
ax.set_ylabel('Probability density')
sTitle=r'Histogram ' + str(len(x)) + ' entries into ' + str(num_bins) + '
Bins: $\mu=' + str(mu) + '$, $\sigma=' + str(sigma) + '$'
ax.set_title(sTitle)

fig.tight_layout()
plt.show()
```

As you can see in Figure 5-3, the binning reduces the 5000 data entries to only 25, with close-to-reality values.

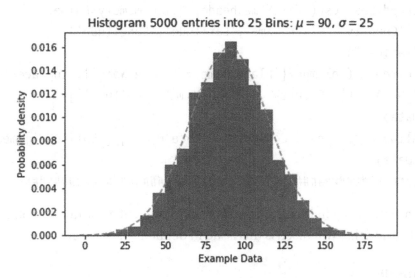

Figure 5-3. *Histogram showing reduced minor observation errors*

Averaging of Data

The use of averaging of features value enables the reduction of data volumes in a control fashion to improve effective data processing.

Example: Open your Python editor and create a file called `DU-Mean.py` in the directory `C:\VKHCG\05-DS\4000-UL\0200-DU`. The following code reduces the data volume from 3562 to 3 data entries, which is a 99.91% reduction.

```
###############################################################
import pandas as pd
###############################################################
InputFileName='IP_DATA_CORE.csv'
OutputFileName='Retrieve_Router_Location.csv'
###############################################################
Base='C:/VKHCG'
sFileName=Base + '/01-Vermeulen/00-RawData/' + InputFileName
print('Loading :',sFileName)
IP_DATA_ALL=pd.read_csv(sFileName,header=0,low_memory=False,
  usecols=['Country','Place Name','Latitude','Longitude'],
encoding="latin-1")
IP_DATA_ALL.rename(columns={'Place Name': 'Place_Name'}, inplace=True)
AllData=IP_DATA_ALL[['Country', 'Place_Name','Latitude']]
print(AllData)
MeanData=AllData.groupby(['Country', 'Place_Name'])['Latitude'].mean()
print(MeanData)
###############################################################
```

This technique also enables the data science to prevent a common issue called overfitting the model.; I will discuss this issue in detail in Chapter 10.

Outlier Detection

Outliers are data that is so different from the rest of the data in the data set that it may be caused by an error in the data source. There is a technique called outlier detection that, with good data science, will identify these outliers.

Example: Open your Python editor and create a file `DU-Outliers.py` in the directory `C:\VKHCG\05-DS\4000-UL\0200-DU`.

```
################################################################
import pandas as pd
################################################################
InputFileName='IP_DATA_CORE.csv'
OutputFileName='Retrieve_Router_Location.csv'
################################################################
Base='C:/VKHCG'
sFileName=Base + '/01-Vermeulen/00-RawData/' + InputFileName
print('Loading :',sFileName)
IP_DATA_ALL=pd.read_csv(sFileName,header=0,low_memory=False,
   usecols=['Country','Place Name','Latitude','Longitude'],
encoding="latin-1")
IP_DATA_ALL.rename(columns={'Place Name': 'Place_Name'}, inplace=True)
LondonData=IP_DATA_ALL.loc[IP_DATA_ALL['Place_Name']=='London']
AllData=LondonData[['Country', 'Place_Name','Latitude']]
print('All Data')
print(AllData)
MeanData=AllData.groupby(['Country', 'Place_Name'])['Latitude'].mean()
StdData=AllData.groupby(['Country', 'Place_Name'])['Latitude'].std()
print('Outliers')
UpperBound=float(MeanData+StdData)
print('Higher than ', UpperBound)
OutliersHigher=AllData[AllData.Latitude>UpperBound]
print(OutliersHigher)
LowerBound=float(MeanData-StdData)
print('Lower than ', LowerBound)
OutliersLower=AllData[AllData.Latitude<LowerBound]
print(OutliersLower)
print('Not Outliers')
OutliersNot=AllData[(AllData.Latitude>=LowerBound) & (AllData.
Latitude<=UpperBound)]
print(OutliersNot)
################################################################
```

Organize Utilities

Utilities for this superstep contain all the processing chains for building the data marts. The organize utilities are mostly used to create data marts against the data science results stored in the data warehouse dimensions and facts. Details on these are provided in Chapter 11.

Report Utilities

Utilities for this superstep contain all the processing chains for building virtualization and reporting of the actionable knowledge. The report utilities are mostly used to create data virtualization against the data science results stored in the data marts. Details on these are provided in Chapter 11.

Maintenance Utilities

The data science solutions you are building are a standard data system and, consequently, require maintenance utilities, as with any other system. Data engineers and data scientists must work together to ensure that the ecosystem works at its most efficient level at all times.

Utilities cover several areas.

Backup and Restore Utilities

These perform different types of database backups and restores for the solution. They are standard for any computer system. For the specific utilities, I suggest you have an in-depth discussion with your own systems manager or the systems manager of your client. I normally provide a wrapper for the specific utility that I can call in my data science ecosystem, without direct exposure to the custom requirements at each customer.

Checks Data Integrity Utilities

These utilities check the allocation and structural integrity of database objects and indexes across the ecosystem, to ensure the accurate processing of the data into knowledge. I will discuss specific utilities in Chapters 7–11.

History Cleanup Utilities

These utilities archive and remove entries in the history tables in the databases. I will discuss specific cleanup utilities in Chapters 7–11.

Note   The "right-to-be-forgotten" statute in various countries around the world imposes multifaceted requirements in this area of data science to be able to implement selective data processing.

Warning   I suggest you look at your information protection laws in detail, because the processing of data now via data science is becoming highly exposed, and in a lot of countries, fines are imposed if you get these processing rules wrong.

Maintenance Cleanup Utilities

These utilities remove artifacts related to maintenance plans and database backup files. I will discuss these utilities in detail in Chapter 6.

Notify Operator Utilities

Utilities that send notification messages to the operations team about the status of the system are crucial to any data science factory. These utilities are discussed in detail in Chapter 6.

Rebuild Data Structure Utilities

These utilities rebuild database tables and views to ensure that all the development is as designed. In Chapters 6–11, I will discuss the specific rebuild utilities.

Reorganize Indexing Utilities

These utilities reorganize indexes in database tables and views, which is a major operational process when your data lake grows at a massive volume and velocity. The variety of data types also complicates the application of indexes to complex data structures. In Chapters 6–11, I will discuss the specific rebuild utilities.

As a data scientist, you must understand when and how your data sources will change. An unclear indexing strategy could slow down algorithms without your taking note, and you could lose data, owing to your not handling the velocity of the data flow.

Shrink/Move Data Structure Utilities

These reduce the footprint size of your database data and associated log artifacts, to ensure an optimum solution is executing. I will discuss specific rebuild utilities in Chapters 6–11.

Solution Statistics Utilities

These utilities update information about the data science artifacts, to ensure that your data science structures are recorded. Call it data science on your data science. These utilities will be discussed in detail in Chapter 6.

The preceding list is a comprehensive, but not all-inclusive. I suggest that you speak to your development and operations organization staff, to ensure that your data science solution fits into the overall data processing structures of your organization.

Processing Utilities

The data science solutions you are building require processing utilities to perform standard system processing. The data science environment requires two basic processing utility types.

Scheduling Utilities

The scheduling utilities I use are based on the basic agile scheduling principles.

Backlog Utilities

Backlog utilities accept new processing requests into the system and are ready to be processed in future processing cycles.

To-Do Utilities

The to-do utilities take a subset of backlog requests for processing during the next processing cycle. They use classification labels, such as priority and parent-child relationships, to decide what process runs during the next cycle.

Doing Utilities

The doing utilities execute the current cycle's requests.

Done Utilities

The done utilities confirm that the completed requests performed the expected processing.

Monitoring Utilities

The monitoring utilities ensure that the complete system is working as expected.

Engineering a Practical Utility Layer

Any source code or other supplementary material referenced by me in this book is available to readers on GitHub, via this book's product page, located at www.apress.com/ 9781484230534.

Note   See the following source code from Chapter 2: ./VKHCG/05-DS/4000-UL.

The utility layer holds all the utilities you share across the data science environment. I suggest that you create three sublayers to help the utility layer support better future use of the utilities.

Maintenance Utility

Collect all the maintenance utilities in this single directory, to enable the environment to handle the utilities as a collection.

Note   See the following source code from Chapter 2: `./VKHCG/05-DS/4000-UL/4000-UL/0100-MU`.

I suggest that you keep a maintenance utilities registry, to enable your entire team to use the common utilities. Include enough documentation for each maintenance utility, to explain its complete workings and requirements.

Data Utility

Collect all the data utilities in this single directory, to enable the environment to handle the utilities as a collection.

Note   See the following source code from Chapter 2: `./VKHCG/05-DS/4000-UL/4000-UL/0200-DU`.

I suggest that you keep a data utilities registry to enable your entire team to use the common utilities. Include enough documentation for each data utility to explain its complete workings and requirements.

Processing Utility

Collect all the processing utilities in this single directory to enable the environment to handle the utilities as a collection.

Note   See the following source code from Chapter 2: `./VKHCG/05-DS/4000-UL/4000-UL/0300-PU`.

I suggest that you keep a processing utilities registry, to enable your entire team to use the common utilities. Include sufficient documentation for each processing utility to explain its complete workings and requirements.

Warning   Ensure that you support your company's processing environment and that the suggested environment supports an agile processing methodology. This may not always match your own environment.

Caution   Remember: These utilities are used by your wider team, if you interrupt them, you will pause other current working processing. Take extra care with this layer's artifacts.

Summary

I have completed the utility layer that supports the common utilities in the data science environment. This layer will evolve as your data science improves. In the beginning of your environment, you will add numerous extra utilities, but in time, you will reuse the artifacts you already own and trust. You must be at ease with the utility layer now and understand that growing it in keeping with your requirements means that you have successfully completed this chapter.

Remember   A good utility layer will continuously evolve. That is normal.

In next chapter, I will move on to the three management layers.

## CHAPTER 6

Three Management Layers

This chapter is about the three management layers that are must-haves for any large-scale data science system. I will discuss them at a basic level. I suggest you scale-out these management capabilities, as your environment grows.

Operational Management Layer

The operational management layer is the core store for the data science ecosystem's complete processing capability. The layer stores every processing schedule and workflow for the all-inclusive ecosystem.

This area enables you to see a singular view of the entire ecosystem. It reports the status of the processing.

The operations management layer is the layer where I record the following.

Processing-Stream Definition and Management

The processing-stream definitions are the building block of the data science ecosystem. I store all my current active processing scripts in this section.

Definition management describes the workflow of the scripts through the system, ensuring that the correct execution order is managed, as per the data scientists' workflow design.

Tip   Keep all your general techniques and algorithms in a source-control-based system, such as GitHub or SVN, in the format of importable libraries. That way, you do not have to verify if they work correctly every time you use them.

© Andreas François Vermeulen 2018
A. F. Vermeulen, *Practical Data Science*, https://doi.org/10.1007/978-1-4842-3054-1_6

Advice   I spend 10% of my time generating new processing building blocks every week and 10% improving existing building blocks. I can confirm that this action easily saves more than 20% of my time on processing new data science projects when I start them, as I already have a base set of tested code to support the activities required. So, please invest in your own and your team's future, by making this a standard practice for the team. You will not regret the investment.

Warning   When you replace existing building blocks, check for impacts downstream. I suggest you use a simple versioning scheme of `mylib_001_01`. That way, you can have 999 versions with 99 sub-versions. This also ensures that your new version can be orderly rolled out into your customer base. The most successful version is to support the process with a good version-control process that can support multiple branched or forked code sets.

Parameters

The parameters for the processing are stored in this section, to ensure a single location for all the system parameters. You will see in all the following examples that there is an ecosystem setup phase.

```
if sys.platform == 'linux':
    Base=os.path.expanduser('~') + '/VKHCG'
else:
    Base='C:/VKHCG'
################################################################
print('##############################')
print('Working Base :',Base, ' using ', sys.platform)
print('##############################')
################################################################
```

```
sFileDir=Base + '/01-Vermeulen/01-Retrieve/01-EDS/02-Python'
if not os.path.exists(sFileDir):
    os.makedirs(sFileDir)
###############################################################
sFileName=Base + '/01-Vermeulen/00-RawData/Country_Currency.xlsx'
```

In my production system, for each customer, we place all these parameters in a single location and then simply call the single location. Two main designs are used:

1. A simple text file that we then import into every Python script

2. A parameter database supported by a standard parameter setup script that we then include into every script

I will also admit to having several parameters that follow the same format as the preceding examples, and I simply collect them in a section at the top of the code.

Advice   Find a way that works best for your team and standardize that method across your team.

Scheduling

The scheduling plan is stored in this section, to enable central control and visibility of the complete scheduling plan for the system. In my solution, I use a Drum-Buffer-Rope (Figure 6-1) methodology. The principle is simple.

Figure 6-1. *Original Drum-Buffer-Rope use*

Similar to a troop of people marching, the Drum-Buffer-Rope methodology is a standard practice to identify the slowest process and then use this process to pace the complete pipeline. You then tie the rest of the pipeline to this process to control the eco-system's speed.

So, you place the "drum" at the slow part of the pipeline, to give the processing pace, and attach the "rope" to the beginning of the pipeline, and the end by ensuring that no processing is done that is not attached to this drum. This ensures that your processes complete more efficiently, as nothing is entering or leaving the process pipe without been recorded by the drum's beat.

I normally use an independent Python program that employs the directed acyclic graph (DAG) that is provided by the network libraries' DiGraph structure. This automatically resolves duplicate dependencies and enables the use of a topological sort, which ensures that tasks are completed in the order of requirement. Following is an example: Open this in your Python editor and view the process.

```
import networkx as nx
```

Here you construct your network in any order.

```
DG = nx.DiGraph([
        ('Start','Retrieve1'),
        ('Start','Retrieve2'),
        ('Retrieve1','Assess1'),
        ('Retrieve2','Assess2'),
        ('Assess1','Process'),
        ('Assess2','Process'),
        ('Process','Transform'),
        ('Transform','Report1'),
        ('Transform','Report2')
        ])
```

Here is your network unsorted:

```
print("Unsorted Nodes")
print(DG.nodes())
```

You can test your network for valid DAG.

```
print("Is a DAG?",nx.is_directed_acyclic_graph(DG))
```

Now you sort the DAG into a correct order.

```
sOrder=nx.topological_sort(DG)
print("Sorted Nodes")
print(sOrder)
```

You can also visualize your network.

```
pos=nx.spring_layout(DG)
nx.draw_networkx_nodes(DG,pos=pos,node_size = 1000)
nx.draw_networkx_edges(DG,pos=pos)
nx.draw_networkx_labels(DG,pos=pos)
```

You can add some extra nodes and see how this resolves the ordering. I suggest that you experiment with networks of different configurations, as this will enable you to understand how the process truly assists with the processing workload.

Tip   I normally store the requirements for the nodes in a common database. That way, I can upload the requirements for multiple data science projects and resolve the optimum requirement with ease.

Monitoring

The central monitoring process is in this section to ensure that there is a single view of the complete system. Always ensure that you monitor your data science from a single point. Having various data science processes running on the same ecosystem without central monitoring is not advised.

Tip   I always get my data science to simply set an active status in a central table when it starts and a not-active when it completes. That way, the entire team knows who is running what and can plan its own processing better.

If you are running on Windows, try the following:

```
conda install -c primer wmi
import wmi
c = wmi.WMI ()
```

```
for process in c.Win32_Process ():
  print (process.ProcessId, process.Name)
```

For Linux, try this

```
import os
pids = [pid for pid in os.listdir('/proc') if pid.isdigit()]

for pid in pids:
    try:
        print open(os.path.join('/proc', pid, 'cmdline'), 'rb').read()
    except IOError: # proc has already terminated
        continue
```

This will give you a full list of all running processes. I normally just load this into a table every minute or so, to create a monitor pattern for the ecosystem.

Communication

All communication from the system is handled in this one section, to ensure that the system can communicate any activities that are happening. We are using a complex communication process via Jira, to ensure we have all our data science tracked. I suggest you look at the Conda install -c conda-forge jira.

I will not provide further details on this subject, as I have found that the internal communication channel in any company is driven by the communication tools it uses. The only advice I will offer is to *communicate*! You would be alarmed if at least once a week, you lost a project owing to someone not communicating what they are running.

Alerting

The alerting section uses communications to inform the correct person, at the correct time, about the correct status of the complete system. I use Jira for alerting, and it works well. If any issue is raised, alerting provides complete details of what the status was and the errors it generated.

I will now discuss each of these sections in more detail and offer practical examples of what to expect or create in each section.

Audit, Balance, and Control Layer

The audit, balance, and control layer controls any processing currently under way. This layer is the engine that ensures that each processing request is completed by the ecosystem as planned. The audit, balance, and control layer is the single area in which you can observe what is currently running within your data scientist environment.

It records

- Process-execution statistics

- Balancing and controls

- Rejects- and error-handling

- Fault codes management

The three subareas are utilized in following manner.

Audit

First, let's define what I mean by *audit*.

An audit is a systematic and independent examination of the ecosystem.

The audit sublayer records the processes that are running at any specific point within the environment. This information is used by data scientists and engineers to understand and plan future improvements to the processing.

Tip   Make sure your algorithms and processing generate a good and complete audit trail.

My experience shows that a good audit trail is extremely crucial. The use of the built-in audit capability of the data science technology stack's components supply you with a rapid and effective base for your auditing. I will discuss what audit statistics are essential to the success of your data science.

In the data science ecosystem, the audit consists of a series of observers that record preapproved processing indicators regarding the ecosystem. I have found the following to be good indicators for audit purposes.

Built-in Logging

I advise you to design your logging into an organized preapproved location, to ensure that you capture every relevant log entry. I also recommend that you do not change the internal or built-in logging process of any of the data science tools, as this will make any future upgrades complex and costly. I suggest that you handle the logs in same manner you would any other data source.

Normally, I build a controlled systematic and independent examination of all the built-in logging vaults. That way, I am sure I can independently collect and process these logs across the ecosystem. I deploy five independent watchers for each logging location, as logging usually has the following five layers.

Debug Watcher

This is the maximum verbose logging level. If I discover any debug logs in my ecosystem, I normally raise an alarm, as this means that the tool is using precise processing cycles to perform low-level debugging.

Warning   Tools running debugging should not be part of a production system.

Information Watcher

The information level is normally utilized to output information that is beneficial to the running and management of a system. I pipe these logs to the central Audit, Balance, and Control data store, using the ecosystem as I would any other data source.

Warning Watcher

Warning is often used for handled "exceptions" or other important log events. Usually this means that the tool handled the issue and took corrective action for recovery. I pipe these logs to the central Audit, Balance, and Control data store, using the ecosystem as I would any other data source. I also add a warning to the Performing a Cause and Effect Analysis System data store. I will discuss this critical tool later in this chapter.

Error Watcher

Error is used to log all unhandled exceptions in the tool. This is not a good state for the overall processing to be in, as it means that a specific step in the planned processing did not complete as expected. Now, the ecosystem must handle the issue and take corrective action for recovery. I pipe these logs to the central Audit, Balance, and Control data store, using the ecosystem as I would any other data source. I also add an error to the Performing a Cause and Effect Analysis System data store. I will discuss this critical tool later in this chapter.

Fatal Watcher

Fatal is reserved for special exceptions/conditions for which it is imperative that you quickly identify these events. This is not a good state for the overall processing to be in, as it means a specific step in the planned processing has not completed as expected. This means the ecosystem must now handle the issue and take corrective action for recovery. Once again, I pipe these logs to the central Audit, Balance, and Control data store, using the ecosystem as I would any other data source. I also add an error to the Performing a Cause and Effect Analysis System data store, which I will discuss later in this chapter.

I have discovered that by simply using built-in logging and a good cause-and-effect analysis system, I can handle more than 95% of all issues that arise in the ecosystem.

Basic Logging

Following is a basic logging process I normally deploy in my data science. Open your Python editor and enter the following logging example. You will require the following libraries:

```
import sys
import os
import logging
import uuid
import shutil
import time
```

You next set up the basic ecosystem, as follows:

```
if sys.platform == 'linux':
    Base=os.path.expanduser('~') + '/VKHCG'
else:
     Base='C:/VKHCG'
```

You need the following constants to cover the ecosystem:

```
sCompanies=['01-Vermeulen','02-Krennwallner','03-Hillman','04-Clark']
sLayers=['01-Retrieve','02-Assess','03-Process','04-Transform','05-
Organise','06-Report']
sLevels=['debug','info','warning','error']
```

You can now build the loops to perform a basic logging run.

```
for sCompany in sCompanies:
    sFileDir=Base + '/' + sCompany
    if not os.path.exists(sFileDir):
        os.makedirs(sFileDir)
    for sLayer in sLayers:
        log = logging.getLogger()  # root logger
        for hdlr in log.handlers[:]:  # remove all old handlers
            log.removeHandler(hdlr)
        sFileDir=Base + '/' + sCompany + '/' + sLayer + '/Logging'
        if os.path.exists(sFileDir):
            shutil.rmtree(sFileDir)
        time.sleep(2)
        if not os.path.exists(sFileDir):
            os.makedirs(sFileDir)
        skey=str(uuid.uuid4())
        sLogFile=Base + '/' + sCompany + '/' + sLayer + '/Logging/
        Logging_'+skey+'.log'
        print('Set up:',sLogFile)
```

You set up logging to file, as follows:

```
logging.basicConfig(level=logging.DEBUG,
                    format='%(asctime)s %(name)-12s %(levelname)-8s
                    %(message)s',
                    datefmt='%m-%d %H:%M',
                    filename=sLogFile,
                    filemode='w')
```

You define a handler, which writes all messages to sys.stderr.

```
console = logging.StreamHandler()
console.setLevel(logging.INFO)
```

You set a format for console use.

```
formatter = logging.Formatter('%(name)-12s: %(levelname)-8s
%(message)s')
```

You activate the handler to use this format.

```
console.setFormatter(formatter)
```

Now, add the handler to the root logger.

```
logging.getLogger('').addHandler(console)
```

Test your root logging.

```
logging.info('Practical Data Science is fun!.')
```

Test all the other levels.

```
for sLevel in sLevels:
    sApp='Application-'+ sCompany + '-' + sLayer + '-' + sLevel
    logger = logging.getLogger(sApp)

    if sLevel == 'debug':
        logger.debug('Practical Data Science logged a debugging
        message.')
```

```
    if sLevel == 'info':
        logger.info('Practical Data Science logged information
        message.')

    if sLevel == 'warning':
        logger.warning('Practical Data Science logged a warning
        message.')

    if sLevel == 'error':
        logger.error('Practical Data Science logged an error
        message.')
```

This logging enables you to log everything that occurs in your data science processing to a central file, for each run of the process.

Process Tracking

I normally build a controlled systematic and independent examination of the process for the hardware logging. There is numerous server-based software that monitors temperature sensors, voltage, fan speeds, and load and clock speeds of a computer system. I suggest you go with the tool with which you and your customer are most comfortable. I do, however, advise that you use the logs for your cause-and-effect analysis system.

Data Provenance

Keep records for every data entity in the data lake, by tracking it through all the transformations in the system. This ensures that you can reproduce the data, if needed, in the future and supplies a detailed history of the data's source in the system.

Data Lineage

Keep records of every change that happens to the individual data values in the data lake. This enables you to know what the exact value of any data record was in the past. It is normally achieved by a valid-from and valid-to audit entry for each data set in the data science environment.

Balance

The balance sublayer ensures that the ecosystem is balanced across the accessible processing capability or has the capability to top up capability during periods of extreme processing. The processing on-demand capability of a cloud ecosystem is highly desirable for this purpose.

Tip   Plan your capability as a combination of always-on and top-up processing.

By using the audit trail, it is possible to adapt to changing requirements and forecast what you will require to complete the schedule of work you submitted to the ecosystem. I have found that deploying a deep reinforced learning algorithm against the cause-and-effect analysis system can handle any balance requirements dynamically.

Note   In my experience, even the best pre-plan solution for processing will disintegrate against a good deep-learning algorithm, with reinforced learning capability handling the balance in the ecosystem.

Control

The control sublayer controls the execution of the current active data science. The control elements are a combination of the control element within the Data Science Technology Stack's individual tools plus a custom interface to control the overarching work.

The control sublayer also ensures that when processing experiences an error, it can try a recovery, as per your requirements, or schedule a clean-up utility to undo the error. The cause-and-effect analysis system is the core data source for the distributed control system in the ecosystem.

I normally use a distributed yoke solution to control the processing. I create an independent process that is created solely to monitor a specific portion of the data processing ecosystem control. So, the control system consists of a series of yokes at each control point that uses Kafka messaging to communicate the control requests. The yoke then converts the requests into a process to execute and manage in the ecosystem.

The yoke system ensures that the distributed tasks are completed, even if it loses contact with the central services. The yoke solution is extremely useful in the Internet of things environment, as you are not always able to communicate directly with the data source.

Yoke Solution

The yoke solution is a custom design I have worked on over years of deployments.

Apache Kafka is an open source stream processing platform developed to deliver a unified, high-throughput, low-latency platform for handling real-time data feeds.

Kafka provides a publish-subscribe solution that can handle all activity-stream data and processing.

The Kafka environment enables you to send messages between producers and consumers that enable you to transfer control between different parts of your ecosystem while ensuring a stable process.

I will give you a simple example of the type of information you can send and receive.

Producer

The producer is the part of the system that generates the requests for data science processing, by creating structures messages for each type of data science process it requires. The producer is the end point of the pipeline that loads messages into Kafka.

Note   This is for your information only. You do not have to code this and make it run.

```
from kafka import KafkaProducer
producer = KafkaProducer(bootstrap_servers='localhost:1234')
for _ in range(100):
    producer.send('Retrieve', b'Person.csv')
# Block until a single message is sent (or timeout)
future = producer.send('Retrieve', b'Last_Name.json')
result = future.get(timeout=60)
# Block until all pending messages are at least put on the network
# NOTE: This does not guarantee delivery or success! It is really
# only useful if you configure internal batching using linger_ms
```

```
producer.flush()
# Use a key for hashed-partitioning
producer.send('York', key=b'Retrieve', value=b'Run')
# Serialize json messages
import json
producer = KafkaProducer(value_serializer=lambda v: json.dumps(v).
encode('utf-8'))
producer.send('Retrieve', {'Retrieve': 'Run'})
# Serialize string keys
producer = KafkaProducer(key_serializer=str.encode)
producer.send('Retrieve', key='ping', value=b'1234')
# Compress messages
producer = KafkaProducer(compression_type='gzip')
for i in range(1000):
    producer.send('Retrieve', b'msg %d' % i)
```

Consumer

The consumer is the part of the process that takes in messages and organizes them for processing by the data science tools. The consumer is the end point of the pipeline that offloads the messages from Kafka.

Note   This is for your information only. You do not have to code this and make it run.

```
from kafka import KafkaConsumer
import msgpack
consumer = KafkaConsumer('Yoke')
for msg in consumer:
    print (msg)
# join a consumer group for dynamic partition assignment and offset commits
from kafka import KafkaConsumer
consumer = KafkaConsumer('Yoke', group_id='Retrieve')
for msg in consumer:
    print (msg)
```

```
# manually assign the partition list for the consumer
from kafka import TopicPartition
consumer = KafkaConsumer(bootstrap_servers='localhost:1234')
consumer.assign([TopicPartition('Retrieve', 2)])
msg = next(consumer)
# Deserialize msgpack-encoded values
consumer = KafkaConsumer(value_deserializer=msgpack.loads)
consumer.subscribe(['Yoke'])
for msg in consumer:
    assert isinstance(msg.value, dict)
```

Directed Acyclic Graph Scheduling

This solution uses a combination of graph theory and publish-subscribe stream data processing to enable scheduling. You can use the Python NetworkX library to resolve any conflicts, by simply formulating the graph into a specific point before or after you send or receive messages via Kafka. That way, you ensure an effective and an efficient processing pipeline.

Tip   I normally publish the request onto three different message queues, to ensure that the pipeline is complete. The extra redundancy outweighs the extra processing, as the message is typically every small.

Yoke Example

Following is a simple simulation of what I suggest you perform with your processing. Open your Python editor and create the following three parts of the yoke processing pipeline:

Create a file called Run-Yoke.py in directory ..\VKHCG\77-Yoke.

Enter this code into the file and save it.

```
################################################################
# -*- coding: utf-8 -*-
################################################################
import sys
```

```
import os
import shutil
################################################################
def prepecosystem():
    if sys.platform == 'linux':
        Base=os.path.expanduser('~') + '/VKHCG'
    else:
        Base='C:/VKHCG'
    ################################################################
    sFileDir=Base + '/77-Yoke'
    if not os.path.exists(sFileDir):
        os.makedirs(sFileDir)
    ################################################################
    sFileDir=Base + '/77-Yoke/10-Master'
    if not os.path.exists(sFileDir):
        os.makedirs(sFileDir)
    ################################################################
    sFileDir=Base + '/77-Yoke/20-Slave'
    if not os.path.exists(sFileDir):
        os.makedirs(sFileDir)
    ################################################################
    return Base
################################################################
def makeslavefile(Base,InputFile):
    sFileNameIn=Base + '/77-Yoke/10-Master/'+InputFile
    sFileNameOut=Base + '/77-Yoke/20-Slave/'+InputFile

    if os.path.isfile(sFileNameIn):
        shutil.move(sFileNameIn,sFileNameOut)
################################################################
if __name__ == '__main__':
    ################################################################
    print('### Start ###################################')
    ################################################################
    Base = prepecosystem()
    sFiles=list(sys.argv)
```

```
    for sFile in sFiles:
        if sFile != 'Run-Yoke.py':
            print(sFile)
            makeslavefile(Base,sFile)
    ################################################################
    print('### Done!! #########################################')
    ################################################################
```

Next, create the Master Producer Script. This script will place nine files in the master message queue simulated by a directory called 10-Master.

Create a file called Master-Yoke.py in directory ..\VKHCG\77-Yoke.

```
################################################################
# -*- coding: utf-8 -*-
################################################################
import sys
import os
import sqlite3 as sq
from pandas.io import sql
import uuid
import re
from multiprocessing import Process
################################################################
def prepecosystem():
    if sys.platform == 'linux':
        Base=os.path.expanduser('~') + '/VKHCG'
    else:
        Base='C:/VKHCG'
    ############################################################
    sFileDir=Base + '/77-Yoke'
    if not os.path.exists(sFileDir):
        os.makedirs(sFileDir)
    ############################################################
    sFileDir=Base + '/77-Yoke/10-Master'
    if not os.path.exists(sFileDir):
        os.makedirs(sFileDir)
    ############################################################
```

```python
        sFileDir=Base + '/77-Yoke/20-Slave'
        if not os.path.exists(sFileDir):
            os.makedirs(sFileDir)
        ################################################################
        sFileDir=Base + '/77-Yoke/99-SQLite'
        if not os.path.exists(sFileDir):
            os.makedirs(sFileDir)
        ################################################################
        sDatabaseName=Base + '/77-Yoke/99-SQLite/Yoke.db'
        conn = sq.connect(sDatabaseName)
        print('Connecting :',sDatabaseName)
        sSQL='CREATE TABLE IF NOT EXISTS YokeData (\
         PathFileName VARCHAR (1000) NOT NULL\
         );'
        sql.execute(sSQL,conn)
        conn.commit()
        conn.close()
        return Base,sDatabaseName
################################################################
def makemasterfile(sseq,Base,sDatabaseName):
    sFileName=Base + '/77-Yoke/10-Master/File_' + sseq +\
    '_' + str(uuid.uuid4()) + '.txt'
    sFileNamePart=os.path.basename(sFileName)
    smessage="Practical Data Science Yoke \n File: " + sFileName
    with open(sFileName, "w") as txt_file:
        txt_file.write(smessage)

    connmerge = sq.connect(sDatabaseName)
    sSQLRaw="INSERT OR REPLACE INTO YokeData(PathFileName)\
            VALUES\
            ('" + sFileNamePart + "');"
    sSQL=re.sub('\s{2,}', ' ', sSQLRaw)
    sql.execute(sSQL,connmerge)
    connmerge.commit()
    connmerge.close()
################################################################
```

```
if __name__ == '__main__':
    ################################################################
    print('### Start ####################################')
    ################################################################
    Base,sDatabaseName = prepecosystem()
    for t in range(1,10):
        sFile='{num:06d}'.format(num=t)
        print('Spawn:',sFile)
        p = Process(target=makemasterfile, args=(sFile,Base,sDatabaseName))
        p.start()
        p.join()
    ################################################################
    print('### Done!! ####################################')
    ################################################################
```

Execute the master script to load the messages into the yoke system.

Next, create the Slave Consumer Script. This script will place nine files in the master message queue simulated by a directory called 20-Slave.

Create a file called Slave-Yoke.py in directory ..\VKHCG\77-Yoke.

```
################################################################
# -*- coding: utf-8 -*-
################################################################
import sys
import os
import sqlite3 as sq
from pandas.io import sql
import pandas as pd
from multiprocessing import Process
################################################################
def prepecosystem():
    if sys.platform == 'linux':
        Base=os.path.expanduser('~') + '/VKHCG'
    else:
        Base='C:/VKHCG'
    ################################################################
    sFileDir=Base + '/77-Yoke'
```

```python
    if not os.path.exists(sFileDir):
        os.makedirs(sFileDir)
    ################################################################
    sFileDir=Base + '/77-Yoke/10-Master'
    if not os.path.exists(sFileDir):
        os.makedirs(sFileDir)
    ################################################################
    sFileDir=Base + '/77-Yoke/20-Slave'
    if not os.path.exists(sFileDir):
        os.makedirs(sFileDir)
    ################################################################
    sFileDir=Base + '/77-Yoke/99-SQLite'
    if not os.path.exists(sFileDir):
        os.makedirs(sFileDir)
    ################################################################
    sDatabaseName=Base + '/77-Yoke/99-SQLite/Yoke.db'
    conn = sq.connect(sDatabaseName)
    print('Connecting :',sDatabaseName)
    sSQL='CREATE TABLE IF NOT EXISTS YokeData (\
     PathFileName VARCHAR (1000) NOT NULL\
     );'
    sql.execute(sSQL,conn)
    conn.commit()
    conn.close()
    return Base,sDatabaseName
################################################################
def makeslavefile(Base,InputFile):
    sExecName=Base + '/77-Yoke/Run-Yoke.py'
    sExecLine='python ' + sExecName + ' ' + InputFile
    os.system(sExecLine)
################################################################
if __name__ == '__main__':
    ################################################################
    print('### Start #######################################')
    ################################################################
    Base,sDatabaseName = prepecosystem()
```

```
connslave = sq.connect(sDatabaseName)
sSQL="SELECT PathFileName FROM YokeData;"
SlaveData=pd.read_sql_query(sSQL, connslave)
for t in range(SlaveData.shape[0]):
    sFile=str(SlaveData['PathFileName'][t])
    print('Spawn:',sFile)
    p = Process(target=makeslavefile, args=(Base,sFile))
    p.start()
    p.join()
##################################################################
print('### Done!! ######################################')
##################################################################
```

Execute the script and observe the slave script retrieving the messages and then using the Run-Yoke to move the files between the 10-Master and 20-Slave directories. This is a simulation of how you could use systems such as Kafka to send messages via a producer and, later, via a consumer to complete the process, by retrieving the messages and executing another process to handle the data science processing.

Well done, you just successfully simulated a simple message system.

Tip   In this manner, I have successfully passed messages between five data centers across four time zones. It is worthwhile to invest time and practice to achieve a high level of expertise with using messaging solutions.

Cause-and-Effect Analysis System

The cause-and-effect analysis system is the part of the ecosystem that collects all the logs, schedules, and other ecosystem-related information and enables data scientists to evaluate the quality of their system.

Advice   Apply the same data science techniques to this data set, to uncover the insights you need to improve your data science ecosystem.

You have now successfully completed the management sections of the ecosystem. I will now introduce the core data science process for this book.

Functional Layer

The functional layer of the data science ecosystem is the largest and most essential layer for programming and modeling. In Chapters 7–11, I will cover the complete functional layer in detail.

Any data science project must have processing elements in this layer. The layer performs all the data processing chains for the practical data science.

Before I officially begin the discussion of the functional layer, I want to share my successful fundamental data science process.

Data Science Process

Following are the five fundamental data science process steps that are the core of my approach to practical data science.

Start with a What-if Question

Decide what you want to know, even if it is only the subset of the data lake you want to use for your data science, which is a good start.

For example, let's consider the example of a small car dealership. Suppose I have been informed that Bob was looking at cars last weekend. Therefore, I ask: "What if I know what car my customer Bob will buy next?"

Take a Guess at a Potential Pattern

Use your experience or insights to guess a pattern you want to discover, to uncover additional insights from the data you already have. For example, I guess Bob will buy a car every three years, and as he currently owns a three-year-old Audi, he will likely buy another Audi. I have no proof; it's just a guess or so-called gut feeling. Something I could prove via my data science techniques.

Gather Observations and Use Them to Produce a Hypothesis

So, I start collecting car-buying patterns on Bob and formulate a hypothesis about his future behavior. For those of you who have not heard of a hypothesis, it is a proposed explanation, prepared on the basis of limited evidence, as a starting point for further investigation.

"I saw Bob looking at cars last weekend in his Audi" then becomes "Bob will buy an Audi next, as his normal three-year buying cycle is approaching."

Use Real-World Evidence to Verify the Hypothesis

Now, we verify our hypothesis with real-world evidence. On our CCTV, I can see that Bob is looking only at Audis and returned to view a yellow Audi R8 five times over last two weeks.

On the sales ledger, I see that Bob bought an Audi both three years previous and six previous. Bob's buying pattern, then, is every three years.

So, our hypothesis is verified. Bob wants to buy my yellow Audi R8.

Collaborate Promptly and Regularly with Customers and Subject Matter Experts As You Gain Insights

The moment I discover Bob's intentions, I contact the salesperson, and we successfully sell Bob the yellow Audi R8.

These five steps work, but I will acknowledge that they serve only as my guide while prototyping. Once you start working with massive volumes, velocities, and variance in data, you will need a more structured framework to handle the data science.

In Chapters 7–11, I will return to the data science process and explain how the preceding five steps fit into the broader framework.

So, let's discuss the functional layer of the framework in more detail.

As previously mentioned, the functional layer of the data science ecosystem is the largest and most essential layer for programming and modeling. The functional layer is the part of the ecosystem that runs the comprehensive data science ecosystem.

Warning   When database administrators refer to a *data model*, they are referring to data schemas and data formats in the data science world. Data science views data models as a set of algorithms and processing rules applied as part of data processing pipelines. So, make sure that when you talk to people, they are clear on what you are talking about in all communications channels.

It consists of several structures, as follows:

- *Data schemas and data formats*: Functional data schemas and data formats deploy onto the data lake's raw data, to perform the required schema-on-query via the functional layer.

- *Data models*: These form the basis for future processing to enhance the processing capabilities of the data lake, by storing already processed data sources for future use by other processes against the data lake.

- *Processing algorithms*: The functional processing is performed via a series of well-designed algorithms across the processing chain.

- *Provisioning of infrastructure*: The functional infrastructure provision enables the framework to add processing capability to the ecosystem, using technology such as Apache Mesos, which enables the dynamic previsioning of processing work cells.

The processing algorithms and data models are spread across six supersteps for processing the data lake.

1. Retrieve

    - This superstep contains all the processing chains for retrieving data from the raw data lake into a more structured format. The details of this superstep are discussed in Chapter 7.

2. Assess

    - This superstep contains all the processing chains for quality assurance and additional data enhancements. The details of this superstep are discussed in Chapter 8.

3. Process

- This superstep contains all the processing chains for building the data vault. The details of this superstep are discussed in Chapter 9.

4. Transform

- This superstep contains all the processing chains for building the data warehouse from the core data vault. The details of this superstep are discussed in Chapter 10.

5. Organize

- This superstep contains all the processing chains for building the data marts from the core data warehouse. The details of this superstep are discussed in Chapter 11.

6. Report

- This superstep contains all the processing chains for building virtualization and reporting of the actionable knowledge. The details of this superstep are discussed in Chapter 11.

These six supersteps, discussed in detail in individual chapters devoted to them, enable the reader to master both them and the relevant tools from the Data Science Technology Stack.

Summary

This chapter covered the three management layers of any data science ecosystem. Operational management ensures that your system has everything ready to process the data with the data techniques you require for your data science. It stores and has all the processing capacity of your solution ready to process data. The audit, balance, and control layer controls any processing that is currently being performed and keeps records of what happened and is happening in the system. Note this area is the biggest source of information for optimizing your own solutions for improving the performance of the ecosystem. The functional layer of the data science ecosystem is the largest and most essential layer for programming and modeling. These are the specific combinations of the techniques, algorithms, and methods deployed against each customer's data sets. This single layer is the biggest part of the ecosystem.

I also discussed other important concepts. I discussed the yoke solution that I personally use for my data science, as it supports distributed processing, using a messaging ecosystem. This is useful, when employing data from various distributed data lakes, to formulate a holistic data science solution. I related the basic data science process that I follow when completing a data science project, regardless of the scale or complexity of the data. It is my guide to good practical data science.

In the next chapter, I will expand on the techniques, algorithms. and methods used in the functional layer, to enable you to process data lakes to gain business insights.

CHAPTER 7

Retrieve Superstep

The Retrieve superstep is a practical method for importing completely into the processing ecosystem a data lake consisting of various external data sources. The Retrieve superstep is the first contact between your data science and the source systems. I will guide you through a methodology of how to handle this discovery of the data up to the point you have all the data you need to evaluate the system you are working with, by deploying your data science skills.

The successful retrieval of the data is a major stepping-stone to ensuring that you are performing good data science. Data lineage delivers the audit trail of the data elements at the lowest granular level, to ensure full data governance. Data governance supports metadata management for system guidelines, processing strategies, policies formulation, and implementation of processing. Data quality and master data management helps to enrich the data lineage with more business values, if you provide complete data source metadata.

Tip I advise importing all the data passing through this layer, as this will ensure data lineage and governance.

The Retrieve superstep supports the edge of the ecosystem, where your data science makes direct contact with the outside data world. I will recommend a current set of data structures that you can use to handle the deluge of data you will need to process to uncover critical business knowledge.

© Andreas François Vermeulen 2018
A. F. Vermeulen, *Practical Data Science*, https://doi.org/10.1007/978-1-4842-3054-1_7

Data Lakes

The Pentaho CTO James Dixon is credited with coining the term *data lake*. As he characterizes it in his now famous blog,[1]

> *If you think of a datamart as a store of bottled water—cleansed and pack-aged and structured for easy consumption—the data lake is a large body of water in a more natural state.*

There is a second part to the blog that is not referred to as frequently.

> *The contents of the data lake stream in from a source to fill the lake, and various users of the lake can come to examine, dive in, or take samples.*

Since this blog, a significant categorization of research and commercial activity have made the term famous and even notorious. Unfortunately, the term is so all-encompassing that you could describe any data that is on the edge of the ecosystem and your business as part of your data lake. I have spent hours and even days in think tanks, theoretical and scientific debates on the complexities and nuances of the data lake.

So, for the practices and procedures of this book, I describe the data lake as follows: a company's data lake covers all data that your business is authorized to process, to attain an improved profitability of your business's core accomplishments. The data lake is the complete data world your company interacts with during its business life span. In simple terms, if you generate data or consume data to perform your business tasks, that data is in your company's data lake.

I will use terms related to a natural lake to explain the specific data lake concepts. So, as a lake needs rivers and streams to feed it, the data lake will consume an unavoidable deluge of data sources from upstream and deliver it to downstream partners.

Tip   Do not fail to appreciate the worth of an unprocessed data source.

I had a business customer who received the entire quarterly shipping manifests of his logistics company on a DVD but merely collected them in a plastic container for five years, because he assumed they were superfluous to requirements, as he had the

---

[1]James Dixon, "Pentaho, Hadoop, and Data Lakes," James Dixon's Blog, https://jamesdixon. wordpress.com/2010/10/14/pentaho-hadoop-and-data-lakes/, October 14, 2010.

electronic e-mails for each shipment online. Fortunately, one of my interns took the time to load a single DVD and noticed that the complete supply chain routing path and coding procedure were also on the DVD. This new information enriched the data lake by more than 30% in real-value logistics costs, by adding this new data to the company's existing data lake. Do not limit the catchment area of your data lake, as the most lucrative knowledge is usually in the merging of data you never before accepted as important.

Note   There are numerous snippets of code that I use to demonstrate concepts throughout this book. I am presuming that you have an R- and Python-enabled environment ready to try these examples. Performing the processes enriches knowledge transfer.

Data Swamps

As I spoke of data lakes and the importance of loading business data to uncover unknown value, I must also explain the data swamp. I have heard and seen various demonstrations asserting data lakes as evil. I have a different point of view. Any unmanaged data structure is evil. On the other hand, I am comfortable admitting that I have billed for many hours for repairing structures containing these evil data ecosystems.

Data swamps are simply data lakes that are not managed. They are not to be feared. They need to be tamed.

Following are my four critical steps to avoid a data swamp.

Start with Concrete Business Questions

Simply dumping a horde of data into a data lake, with no tangible purpose in mind, will result in a big business risk. Countless times, I have caught sight of data-loading agreements that state that they want to load all data. Please see the section on "Ambiguity" in Chapter 4, on business requirements.

My recommendation is to perform a comprehensive analysis of the entire set of data you have and then apply a metadata classification for the data, stating full data lineage for allowing it into the data lake. The data lake must be enabled to collect the data required to answer your business questions.

If, for example, you wanted to calculate the shipping patterns for the last five years, you would require all the data for the last five years related to shipping to be available in the data lake. You would also need to start a process of collecting the relevant data in the future, at agreed intervals, to ensure that you could use the same data science in a year's time, to evaluate the previous five years' data. The data lineage ensures that you can reference the data sources once you have your results.

Tip   At minimum, assert the true source of the data and a full timestamp of when it entered the data lake. The more data lineage you can provide, the better. Ensure that you have accredited sources.

Example:

It is better to say "I have found that your shipping pattern is good, and I have ensured that we have all the data for the last five years, by getting it signed off by our logistics managers as true and complete." That statement gives your data science a solid foundation and accredited lineage.

Last year, I had a junior data scientist try to proof his data science results by convincing a CFO that the source unknown, date unknown DVDs with shipping invoices that he found in the archives were a good data lineage for his four weeks of data science. Sadly, I must report, that results document plus the DVDs are now part of my "How not to perform data science" presentation!

The same data scientist, ten days later, with the same data science algorithms but accredited lineage, presented a major cost saving to my customer.

Data Governance

The role of data governance, data access, and data security does not go away with the volume of data in the data lake. It simply collects together into a worse problem, if not managed.

Warning   *Here be dragons!* Finding a piece of unmanaged data in a petabyte, or even terabyte, data lake will cost you hours of time.

Spend the time on data governance up front, as recovery is not easy. I typically spend 70%+ of the project's time on governance and classification of data sources. I will demonstrate in Chapters 8–11 how a well-established data lake from this chapter makes processing data so much easier.

Data Source Catalog

Metadata that link ingested data-to-data sources are a must-have for any data lake. I suggest you note the following as general rules for the data you process.

- *Unique data catalog number*: I normally use YYYYMMDD/ NNNNNN/NNN. E.g. 20171230/000000001/001 for data first registered into the metadata registers on December 30, 2017, as data source 1 of data type 1. This is a critical requirement.

- *Short description (keep it under 100 characters)*: Country codes and country names (Country Codes—ISO 3166)

- *Long description (keep it as complete as possible)*: Country codes and country names used by VKHC as standard for country entries

- *Contact information for external data source*: ISO 3166-1:2013 code lists from `www.iso.org/iso-3166-country-codes.html`

- *Expected frequency*: Irregular (i.e., no fixed frequency, also known as ad hoc). Other options are near-real-time, every 5 seconds, every minute, hourly, daily, weekly, monthly, or yearly.

- *Internal business purpose*: Validate country codes and names.

I have found that if your data source catalog is up to date, the rest of the processing is stress-free.

Business Glossary

The business glossary maps the data-source fields and classifies them into respective lines of business. This glossary is a must-have for any good data lake.

Create a data-mapping registry with the following information:

- *Unique data catalog number*: I normally use YYYYMMDD/ NNNNNN/NNN.

- *Unique data mapping number*: I normally use NNNNNNN/ NNNNNNNNN. E.g., 0000001/000000001 for field 1 mapped to internal field 1

- *External data source field name*: States the field as found in the raw data source

- *External data source field type*: Records the full set of the field's data types when loading the data lake

- *Internal data source field name*: Records every internal data field name to use once loaded from the data lake

- *Internal data source field type*: Records the full set of the field's types to use internally once loaded

- *Timestamp of last verification of the data mapping*: I normally use YYYYMMDD-HHMMSS-SSS that supports timestamp down to a thousandth of a second.

Note   There can be a many-to-many relationship between external and internal mappings. Keep the mappings to a controlled level of consolation. However, balance the time you spend resolving them against the gain in efficiency.

The business glossary records the data sources ready for the retrieve processing to load the data. In several of my customer's systems, I have had to perform this classification process, using machine learning with success.

This simply points the classification bots at a data lake and finds any new or changed metadata in that manner. Once newly identified metadata is collected, the human operators have only to deal with the exceptions.

I have several bots that can perform these discovery tasks, but I will not spend more time in this book on how I accomplished this, as, in general, every system requires a custom build for each customer's ecosystem. I recommend that you invest time automating your data lake, as this allows you more time to perform the stimulating data analytics part of the data science.

Analytical Model Usage

Data tagged in respective analytical models define the profile of the data that requires loading and guides the data scientist to what additional processing is required. I perform the following data analytical models on every data set in my data lake by default.

Let's start by loading a sample data set.

Note   I will use a sample data set, to show each of the steps of this process. So, you will require either R or Python, to perform some data processing.

Disclaimer   As I introduce you to several processing R and Python packages in the course of this book, I will provide you with only a basic introduction to these tools. I suggest that you investigate the other details of these packages, to improve your overall understanding and capabilities.

For R, perform the following commands:
Start your RStudio editor.

Note   For the rest of this chapter, I will use `setwd(C:/VKHCG)` for Windows and `setwd(/home/<youruser>/VKHCG)` for Linux ecosystems.

Execute the following to import a data loader package:

```
library(readr)
```

Note   See `https://cran.r-project.org/web/packages/readr/index.html` for more details.

This should load an amusing R data loading package. So, let's load the data for the examples for Vermeulen company.

Let's load a table named IP_DATA_ALL.csv.

```
Base=getwd()
FileName=paste0(Base,'/01-Vermeulen/00-RawData/IP_DATA_ALL.csv')
IP_DATA_ALL <- read_csv(FileName)
```

You should have successfully loaded a data set with eight columns.

Using the spec(IP_DATA_ALL) command should yield the following results:

```
cols(
  ID = col_integer(),
  Country = col_character(),
  'Place Name' = col_character(),
  'Post Code' = col_character(),
  Latitude = col_double(),
  Longitude = col_double(),
  'First IP Number' = col_integer(),
  'Last IP Number' = col_integer()
)
```

This informs you that you have the following eight columns:

- ID of type integer

- Place name of type character

- Post code of type character

- Latitude of type numeric double

- Longitude of type numeric double

- First IP number of type integer

- Last IP number of type integer

Now you can look at the data by using View(IP_DATA_ALL). You will see a data grid appear with your data.

Data Field Name Verification

I use this to validate and verify the data field's names in the retrieve processing in an easy manner. To add an extra library, you will have to run the following commands:

Warning   Ensure that you have the latest package, by performing an `install.packages(tibble)`. So, if one of my proofreaders reports that the `set_tidy_names()` command is not present, I use R 3.4.1 and `tibble` version 1.3.4.

```
library(tibble)
```

(See https://cran.r-project.org/web/packages/tibble/.)

```
set_tidy_names(IP_DATA_ALL, syntactic = TRUE, quiet = FALSE)
```

You will detect that some field names are not easy to use.

```
New names:
Place Name -> Place.Name
Post Code -> Post.Code
First IP Number -> First.IP.Number
Last IP Number -> Last.IP.Number
```

This informs you that four of the field names are not valid and suggests new field names that are valid.

Tip   I always use existing tools that are available, as this saves you time to verify that they work as expected. Good data scientists learn from others.

You can fix any detected invalid column names by executing

```
IP_DATA_ALL_FIX=set_tidy_names(IP_DATA_ALL, syntactic = TRUE, quiet = TRUE)
```

By using command View(IP_DATA_ALL_FIX), you can check that you have fixed the columns.

Well done! You just fixed the data field names. So now, let's store your progress.

```
Base=getwd()
FileDir= paste0(Base,'/01-Vermeulen/01-Retrieve/01-EDS)
dir.create(FileDir)
FileDir= paste0(Base,'/01-Vermeulen/01-Retrieve/01-EDS/01-R')
dir.create(FileDir)

FileName=paste0(FileDir,'/IP_DATA_ALL_FIX.csv')
write.csv(IP_DATA_ALL_FIX, FileName)
```

You should now have a file named IP_DATA_ALL_FIX.csv.

Unique Identifier of Each Data Entry

Allocate a unique identifier within the system that is independent of the given file name. This ensures that the system can handle different files from different paths and keep track of all data entries in an effective manner. Then allocate a unique identifier for each record or data element in the files that are retrieved.

Tip   I have found that the extra time spent to perform this task prevents major data referential issues. At scale, the probability of getting similar or even the same data loaded into the data lake more than once is high.

Now, investigate your sample data.

For R, see the following.

To add the unique identifier, run the following command:

```
IP_DATA_ALL_with_ID=rowid_to_column(IP_DATA_ALL_FIX, var = "RowID")
```

Check if you successfully added the unique identifier, as follows:

```
View(IP_DATA_ALL_with_ID)
```

This will show you that every record has a unique ID named "Row ID."

So, let's store your progress.

```
Base=getwd()
FileName=paste0(FileDir,'/IP_DATA_ALL_with_ID.csv')
write.csv(IP_DATA_ALL_with_ID,FileName)
```

You should now have a file named IP_DATA_ALL_with_ID.csv.

Data Type of Each Data Column

Determine the best data type for each column, to assist you in completing the business glossary, to ensure that you record the correct import processing rules.

Use the following R command:

```
sapply(IP_DATA_ALL_with_ID, typeof)
```

You should see a complete data type analysis of the data you've loaded.

RowId	ID	Country	Place. Name	Post. Code	Latitude	Longitude	First. IP.Number	Last. IP.Number
integer	integer	character	character	character	double	double	Integer	integer

---

Warning   Keep your eye on data types, such as account numbers or identification numbers, that naturally appear to be numeric but are, in reality, text columns.

Histograms of Each Column

I always generate a histogram across every column, to determine the spread of the data value. In the practical section of this chapter, I will generate a histogram against the sample data set, to demonstrate how to perform this action.

Now we translate the data set to a data table, which will assist us to investigate the data in an enhanced manner. I will show you in R how to find the histogram for the country data, as follows:

```
library(data.table)
```

```
hist_country=data.table(Country=unique(IP_DATA_ALL_with_ID[is.na(IP_DATA_
ALL_with_ID ['Country']) == 0, ]$Country))
setorder(hist_country,'Country')
hist_country_with_id=rowid_to_column(hist_country, var = "RowIDCountry")
View(hist_country_with_id)
```

```
IP_DATA_COUNTRY_FREQ=data.table(with(IP_DATA_ALL_with_ID, table(Country)))
View(IP_DATA_COUNTRY_FREQ)
```

Top-Two Country Codes	Frequency
US	512714
GB	127069

Business Insights:

- The two biggest subset volumes are from the US and GB.

- The US has just over four times the data as GB.

This information is useful for designing parallel processing of loaded data.

During parallel processing, the skew of data can result in unwanted wait times or overloading of processors, owing to specific split criteria for the parallel processing. In the case just discussed, if you spread the data, for example, for one processor per country code, you would have a result of US and GB running longer than the other country codes, owing to the volumes. Even the mismatch in volumes between GB and US would result in more skew processing.

Caution   Parallel processing is an essential requirement for data science, and it is important that you understand the behaviors of your data model and/or algorithms when dealing with nonconforming data spreads. Sometimes, you will be better off not going parallel, as the split process may take longer than the savings it brings to the overall process.

So, let's store your progress.

```
Base=getwd()
FileName=paste0(FileDir,'/IP_DATA_COUNTRY_FREQ.csv')
IP_DATA_COUNTRY_FREQ.to_csv(FileName)
```

You should now have a file named IP_DATA_COUNTRY_FREQ.csv.

Tip   As you start to understand the analytic characteristics of the data lake, you will start to formulate your plans on how you would process it. I normally keep a few key notes on the data profile of the data lake.

Profiling the data lake is normally a structured approach. My team and I keep a simple cheat sheet on each data source on which we note what the profile of the data is. We share this via our GitHub source control, to ensure that any new discoveries are circulated to everyone involved.

Let's look at some more key characteristics you may want to investigate.

Histograms explain how the complete data set is spread over a set of values. I will show you in R how to find the histogram for the latitude data field.

```
hist_latitude =data.table(Latitude=unique(IP_DATA_ALL_with_ID [is.na(IP_
DATA_ALL_with_ID ['Latitude']) == 0, ]$Latitude))
setkeyv(hist_latitude, 'Latitude')
setorder(hist_latitude)
hist_latitude_with_id=rowid_to_column(hist_latitude, var = "RowID")

View(hist_latitude_with_id)

IP_DATA_Latitude_FREQ=data.table(with(IP_DATA_ALL_with_ID,
table(Latitude)))

View(IP_DATA_Latitude_FREQ)
```

Top-Three Latitudes	Frequency
35.6427	4838
51.5092	4442
48.8628	3230

Business Insights:

- The two biggest data volumes are from latitudes 35.6427 and 51.5092.

- The spread appears to be nearly equal between the top-two latitudes.

So, let's store your progress.

```
Base=getwd()
FileName=paste0(FileDir,'/IP_DATA_Latitude_FREQ.csv')
IP_DATA_Latitude_FREQ.to_csv(FileName)
```

You should now have a file named IP_DATA_Latitude_FREQ.csv.

I will show you in R how to find the histogram for the longitude data.

```
hist_longitude =data.table(Longitude=unique(IP_DATA_ALL_with_ID [is.na(IP_
DATA_ALL_with_ID ['Longitude']) == 0, ]$Longitude))
setkeyv(hist_longitude, 'Longitude')
setorder(hist_longitude,'Longitude')
hist_longitude_with_id=rowid_to_column(hist_longitude, var = "RowID")
View(hist_longitude_with_id)

IP_DATA_Longitude_FREQ=data.table(with(IP_DATA_ALL_with_ID,
table(Longitude)))

View(IP_DATA_Longitude_FREQ)
```

Top-Three Longitudes	Frequency
139.7677	4837
-0.0955	4436
2.3292	3230

So, let's store your progress.

```
Base=getwd()
FileName=paste0(FileDir,'/IP_DATA_Longitude_FREQ.csv')
IP_DATA_Longitude_FREQ.to_csv(FileName)
```

You should now have a file named IP_DATA_Longitude_FREQ.csv.

Minimum Value

Determine the minimum value in a specific column.

In R, use the following. I suggest that you inspect the country.

```
min(hist_country$Country)
```

You should return AD. AD is the country code for Andorra (an independent principality situated between France and Spain in the Pyrenees Mountains).

Tip   I will be using a common processing function in R named `sapply()`. If you have not used it before, I suggest you look at www.r-bloggers.com/using-apply-sapply-lapply-in-r/. In simple terms, it applies a function to a data set. Please read up on it, as you will use it frequently in your data science.

I will now apply the function `min` to the data set in `hist_country`'s country field. You can also use

```
sapply(hist_country[,'Country'], min, na.rm=TRUE)
```

You should return AD.

I suggest you examine latitude next.

In R, execute

```
sapply(hist_latitude_with_id[,'Latitude'], min, na.rm=TRUE)
```

The result is -54.2767. What does this tell you?

Fact: The range of latitude for the Southern Hemisphere is from -90 to 0. So, if you do not have any latitudes farther south than -54.2767, you can improve your retrieve routine.

Calculate the following in R:

```
round(1-(-54.2767/-90),4)*100
```

The result is 39.69%. A 39% minimum saving in processing can be achieved using this discovery.

I suggest you examine longitude next.

In R, execute

```
sapply(hist_longitude_with_id[,'Longitude'], min, na.rm=TRUE)
```

161

The result is -176.5. What does this tell you?

Fact: The range of longitude for the Western Hemisphere is from -180 to 0.

So, if you do not have any latitudes more westerly than -176.5, you can improve your retrieve routine.

Calculate the following in R:

```
round(1-(-176.5/-180),4)*100
```

The result is 1.94%. A small 1% saving in processing can be achieved using this discovery.

Business Insights:

- The insight on Andorra is interesting but does not impact the processing plan for retrieval of the data.

- The range in latitude for the Southern Hemisphere is from -90 to 0, with the minimum latitude confirmed at greater than -55. Knowing this, you do not now have to process values for +/-39% of the Southern Hemisphere. That is a major gain in efficiencies.

- The range in longitude for the Western Hemisphere is -180 to 0, with the minimum longitude confirmed at greater than -177. Therefore, you will not achieve a similar gain in efficiencies, as you can only not load less than 2% of total range of the Eastern Hemisphere.

You have discovered that when you planned for 180 processes to load the individual bands of latitudes, you already removed the need for 35 of those processes. You also discovered that if you have loaded via bands based on longitudes, the improvements are not so remarkable.

Maximum Value

Determine the maximum value in a specific column.

In R, use

```
max(hist_country$Country)
```

You should return ZW. You can also use

```
sapply(hist_country[,'Country'], max, na.rm=TRUE)
```

You should return ZW. I suggest that you investigate latitude next.

In R, execute

```
sapply(hist_latitude_with_id[,'Latitude'], max, na.rm=TRUE)
```

The result is 78.2167. What does this tell you?

Fact: The range in latitude for the Northern Hemisphere is from 0 to 90. So, if you do not have any latitudes more northerly than 78.2167, you can improve your retrieve routine.

Calculate the following in R:

```
round(1-(78.2167/90),4)*100
```

The result is 13.09%. A 13% minimum saving in processing can be achieved using this discovery.

In R, execute

```
sapply(hist_longitude_with_id[,'Longitude'], max, na.rm=TRUE)
```

The result is 179.2167. What does this tell you?

Fact: The range in longitude for the Eastern Hemisphere is from 0 to 180. So, if you do not have any longitudes more easterly than 179.2167, you can improve your retrieve routine.

Calculate the following in R:

```
round(1-(179.2167/180),4)*100
```

The result is 0.44%. A small 0.4% saving in processing can be achieved using this discovery.

Mean

If the column is numeric in nature, determine the average value in a specific column.

Let's examine latitude.

In R, use

```
sapply(hist_latitude_with_id[,'Latitude'], mean, na.rm=TRUE)
```

The result is 38.23803.

Let's examine longitude.

In R, use

```
sapply(hist_longitude_with_id[,'Longitude'], mean, na.rm=TRUE)
```

The result is -13.87398.

Median

Determine the value that splits the data set into two parts in a specific column.

In R, use the following command against the latitude column:

```
sapply(hist_latitude_with_id[,'Latitude'], median, na.rm=TRUE)
```

The result is 43.2126.

In R, use the following command against the longitude column:

```
sapply(hist_longitude_with_id[,'Longitude'], median, na.rm=TRUE)
```

The result is 1.8181.

Mode

Determine the value that appears most in a specific column.

In R, use the following command against the country column:

```
IP_DATA_COUNTRY_FREQ=data.table(with(IP_DATA_ALL_with_ID, table(Country)))
setorder(IP_DATA_COUNTRY_FREQ,-N)
IP_DATA_COUNTRY_FREQ[1,'Country']
```

The result is US.

In R, use the following command against the latitude column:

```
IP_DATA_Latitude_FREQ=data.table(with(IP_DATA_ALL_with_ID,
table(Latitude)))
setorder(IP_DATA_Latitude_FREQ,-N)
IP_DATA_Latitude_FREQ[1,'Latitude']
```

The result is 35.6427.

In R, use the following command against the longitude column:

```
IP_DATA_Longitude_FREQ=data.table(with(IP_DATA_ALL_with_ID,
table(Longitude)))
```

```
setorder(IP_DATA_Longitude_FREQ,-N)
IP_DATA_Longitude_FREQ[1,'Longitude']
```

The result is 139.7677.

Range

For numeric values, you determine the range of the values by taking the maximum value and subtracting the minimum value.

In R, use the following command against the latitude column:

```
sapply(hist_latitude_with_id[,'Latitude'], range, na.rm=TRUE)
```

The range in latitude is -54.2767 to 78.2167.

In R, use the following command against the longitude column:

```
sapply(hist_longitude_with_id[,'Longitude'], range, na.rm=TRUE)
```

The range in longitude is -176.5000 to 179.2167.

In R, use the following command against the country column:

```
sapply(hist_country_with_id[,'Country'], range, na.rm=TRUE)
Country Range is: "AD" to "ZW"
```

Quartiles

Quartiles are the base values dividing a data set into quarters. Simply sort the data column and split it into four groups that are of four equal parts.

In R, use the following command against the latitude column:

```
sapply(hist_latitude_with_id[,'Latitude'], quantile, na.rm=TRUE)
```

The results are as follows:

	Latitude
0%	-54.2767
25%	36.4151
50%	43.2126
75%	48.6219
100%	78.2167

Knowledge Checkpoint   Could you perform the process for the longitude? Is there any other insight you could achieve? How would you perform the task?

Here is my attempt:

In R, use the following command against the longitude column:

```
sapply(hist_longitude_with_id[,'Longitude'], quantile, na.rm=TRUE)
```

The results are as follows:

	Latitude
0%	-176.50000
25%	-80.20185
50%	1.81810
75%	12.69480
100%	179.21670

Standard Deviation

In R, use the following command against the latitude column:

```
sapply(hist_latitude_with_id[,'Latitude'], sd, na.rm=TRUE)
```

The result is 19.86071.

In R, use the following command against the longitude column:

```
sapply(hist_longitude_with_id[,'Longitude'], sd, na.rm=TRUE)
```

The result is 68.981.

Skewness

Skewness describes the shape or profile of the distribution of the data in the column.

Remember   This is the characteristic that gives you a profile for doing parallel processing.

In R, use the following command against the latitude column:

Warning   You will install a new package now: `install.packges('e1071')`.
This is a set of useful functions known as E1071, elaborated by the Vienna
University of Technology's Department of Statistics' Probability Theory Group. There
are thousands of R research groups worldwide that are achieving extremely useful
progress in data processing. I am simply enriching our processing capability by
adding their already proven processing capability.

```
library(e1071)
skewness(hist_latitude_with_id$Latitude, na.rm = FALSE, type = 2)
```

The result is -2.504929.

Negative skew: The mass of the distribution is concentrated on the right, and the distribution is said to be left-skewed. The result is only an indicator of how the data is skewed. By combining it with a histogram, you can gather good insights into how the data is spread.

Example:

If you are looking at a distribution of the latitudes ordered in ascending order, from right to left, and found the skew to be negative, it is highly improbable that you will find fewer locations on the left, but more probable that you will find more on the right-hand side of the scale. This is just a quick estimate. The histogram of the sizes will give you a precise count of each size.

Knowledge Checkpoint   Is it true about the longitude also? Using `View(hist_longitude_with_id)`, can you predict what you should find? Can you guess or calculate the skew yourself?

In R, use the following command against the longitude column:

```
skewness(hist_longitude_with_id$Longitude,na.rm = FALSE, type = 2)
```

The result is 0.4852374.

Positive skew: The mass of the distribution is concentrated on the left, and the distribution is said to be right-skewed.

Warning   Missing data or unknown values is a common but unavoidable fact of data science.

Example:

If you have data only about customers in the United States and Germany, does that mean you have no customers in the UK? What if your data lineage infers that you have data, but it is simply not 100% filled in on the data lake? What do you process?

Missing or Unknown Values

Identify if you have missing or unknown values in the data sets.

In R, use the following command against the country column:

```
missing_country=data.table(Country=unique(IP_DATA_ALL_with_ID[is.na(IP_
DATA_ALL_with_ID ['Country']) == 1, ]))

View(missing_country)
```

Data Pattern

I have used the following process for years, to determine a pattern of the data values themselves. Here is my standard version:

Replace all alphabet values with an uppercase case A, all numbers with an uppercase N, and replace any spaces with a lowercase letter b and all other unknown characters with a lowercase u. As a result, "Good Book 101" becomes "AAAAbAAAAbNNNu."

This pattern creation is beneficial for designing any specific assess rules in Chapter 10, as specific unwanted patterns can be processed to remove or repair the invalid pattern. This pattern view of data is a quick way to identify common patterns or determine standard layouts.

Example 1:

If you have two patterns—NNNNuNNuNN and uuNNuNNuNN—for a date column, it is easy to see that you have an issue with the uuNNuNNuNN pattern. It is also easy to create a quality matching grid in regular expressions: (`regexpr()`).

See https://cran.r-project.org/web/packages/stringr/vignettes/regular-expressions.html.

Example 2:

It also helps with determining how to load the data in a parallel manner, as each pattern can be loaded in separate retrieve procedures. If the same two patterns, NNNNuNNuNN and uuNNuNNuNN, are found, you can send NNNNuNNuNN directly to be converted into a date, while uuNNuNNuNN goes through a quality-improvement process to then route back to the same queue as NNNNuNNuNN, once it complies. This is a common use of patterns to separate common standards and structures.

So, let's see how you can achieve this. Let's pattern our first column. I suggest the country column.

In R, use the following commands:

```
library(readr)
library(data.table)
Base=getwd()
FileName=paste0(Base,'/01-Vermeulen/00-RawData/IP_DATA_ALL.csv')
IP_DATA_ALL <- read_csv(FileName)
hist_country=data.table(Country=unique(IP_DATA_ALL$Country))
pattern_country=data.table(Country=hist_country$Country,
PatternCountry=hist_country$Country)
oldchar=c(letters,LETTERS)
newchar=replicate(length(oldchar),"A")
for (r in seq(nrow(pattern_country))){
        s=pattern_country[r,]$PatternCountry;
        for (c in seq(length(oldchar))){
                s=chartr(oldchar[c],newchar[c],s)
        };
        for (n in seq(0,9,1)){
        s=chartr(as.character(n),"N",s)
    };
        s=chartr(" ","b",s)
        s=chartr(".","u",s)
        pattern_country[r,]$PatternCountry=s;
};
View(pattern_country)
```

> **Note**   The process of generating these patterns can be moved to a custom function or even a package. If you are interested in this process, for functions, see www.r-bloggers.com/how-to-write-and-debug-an-r-function/. For packages, see https://cran.r-project.org/doc/manuals/R-exts.html.

So, let's store your progress.

```
Base=getwd()
FileName=paste0(FileDir,'/pattern_country.csv')
pattern_country.to_csv(FileName)
```

You should now have a file named pattern_country.csv.

Let's pattern another column. I suggest the latitude column.

In R, use the following commands:

```
library(readr)
library(data.table)
Base=getwd()
FileName=paste0(Base,'/01-Vermeulen/00-RawData/IP_DATA_ALL.csv')
IP_DATA_ALL <- read_csv(FileName)
hist_latitude=data.table(Latitude=unique(IP_DATA_ALL$Latitude))
pattern_latitude=data.table(latitude=hist_latitude$Latitude,
Patternlatitude=as.character(hist_latitude$Latitude))
oldchar=c(letters,LETTERS)
newchar=replicate(length(oldchar),"A")
for (r in seq(nrow(pattern_latitude))){
        s=pattern_latitude[r,]$Patternlatitude;
        for (c in seq(length(oldchar))){
                s=chartr(oldchar[c],newchar[c],s)
        };
        for (n in seq(0,9,1)){
        s=chartr(as.character(n),"N",s)
};
        s=chartr(" ","b",s)
        s=chartr("+","u",s)
        s=chartr("-","u",s)
```

```
     s=chartr(".","u",s)
     pattern_latitude[r,]$Patternlatitude=s;
};
```

```
setorder(pattern_latitude,latitude)
View(pattern_latitude[1:3])
```

The results are as follows:

	latitude	Patternlatitude
1	-54.2767	uNNuNNNN
2	-54.1561	uNNuNNNN
3	-51.7	uNNuN

So, let's store your progress.

```
Base=getwd()
FileName=paste0(FileDir,'/pattern_country_asc.csv')
pattern_country.to_csv(FileName)
```

You should now have a file named pattern_country_asc.csv.

```
setorder(pattern_latitude,-latitude)
View(pattern_latitude[1:3])
```

The results are as follows:

	latitude	Patternlatitude
1	78.2167	NNuNNNN
2	77.3025	NNuNNNN
3	72.7	NNuN

So, let's store your progress.

```
Base=getwd()
FileName=paste0(FileDir,'/pattern_country_desc.csv')
pattern_country.to_csv(FileName)
```

You should now have a file named `pattern_country_desc.csv`.

Note   At this point, you can save all your work, as I will now discuss various non-programming information.

Data Quality

More data points do not mean that data quality is less relevant. Data quality can cause the invalidation of a complete data set, if not dealt with correctly.

Warning   Bad quality data will damage even the best designed data science model. So, spend the time to ensure that you understand any data issues early in the process.

In Chapter 8, I will discuss in detail how to resolve any data quality issues.

Audit and Version Management

Up to now, I have simply allowed you to save and work with data in a use-once method. That is not the correct way to perform data science. You must always report the following:

- Who used the process?

- When was it used?

- Which version of code was used?

I refer you to Chapter 6, on version management, for processes and audit requirements for processing data. At this point, I want to discuss the other part of a successful data science process: you and your team.

Training the Trainer Model

To prevent a data swamp, it is essential that you train your team also. Data science is a team effort.

People, process, and technology are the three cornerstones to ensure that data is curated and protected. You are responsible for your people; share the knowledge you acquire from this book. The process I teach you, you need to teach them. Alone, you cannot achieve success.

Technology requires that you invest time to understand it fully. We are only at the dawn of major developments in the field of data engineering and data science. Remember: A big part of this process is to ensure that business users and data scientists understand the need to start small, have concrete questions in mind, and realize that there is work to do with all data to achieve success.

Understanding the Business Dynamics of the Data Lake

You now have the basic techniques of how to handle a retrieve step in your data science process. Let's now process the examples I supplied. Now you can measure your own capability against the example.

R Retrieve Solution

First, I will guide you through an R-based solution for the Retrieve superstep.

Vermeulen PLC

I will guide you through the first company. Start your RStudio editor.

Note   For the rest of the chapter, I will use `setwd("C:/VKHCG")` for Windows and `setwd("/home/<youruser>/VKHCG")` for Linux ecosystems. Load `IP_DATA_ALL`.

Now, execute the following command via RStudio:

```
getwd()
```

Test if your workspace directory is correct. You should see the directory you set.

Loading IP_DATA_ALL

This data set contains all the IP address allocations in the world. It will help you to locate your customers when interacting with them online.

Warning   I am using default directory references: for windows, C:/VKHCG;for Linux, /home/<youruser>/VKHCG.

If you have code and data at neither of these locations, you will have to make code changes for every example in this book.

Let's start our first load for the data process. Create a new R script file and save it as Retrieve-IP_DATA_ALL.r in directory C:\VKHCG\01-Vermeulen\01-Retrieve. Now, start copying the following code into the file:

```
###########################################
rm(list=ls()) #will remove ALL objects
###########################################
if (Sys.info()['sysname'] == "Windows")
  {
    BaseSet = "C:/VKHCG"
    setwd(BaseSet)
  } else {
    BaseSet = paste0("/home/", Sys.info()['user'], "/VKHCG")
    setwd(BaseSet)
  }
###########################################
Base=getwd() FileDir= paste0(Base,'/01-Vermeulen/01-Retrieve/01-EDS/01-R')
dir.create(FileDir)
FileDirLog=paste0(FileDir,'/log')
dir.create(FileDirLog)
FileDirRun=paste0(FileDirLog,'/Run0001')
dir.create(FileDirRun)
###########################################
StartTime=Sys.time()
###########################################
```

```
# Set up logging
##########################################
debugLog=paste0(FileDirRun,'/debug.Log')
infoLog=paste0(FileDirRun,'/info.Log')
errorLog=paste0(FileDirRun,'/error.Log')
##########################################
write(paste0('Start Debug Log File ',
    format(StartTime, "%Y/%d/%m %H:%M:%S")),
    file=debugLog,append = FALSE)
write(paste0('Start Information Log File ',
    format(StartTime, "%Y/%d/%m %H:%M:%S")),
    file=infoLog,append = FALSE)
write(paste0('Start Error Log File ',
    format(StartTime, "%Y/%d/%m %H:%M:%S")),
    file=errorLog,append = FALSE)
##########################################
UserName='Practical Data Scientist'
##########################################
write(paste0(UserName,' Load library: ', 'readr'),
    infoLog,append = TRUE)
library(readr)
write(paste0(UserName,' Load library: ', 'data.table'),
    infoLog,append = TRUE)
library(data.table)
write(paste0(UserName,' Load library: ', 'tibble'),
    infoLog,append = TRUE)
library(tibble)
##########################################
FileName=paste0(Base,'/01-Vermeulen/00-RawData/IP_DATA_ALL.csv')
write(paste0(UserName,' Retrieve data file: ', FileName),
    file=infoLog,append = TRUE)
##########################################
IP_DATA_ALL <- read_csv(FileName,
                    col_types = cols(
                        Country = col_character(),
                        ID = col_skip(),
```

175

```
                            'First IP Number' = col_skip(),
                            'Last IP Number' = col_skip(),
                            Latitude = col_double(),
                            Longitude = col_double(),
                            'Place Name' = col_character(),
                            'Post Code' = col_character()
                      ), na = "empty")
##############################################
IP_DATA_ALL_FIX=set_tidy_names(IP_DATA_ALL, syntactic = TRUE, quiet = TRUE)
##############################################
IP_DATA_ALL_with_ID=rowid_to_column(IP_DATA_ALL_FIX, var = "RowID")
##############################################
FileNameOut=paste0(FileDir,'/Retrieve_IP_DATA.csv')
fwrite(IP_DATA_ALL_with_ID, FileNameOut)
write(paste0(UserName,' Stores Retrieve data file: ', FileNameOut),
     file=infoLog,append = TRUE)
##############################################
StopTime=Sys.time()
##############################################
write(paste0('Stop Debug Log File ',
     format(StopTime, "%Y/%d/%m %H:%M:%S")),
     file=debugLog,append = TRUE)
write(paste0('Stop Information Log File ',
     format(StopTime, "%Y/%d/%m %H:%M:%S")),
     file=infoLog,append = TRUE)
write(paste0('Stop Error Log File ',
     format(StopTime, "%Y/%d/%m %H:%M:%S")),
     file=errorLog,append = TRUE)
##############################################
View(IP_DATA_ALL_with_ID)
##############################################
```

Save and execute the script.

If you completed the code correctly, you should see four outcomes, as follows:

1. In directory C:\VKHCG\01-Vermeulen\01-Retrieve, you should have a file named Retrieve-IP_DATA_ALL.r.

2. In directory C:\VKHCG\01-Vermeulen\01-Retrieve\01-EDS\01-R\log\Run0001, you should have three files: debug.Log, error.Log, and info.Log. If you open info.Log in a text editor, you should see the following:

```
Start Information Log File 2017/05/08 03:35:54
Practical Data Scientist Load library: readr
Practical Data Scientist Load library: data.table
Practical Data Scientist Load library: tibble
Practical Data Scientist Retrieve data file: C:/VKHCG/01-Vermeulen/00-
RawData/IP_DATA_ALL.csv
Practical Data Scientist Stores Retrieve data file: C:/VKHCG/01-
Vermeulen/01-Retrieve/01-EDS/01-R/Retrieve_IP_DATA.csv
Stop Information Log File 2017/05/08 03:35:57
```

3. In directory C:\VKHCG\01-Vermeulen\01-Retrieve\01-EDS\01-R, You should have a file named Retrieve_IP_DATA.csv.

4. You will also see a display of the file you loaded, as follows:

Row ID	Country	Place.Name	Post.Code	Latitude	Longitude
1	BW	Gaborone	NA	-24.6464	25.9119
2	BW	Gaborone	NA	-24.6464	25.9119
3	BW	Gaborone	NA	-24.6464	25.9119
4	BW	Gaborone	NA	-24.6464	25.9119

So, what have you achieved? You've

- Loaded the IP_DATA_ALL.csv into R

- Removed, by using col_skip() for three columns: ID, First IP Number, and Last IP Number

- Set data type, by using col_character() for three columns: Country, Place Name, and Post Code, i.e., set it to be a string or characters

- Converted all missing data to empty

- Changed Place Name and Post Code to Place.Name and Post.Code

- Learned how to log your progress throughout the program

Wow, you have taken the first steps to becoming a practical data scientist. Now, I suggest that you close Retrieve-IP_DATA_ALL.r, and we will move on to next file to load.

Loading IP_DATA_C_VKHCG

This is a specific data load, as it only contains one class C address that VKHCG is using for its data systems. Create a new R script file and save it as Retrieve-IP_DATA_ALL.r in directory C:\VKHCG\01-Vermeulen\01-Retrieve.

Now, you can start copying the following code into the script file:

```
###########################################
rm(list=ls()) #will remove ALL objects
###########################################
###########################################
if (Sys.info()['sysname'] == "Windows")
  {
    BaseSet = "C:/VKHCG"
    setwd(BaseSet)
  } else {
    BaseSet = paste0("/home/", Sys.info()['user'], "/VKHCG")
    setwd(BaseSet)
  }
###########################################
Base=getwd()
FileDir= paste0(Base,'/01-Vermeulen/01-Retrieve/01-EDS/01-R')
dir.create(FileDir)
FileDirLog=paste0(FileDir,'/log')
dir.create(FileDirLog)
FileDirRun=paste0(FileDirLog,'/Run0002')
```

```
dir.create(FileDirRun)
###########################################
StartTime=Sys.time()
###########################################
# Set up logging
###########################################
debugLog=paste0(FileDirRun,'/debug.Log')
infoLog=paste0(FileDirRun,'/info.Log')
errorLog=paste0(FileDirRun,'/error.Log')
###########################################
write(paste0('Start Debug Log File ',
      format(StartTime, "%Y/%d/%m %H:%M:%S")),
      file=debugLog,append = FALSE)
write(paste0('Start Information Log File ',
      format(StartTime, "%Y/%d/%m %H:%M:%S")),
      file=infoLog,append = FALSE)
write(paste0('Start Error Log File ',
      format(StartTime, "%Y/%d/%m %H:%M:%S")),
      file=errorLog,append = FALSE)
###########################################
UserName='Practical Data Scientist'
###########################################
write(paste0(UserName,' Load library: ', 'readr'),
      infoLog,append = TRUE)
library(readr)
write(paste0(UserName,' Load library: ', 'data.table'),
      infoLog,append = TRUE)
library(data.table)
write(paste0(UserName,' Load library: ', 'tibble'),
      infoLog,append = TRUE)
library(tibble)
###########################################
FileName=paste0(Base,'/01-Vermeulen/00-RawData/IP_DATA_C_VKHCG.csv')
write(paste0(UserName,' Retrieve data file: ', FileName),
      file=infoLog,append = TRUE)
```

```
IP_DATA_C_VKHCG <- read_csv (FileName,
     col_types = cols(
   `IP Address` = col_character(),
   `IP Number` = col_double(),
    w = col_integer(),
    x = col_integer(),
    y = col_integer(),
    z = col_integer())))
############################################
IP_DATA_C_VKHCG_FIX=set_tidy_names(IP_DATA_C_VKHCG,
    syntactic = TRUE, quiet = TRUE)
############################################
IP_DATA_C_VKHCG_with_ID=rowid_to_column(IP_DATA_C_VKHCG_FIX, var = "RowID")
############################################
setorderv(IP_DATA_C_VKHCG_with_ID, 'IP Number', order= -1L, na.last=FALSE)
############################################
FileNameOut=paste0(FileDir,'/Retrieve_IP_C_VKHCG.csv')
fwrite(IP_DATA_C_VKHCG_with_ID, FileNameOut)
write(paste0(UserName,' Stores Retrieve data file: ', FileNameOut),
      file=infoLog,append = TRUE)
############################################
StopTime=Sys.time()
############################################
write(paste0('Stop Debug Log File ',
      format(StopTime, "%Y/%d/%m %H:%M:%S")),
      file=debugLog,append = TRUE)
write(paste0('Stop Information Log File ',
      format(StopTime, "%Y/%d/%m %H:%M:%S")),
      file=infoLog,append = TRUE)
write(paste0('Stop Error Log File ',
      format(StopTime, "%Y/%d/%m %H:%M:%S")),
      file=errorLog,append = TRUE)
############################################
View(IP_DATA_C_VKHCG_with_ID)
############################################
```

Save the script file Retrieve-IP_DATA_ALL.r. Now source it to execute.

If completed as required, you should have a new file named Retrieve_IP_C_VKHCG.csv.

Loading IP_DATA_CORE

The next data source supplies all the core router's addresses within VKHCG. You will now load IP_DATA_CORE.csv into Retrieve_IP_DATA_CORE.csv.

Create a new R script file and save it as Retrieve-IP_DATA_CORE.r in directory C:\VKHCG\01-Vermeulen\01-Retrieve. Now, you can start to copy the following code into the script file:

```r
###########################################
rm(list=ls()) #will remove ALL objects
###########################################
setwd("C:/VKHCG")
Base=getwd()
FileDir= paste0(Base,'/01-Vermeulen/01-Retrieve/01-EDS/01-R')
dir.create(FileDir)
FileDirLog=paste0(FileDir,'/log')
dir.create(FileDirLog)
FileDirRun=paste0(FileDirLog,'/Run0003')
dir.create(FileDirRun)
###########################################
StartTime=Sys.time()
###########################################
# Set up logging
###########################################
debugLog=paste0(FileDirRun,'/debug.Log')
infoLog=paste0(FileDirRun,'/info.Log')
errorLog=paste0(FileDirRun,'/error.Log')
###########################################
write(paste0('Start Debug Log File ',
        format(StartTime, "%Y/%d/%m %H:%M:%S")),
        file=debugLog,append = FALSE)
```

```
write(paste0('Start Information Log File ',
        format(StartTime, "%Y/%d/%m %H:%M:%S")),
        file=infoLog,append = FALSE)
write(paste0('Start Error Log File ',
        format(StartTime, "%Y/%d/%m %H:%M:%S")),
        file=errorLog,append = FALSE)
##########################################
UserName='Practical Data Scientist'
##########################################
write(paste0(UserName,' Load library: ', 'readr'),
        infoLog,append = TRUE)
library(readr)
write(paste0(UserName,' Load library: ', 'data.table'),
        infoLog,append = TRUE)
library(data.table)
write(paste0(UserName,' Load library: ', 'tibble'),
        infoLog,append = TRUE)
library(tibble)
##########################################
FileName=paste0(Base,'/01-Vermeulen/00-RawData/IP_DATA_CORE.csv')
write(paste0(UserName,' Retrieve data file: ', FileName),
        file=infoLog,append = TRUE)
IP_DATA_CORE <- read_csv (FileName,
  col_types = cols(
                Country = col_character(),
                `First IP Number` = col_double(),
                ID = col_integer(),
                `Last IP Number` = col_double(),
                Latitude = col_double(),
                Longitude = col_double(),
                `Place Name` = col_character(),
                `Post Code` = col_character()
                ),
  na = "empty"
)
```

```
###########################################
IP_DATA_CORE_FIX=set_tidy_names(IP_DATA_CORE,
    syntactic = TRUE, quiet = TRUE)
###########################################
IP_DATA_CORE_with_ID=rowid_to_column(IP_DATA_CORE_FIX, var = "RowID")
###########################################
setorderv(IP_DATA_CORE_with_ID, c('Country','Place.Name','Post.Code'),
        order= 1L, na.last=FALSE)
###########################################
FileNameOut=paste0(FileDir,'/Retrieve_IP_ DATA_CORE.csv')
fwrite(IP_DATA_CORE_with_ID, FileNameOut)
write(paste0(UserName,' Stores Retrieve data file: ',
      FileNameOut),
      file=infoLog,append = TRUE)
###########################################
StopTime=Sys.time()
###########################################
write(paste0('Stop Debug Log File ',
      format(StopTime, "%Y/%d/%m %H:%M:%S")),
      file=debugLog,append = TRUE)
write(paste0('Stop Information Log File ',
      format(StopTime, "%Y/%d/%m %H:%M:%S")),
      file=infoLog,append = TRUE)
write(paste0('Stop Error Log File ',
      format(StopTime, "%Y/%d/%m %H:%M:%S")),
      file=errorLog,append = TRUE)
###########################################
View(IP_DATA_CORE_with_ID)
###########################################
```

Save the Retrieve-IP_DATA_CORE.r file and source it to execute. Now that you have the file Retrieve_IP_ DATA_CORE.csv completed, you should be able to follow the basic principles of retrieving files.

Loading COUNTRY-CODES

Now, we add a reference file to the main data set, to assist with requirements for the access data processing in Chapter 10. For now, it is only another data file, but I will guide you through the process of how to use this reference data.

Create a new R script file and save it as `Retrieve-Country_Code.r` in directory `C:\VKHCG\01-Vermeulen\01-Retrieve`. Now, you can start to copy the following code into the script file:

```
############################################
rm(list=ls()) #will remove ALL objects
############################################
if (Sys.info()['sysname'] == "Windows")
  {
    BaseSet = "C:/VKHCG"
    setwd(BaseSet)
  } else {
    BaseSet = paste0("/home/",  Sys.info()['user'],  "/VKHCG")
    setwd(BaseSet)
  }
############################################
Base=getwd()
FileDir= paste0(Base,'/01-Vermeulen/01-Retrieve/01-EDS/01-R')
dir.create(FileDir)
FileDirLog=paste0(FileDir,'/log')
dir.create(FileDirLog)
FileDirRun=paste0(FileDirLog,'/Run0004')
dir.create(FileDirRun)
############################################
StartTime=Sys.time()
############################################
# Set up logging
############################################
debugLog=paste0(FileDirRun,'/debug.Log')
infoLog=paste0(FileDirRun,'/info.Log')
errorLog=paste0(FileDirRun,'/error.Log')
############################################
```

```
write(paste0('Start Debug Log File ',
        format(StartTime, "%Y/%d/%m %H:%M:%S")),
        file=debugLog,append = FALSE)
write(paste0('Start Information Log File ',
        format(StartTime, "%Y/%d/%m %H:%M:%S")),
        file=infoLog,append = FALSE)
write(paste0('Start Error Log File ',
        format(StartTime, "%Y/%d/%m %H:%M:%S")),
        file=errorLog,append = FALSE)
############################################
UserName='Practical Data Scientist'
############################################
write(paste0(UserName,' Load library: ', 'readr'),
        infoLog,append = TRUE)
library(readr)
write(paste0(UserName,' Load library: ', 'data.table'),
        infoLog,append = TRUE)
library(data.table)
write(paste0(UserName,' Load library: ', 'tibble'),
        infoLog,append = TRUE)
library(tibble)
############################################
FileName=paste0(Base,'/01-Vermeulen/00-RawData/Country_Code.csv')
write(paste0(UserName,' Retrieve data file: ', FileName),
        file=infoLog,append = TRUE)
Country_Code <- read_csv (FileName,
  col_types = cols(Country = col_character(),
                    'ISO-2-CODE' = col_character(),
                    'ISO-3-Code' = col_character(),
                    'ISO-M49' = col_integer()
                    ),
  na = "empty"
)
############################################
```

```
Country_Code_with_ID=rowid_to_column(Country_Code, var = "RowID")
#############################################
setorderv(Country_Code_with_ID, 'ISO-2-CODE',order= 1L, na.last=FALSE)
#############################################
FileNameOut=paste0(FileDir,'/Retrieve_Country_Code.csv')
fwrite(Country_Code_with_ID, FileNameOut)
write(paste0(UserName,' Stores Retrieve data file: ', FileNameOut),
        file=infoLog,append = TRUE)
#############################################
StopTime=Sys.time()
#############################################
write(paste0('Stop Debug Log File ',
        format(StopTime, "%Y/%d/%m %H:%M:%S")),
        file=debugLog,append = TRUE)
write(paste0('Stop Information Log File ',
        format(StopTime, "%Y/%d/%m %H:%M:%S")),
        file=infoLog,append = TRUE)
write(paste0('Stop Error Log File ',
        format(StopTime, "%Y/%d/%m %H:%M:%S")),
        file=errorLog,append = TRUE)
#############################################
View(Country_Code_with_ID)
#############################################
```

Save `Retrieve-Country_Code.r` and source it to execute.

Note   From this point, I have removed the logging code, as I want you to get to the core data retrieve process. Logging must be implemented when you deploy your new skills against real data lakes.

Krennwallner AG

Remember that this company markets our customers' products on billboards across the country.

> **Note**   I am loading from a new location named `C:\VKHCG\02-Krennwallner\`
> `00-RawData`.

Loading DE_Billboard_Locations

I will next guide you to retrieve the billboard locations for Germany. First, load
`DE_Billboard_Locations.csv` into `Retrieve_DE_Billboard_Locations.csv`.

   Create a new R script file and save it as `Retrieve-DE_Billboard_Locations.r` in
directory `C:\VKHCG\02-Krennwallner\01-Retrieve`. Now, you can start copying the
following code into the script file:

```
##########################################
rm(list=ls()) #will remove ALL objects
##########################################
if (Sys.info()['sysname'] == "Windows")
  {
    BaseSet = "C:/VKHCG"
    setwd(BaseSet)
  } else {
    BaseSet = paste0("/home/", Sys.info()['user'], "/VKHCG")
    setwd(BaseSet)
  }
##########################################
Base=getwd()
FileDir= paste0(Base,'/02-Krennwallner/01-Retrieve/01-EDS/01-R')
##########################################
library(readr)
library(data.table)
##########################################
FileName=paste0(Base,'/02-Krennwallner/00-RawData/DE_Billboard_Locations.csv')
##########################################
DE_Billboard_Locations <- read_csv(FileName,
     col_types = cols(
       Country = col_character(),
```

```
      ID = col_integer(),
      Latitude = col_double(),
      Longitude = col_double(),
      PlaceName = col_character()
   ), na = "empty")
###########################################
setnames(DE_Billboard_Locations,'PlaceName','Place.Name')
###########################################
setorder(DE_Billboard_Locations,Latitude,Longitude)
###########################################
FileNameOut=paste0(FileDir,'/Retrieve_DE_Billboard_Locations.csv')
fwrite(DE_Billboard_Locations, FileNameOut)
###########################################
View(DE_Billboard_Locations)
###########################################
```

Save `Retrieve-DE_Billboard_Locations.r` and source it to execute.

Tip   Did you spot the data issues in column `Place.Name`? I will help you to fix them in Chapter 10.

The data error you spotted is owing to the special characters in German names. The export program that delivered the data to the data lake has caused an issue. This is a common problem that a data scientist must handle to translate the data lake into business knowledge.

The Krennwallner company's data is loaded, and we can move on to the next company.

Hillman Ltd

This is our logistics company.

Note   I am loading from a new location: `..\VKHCG\03-Hillman\00-RawData`.

Loading GB_Postcode_Full

This data source contains all the postal code information for the United Kingdom. First, load DE_Billboard_Locations.csv into Retrieve_GB_Postcode_Full.csv. Create a new RsScript file and save it as Retrieve-GB_Postcode_Full.r in directory C:/VKHCG/ 03-Hillman/01-Retrieve/01-EDS/01-R.

```
###########################################
rm(list=ls()) #will remove ALL objects
###########################################
library(readr)
library(data.table)
library(tibble)
###########################################
if (Sys.info()['sysname'] == "Windows")
  {
    BaseSet = "C:/VKHCG"
    setwd(BaseSet)
  } else {
    BaseSet = paste0("/home/", Sys.info()['user'], "/VKHCG")
    setwd(BaseSet)
  }
###########################################
Base=getwd()
FileDir= paste0(Base,'/03-Hillman/01-Retrieve/01-EDS/01-R')
FileName=paste0(Base,'/03-Hillman/00-RawData/GB_Postcode_Full.csv')
GB_Postcode_Full <- read_csv(FileName,
                        col_types = cols(
                          AreaName = col_character(),
                          Country = col_character(),
                          ID = col_integer(),
                          PlaceName = col_character(),
                          PostCode = col_character(),
                          Region = col_character(),
                          RegionCode = col_character()
                        ),
                        na = "empty")
```

```
############################################
FileNameOut=paste0(FileDir,'/Retrieve_GB_Postcode_Full.csv')
fwrite(GB_Postcode_Full, FileNameOut)
############################################
View(GB_Postcode_Full)
############################################
```

Save `Retrieve-GB_Postcode_Full.r` and source it to execute.

Loading GB_Postcode_Warehouse

The following data holds all the locations for the warehouses. First, load `GB_Postcode_Warehouse.csv` into `Retrieve_GB_Postcode_Warehouse.csv`. I will then guide you to create and read back the data in JSON format. I will also show you how to use `library(jsonlite)` to generate two files: `Retrieve_GB_Postcode_Warehouse_A.json` and `Retrieve_GB_Postcode_Warehouse_B.json`.

Warning   I am assuming that you know the JSON data format. If you are not familiar with it, please read the next paragraph.

JSON (JavaScript Object Notation) is a lightweight data-interchange format used by source systems to generate data to transport more easily.

Create a new R script file and save it as `Retrieve-GB_Postcode_Warehouse.r` in directory `../VKHCG/03-Hillman/01-Retrieve/01-EDS/01-R`.

```
############################################
rm(list=ls()) #will remove ALL objects
############################################
library(readr)
library(data.table)
library(jsonlite)
############################################
if (Sys.info()['sysname'] == "Windows")
  {
    BaseSet = "C:/VKHCG"
    setwd(BaseSet)
```

```r
  } else {
    BaseSet = paste0("/home/",  Sys.info()['user'],  "/VKHCG")
    setwd(BaseSet)
  }
###########################################

Base=getwd()
FileDir= paste0(Base,'/03-Hillman/01-Retrieve/01-EDS/01-R')
FileName=paste0(Base,'/03-Hillman/00-RawData/GB_Postcode_Warehouse.csv')
GB_Postcode_Warehouse <- read_csv(FileName,
                        col_types = cols(
                           id = col_integer(),
                           latitude = col_double(),
                           longitude = col_double(),
                           postcode = col_character()
                        ),
                        na = "empty")
###########################################
FileNameOut=paste0(FileDir,'/Retrieve_GB_Postcode_Warehouse.csv')
fwrite(GB_Postcode_Warehouse, FileNameOut)
###########################################
GB_Postcode_Warehouse_JSON_A=toJSON(GB_Postcode_Warehouse, pretty=TRUE)
###########################################
FileNameJSONA=paste0(FileDir,'/Retrieve_GB_Postcode_Warehouse_A.json')
write(GB_Postcode_Warehouse_JSON_A, FileNameJSONA)
###########################################
GB_Postcode_Warehouse_JSON_B=toJSON(GB_Postcode_Warehouse, pretty=FALSE)
###########################################
FileNameJSONB=paste0(FileDir,'/Retrieve_GB_Postcode_Warehouse_B.json')
write(GB_Postcode_Warehouse_JSON_B, FileNameJSONB)
###########################################
View(GB_Postcode_Warehouse)
###########################################
GB_Postcode_Warehouse_A=json_data <- fromJSON(paste(readLines(FileNameJSONA),
collapse=""))
###########################################
```

```
View(GB_Postcode_Warehouse_A)
###########################################
GB_Postcode_Warehouse_B=json_data <- fromJSON(paste(readLines
(FileNameJSONB), collapse=""))
###########################################
View(GB_Postcode_Warehouse_B)
```

Save Retrieve- GB_Postcode_Warehouse.r and source it to execute.

You can create and load JSON files. This is another skill you will find useful in your interactions with a data lake.

Loading GB_Postcode_Shops

The following data sets the locations of all the shops. First, load GB_Postcodes_Shops.csv into Retrieve_GB_Postcodes_Shops.csv.

I will also guide you to create and read back the data for five shops in XML format. I will also show you how to use library(XML) to generate file Retrieve_GB_Postcodes_Shops.xml.

Warning   I am assuming that you know the XML data format. If you are not familiar with it, please read the next paragraph.

XML stands for EXtensible Markup Language. XML is a markup language similar to HTML, but XML is used to transport data across the Internet. It is a protocol that is used to transport data using standard HTTP-capable technology.

Tip   Generating XML files is time-consuming and must be used at a minimum inside the data lake ecosystem. Convert these to less complex and less verbose formats as soon as possible.

Create a new R script file and save it as Retrieve-GB_Postcode_Shops.r in directory ../VKHCG/03-Hillman/01-Retrieve/01-EDS/01-R.

```
###########################################
rm(list=ls()) #will remove ALL objects
###########################################
```

```r
library(readr)
library(data.table)
############################################
if (Sys.info()['sysname'] == "Windows")
  {
    BaseSet = "C:/VKHCG"
    setwd(BaseSet)
  } else {
    BaseSet = paste0("/home/", Sys.info()['user'],  "/VKHCG")
    setwd(BaseSet)
  }
############################################
Base=getwd()
FileDir= paste0(Base,'/03-Hillman/01-Retrieve/01-EDS/01-R')
FileName=paste0(Base,'/03-Hillman/00-RawData/GB_Postcodes_Shops.csv')
GB_Postcodes_Shops <- read_csv(FileName,
                               col_types = cols(
                                 id = col_integer(),
                                 latitude = col_double(),
                                 longitude = col_double(),
                                 postcode = col_character()
                               ),
                               na = "empty")
############################################
FileNameOut=paste0(FileDir,'/Retrieve_GB_Postcodes_Shops.csv')
fwrite(GB_Postcodes_Shops, FileNameOut)
############################################
View(GB_Postcodes_Shops)
############################################
GB_Five_Shops=GB_Postcodes_Shops[1:5,]
############################################
library(XML)
############################################
xml <- xmlTree()
xml$addTag("document", close=FALSE)
```

```
for (i in 1:nrow(GB_Five_Shops)) {
  xml$addTag("row", close=FALSE)
  for (j in 1:length(names(GB_Five_Shops))) {
    xml$addTag(names(GB_Five_Shops)[j], GB_Five_Shops[i, j])
  }
  xml$closeTag()
}
xml$closeTag()
GB_Postcodes_Shops_XML=saveXML(xml)
##########################################
FileNameXML=paste0(FileDir,'/Retrieve_GB_Postcodes_Shops.xml')
write(GB_Postcodes_Shops_XML, FileNameXML)
##########################################
xmlFile<-xmlTreeParse(FileNameXML)
class(xmlFile)
xmlTop = xmlRoot(xmlFile)
xmlText<- xmlSApply(xmlTop, function(x) xmlSApply(x, xmlValue))
GB_Postcodes_Shops_XML_A <- data.frame(t(xmlText),row.names=NULL)
##########################################
View(GB_Postcodes_Shops_XML_A)
```

Save Retrieve-GB_Postcode_Shops.r and source it to execute.

You have now generated an XML document and reloaded it into a data frame. This is a skill you will use several times in your work as data scientist.

Clark Ltd

Clark is the source of the financial information in the group.

Note   I am loading from a new location: ../VKHCG/04-Clark/00-RawData.

Loading Euro_ExchangeRates

First, load Euro_ExchangeRates.csv into Retrieve_Euro_ExchangeRates_Pivot.csv.
Create a new R script file and save it as Retrieve-Euro_ExchangeRates_Pivot.r in
directory C:/VKHCG/04-Clark/00-RawData/01-Retrieve/01-EDS/01-R.

```r
##############################################
rm(list=ls()) #will remove ALL objects
##############################################
library(readr)
library(data.table)
library(stringi)
##############################################
if (Sys.info()['sysname'] == "Windows")
  {
    BaseSet = "C:/VKHCG"
    setwd(BaseSet)
  } else {
    BaseSet = paste0("/home/", Sys.info()['user'], "/VKHCG")
    setwd(BaseSet)
  }
##############################################
Base=getwd()
FileDir= paste0(Base,'/04-Clark/01-Retrieve/01-EDS/01-R')
FileName=paste0(Base,'/04-Clark/00-RawData/Euro_ExchangeRates.csv')
Euro_ExchangeRates <- read_csv(FileName,
                  col_types = cols(
                    .default = col_double(),
                    Date = col_date(format = "%d/%m/%Y")
                  ),
                  locale = locale(asciify = TRUE), na = "empty")
##############################################
FileNameOut=paste0(FileDir,'/Retrieve_Euro_ExchangeRates_Pivot.csv')
fwrite(Euro_ExchangeRates, FileNameOut)
##############################################
View(Euro_ExchangeRates)
##############################################
```

The file you have for the data structure is known as a pivot table. This is a common data output that financial staff generate. This is, however, not the format in which data scientists want the data to be loaded. So, I must guide you through a process that you can use to resolve this pivot structure.

Reload Euro_ExchangeRates.csv into Retrieve_Euro_ExchangeRates.csv, create a new R script file, and save it as Retrieve-Euro_ExchangeRates.r in directory C:/VKHCG/04-Clark/00-RawData/01-Retrieve/01-EDS/01-R.

```r
###########################################
rm(list=ls()) #will remove ALL objects
###########################################
library("data.table")
library("readr")
###########################################
if (Sys.info()['sysname'] == "Windows")
  {
    BaseSet = "C:/VKHCG"
    setwd(BaseSet)
  } else {
    BaseSet = paste0("/home/", Sys.info()['user'], "/VKHCG")
    setwd(BaseSet)
  }
###########################################
Base=getwd()
FileDir= paste0(Base,'/04-Clark/01-Retrieve/01-EDS/01-R')
FileName=paste0(Base,'/04-Clark/00-RawData/Euro_ExchangeRates.csv')
Euro_ExchangeRates <- read_csv(FileName,
                  col_types = cols(
                    Date = col_date(format = "%d/%m/%Y"),
                    .default = col_character()
          ), locale = locale(encoding = "ASCII", asciify = TRUE)
)
###########################################
### Get the list of headings
###########################################
CA=as.vector(names(Euro_ExchangeRates))
```

```
##########################################
### Remove Date from the vector to get Exchange Codes
##########################################
C=CA [! CA %in% "Date"]
##########################################
### Create a default table structure
##########################################
Exchange=data.table(Date="1900-01-01",Code="Code",Rate=0,Base="Base")
##########################################
### Add the data for Exchange Code Euro pairs
##########################################
for (R in C) {
  ExchangeRates=data.table(subset(Euro_ExchangeRates[c('Date',R)],
                     is.na(Euro_ExchangeRates[R])==FALSE),R,Base="EUR")
  colnames(ExchangeRates)[colnames(ExchangeRates)==R] <- "Rate"
  colnames(ExchangeRates)[colnames(ExchangeRates)=="R"] <- "Code"
  if(nrow(Exchange)==1) Exchange=ExchangeRates
  if(nrow(Exchange)>1) Exchange=data.table(rbind(Exchange,ExchangeRates))
}
##########################################
Exchange2=Exchange
##########################################
colnames(Exchange)[colnames(Exchange)=="Rate"] <- "RateIn"
colnames(Exchange)[colnames(Exchange)=="Code"] <- "CodeIn"
colnames(Exchange2)[colnames(Exchange2)=="Rate"] <- "RateOut"
colnames(Exchange2)[colnames(Exchange2)=="Code"] <- "CodeOut"
ExchangeRate=merge(Exchange, Exchange2, by=c("Base","Date"), all=TRUE,
                   sort=FALSE,allow.cartesian=TRUE)
ExchangeRates <- data.table(ExchangeRate,
                Rate=with(
                  ExchangeRate,
                  round((as.numeric(RateOut) / as.numeric(RateIn)),9)
                ))
##########################################
### Remove work columns
```

```
##########################################
ExchangeRates$Base <-NULL
ExchangeRates$RateIn <-NULL
ExchangeRates$RateOut  <-NULL
##########################################
### Make entries unique
##########################################
ExchangeRate=unique(ExchangeRates)
##########################################
### Sort the results
##########################################
setorderv(ExchangeRate, c('Date','CodeIn','CodeOut'),
          order= c(-1L,1L,1L), na.last=FALSE)
##########################################
### Write Results
##########################################
FileNameOut=paste0(FileDir,'/Retrieve_Euro_ExchangeRates.csv')
fwrite(ExchangeRate, FileNameOut)
##########################################
### View Results
##########################################
View(ExchangeRate)
##########################################
```

Save the script `Retrieve-Euro_ExchangeRates.r` and source it to execute. You have resolved the pivot by converting it to a relationship between two exchange rates for a specific date. This will be useful during the data science in Chapters 10 and 11.

Load: Profit_And_Loss

First, load `Profit_And_Loss.csv` into `Retrieve_Profit_And_Loss.csv`. Create a new R script file and save it as `Retrieve-Profit_And_Loss.r` in directory `../VKHCG/04-Clark/00-RawData/01-Retrieve/01-EDS/01-R`.

```
##########################################

##########################################
rm(list=ls()) #will remove ALL objects
##########################################
```

```r
library(readr)
library(data.table)
library(tibble)
###########################################
if (Sys.info()['sysname'] == "Windows")
        {
                BaseSet = "C:/VKHCG"
                setwd(BaseSet)
        } else {
                BaseSet = paste0("/home/", Sys.info()['user'], "/VKHCG")
                setwd(BaseSet)
        }
###########################################
Base=getwd()
FileDir= paste0(Base,'/04-Clark/01-Retrieve/01-EDS/01-R')
FileName=paste0(Base,'/04-Clark/00-RawData/Profit_And_Loss.csv')
###########################################
Profit_And_Loss <- read_csv(FileName,
                    col_types = cols(
                       Amount = col_double(),
                       ProductClass1 = col_character(),
                       ProductClass2 = col_character(),
                       ProductClass3 = col_character(),
                       QTR = col_character(),
                       QTY = col_double(),
                       TypeOfEntry = col_character()
                    ), na = "empty")
###########################################
### Sort the results
###########################################
keyList=c('QTR','TypeOfEntry','ProductClass1','ProductClass2','ProductClass3')
setorderv(Profit_And_Loss, keyList, order= c(-1L,1L,1L,1L,1L),
na.last=FALSE)
###########################################
FileNameOut=paste0(FileDir,'/Retrieve_Profit_And_Loss.csv')
fwrite(Profit_And_Loss, FileNameOut)
```

```
############################################
View(Profit_And_Loss)
############################################
```

Save `Retrieve_Profit_And_Loss.csv` and source it to execute.

Python Retrieve Solution

I will now guide you through the Python solution for the Retrieve superstep.

Note   I will repeat an already complete R step in Python, to illustrate that you can interchange technology stacks and still accomplish similar good data science principles.

Vermeulen PLC

I will guide you through the first company using Python. I just want to show you that with Python, you can achieve the same results as with R.

Note   I am only showing the first data load, to make my point.

Start your Python editor. Create a text file named `Retrieve-IP_DATA_ALL.py` in directory `..\VKHCG\01-Vermeulen\01-Retrieve`.

Here is the Python code you must copy into the file:

```
################################################################
# -*- coding: utf-8 -*-
################################################################
import os
import pandas as pd
################################################################
if sys.platform == 'linux':
    Base=os.path.expanduser('~') + '/VKHCG'
else:
    Base='C:/VKHCG'
```

```python
###############################################################
sFileName=Base + '/01-Vermeulen/00-RawData/IP_DATA_ALL.csv'
print('Loading :',sFileName)
IP_DATA_ALL=pd.read_csv(sFileName,header=0,low_memory=False)
###############################################################
sFileDir=Base + '/01-Vermeulen/01-Retrieve/01-EDS/02-Python'
if not os.path.exists(sFileDir):
    os.makedirs(sFileDir)

print('Rows:', IP_DATA_ALL.shape[0])
print('Columns:', IP_DATA_ALL.shape[1])
print('### Raw Data Set ###################################')
for i in range(0,len(IP_DATA_ALL.columns)):
    print(IP_DATA_ALL.columns[i],type(IP_DATA_ALL.columns[i]))
print('### Fixed Data Set ###################################')
IP_DATA_ALL_FIX=IP_DATA_ALL
for i in range(0,len(IP_DATA_ALL.columns)):
    cNameOld=IP_DATA_ALL_FIX.columns[i] + '    '
    cNameNew=cNameOld.strip().replace(" ", ".")
    IP_DATA_ALL_FIX.columns.values[i] = cNameNew
    print(IP_DATA_ALL.columns[i],type(IP_DATA_ALL.columns[i]))
###############################################################
#print(IP_DATA_ALL_FIX.head())
###############################################################
print('Fixed Data Set with ID')
IP_DATA_ALL_with_ID=IP_DATA_ALL_FIX
IP_DATA_ALL_with_ID.index.names = ['RowID']
#print(IP_DATA_ALL_with_ID.head())

sFileName2=sFileDir + '/Retrieve_IP_DATA.csv'
IP_DATA_ALL_with_ID.to_csv(sFileName2, index = True)

###############################################################
print('### Done!! ###################################')
###############################################################
```

Now, save the text file. Execute the `Retrieve-IP_DATA_ALL.py` with your preferred Python compiler.

Note   I use the Spyder 3 environment from Anaconda and Python 3.

On completion of this task, you have proven that with different tools and a good data scientist, you can process most data with ease.

Actionable Business Knowledge from Data Lakes

I will guide you through several actionable business processes that you can formulate directly from the data in the sample data set.

Engineering a Practical Retrieve Superstep

Any source code or other supplementary material referenced by me in this book is available to readers on GitHub, via this book's product page, located at `www.apress.com/9781484230534`. Please note that this source code assumes that you have completed the source code setup from Chapter 2.

Now that I have explained the various aspects of the Retrieve superstep, I will explain how to assist our company with its processing.

The means are as follows:

- Identify the data sources required.

- Identify source data format (CSV, XML, JSON, or database).

- Data profile the data distribution (Skew, Histogram, Min, Max).

- Identify any loading characteristics (Columns Names, Data Types, Volumes).

- Determine the delivery format (CSV, XML, JSON, or database).

Warning   I will be using Python as the processing engine for the rest of this book and will add and explain new libraries and functions as we progress. I am using an Anaconda ecosystem, with a Spyder 3 editor and Python 3 compiler.

Vermeulen PLC

The company has two main jobs on which to focus your attention:

- Designing a routing diagram for company

- Planning a schedule of jobs to be performed for the router network

To perform these tasks, I will guide you through two separate data retrieve actions. For this practical data science, you will need to have your Python editor ready.

Designing a Routing Diagram for the Company

A network routing diagram is a map of all potential routes through the company's network, like a road map. I will help you to create a set of start and end locations, using longitude and latitude. I will implement feature extraction, by calculating the distance between locations in kilometers and miles.

Start your Python editor and create a text file named Retrieve-IP_Routing.py in directory ..\VKHCG\01-Vermeulen\01-Retrieve.

Following is the Python code you must copy into the file.

Let's retrieve the core IP data structure for the company.

```
###############################################################
# -*- coding: utf-8 -*-
###############################################################
import os
import pandas as pd
from math import radians, cos, sin, asin, sqrt
###############################################################
def haversine(lon1, lat1, lon2, lat2,stype):
    ### convert decimal degrees to radians
    lon1, lat1, lon2, lat2 = map(radians, [lon1, lat1, lon2, lat2])
```

```
    ### haversine formula
    dlon = lon2 - lon1
    dlat = lat2 - lat1
    a = sin(dlat/2)**2 + cos(lat1) * cos(lat2) * sin(dlon/2)**2
    c = 2 * asin(sqrt(a))
    ### Type of Distance (Kilometers/Miles)
    if stype == 'km':
        r = 6371 # Radius of earth in kilometers
    else:
        r = 3956 # Radius of earth in miles
    d=round(c * r,3)
    return d
################################################################
Base='C:/VKHCG'
sFileName=Base + '/01-Vermeulen/00-RawData/IP_DATA_CORE.csv'
print('Loading :',sFileName)
IP_DATA_ALL=pd.read_csv(sFileName,header=0,low_memory=False,
  usecols=['Country','Place Name','Latitude','Longitude'])
################################################################
sFileDir=Base + '/01-Vermeulen/01-Retrieve/01-EDS/02-Python'
if not os.path.exists(sFileDir):
    os.makedirs(sFileDir)

IP_DATA = IP_DATA_ALL.drop_duplicates(subset=None, keep='first',
inplace=False)
IP_DATA.rename(columns={'Place Name': 'Place_Name'}, inplace=True)
IP_DATA1 = IP_DATA
IP_DATA1.insert(0, 'K', 1)
IP_DATA2 = IP_DATA1

print(IP_DATA1.shape)

IP_CROSS=pd.merge(right=IP_DATA1,left=IP_DATA2,on='K')
IP_CROSS.drop('K', axis=1, inplace=True)
IP_CROSS.rename(columns={'Longitude_x': 'Longitude_from', 'Longitude_y':
'Longitude_to'}, inplace=True)
```

```
IP_CROSS.rename(columns={'Latitude_x': 'Latitude_from', 'Latitude_y':
'Latitude_to'}, inplace=True)
IP_CROSS.rename(columns={'Place_Name_x': 'Place_Name_from', 'Place_Name_y':
'Place_Name_to'}, inplace=True)
IP_CROSS.rename(columns={'Country_x': 'Country_from', 'Country_y':
'Country_to'}, inplace=True)

IP_CROSS['DistanceBetweenKilometers'] = IP_CROSS.apply(lambda row:
    haversine(
            row['Longitude_from'],
            row['Latitude_from'],
            row['Longitude_to'],
            row['Latitude_to'],
            'km')
            ,axis=1)

IP_CROSS['DistanceBetweenMiles'] = IP_CROSS.apply(lambda row:
    haversine(
            row['Longitude_from'],
            row['Latitude_from'],
            row['Longitude_to'],
            row['Latitude_to'],
            'miles')
            ,axis=1)
print(IP_CROSS.shape)
sFileName2=sFileDir + '/Retrieve_IP_Routing.csv'
IP_CROSS.to_csv(sFileName2, index = False)

################################################################
print('### Done!! #######################################')
################################################################
```

Now, save the text file.

Execute the `Retrieve-IP_Routing.py` with your preferred Python compiler. You will see a file named `Retrieve_IP_Routing.csv` in `C:\VKHCG\01-Vermeulen\01-Retrieve\01-EDS\02-Python`. Open this file, and you will see a new data set similar in layout to the following:

Country_ from	Place_ Name_ from	Latitude_ from	Longitude_ from	Country_ to	Place_ Name_ to	Latitude_ to	Longitude_ to	Distance Between Kilometers	Distance Between Miles
US	New York	40.7528	-73.9725	US	New York	40.7528	-73.9725	0	0
US	New York	40.7528	-73.9725	US	New York	40.7214	-74.0052	4.448	2.762

So, the distance between New York (40.7528, -73.9725) to New York (40.7214, -74.0052) is 4.45 kilometers, or 2.77 miles.

Well done! You have taken a data set of IP addresses and extracted a feature and a hidden feature of the distance between the IP addresses.

Building a Diagram for the Scheduling of Jobs

You can extract core routers locations to schedule maintenance jobs. Now that we know where the routers are, we must set up a schedule for maintenance jobs that have to be completed every month. To accomplish this, we must retrieve the location of each router, to prepare a schedule for the staff maintaining them.

Start your Python editor and create a text file named `Retrieve-Router-Location.py` in directory `..\VKHCG\01-Vermeulen\01-Retrieve`. Here is the Python code you must copy into the file:

```
################################################################
# -*- coding: utf-8 -*-
################################################################
import os
import pandas as pd
################################################################
InputFileName='IP_DATA_CORE.csv'
OutputFileName='Retrieve_Router_Location.csv'
################################################################
```

```python
if sys.platform == 'linux':
    Base=os.path.expanduser('~') + '/VKHCG'
else:
    Base='C:/VKHCG'
################################################################
sFileName=Base + '/01-Vermeulen/00-RawData/' + InputFileName
print('Loading :',sFileName)
IP_DATA_ALL=pd.read_csv(sFileName,header=0,low_memory=False,
  usecols=['Country','Place Name','Latitude','Longitude'])

IP_DATA_ALL.rename(columns={'Place Name': 'Place_Name'}, inplace=True)
################################################################
sFileDir=Base + '/01-Vermeulen/01-Retrieve/01-EDS/02-Python'
if not os.path.exists(sFileDir):
    os.makedirs(sFileDir)

ROUTERLOC = IP_DATA_ALL.drop_duplicates(subset=None, keep='first',
inplace=False)

print('Rows :',ROUTERLOC.shape[0])
print('Columns :',ROUTERLOC.shape[1])

sFileName2=sFileDir + '/' + OutputFileName
ROUTERLOC.to_csv(sFileName2, index = False)

################################################################
print('### Done!! ########################################')
################################################################
```

Now save the Python file. Execute the Retrieve-Router-Location.py with your preferred Python compiler. You will see a file named Retrieve_Router_Location.csv in C:\VKHCG\01-Vermeulen\01-Retrieve\01-EDS\02-Python.

Open this file, and you should see a data set like the following:

Country	Place_Name	Latitude	Longitude
US	New York	40.7528	-73.9725
US	New York	40.7214	-74.0052
US	New York	40.7662	-73.9862

You have now successfully retrieved the location of 150 routers.

There is another set of data you can try your newfound skills against. First, modify the code you just created, by changing the following code values:

```
InputFileName='IP_DATA_ALL.csv'
OutputFileName='Retrieve_All_Router_Location.csv'
```

Now, save the Python file as `Retrieve-Router-All-Location.py` and execute the new code with your preferred Python compiler. You will see a file named `Retrieve_All_Router_Location.csv` in `C:\VKHCG\01-Vermeulen\01-Retrieve\01-EDS\02-Python`.

Note   You may recall the discussion in Chapter 5 regarding utilities. This code is the start of a utility.

Well done, you just completed the retrieve for Vermeulen.

Let's review what you have achieved.

- You can now use Python to extract data from a CSV file.

- You know how to extract hidden features from a data set, i.e., the distance between two locations.

- You can save programming time with minor code changes, using common variables.

You have now succeeded in generating super data science code.

Krennwallner AG

The company has two main jobs in need of your attention:

- *Picking content for billboards*: I will guide you through the data science required to pick advertisements for each billboard in the company.

- *Understanding your online visitor data*: I will guide you through the evaluation of the web traffic to the billboard's online web servers.

Picking Content for Billboards

Let's retrieve the billboard data we have for Germany. Start your Python editor and create a text file named `Retrieve-DE-Billboard-Locations.py` in directory `..\ VKHCG\02-Krennwallner\01-Retrieve`.

Following is the Python code you must copy into the file:

```
###############################################################
# -*- coding: utf-8 -*-
###############################################################
import os
import pandas as pd
###############################################################
InputFileName='DE_Billboard_Locations.csv'
OutputFileName='Retrieve_DE_Billboard_Locations.csv'
Company='02-Krennwallner'
###############################################################
if sys.platform == 'linux':
    Base=os.path.expanduser('~') + '/VKHCG'
else:
    Base='C:/VKHCG'
###############################################################
sFileName=Base + '/' + Company + '/00-RawData/' + InputFileName
print('Loading :',sFileName)
IP_DATA_ALL=pd.read_csv(sFileName,header=0,low_memory=False,
  usecols=['Country','PlaceName','Latitude','Longitude'])
```

```
IP_DATA_ALL.rename(columns={'PlaceName': 'Place_Name'}, inplace=True)
##################################################################
sFileDir=Base + '/' + Company + '/01-Retrieve/01-EDS/02-Python'
if not os.path.exists(sFileDir):
    os.makedirs(sFileDir)

ROUTERLOC = IP_DATA_ALL.drop_duplicates(subset=None, keep='first',
inplace=False)

print('Rows :',ROUTERLOC.shape[0])
print('Columns :',ROUTERLOC.shape[1])

sFileName2=sFileDir + '/' + OutputFileName
ROUTERLOC.to_csv(sFileName2, index = False)

##################################################################
print('### Done!! #######################################')
##################################################################
```

Now save the Python file and execute the `Retrieve-DE-Billboard-Locations. py` with your preferred Python compiler. You will see a file named `Retrieve_Router_ Location.csv` in `C:\VKHCG\02-Krennwallner\01-Retrieve\01-EDS\02-Python`.

Open this file, and you should see a data set like this:

Country	Place_Name	Latitude	Longitude
DE	Lake	51.7833	8.5667
DE	Horb	48.4333	8.6833
DE	Hardenberg	51.1	7.7333
DE	Horn-bad Meinberg	51.9833	8.9667

Understanding Your Online Visitor Data

Let's retrieve the visitor data for the billboard we have in Germany.

Issue   We have no visitor data in Krennwallner.

I suggest that you look at all available data sources.

Resolved   The data is embedded in Vermeulen's data sources. See `IP_DATA_ALL.csv`.

I have found that if you are careful about conducting your business information gathering, which I covered in Chapter 4, you will have an entry on the data mapping matrix (see Chapter 4) for each data source. This cross use of data sources can be a reliable solution in your data science.

Several times in my engagements with customers, I find that common and important information is buried somewhere in the company's various data sources. I recommend investigating any direct suppliers or consumers' upstream or downstream data sources attached to the specific business process. That is part of your skills that you are applying to data science. I have personally found numerous insightful fragments of information in the data sources surrounding a customer's business processes.

Now that we know where the visitor data is, I will guide you in loading it. Start your Python editor and create a text file named `Retrieve-Online-Visitor.py` in directory `.\VKHCG\02-Krennwallner\01-Retrieve`.

Following is the Python code you must copy into the file:

```python
################################################################
# -*- coding: utf-8 -*-
################################################################
import sys
import os
import pandas as pd
import gzip as gz
################################################################
InputFileName='IP_DATA_ALL.csv'
OutputFileName='Retrieve_Online_Visitor'
CompanyIn= '01-Vermeulen'
CompanyOut= '02-Krennwallner'
################################################################
if sys.platform == 'linux':
```

```
    Base=os.path.expanduser('~') + '/VKHCG'
else:
    Base='C:/VKHCG'
print('##############################')
print('Working Base :',Base, ' using ', sys.platform)
print('##############################')
################################################################
Base='C:/VKHCG'
sFileName=Base + '/' + CompanyIn + '/00-RawData/' + InputFileName
print('Loading :',sFileName)
IP_DATA_ALL=pd.read_csv(sFileName,header=0,low_memory=False,
  usecols=['Country','Place Name','Latitude','Longitude','First IP
Number','Last IP Number'])

IP_DATA_ALL.rename(columns={'Place Name': 'Place_Name'}, inplace=True)
IP_DATA_ALL.rename(columns={'First IP Number': 'First_IP_Number'},
inplace=True)
IP_DATA_ALL.rename(columns={'Last IP Number': 'Last_IP_Number'},
inplace=True)
################################################################
sFileDir=Base + '/' + CompanyOut + '/01-Retrieve/01-EDS/02-Python'
if not os.path.exists(sFileDir):
    os.makedirs(sFileDir)

visitordata = IP_DATA_ALL.drop_duplicates(subset=None, keep='first',
inplace=False)
visitordata10=visitordata.head(10)

print('Rows :',visitordata.shape[0])
print('Columns :',visitordata.shape[1])

print('Export CSV')
sFileName2=sFileDir + '/' + OutputFileName + '.csv'
visitordata.to_csv(sFileName2, index = False)
print('Store All:',sFileName2)

sFileName3=sFileDir + '/' + OutputFileName + '_10.csv'
visitordata10.to_csv(sFileName3, index = False)
print('Store 10:',sFileName3)
```

```
for z in ['gzip', 'bz2', 'xz']:
    if z == 'gzip':
        sFileName4=sFileName2 + '.gz'
    else:
        sFileName4=sFileName2 + '.' + z
    visitordata.to_csv(sFileName4, index = False, compression=z)
    print('Store :',sFileName4)
#################################################################
print('Export JSON')
for sOrient in ['split','records','index', 'columns','values','table']:
    sFileName2=sFileDir + '/' + OutputFileName + '_' + sOrient + '.json'
    visitordata.to_json(sFileName2,orient=sOrient,force_ascii=True)
    print('Store All:',sFileName2)

    sFileName3=sFileDir + '/' + OutputFileName + '_10_' + sOrient + '.json'
    visitordata10.to_json(sFileName3,orient=sOrient,force_ascii=True)
    print('Store 10:',sFileName3)

    sFileName4=sFileName2 + '.gz'
    file_in = open(sFileName2, 'rb')
    file_out = gz.open(sFileName4, 'wb')
    file_out.writelines(file_in)
    file_in.close()
    file_out.close()
    print('Store GZIP All:',sFileName4)

    sFileName5=sFileDir + '/' + OutputFileName + '_' + sOrient + '_UnGZip.json'
    file_in = gz.open(sFileName4, 'rb')
    file_out = open(sFileName5, 'wb')
    file_out.writelines(file_in)
    file_in.close()
    file_out.close()
    print('Store UnGZIP All:',sFileName5)
#################################################################
print('### Done!! #####################################')
#################################################################
```

Now save the Python file. Execute the Retrieve-Online-Visitor.py with your preferred Python compiler. You will see a file named Retrieve_Online_Visitor.csv in C:\VKHCG\02-Krennwallner\01-Retrieve\01-EDS\02-Python. Open this file, and you should see a data set like this:

Country	Place_Name	Latitude	Longitude	First_IP_Number	Last_IP_Number
BW	Gaborone	-24.6464	25.9119	692781056	692781567
BW	Gaborone	-24.6464	25.9119	692781824	692783103
BW	Gaborone	-24.6464	25.9119	692909056	692909311
BW	Gaborone	-24.6464	25.9119	692909568	692910079

I have demonstrated that you can export the same data in various formats of JSON.

Warning   Remember: Being JSON-compliant is not sufficient to read the files. The orientation formatting is vital to processing the files. If you use a different layout orientation than the source system's, the data file will not load.

You can also see the following JSON files of only ten records. Open them in a text editor, as follows:

split : uses the format {index -> [index], columns -> [columns], data -> [values]}. See Retrieve_Online_Visitor_10_split.json.

```
{
    "columns":

        "Country","Place_Name","Latitude","Longitude","First_IP_
        Number","Last_IP_Number"
    ],
    "index":
    [
        0,1
    ],
    "data":[
        ["BW","Gaborone",-24.6464,25.9119,692781056,692781567],
        ["BW","Gaborone",-24.6464,25.9119,692781824,692783103]
```

214

```
    ]
}
```

records : uses the format [{column -> value}, . . ., {column -> value}]. See `Retrieve_Online_Visitor_10_records.json`.

```
[
    {
        "Country":"BW",
        "Place_Name":"Gaborone",
        "Latitude":-24.6464,
        "Longitude":25.9119,
        "First_IP_Number":692781056,
        "Last_IP_Number":692781567
    }
,
    {
        "Country":"BW",
        "Place_Name":"Gaborone",
        "Latitude":-24.6464,
        "Longitude":25.9119,
        "First_IP_Number":692781824,
        "Last_IP_Number":692783103
    }
]
```

index : uses the format {index -> {column -> value}}. See `Retrieve_Online_Visitor_10_index.json`.

```
{
    "0":
    {
        "Country":"BW",
        "Place_Name":"Gaborone",
        "Latitude":-24.6464,
        "Longitude":25.9119,
        "First_IP_Number":692781056,
```

```
        "Last_IP_Number":692781567
    }
,
    "1":
    {
        "Country":"BW",
        "Place_Name":"Gaborone",
        "Latitude":-24.6464,
        "Longitude":25.9119,
        "First_IP_Number":692781824,
        "Last_IP_Number":692783103
    }
}
```

columns : uses the format {column -> {index -> value}}. See `Retrieve_Online_Visitor_10_columns.json`.

```
{
    "Country":
        {"0":"BW","1":"BW"},
    "Place_Name":
        {"0":"Gaborone","1":"Gaborone"},
    "Latitude":
        {"0":-24.6464,"1":-24.6464},
    "Longitude":
        {"0":25.9119,"1":25.9119},
    "First_IP_Number":
        {"0":692781056,"1":692781824},
    "Last_IP_Number":
        {"0":692781567,"1":692783103}
}
```

values : uses the format of a simple values array. See `Retrieve_Online_Visitor_10_values.json`.

```
[
    ["BW","Gaborone",-24.6464,25.9119,692781056,692781567]
,
    ["BW","Gaborone",-24.6464,25.9119,692781824,692783103]
]
```

Warning   This JSON format causes you to lose all the metadata, i.e., no column name and data types are passed to the next part of the processing. You will have to remap and rename all the data items at each step in the process.

---

table : uses the format {'schema': {schema}, 'data': {data}}. See Retrieve_Online_Visitor_10_table.json.

```
{"schema":
    {
    "fields":
        [
            {"name":"index","type":"integer"},
            {"name":"Country","type":"string"},
            {"name":"Place_Name","type":"string"},
            {"name":"Latitude","type":"number"},
            {"name":"Longitude","type":"number"},
            {"name":"First_IP_Number","type":"integer"},
            {"name":"Last_IP_Number","type":"integer"}
        ]
    ,
    "primaryKey":["index"],
    "pandas_version":"0.20.0"},
    "data":
    [
        {
            "index":0,
            "Country":"BW",
            "Place_Name":"Gaborone",
            "Latitude":-24.6464,
```

```
                "Longitude":25.9119,
                "First_IP_Number":692781056,
                "Last_IP_Number":692781567
        }
        ,
        {

                "index":1,
                "Country":"BW",
                "Place_Name":"Gaborone",
                "Latitude":-24.6464,
                "Longitude":25.9119,
                "First_IP_Number":692781824,
                "Last_IP_Number":692783103
        }
    ]
}
```

Note   Table JSON formatting passes full metadata to the next data process. If you have a choice, I suggest that you always support this JSON format.

For visitor data, I suggest we also look at using XML formats for the data retrieve step.

Warning   XML can have complex structures in real-world applications. The following code can be modified to work on more complex structures. This is normally a time-consuming process of discovery. I suggest you experiment with the XML package, which I suggest we use for this task.

Following is a simple but effective introduction to XML processing. Start your Python editor and create a text file named `Retrieve-Online-Visitor-XML.py` in directory `.\VKHCG\02-Krennwallner\01-Retrieve`.

Here is the Python code you must copy into the file:

```python
################################################################
# -*- coding: utf-8 -*-
################################################################
import os
import pandas as pd
import xml.etree.ElementTree as ET
################################################################
def df2xml(data):
    header = data.columns
    root = ET.Element('root')
    for row in range(data.shape[0]):
        entry = ET.SubElement(root,'entry')
        for index in range(data.shape[1]):
            schild=str(header[index])
            child = ET.SubElement(entry, schild)
            if str(data[schild][row]) != 'nan':
                child.text = str(data[schild][row])
            else:
                child.text = 'n/a'
            entry.append(child)
    result = ET.tostring(root)
    return result
################################################################
def xml2df(xml:data):
    root = ET.XML(xml:data)
    all_records = []
    for i, child in enumerate(root):
        record = {}
        for subchild in child:
            record[subchild.tag] = subchild.text
        all_records.append(record)
    return pd.DataFrame(all_records)
################################################################
```

```
InputFileName='IP_DATA_ALL.csv'
OutputFileName='Retrieve_Online_Visitor.xml'
CompanyIn= '01-Vermeulen'
CompanyOut= '02-Krennwallner'
################################################################
if sys.platform == 'linux':
    Base=os.path.expanduser('~') + '/VKHCG'
else:
    Base='C:/VKHCG'
print('################################')
print('Working Base :',Base, ' using ', sys.platform)
print('################################')
################################################################
sFileName=Base + '/' + CompanyIn + '/00-RawData/' + InputFileName
print('Loading :',sFileName)
IP_DATA_ALL=pd.read_csv(sFileName,header=0,low_memory=False)

IP_DATA_ALL.rename(columns={'Place Name': 'Place_Name'}, inplace=True)
IP_DATA_ALL.rename(columns={'First IP Number': 'First_IP_Number'},
inplace=True)
IP_DATA_ALL.rename(columns={'Last IP Number': 'Last_IP_Number'},
inplace=True)
IP_DATA_ALL.rename(columns={'Post Code': 'Post_Code'}, inplace=True)
################################################################
sFileDir=Base + '/' + CompanyOut + '/01-Retrieve/01-EDS/02-Python'
if not os.path.exists(sFileDir):
    os.makedirs(sFileDir)

visitordata = IP_DATA_ALL.head(10000)

print('Original Subset Data Frame')
print('Rows :',visitordata.shape[0])
print('Columns :',visitordata.shape[1])
print(visitordata)
```

```
print('Export XML')
sXML=df2xml(visitordata)

sFileName=sFileDir + '/' + OutputFileName
file_out = open(sFileName, 'wb')
file_out.write(sXML)
file_out.close()
print('Store XML:',sFileName)

xml:data = open(sFileName).read()
unxmlrawdata=xml2df(xml:data)

print('Raw XML Data Frame')
print('Rows :',unxmlrawdata.shape[0])
print('Columns :',unxmlrawdata.shape[1])
print(unxmlrawdata)

unxmldata = unxmlrawdata.drop_duplicates(subset=None, keep='first',
inplace=False)


print('Deduplicated XML Data Frame')
print('Rows :',unxmldata.shape[0])
print('Columns :',unxmldata.shape[1])
print(unxmldata)
#################################################################
#print('### Done!! #####################################')
#################################################################
```

Now save the Python file and execute the `Retrieve-Online-Visitor-XML.py` with your preferred Python compiler. You will see a file named `Retrieve_Online_Visitor.xml` in `C:\VKHCG\02-Krennwallner\01-Retrieve\01-EDS\02-Python`. This enables you to deliver XML format data as part of the retrieve step.

I have assisted you with many formats to process and pass data between steps in your processing pipeline. For the rest of this chapter, I am only demonstrating a smaller subset. Remember: The processing methods and techniques can be used on all the data formats. So, let's see what our Hillman data brings to our knowledge base.

Hillman Ltd

The company has four main jobs requiring your attention:

- *Planning the locations of the warehouses*: Hillman has countless UK warehouses, but owing to financial hardships, the business wants to shrink the quantity of warehouses by 20%.

- *Planning the shipping rules for best-fit international logistics*: At Hillman Global Logistics' expense, the company has shipped goods from its international warehouses to its UK shops. This model is no longer sustainable. The co-owned shops now want more feasibility regarding shipping options.

- *Adopting the best packing option for shipping in containers*: Hillman has introduced a new three-size-shipping-container solution. It needs a packing solution encompassing the warehouses, shops, and customers.

- *Creating a delivery route*: Hillman needs to preplan a delivery route for each of its warehouses to shops, to realize a 30% savings in shipping costs.

Plan the locations of the warehouses. I will assist you in retrieving the warehouse locations.

Warning   Stop! This data is already loaded.

Earlier in the chapter, you have created the load, as part of your R processing. The file is named `Retrieve_GB_Postcode_Warehouse.csv` in directory `C:\VKHCG\03-Hillman\01-Retrieve\01-EDS\01-R`.

I suggest that you update the data mapping matrix (see Chapter 4) on a regular basis. An up-to-date data mapping matrix will save you from having to code a new solution for the data retrieve data sources.

Planning Shipping Rules for Best-Fit International Logistics

I will now guide you through the retrieve process for shipping rules. Yes, you must understand the business to map its business processes into your data discovery.

> **Note**   I do not expect you to grasp all the rules on first read. However, as a data scientist, you will have to understand the business fully, to create the data sets that are required.

Important Shipping Information

In the world of shipping, there is a short list of important terms you must understand first.

Shipping Terms

These determine the rules of the shipment, the conditions under which it is made. Normally, these are stated on the shipping manifest. Currently, Hillman is shipping everything as DDP. Not what we want! I will discuss this shipping term in detail within the next pages.

Seller

The person/company sending the products on the shipping manifest is the *seller*. In our case, there will be warehouses, shops, and customers. Note that this is not a location but a legal entity sending the products.

Carrier

The person/company that physically carries the products on the shipping manifest is the *carrier*. Note that this is not a location but a legal entity transporting the products.

Port

A *port* is any point from which you have to exit or enter a country. Normally, these are shipping ports or airports but can also include border crossings via road. Note that there are two ports in the complete process. This is important. There is a port of exit and a port of entry.

Ship

Ship is the general term for the physical transport method used for the goods. This can refer to a cargo ship, airplane, truck, or even person, but it must be identified by a unique allocation number.

223

Terminal

A *terminal* is the physical point at which the goods are handed off for the next phase of the physical shipping.

Named Place

This is the location where the ownership is legally changed from seller to buyer. This is a specific location in the overall process. Remember this point, as it causes many legal disputes in the logistics industry.

Buyer

The person/company receiving the products on the shipping manifest is the *buyer*. In our case, there will be warehouses, shops, and customers. Note that this is not a location but a legal entity receiving the products.

In general, the complete end-to-end process in a real shipping route consists of several chains of shipping manifests. Yes, you read correctly, there is more than one shipping manifest per shipping route!

Incoterm 2010

Let's first tackle the standard shipping terms. Following is a summary of the basic options, as determined by a standard board and published as Incoterm 2010.

Here they are in a simple grid:

Shipping Term	Seller	Carrier	Port	Ship	Port	Terminal	Named Place	Buyer
EXW	Seller	Buyer	Buyer	Buyer	Buyer	Buyer	Buyer	Buyer
FCA	Seller	Seller	Buyer	Buyer	Buyer	Buyer	Buyer	Buyer
CPT	Seller	Seller	Buyer	Buyer	Buyer	Buyer	Buyer	Buyer
CIP	Seller	Seller	Insurance	Insurance	Insurance	Insurance	Insurance	Buyer
DAT	Seller	Seller	Seller	Seller	Seller	Seller	Buyer	Buyer
DAP	Seller	Seller	Seller	Seller	Seller	Seller	Seller	Buyer
DDP	Seller	Seller	Seller	Seller	Seller	Seller	Seller	Seller

If right now you feel the same as I did the first time I saw this grid—lost, lonely, in need of coffee—don't close this book. I will quickly guide you from beginning to end. The first item of discussion is the shipping terms.

EXW—Ex Works (Named Place of Delivery)

By this term, the seller makes the goods available at its premises or at another named place. This term places the maximum obligation on the buyer and minimum obligations on the seller.

Here is the data science version: If I were to buy *Practical Data Science* at a local bookshop and take it home, and the shop has shipped it EXW—Ex Works, the moment I pay at the register, the ownership is transferred to me. If anything happens to the book, I would have to pay to replace it.

So, let's see what the data science reveals, using our sample company. Start your Python editor and create a text file named `Retrieve-Incoterm-EXW.py` in directory `.\VKHCG\03-Hillman\01-Retrieve`. Following is the Python code that you must copy into the file:

```
################################################################
# -*- coding: utf-8 -*-
################################################################
import os
import pandas as pd
################################################################
IncoTerm='EXW'
InputFileName='Incoterm_2010.csv'
OutputFileName='Retrieve_Incoterm_' + IncoTerm + '_RuleSet.csv'
Company='03-Hillman'
################################################################
if sys.platform == 'linux':
    Base=os.path.expanduser('~') + '/VKHCG'
else:
    Base='C:/VKHCG'
print('###############################')
print('Working Base :',Base, ' using ', sys.platform)
print('###############################')
```

```
################################################################
sFileDir=Base + '/' + Company + '/01-Retrieve/01-EDS/02-Python'
if not os.path.exists(sFileDir):
    os.makedirs(sFileDir)
################################################################
### Import Incoterms
################################################################
sFileName=Base + '/' + Company + '/00-RawData/' + InputFileName
print('###########')
print('Loading :',sFileName)
IncotermGrid=pd.read_csv(sFileName,header=0,low_memory=False)
IncotermRule=IncotermGrid[IncotermGrid.Shipping_Term == IncoTerm]
print('Rows :',IncotermRule.shape[0])
print('Columns :',IncotermRule.shape[1])
print('###########')

print(IncotermRule)
################################################################

sFileName=sFileDir + '/' + OutputFileName
IncotermRule.to_csv(sFileName, index = False)

################################################################
print('### Done!! #########################################')
################################################################
```

Now save the Python file and execute the Retrieve-Incoterm-EXW.py file with your preferred Python compiler. You will see a file named Retrieve_Incoterm_EXW.csv in C:\VKHCG\03-Hillman\01-Retrieve\01-EDS\02-Python. Open this file, and you should see a data set like this:

Shipping Term	Seller	Carrier	Port	Ship	Port	Terminal	Named Place	Buyer
EXW	Seller	Buyer	Buyer	Buyer	Buyer	Buyer	Buyer	Buyer

Note   The ownership changes prematurely in the supply chain. All the costs and risks are with the buyer.

FCA—Free Carrier (Named Place of Delivery)

Under this condition, the seller delivers the goods, cleared for export, at a named place. Following is the data science version.

If I were to buy *Practical Data Science* at an overseas duty-free shop and then pick it up at the duty-free desk before taking it home, and the shop has shipped it FCA—Free Carrier—to the duty-free desk, the moment I pay at the register, the ownership is transferred to me, but if anything happens to the book between the shop and the duty-free desk, the shop will have to pay. It is only once I pick it up at the desk that I will have to pay, if anything happens. So, the moment I take the book, the transaction becomes EXW, so I have to pay any necessary import duties on arrival in my home country.

Let's see what the data science finds. Start your Python editor and create a text file named `Retrieve-Incoterm-FCA.py` in directory `.\VKHCG\03-Hillman\01-Retrieve`. Here is the Python code you must copy into the file:

```
###################################################################
# -*- coding: utf-8 -*-
###################################################################
import os
import pandas as pd
###################################################################
IncoTerm='FCA'
InputFileName='Incoterm_2010.csv'
OutputFileName='Retrieve_Incoterm_' + IncoTerm + '_RuleSet.csv'
Company='03-Hillman'
###################################################################
if sys.platform == 'linux':
    Base=os.path.expanduser('~') + '/VKHCG'
else:
    Base='C:/VKHCG'
print('###############################')
print('Working Base :',Base, ' using ', sys.platform)
print('###############################')
###################################################################
sFileDir=Base + '/' + Company + '/01-Retrieve/01-EDS/02-Python'
if not os.path.exists(sFileDir):
    os.makedirs(sFileDir)
```

```
################################################################
### Import Incoterms
################################################################
sFileName=Base + '/' + Company + '/00-RawData/' + InputFileName
print('###########')
print('Loading :',sFileName)
IncotermGrid=pd.read_csv(sFileName,header=0,low_memory=False)
IncotermRule=IncotermGrid[IncotermGrid.Shipping_Term == IncoTerm]
print('Rows :',IncotermRule.shape[0])
print('Columns :',IncotermRule.shape[1])
print('###########')

print(IncotermRule)
################################################################

sFileName=sFileDir + '/' + OutputFileName
IncotermRule.to_csv(sFileName, index = False)

################################################################
print('### Done!! #######################################')
################################################################
```

Now save the Python file and execute the Retrieve-Incoterm-FCA.py file with your preferred Python compiler. You will see a file named Retrieve_Incoterm_FCA.csv in C:\ VKHCG\03-Hillman\01-Retrieve\01-EDS\02-Python. Open this file, and you should see a data set like this:

Shipping Term	Seller	Carrier	Port	Ship	Port	Terminal	Named Place	Buyer
FCA	Seller	Seller	Buyer	Buyer	Buyer	Buyer	Buyer	Buyer

Note   The ownership changes prematurely, at the port in the supply chain. Most of the risks and costs are again with the buyer.

CPT—Carriage Paid To (Named Place of Destination)

The seller, under this term, pays for the carriage of the goods up to the named place of destination. However, the goods are considered to be delivered when they have been handed over to the first carrier, so that the risk transfers to the buyer upon handing the goods over to the carrier at the place of shipment in the country of export.

The seller is responsible for origin costs, including export clearance and freight costs for carriage to the named place of destination. (This is either the final destination, such as the buyer's facilities, or a port of destination. This must be agreed upon by both seller and buyer, however.)

Now, here is the data science version: If I were to buy *Practical Data Science* at an overseas bookshop and then pick it up at the export desk before taking it home and the shop shipped it CPT—Carriage Paid To—the duty desk for free, the moment I pay at the register, the ownership is transferred to me, but if anything happens to the book between the shop and the duty desk of the shop, I will have to pay. It is only once I have picked up the book at the desk that I have to pay if anything happens. So, the moment I take the book, the transaction becomes EXW, so I must pay any required export and import duties on arrival in my home country.

Let's see what the data science finds.

Start your Python editor and create a text file named `Retrieve-Incoterm-CPT.py` in directory `.\VKHCG\03-Hillman\01-Retrieve`. Here is the Python code you must copy into the file:

```python
################################################################
# -*- coding: utf-8 -*-
################################################################
import os
import pandas as pd
################################################################
IncoTerm='CPT'
InputFileName='Incoterm_2010.csv'
OutputFileName='Retrieve_Incoterm_' + IncoTerm + '_RuleSet.csv'
Company='03-Hillman'
################################################################
```

```
if sys.platform == 'linux':
    Base=os.path.expanduser('~') + '/VKHCG'
else:
    Base='C:/VKHCG'
print('###############################')
print('Working Base :',Base, ' using ', sys.platform)
print('###############################')
################################################################
################################################################
sFileDir=Base + '/' + Company + '/01-Retrieve/01-EDS/02-Python'
if not os.path.exists(sFileDir):
    os.makedirs(sFileDir)
################################################################
### Import Incoterms
################################################################
sFileName=Base + '/' + Company + '/00-RawData/' + InputFileName
print('###########')
print('Loading :',sFileName)
IncotermGrid=pd.read_csv(sFileName,header=0,low_memory=False)
IncotermRule=IncotermGrid[IncotermGrid.Shipping_Term == IncoTerm]
print('Rows :',IncotermRule.shape[0])
print('Columns :',IncotermRule.shape[1])
print('###########')

print(IncotermRule)
################################################################

sFileName=sFileDir + '/' + OutputFileName
IncotermRule.to_csv(sFileName, index = False)

################################################################
print('### Done!! ##########################################')
################################################################
```

Now save the Python file and execute the Retrieve-Incoterm-CPT.py file with your preferred Python compiler. You will see a file named: Retrieve_Incoterm_CPT.csv in

`C:\VKHCG\03-Hillman\01-Retrieve\01-EDS\02-Python`. Open this file, and you should see a data set like this:

Shipping Term	Seller	Carrier	Port	Ship	Port	Terminal	Named Place	Buyer
CPT	Seller	Seller	Buyer	Buyer	Buyer	Buyer	Buyer	Buyer

Note   The ownership changes prematurely, at the port in the supply chain. Most of the risks and cost are with the buyer. This is bad if we are buying but good if we are selling.

CIP—Carriage and Insurance Paid To (Named Place of Destination)

This term is generally similar to the preceding CPT, with the exception that the seller is required to obtain insurance for the goods while in transit. Following is the data science version.

If I buy *Practical Data Science* in an overseas bookshop and then pick it up at the export desk before taking it home, and the shop has shipped it CPT—Carriage Paid To— to the duty desk for free, the moment I pay at the register, the ownership is transferred to me. However, if anything happens to the book between the shop and the duty desk at the shop, I have to take out insurance to pay for the damage. It is only once I have picked it up at the desk that I have to pay if anything happens. So, the moment I take the book, it becomes EXW, so I have to pay any export and import duties on arrival in my home country. Note that insurance only covers that portion of the transaction between the shop and duty desk.

Let's see what the data science finds. Start your Python editor and create a text file named `Retrieve-Incoterm-CIP.py` in directory `.\VKHCG\03-Hillman\01-Retrieve`. Here is the Python code you must copy into the file:

```
################################################################
# -*- coding: utf-8 -*-
################################################################
import os
import pandas as pd
```

```
################################################################
IncoTerm='CIP'
InputFileName='Incoterm_2010.csv'
OutputFileName='Retrieve_Incoterm_' + IncoTerm + '_RuleSet.csv'
Company='03-Hillman'
################################################################
if sys.platform == 'linux':
    Base=os.path.expanduser('~') + '/VKHCG'
else:
    Base='C:/VKHCG'
print('##############################')
print('Working Base :',Base, ' using ', sys.platform)
print('##############################')
################################################################
sFileDir=Base + '/' + Company + '/01-Retrieve/01-EDS/02-Python'
if not os.path.exists(sFileDir):
    os.makedirs(sFileDir)
################################################################
### Import Incoterms
################################################################
sFileName=Base + '/' + Company + '/00-RawData/' + InputFileName
print('##########')
print('Loading :',sFileName)
IncotermGrid=pd.read_csv(sFileName,header=0,low_memory=False)
IncotermRule=IncotermGrid[IncotermGrid.Shipping_Term == IncoTerm]
print('Rows :',IncotermRule.shape[0])
print('Columns :',IncotermRule.shape[1])
print('##########')

print(IncotermRule)
################################################################

sFileName=sFileDir + '/' + OutputFileName
IncotermRule.to_csv(sFileName, index = False)
```

```
################################################################
print('### Done!! ######################################')
################################################################
```

Now save the Python file and execute the `Retrieve-Incoterm-CIP.py` file with your preferred Python compiler. You will see a file named `Retrieve_Incoterm_CIP.csv` in `C:\VKHCG\03-Hillman\01-Retrieve\01-EDS\02-Python`. Open this file, and you should see a data set like this:

Shipping Term	Seller	Carrier	Port	Ship	Port	Terminal	Named Place	Buyer
CIP	Seller	Seller	Insurance	Insurance	Insurance	Insurance	Insurance	Buyer

Note   The ownership changes prematurely, at the port in the supply chain. Most of the risks and cost are with the buyer, with some of it covered by insurance. Risk is lower, compared to that calculated for house or car insurance. This is not so bad if you are buying but not so good if you are selling!

DAT—Delivered at Terminal (Named Terminal at Port or Place of Destination)

This Incoterm requires that the seller deliver the goods, unloaded, at the named terminal. The seller covers all the costs of transport (export fees, carriage, unloading from the main carrier at destination port, and destination port charges) and assumes all risks until arrival at the destination port or terminal.

The terminal can be a port, airport, or inland freight interchange, but it must be a facility with the capability to receive the shipment. If the seller is not able to organize unloading, it should consider shipping under DAP terms instead. All charges after unloading (for example, import duty, taxes, customs and on-carriage costs) are to be borne by buyer.

Following is the data science version. If I were to buy *Practical Data Science* overseas and then pick it up at a local bookshop before taking it home, and the overseas shop shipped it—Delivered At Terminal (Local Shop)—the moment I pay at the register, the ownership is transferred to me. However, if anything happens to the book between the payment and the pickup, the local shop pays. It is picked up only once at the local shop. I have to pay if anything happens. So, the moment I take it, the transaction becomes EXW, so I have to pay any import duties on arrival in my home.

Let's see what the data science finds. Start your Python editor and create a text file named `Retrieve-Incoterm-DAT.py` in directory `".\VKHCG\03-Hillman\01-Retrieve`. Following is the Python code you must copy into the file:

```python
###################################################################
# -*- coding: utf-8 -*-
###################################################################
import os
import pandas as pd
###################################################################
IncoTerm='DAT'
InputFileName='Incoterm_2010.csv'
OutputFileName='Retrieve_Incoterm_' + IncoTerm + '_RuleSet.csv'
Company='03-Hillman'
###################################################################
if sys.platform == 'linux':
    Base=os.path.expanduser('~') + '/VKHCG'
else:
    Base='C:/VKHCG'
print('#############################')
print('Working Base :',Base, ' using ', sys.platform)
print('#############################')
###################################################################
sFileDir=Base + '/' + Company + '/01-Retrieve/01-EDS/02-Python'
if not os.path.exists(sFileDir):
    os.makedirs(sFileDir)
###################################################################
### Import Incoterms
###################################################################
sFileName=Base + '/' + Company + '/00-RawData/' + InputFileName
print('###########')
print('Loading :',sFileName)
IncotermGrid=pd.read_csv(sFileName,header=0,low_memory=False)
IncotermRule=IncotermGrid[IncotermGrid.Shipping_Term == IncoTerm]
print('Rows :',IncotermRule.shape[0])
```

```
print('Columns :',IncotermRule.shape[1])
print('##########')

print(IncotermRule)
################################################################

sFileName=sFileDir + '/' + OutputFileName
IncotermRule.to_csv(sFileName, index = False)


################################################################
print('### Done!! ####################################')
################################################################
```

Now save the Python file and execute the Retrieve-Incoterm-DAT.py file with your preferred Python compiler. You will see a file named Retrieve_Incoterm_DAT.csv in C:\ VKHCG\03-Hillman\01-Retrieve\01-EDS\02-Python. Open this file, and you should see a data set like this:

Shipping Term	Seller	Carrier	Port	Ship	Port	Terminal	Named Place	Buyer
DAT		Seller	Seller	Seller	Seller	Seller	Seller	Buyer		Buyer

Note   Now the ownership changes later at the named place in the supply chain. Most of the risk and cost are borne by the seller. This is positive, if you are buying, less so, if you are selling.

DAP—Delivered at Place (Named Place of Destination)

According to Incoterm 2010's definition, DAP—Delivered at Place—means that, at the disposal of the buyer, the seller delivers when the goods are placed on the arriving means of transport, ready for unloading at the named place of destination. Under DAP terms, the risk passes from seller to buyer from the point of destination mentioned in the contract of delivery.

Once goods are ready for shipment, the necessary packing is carried out by the seller at his own cost, so that the goods reach their final destination safely. All necessary legal formalities in the exporting country are completed by the seller at his own cost and risk to clear the goods for export.

After arrival of the goods in the country of destination, the customs clearance in the importing country must be completed by the buyer at his own cost and risk, including all customs duties and taxes. However, as with DAT terms, any delay or demurrage charges are to be borne by the seller.

Under DAP terms, all carriage expenses with any terminal expenses are paid by seller, up to the agreed destination point. The required unloading cost at the final destination has to be accepted by the buyer under DAP terms.

Here is the data science version. If I were to buy 100 copies of *Practical Data Science* from an overseas web site and then pick up the copies at a local bookshop before taking them home, and the shop shipped the copies DAP—Delivered At Place (Local Shop)— the moment I paid at the register, the ownership would be transferred to me. However, if anything happened to the books between the payment and the pickup, the web site owner pays. Once the copies are picked up at the local shop, I have to pay to unpack them at bookshop. So, the moment I take the copies, the transaction becomes EXW, so I will have to pay costs after I take the copies.

Let's see what the data science finds. Start your Python editor and create a text file named `Retrieve-Incoterm-DAP.py` in directory `.\VKHCG\03-Hillman\01-Retrieve`. Here is the Python code you must copy into the file:

```python
###############################################################
# -*- coding: utf-8 -*-
###############################################################
import os
import pandas as pd
###############################################################
IncoTerm='DAP'
InputFileName='Incoterm_2010.csv'
OutputFileName='Retrieve_Incoterm_' + IncoTerm + '_RuleSet.csv'
Company='03-Hillman'
###############################################################
if sys.platform == 'linux':
    Base=os.path.expanduser('~') + '/VKHCG'
else:
    Base='C:/VKHCG'
print('###############################')
print('Working Base :',Base, ' using ', sys.platform)
```

```python
print('###############################')
#################################################################
sFileDir=Base + '/' + Company + '/01-Retrieve/01-EDS/02-Python'
if not os.path.exists(sFileDir):
    os.makedirs(sFileDir)
#################################################################
### Import Incoterms
#################################################################
sFileName=Base + '/' + Company + '/00-RawData/' + InputFileName
print('###########')
print('Loading :',sFileName)
IncotermGrid=pd.read_csv(sFileName,header=0,low_memory=False)
IncotermRule=IncotermGrid[IncotermGrid.Shipping_Term == IncoTerm]
print('Rows :',IncotermRule.shape[0])
print('Columns :',IncotermRule.shape[1])
print('###########')

print(IncotermRule)
#################################################################

sFileName=sFileDir + '/' + OutputFileName
IncotermRule.to_csv(sFileName, index = False)

#################################################################
print('### Done!! #########################################')
#################################################################
```

Now save the Python file.

Execute the Retrieve-Incoterm-DAP.py file with your preferred Python compiler. You will see a file named Retrieve_Incoterm_DAP.csv in C:\VKHCG\03-Hillman\01-Retrieve\01-EDS\02-Python. Open this file, and you should see a data set like this:

Shipping Term	Seller	Carrier	Port	Ship	Port	Terminal	Named Place	Buyer
DAP	Seller	Seller	Seller	Seller	Seller	Seller	Seller	Buyer

> **Note**   Now, the ownership changes later, at the buyer's place in the supply chain. All of the risks and cost are with the seller. This is very good if you are buying but really bad if you are selling. You would be responsible for any mishaps with the shipping. The only portion you would not need to cover is offloading.

DDP—Delivered Duty Paid (Named Place of Destination)

By this term, the seller is responsible for delivering the goods to the named place in the country of the buyer and pays all costs in bringing the goods to the destination, including import duties and taxes. The seller is not responsible for unloading. This term places the maximum obligations on the seller and minimum obligations on the buyer. No risk or responsibility is transferred to the buyer until delivery of the goods at the named place of destination.

The most important consideration regarding DDP terms is that the seller is responsible for clearing the goods through customs in the buyer's country, including both paying the duties and taxes, and obtaining the necessary authorizations and registrations from the authorities in that country.

Here is the data science version. If I were to buy 100 copies of *Practical Data Science* on an overseas web site and then pick them up at a local bookshop before taking them home, and the shop shipped DDP—Delivered Duty Paid (my home)—the moment I pay at the till, the ownership is transferred to me. However, if anything were to happen to the books between the payment and the delivery at my house, the bookshop must replace the books as the term covers the delivery to my house.

Let's see what the data science finds. Start your Python editor and create a text file named Retrieve-Incoterm-DDP.py in directory .\VKHCG\03-Hillman\01-Retrieve. Following is the Python code you must copy into the file:

```
###############################################################
# -*- coding: utf-8 -*-
###############################################################
import os
import pandas as pd
###############################################################
IncoTerm='DDP'
InputFileName='Incoterm_2010.csv'
```

```
OutputFileName='Retrieve_Incoterm_' + IncoTerm + '_RuleSet.csv'
Company='03-Hillman'
################################################################
if sys.platform == 'linux':
    Base=os.path.expanduser('~') + '/VKHCG'
else:
    Base='C:/VKHCG'
print('##############################')
print('Working Base :',Base, ' using ', sys.platform)
print('##############################')
################################################################
sFileDir=Base + '/' + Company + '/01-Retrieve/01-EDS/02-Python'
if not os.path.exists(sFileDir):
    os.makedirs(sFileDir)
################################################################
### Import Incoterms
################################################################
sFileName=Base + '/' + Company + '/00-RawData/' + InputFileName
print('###########')
print('Loading :',sFileName)
IncotermGrid=pd.read_csv(sFileName,header=0,low_memory=False)
IncotermRule=IncotermGrid[IncotermGrid.Shipping_Term == IncoTerm]
print('Rows :',IncotermRule.shape[0])
print('Columns :',IncotermRule.shape[1])
print('###########')

print(IncotermRule)
################################################################

sFileName=sFileDir + '/' + OutputFileName
IncotermRule.to_csv(sFileName, index = False)

################################################################
print('### Done!! ######################################')
################################################################
```

Now save the Python file and execute the `Retrieve-Incoterm-DDP.py` file with your preferred Python compiler. You will see a file named `Retrieve_Incoterm_DDP.csv` in `C:\ VKHCG\03-Hillman\01-Retrieve\01-EDS\02-Python`. Open this file, and you should see a data set like this:

Shipping Term	Seller	Carrier	Port	Ship	Port	Terminal	Named Place	Buyer
DDP	Seller	Seller	Seller	Seller	Seller	Seller	Seller	Seller

Note   Now, the ownership changes later, at the buyer's place in the supply chain. All risks and cost are with the seller. This is very good, if you are buying but really bad if you are selling. You would be responsible for any mishaps with the shipping.

Shipping Chains

The code works between any two points on a shipping route. These chains can contain numerous different Incoterm options, depending on the agreements between buyer and seller.

Warning   There can be several shipping chains. Numerous combinations and permutations are legally conceivable.

Call an Expert

So, here is my important proven advice when dealing with a complex business process: find an expert in the field. It is essential to get experts with domain knowledge as involved in the process as you can.

In the following scenario, I will assume the role of the expert Mr. Maximillian Logistics. Max suggests that you

- Stop shipping DDP everywhere.

- Ship CIP between warehouses only.

- Ship CIP or DDP between warehouses and shops.

- Ship CIP or EXW between shops.

- Ship EXW to customers.

Max confirms Hillman's policy and states currently that

- Shipping costs are £10 per kilometer.

- Insurance costs are £5 per kilometer.

- Import and export duties are a flat rate of £2 per item.

- You can only ship to a shop from the closest three warehouses.

- You can only ship between in-country shops, i.e., shops only in the same country.

- You ship to a customer only from the closest in-county shop.

- You cannot ship to a customer farther than 20 kilometers away.

What does this mean to you as a data scientist at the retrieve superstep level?

- *You must know the locations of the warehouses.* Earlier in the chapter, you created the load as part of your R processing. This is the file named `Retrieve_GB_Postcode_Warehouse.csv` in directory `..\VKHCG\03-Hillman\01-Retrieve\01-EDS\01-R`. So, that data is ready.

- *You must know the locations of the UK shops.* Earlier in the chapter, you created the load as part of your R processing. The file is named `Retrieve_GB_Postcodes_Shops.csv` in directory `..\VKHCG\03-Hillman\01-Retrieve\01-EDS\01-R`. So that data is ready.

- *You must know where the customers are located.* Earlier in the chapter, you created the load as part of your R processing. This is the file named `Retrieve_IP_DATA.csv` in directory `..\VKHCG\01-Vermeulen\01-Retrieve\01-EDS\01-R`. So the required data is ready.

The rest of Max's (my) advice will be used in Chapters 8–11, to formulate the required information to achieve knowledge of the logistics process.

Possible Shipping Routes

There are numerous potential shipping routes available to the company. The retrieve step can generate the potential set, by using a route combination generator. This will give you a set of routes, but it is highly unlikely that you will ship along all of them. It is simply a population of routes that can be used by the data science to find the optimum solution.

Start your Python editor and create a text file named `Retrieve-Warehouse-Incoterm-Chains.py` in directory `.\VKHCG\03-Hillman\01-Retrieve`. Here is the Python code you must copy into the file:

```python
################################################################
# -*- coding: utf-8 -*-
################################################################
import os
import pandas as pd
from math import radians, cos, sin, asin, sqrt
################################################################
```

In navigation, the haversine formula determines the great-circle distance between two points on a sphere, given their longitudes and latitudes. Here is my implementation:

```python
def haversine(lon1, lat1, lon2, lat2,stype):
    """
    Calculate the great circle distance between two given GPS points
    on the surface the earth (specified in decimal degrees)
    """
    # convert decimal degrees to radians
    lon1, lat1, lon2, lat2 = map(radians, [lon1, lat1, lon2, lat2])

    # haversine formula
    dlon = lon2 - lon1
    dlat = lat2 - lat1
    a = sin(dlat/2)**2 + cos(lat1) * cos(lat2) * sin(dlon/2)**2
    c = 2 * asin(sqrt(a))
    if stype == 'km':
        r = 6371 # Radius of earth in kilometers
```

```
    else:
        r = 3956 # Radius of earth in miles
    d=round(c * r,3)
    return d
```

Now that you have the formula, I suggest you load the following data:

```
################################################################
InputFileName='GB_Postcode_Warehouse.csv'
OutputFileName='Retrieve_Incoterm_Chain_GB_Warehouse.csv'
Company='03-Hillman'
################################################################
if sys.platform == 'linux':
    Base=os.path.expanduser('~') + '/VKHCG'
else:
    Base='C:/VKHCG'
print('###############################')
print('Working Base :',Base, ' using ', sys.platform)
print('###############################')
################################################################
sFileDir=Base + '/' + Company + '/01-Retrieve/01-EDS/02-Python'
if not os.path.exists(sFileDir):
    os.makedirs(sFileDir)
```

We need the warehouse locations. These are important, as they determine from where to where you should ship products.

```
################################################################
### Import Warehouses
################################################################
sFileName=Base + '/' + Company + '/00-RawData/' + InputFileName
print('##########')
print('Loading :',sFileName)
Warehouse=pd.read_csv(sFileName,header=0,low_memory=False)
```

243

You now select the entire valid warehouse list, which includes known locations. Remember the discussion about handling missing data? In this case, we are simply removing unknown data.

```
WarehouseGood=Warehouse[Warehouse.latitude != 0]
```

Your warehouse is named WH- plus the post code. This is a standard pattern, so impose it on the data, as follows:

```
WarehouseGood['Warehouse_Name']=WarehouseGood.apply(lambda row:
        'WH-' + row['postcode']
        ,axis=1)
```

You have irrelevant data in the data set. Remove it now, with the following code:

```
WarehouseGood.drop('id', axis=1, inplace=True)
WarehouseGood.drop('postcode', axis=1, inplace=True)
```

You will now limit the warehouses to 100. This is purely to speed up the progress of the example. You can experiment with different numbers of warehouses, to see the impact of the selection.

```
###############################################################
WarehouseFrom=WarehouseGood.head(100)
###############################################################
```

You will now construct two data sets to model the shipping between warehouses. Any warehouse can ship to any other warehouse.

```
###############################################################
for i in range(WarehouseFrom.shape[1]):
    oldColumn=WarehouseFrom.columns[i]
    newColumn=oldColumn + '_from'
    WarehouseFrom.rename(columns={oldColumn: newColumn}, inplace=True)
WarehouseFrom.insert(3,'Keys', 1)
###############################################################
WarehouseTo=WarehouseGood.head(100)
for i in range(WarehouseTo.shape[1]):
```

```
    oldColumn=WarehouseTo.columns[i]
    newColumn=oldColumn + '_to'
    WarehouseTo.rename(columns={oldColumn: newColumn}, inplace=True)
WarehouseTo.insert(3,'Keys', 1)
###################################################################
WarehouseCross=pd.merge(right=WarehouseFrom,
                        left=WarehouseTo,
                         how='outer',
                         on='Keys')
```

Congratulations! You have just successfully modeled a matrix model for the intra-warehouse shipping. You should now remove the excess key you used for the join.

```
WarehouseCross.drop('Keys', axis=1, inplace=True)
```

For these shipments, your Incoterm is DDP. Can you still remember what DDP shipping means? It means that the seller's warehouse pays everything.

```
WarehouseCross.insert(0,'Incoterm', 'DDP')
```

You will now use the distance formula to calculate a distance in kilometers between the two warehouses involved in the shipping.

```
WarehouseCross['DistanceBetweenKilometers'] = WarehouseCross.apply(lambda
row:
    haversine(
            row['longitude_from'],
            row['latitude_from'],
            row['longitude_to'],
            row['latitude_to'],
            'km')
            ,axis=1)
```

You must also know the distance in miles. Having both distances allows the option of using either distance, depending on the cost models. Always supply more features, as this enriches your results.

245

```
WarehouseCross['DistanceBetweenMiles'] = WarehouseCross.apply(lambda row:
    haversine(
            row['longitude_from'],
            row['latitude_from'],
            row['longitude_to'],
            row['latitude_to'],
            'miles')
            ,axis=1)
```

Clean up the now excess fields in the model.

```
WarehouseCross.drop('longitude_from', axis=1, inplace=True)
WarehouseCross.drop('latitude_from', axis=1, inplace=True)
WarehouseCross.drop('longitude_to', axis=1, inplace=True)
WarehouseCross.drop('latitude_to', axis=1, inplace=True)
```

On a validation check, we have found that our model supports the same warehouse shipments that are not in the real-world ecosystem. So, remove them from your simulated model.

```
WarehouseCrossClean=WarehouseCross[WarehouseCross.DistanceBetweenKilometers
!=0]
```

Congratulations! You have successfully built a model for the shipping of products. Now, you can look at some quick high-level check for the model and save the model to disk.

```
print('###########')
print('Rows :',WarehouseCrossClean.shape[0])
print('Columns :',WarehouseCrossClean.shape[1])
print('###########')
################################################################

sFileName=sFileDir + '/' + OutputFileName
WarehouseCrossClean.to_csv(sFileName, index = False)

################################################################
print('### Done!! #####################################')
################################################################
```

Now save the Python file.

Execute the `Retrieve-Warehouse-Incoterm-Chains.py` file with your preferred Python compiler.

You have just run your model end-to-end. Well done! You will see a file named `Retrieve_Incoterm_Chain_GB_Warehouse.csv` in `C:\VKHCG\03-Hillman\01-Retrieve\01-EDS\02-Python`. Open this file, and you should see a data set similar to the following:

Incoterm	Warehouse_ Name_to	Warehouse_ Name_from	Distance Between Kilometers	Distance Between Miles
DDP	WH-AB10	WH-AB11	1.644	1.021
DDP	WH-AB10	WH-AB12	3.818	2.371
DDP	WH-AB10	WH-AB13	7.872	4.888
DDP	WH-AB10	WH-AB14	10.02	6.222

I will point out that the current data still ships the product as DDP. You will use your new data science skills to resolve this, as I guide you through the book.

Adopt New Shipping Containers

Adopting the best packing option for shipping in containers will require that I introduce a new concept. Shipping of containers is based on a concept reducing the packaging you use down to an optimum set of sizes having the following requirements:

- The product must fit within the box formed by the four sides of a cube.

- The product can be secured using packing foam, which will fill any void volume in the packaging.

- Packaging must fit in shipping containers with zero space gaps.

- Containers can only hold product that is shipped to a single warehouse, shop, or customer.

Note   As this is a new concept for Hillman, you will have to generate data for use by your data science.

So, I will suggest a well-designed set of generated data that is an acceptable alternative for real data, to build the model as you expect it to work in future. However, once you start collecting the results of the business process, you always swap the generated data for real production data.

I will now guide you to formulate a set of seed data for your data science. Start your Python editor and create a text file named `Retrieve-Container-Plan.py` in directory `.\VKHCG\03-Hillman\01-Retrieve`. Here is the Python code you must copy into the file.

Note   Please follow the subsequent code entry. I will discuss important parts of the coding, as required.

```
################################################################
# -*- coding: utf-8 -*-
################################################################
import os
import pandas as pd
################################################################
ContainerFileName='Retrieve_Container.csv'
BoxFileName='Retrieve_Box.csv'
ProductFileName='Retrieve_Product.csv'
Company='03-Hillman'
################################################################
if sys.platform == 'linux':
    Base=os.path.expanduser('~') + '/VKHCG'
else:
    Base='C:/VKHCG'
print('###############################')
print('Working Base :',Base, ' using ', sys.platform)
print('###############################')
################################################################
sFileDir=Base + '/' + Company + '/01-Retrieve/01-EDS/02-Python'
if not os.path.exists(sFileDir):
    os.makedirs(sFileDir)
```

```
################################################################
### Create the Containers
################################################################
```

The business requires a model that simulates shipping containers of dimensions 1 meter × 1 meter × 1 meter to 21 meters × 10 meters × 6 meters. So, let's simulate these.

```python
containerLength=range(1,21)
containerWidth=range(1,10)
containerHeigth=range(1,6)
containerStep=1
c=0
for l in containerLength:
    for w in containerWidth:
        for h in containerHeigth:
containerVolume=(l/containerStep)*(w/containerStep)*(h/containerStep)
            c=c+1
            ContainerLine=[('ShipType', ['Container']),
                    ('UnitNumber', ('C'+format(c,"06d"))),
                    ('Length',(format(round(l,3),".4f"))),
                    ('Width',(format(round(w,3),".4f"))),
                    ('Height',(format(round(h,3),".4f"))),
                    ('ContainerVolume',(format(round(containerVolume,6),
                    ".6f")))]
            if c==1:
                ContainerFrame = pd.DataFrame.from_items(ContainerLine)
            else:
                ContainerRow = pd.DataFrame.from_items(ContainerLine)
                ContainerFrame = ContainerFrame.append(ContainerRow)
                ContainerFrame.index.name = 'IDNumber'

print('###############')
print('## Container')
print('###############')
print('Rows :',ContainerFrame.shape[0])
print('Columns :',ContainerFrame.shape[1])
```

```
print('################')
################################################################
sFileContainerName=sFileDir + '/' + ContainerFileName
ContainerFrame.to_csv(sFileContainerName, index = False)
################################################################
```

Well done! You have just completed a simulated model, based on the requirement of a future system. So, when the "real" containers arrive, you simply replace this simulated data set with the "real" data and none of the downstream data science needs changing.

Your second simulation is the cardboard boxes for the packing of the products. The requirement is for boxes having a dimension of 100 centimeters × 100 centimeters × 100 centimeters to 2.1 meters × 2.1 meters × 2.1 meters. You can also use between zero and 600 centimeters of packing foam to secure any product in the box.

You can simulate these requirements as follows:

```
################################################################
## Create valid Boxes with packing foam
################################################################
boxLength=range(1,21)
boxWidth=range(1,21)
boxHeigth=range(1,21)
packThick=range(0,6)
boxStep=10
b=0
for l in boxLength:
    for w in boxWidth:
        for h in boxHeigth:
            for t in packThick:
                boxVolume=round((l/boxStep)*(w/boxStep)*(h/boxStep),6)
                productVolume=round(((l-t)/boxStep)*((w-t)/boxStep)*((h-t)/
                boxStep),6)
                if productVolume > 0:
                    b=b+1
                    BoxLine=[('ShipType', ['Box']),
                        ('UnitNumber', ('B'+format(b,"06d"))),
                        ('Length',(format(round(l/10,6),".6f"))),
```

```
                ('Width',(format(round(w/10,6),".6f"))),
                ('Height',(format(round(h/10,6),".6f"))),
                ('Thickness',(format(round(t/5,6),".6f"))),
                ('BoxVolume',(format(round(boxVolume,9),".9f"))),
                ('ProductVolume',(format(round(productVolume,9),
                ".9f")))]
        if b==1:
            BoxFrame = pd.DataFrame.from_items(BoxLine)
        else:
            BoxRow = pd.DataFrame.from_items(BoxLine)
            BoxFrame = BoxFrame.append(BoxRow)
        BoxFrame.index.name = 'IDNumber'

print('################')
print('## Box')
print('################')
print('Rows :',BoxFrame.shape[0])
print('Columns :',BoxFrame.shape[1])
print('################')
################################################################
sFileBoxName=sFileDir + '/' + BoxFileName
BoxFrame.to_csv(sFileBoxName, index = False)
################################################################
```

Two simulations are done, you are making great progress. Now, your third model is for the products. The requirement is for products having a dimension of 100 centimeters × 100 centimeters × 100 centimeters to 2.1 meters × 2.1 meters × 2.1 meters. You can build this as follows:

```
################################################################
## Create valid Product
################################################################
productLength=range(1,21)
productWidth=range(1,21)
productHeigth=range(1,21)
productStep=10
p=0
```

```python
for l in productLength:
    for w in productWidth:
        for h in productHeigth:
            productVolume=round((l/productStep)*(w/productStep)*(h/productStep),6)
            if productVolume > 0:
                p=p+1
                ProductLine=[('ShipType', ['Product']),
                    ('UnitNumber', ('P'+format(p,"06d"))),
                    ('Length',(format(round(l/10,6),".6f"))),
                    ('Width',(format(round(w/10,6),".6f"))),
                    ('Height',(format(round(h/10,6),".6f"))),
                    ('ProductVolume',(format(round(productVolume,9),".9f")))]
                if p==1:
                    ProductFrame = pd.DataFrame.from_items(ProductLine)
                else:
                    ProductRow = pd.DataFrame.from_items(ProductLine)
                    ProductFrame = ProductFrame.append(ProductRow)
                BoxFrame.index.name = 'IDNumber'

print('################')
print('## Product')
print('################')
print('Rows :',ProductFrame.shape[0])
print('Columns :',ProductFrame.shape[1])
print('################')
################################################################
sFileProductName=sFileDir + '/' + ProductFileName
ProductFrame.to_csv(sFileProductName, index = False)
################################################################
################################################################
print('### Done!! #####################################')
################################################################
################################################################
```

Now save the Python file. Execute the `Retrieve-Container-Plan.py` file with your preferred Python compiler.

You have just simulated a complex packing solution from simple requirements, and you are ready to apply data science to the model in Chapters 8–11. You will see a container data file in a format similar to the following: `Retrieve_Container.csv` in `C:\VKHCG\03-Hillman\01-Retrieve\01-EDS\02-Python`. Open this file, and you should see a data set like this:

ShipType	UnitNumber	Length	Width	Height	ContainerVolume
Container	C000001	1	1	1	1
Container	C000002	1	1	2	2
Container	C000003	1	1	3	3

You will see a box packing file in a format similar to `Retrieve_Box.csv`, in `C:\VKHCG\03-Hillman\01-Retrieve\01-EDS\02-Python`.

ShipType	UnitNumber	Length	Width	Height	Thickness	BoxVolume	ProductVolume
Box	B000001	0.1	0.1	0.1	0	0.001	0.001
Box	B000002	0.1	0.1	0.2	0	0.002	0.002
Box	B000003	0.1	0.1	0.3	0	0.003	0.003

You will see a box packing file similar in format to `'Retrieve_Box.csv'` in `C:\VKHCG\03-Hillman\01-Retrieve\01-EDS\02-Python`.

ShipType	UnitNumber	Length	Width	Height	ProductVolume
Product	P034485	0.1	0.1	0.1	0.001
Product	P034485	0.1	0.1	0.2	0.002
Product	P034485	0.1	0.1	0.3	0.003

I have now successfully assisted you in creating a data set with all supported packing seed data for your data science. Now, I will point out that if you change any of the following parameters in the code

```
containerLength=range(1,21)
containerWidth=range(1,10)
containerHeigth=range(1,6)
containerStep=1

boxLength=range(1,21)
boxWidth=range(1,21)
boxHeigth=range(1,21)
packThick=range(0,6)
boxStep=10

productLength=range(1,21)
productWidth=range(1,21)
productHeigth=range(1,21)
productStep=10
```

you will produce a different packing solution that will produce interesting results during the data science investigations. I will leave that to you to experiment with, once you understand your skills.

Hillman now has data on its containers. Well done.

Create a Delivery Route

I will guide you through the process of creating a delivery route for a customer. Doing so requires that I lead you through a remarkable model that I discovered for shipping items.

I want to introduce you to the useful Python package named geopy. This package is the Python Geocoding Toolbox. It supports most location-related queries.

The model enables you to generate a complex routing plan for the shipping routes of the company. Start your Python editor and create a text file named Retrieve-Route-Plan.py in directory .\VKHCG\03-Hillman\01-Retrieve. Following is the Python code you must copy into the file:

```
###############################################################
# -*- coding: utf-8 -*-
###############################################################
```

```
import os
import pandas as pd
from geopy.distance import vincenty
################################################################
InputFileName='GB_Postcode_Warehouse.csv'
OutputFileName='Retrieve_GB_Warehouse.csv'
Company='03-Hillman'
################################################################
if sys.platform == 'linux':
    Base=os.path.expanduser('~') + '/VKHCG'
else:
    Base='C:/VKHCG'
print('################################')
print('Working Base :',Base, ' using ', sys.platform)
print('################################') ################################
################################
sFileDir=Base + '/' + Company + '/01-Retrieve/01-EDS/02-Python'
if not os.path.exists(sFileDir):
    os.makedirs(sFileDir)
################################################################
sFileName=Base + '/' + Company + '/00-RawData/' + InputFileName
print('###########')
print('Loading :',sFileName)
Warehouse=pd.read_csv(sFileName,header=0,low_memory=False)

WarehouseClean=Warehouse[Warehouse.latitude != 0]
WarehouseGood=WarehouseClean[WarehouseClean.longitude != 0]

WarehouseGood.drop_duplicates(subset='postcode', keep='first',
inplace=True)

WarehouseGood.sort_values(by='postcode', ascending=1)
################################################################
sFileName=sFileDir + '/' + OutputFileName
WarehouseGood.to_csv(sFileName, index = False)
################################################################
```

```
WarehouseLoop = WarehouseGood.head(20)

for i in range(0,WarehouseLoop.shape[0]):
    print('Run :',i,' =======>>>>>>>>>>',WarehouseLoop['postcode'][i])
    WarehouseHold = WarehouseGood.head(10000)
    WarehouseHold['Transaction']=WarehouseHold.apply(lambda row:
              'WH-to-WH'
              ,axis=1)
    OutputLoopName='Retrieve_Route_' + 'WH-' + WarehouseLoop['postcode'][i]
    + '_Route.csv'

    WarehouseHold['Seller']=WarehouseHold.apply(lambda row:
              'WH-' + WarehouseLoop['postcode'][i]
              ,axis=1)

    WarehouseHold['Seller_Latitude']=WarehouseHold.apply(lambda row:
              WarehouseHold['latitude'][i]
              ,axis=1)
    WarehouseHold['Seller_Longitude']=WarehouseHold.apply(lambda row:
              WarehouseLoop['longitude'][i]
              ,axis=1)

    WarehouseHold['Buyer']=WarehouseHold.apply(lambda row:
              'WH-' + row['postcode']
              ,axis=1)

    WarehouseHold['Buyer_Latitude']=WarehouseHold.apply(lambda row:
              row['latitude']
              ,axis=1)
    WarehouseHold['Buyer_Longitude']=WarehouseHold.apply(lambda row:
              row['longitude']
              ,axis=1)

    WarehouseHold['Distance']=WarehouseHold.apply(lambda row: round(
         vincenty((WarehouseLoop['latitude'][i],WarehouseLoop['longitude'][i]),
              (row['latitude'],row['longitude'])).miles,6)
              ,axis=1)
```

```
WarehouseHold.drop('id', axis=1, inplace=True)
WarehouseHold.drop('postcode', axis=1, inplace=True)
WarehouseHold.drop('latitude', axis=1, inplace=True)
WarehouseHold.drop('longitude', axis=1, inplace=True)
###############################################################
sFileLoopName=sFileDir + '/' + OutputLoopName
WarehouseHold.to_csv(sFileLoopName, index = False)
#################################################################
print('### Done!! #######################################')
#################################################################
```

Now save the Python file. Execute the `Retrieve-Route-Plan.py` file with your preferred Python compiler.

You will see a collection of files similar in format to `Retrieve_Route_WH-AB10_Route.csv` in `C:\VKHCG\03-Hillman\01-Retrieve\01-EDS\02-Python`. Open this file, and you should see a data set like this:

Transaction	Seller	Seller_ Latitude	Seller_ Longitude	Buyer	Buyer_ Latitude	Buyer_ Longitude	Distance
WH-to-WH	WH-AB10	57.13514	-2.11731	WH-AB10	57.13514	-2.11731	0
WH-to-WH	WH-AB10	57.13514	-2.11731	WH-AB11	57.13875	-2.09089	1.024914
WH-to-WH	WH-AB10	57.13514	-2.11731	WH-AB12	57.101	-2.1106	2.375854
WH-to-WH	WH-AB10	57.13514	-2.11731	WH-AB13	57.10801	-2.23776	4.906919

The preceding code enables you to produce numerous routing plans for the business. For example, changing the WarehouseLoop = `WarehouseGood.head(20)` to `WarehouseGood.head(200)` will produce 200 new route plans. The code will produce a route map for every warehouse, shop, or customer you load.

Global Post Codes

I will guide you through a global post code data set that will form the base for most of the non-company address and shipping requirements.

Tip   Normally, I find the best post code information at the local post office or shipping agent. Remember what I said about experts. The Post Office experts ship items every minute of the day. They are the people to contact for post code information.

Warning   You must perform some data science in your R environment.

Note   If you have code ready to perform a task, use it!

Now, open your RStudio and use R to process the following R script: Retrieve-Postcode-Global.r.

```r
library(readr)
All_Countries <- read_delim("C:/VKHCG/03-Hillman/00-RawData/
All_Countries.txt",
"\t", col_names = FALSE,
col_types = cols(
X12 = col_skip(),
X6 = col_skip(),
X7 = col_skip(),
X8 = col_skip(),
X9 = col_skip()),
na = "null", trim_ws = TRUE)
write.csv(All_Countries,
file = "C:/VKHCG/03-Hillman/01-Retrieve/01-EDS/01-R/Retrieve_All_Countries.csv")
```

You have just successfully uploaded a new file named Retrieve_All_Countries. csv.' That was the last of the Hillman retrieve tasks.

Clark Ltd

Clark is the financial powerhouse of the group. It must process all the money-related data sources.

Forex

The first financial duty of the company is to perform any foreign exchange trading.

Forex Base Data

Previously, you found a single data source (`Euro_ExchangeRates.csv`) for forex rates in Clark. Earlier in the chapter, I helped you to create the load, as part of your R processing. The relevant file is `Retrieve_Retrieve_Euro_ExchangeRates.csv` in directory `C:\ VKHCG\04-Clark\01-Retrieve\01-EDS\01-R`. So, that data is ready.

Financials

Clark generates the financial statements for all the group's companies.

Financial Base Data

You found a single data source (`Profit_And_Loss.csv`) in Clark for financials and, as mentioned previously, a single data source (`Euro_ExchangeRates.csv`) for forex rates. Earlier in the chapter, I helped you to create the load, as part of your R processing. The file relevant file is `Retrieve_Profit_And_Loss.csv` in directory `C:\VKHCG\04-Clark\01-Retrieve\01-EDS\01-R`.

Person Base Data

Now, I will explain how to extract personal information from three compressed data structures and then join the separate files together to form the base for the personal information.

Start your Python editor and create a text file named Retrieve-PersonData.py in directory .\VKHCG\04-Clark\01-Retrieve. Following is the Python code you must copy into the file:

```python
################################################################
# -*- coding: utf-8 -*-
################################################################
import sys
import os
import shutil
import zipfile
import pandas as pd
################################################################
if sys.platform == 'linux':
    Base=os.path.expanduser('~') + 'VKHCG'
else:
    Base='C:/VKHCG'
print('################################')
print('Working Base :',Base, ' using ', sys.platform)
print('################################')
################################################################
Company='04-Clark'
ZIPFiles=['Data_female-names','Data_male-names','Data_last-names']
for ZIPFile in ZIPFiles:
    InputZIPFile=Base+'/'+Company+'/00-RawData/' + ZIPFile + '.zip'
    OutputDir=Base+'/'+Company+'/01-Retrieve/01-EDS/02-Python/' + ZIPFile
    OutputFile=Base+'/'+Company+'/01-Retrieve/01-EDS/02-Python/Retrieve-'+ZIPFile+'.csv'
    zip_file = zipfile.ZipFile(InputZIPFile, 'r')
    zip_file.extractall(OutputDir)
    zip_file.close()
    t=0
    for dirname, dirnames, filenames in os.walk(OutputDir):
        for filename in filenames:
            sCSVFile = dirname + '/' + filename
```

```
        t=t+1
        if t==1:
          NameRawData=pd.read_csv(sCSVFile,header=None,low_memory=False)
            NameData=NameRawData
        else:
          NameRawData=pd.read_csv(sCSVFile,header=None,low_memory=False)
            NameData=NameData.append(NameRawData)
    NameData.rename(columns={0 : 'Name'},inplace=True)
    NameData.to_csv(OutputFile, index = False)
    shutil.rmtree(OutputDir)
    print('Process: ',InputZIPFile)
################################################################
print('### Done!! ###########################################')
################################################################
```

Now save the Python file and execute the `Retrieve-PersonData.py` file with your preferred Python compiler. This generates three files named `Retrieve-Data_female-names.csv`, `Retrieve-Data_male-names.csv`, and `Retrieve-Data_last-names.csv`.

You have now completed all the retrieve tasks. Clarks's retrieve data is ready.

Connecting to Other Data Sources

I will now offer a high-level explanation of how to handle non-CSV data sources. The basic data flow and processing stays the same; hence, the data science stays the same. It is only the connection methods that change.

The following sections showcase a few common data stores I have used on a regular basis.

SQLite

This requires a package named `sqlite3`. Let's load the `IP_DATA_ALL.csv` data in a SQLite database: `vermeulen.db` at `.\VKHCG\01-Vermeulen\00-RawData\SQLite\`. Open your Python editor and create a text file named `Retrieve-IP_DATA_ALL_2_SQLite.py` in directory `.\VKHCG\03-Hillman\01-Retrieve`.

Here is the Python code you must copy into the file:

```python
###################################################################
# -*- coding: utf-8 -*-
###################################################################
import sqlite3 as sq
import pandas as pd
###################################################################
Base='C:/VKHCG'
sDatabaseName=Base + '/01-Vermeulen/00-RawData/SQLite/vermeulen.db'
conn = sq.connect(sDatabaseName)
###################################################################
sFileName=Base + '/01-Vermeulen/00-RawData/IP_DATA_ALL.csv'
print('Loading :',sFileName)
IP_DATA_ALL=pd.read_csv(sFileName,header=0,low_memory=False)

IP_DATA_ALL.index.names = ['RowIDCSV']
sTable='IP_DATA_ALL'

print('Storing :',sDatabaseName,' Table:',sTable)
IP_DATA_ALL.to_sql(sTable, conn, if_exists="replace")

print('Loading :',sDatabaseName,' Table:',sTable)
TestData=pd.read_sql_query("select * from IP_DATA_ALL;", conn)

print('################')
print('## Data Values')
print('################')
print(TestData)
print('################')
print('## Data Profile')
print('################')
print('Rows :',TestData.shape[0])
print('Columns :',TestData.shape[1])
print('################')
###################################################################
print('### Done!! #######################################')
###################################################################
```

Now save the Python file and execute the `Retrieve-IP_DATA_ALL_2_SQLite.py` file with your preferred Python compiler. The data will now be loaded into SQLite.

Now, to transfer data within SQLite, you must perform the following. Open your Python editor and create a text file named `Retrieve-IP_DATA_ALL_SQLite.py` in directory `.\VKHCG\03-Hillman\01-Retrieve`.

Following is the Python code you must copy into the file:

```
################################################################
import pandas as pd
import sqlite3 as sq
################################################################
Base='C:/VKHCG'
sDatabaseName=Base + '/01-Vermeulen/00-RawData/SQLite/Vermeulen.db'
conn = sq.connect(sDatabaseName)
print('Loading :',sDatabaseName)
IP_DATA_ALL=pd.read_sql_query("select * from IP_DATA_ALL;", conn)
################################################################
print('Rows:', IP_DATA_ALL.shape[0])
print('Columns:', IP_DATA_ALL.shape[1])
print('### Raw Data Set #################################')
for i in range(0,len(IP_DATA_ALL.columns)):
    print(IP_DATA_ALL.columns[i],type(IP_DATA_ALL.columns[i]))
print('### Fixed Data Set #################################')
IP_DATA_ALL_FIX=IP_DATA_ALL
for i in range(0,len(IP_DATA_ALL.columns)):
    cNameOld=IP_DATA_ALL_FIX.columns[i] + '        '
    cNameNew=cNameOld.strip().replace(" ", ".")
    IP_DATA_ALL_FIX.columns.values[i] = cNameNew
    print(IP_DATA_ALL.columns[i],type(IP_DATA_ALL.columns[i]))
################################################################
#print(IP_DATA_ALL_FIX.head())
################################################################
print('################')
print('Fixed Data Set with ID')
```

```
print('###############')
IP_DATA_ALL_with_ID=IP_DATA_ALL_FIX
print('###############')
print(IP_DATA_ALL_with_ID.head())
print('###############')

sTable2='Retrieve_IP_DATA'
IP_DATA_ALL_with_ID.to_sql(sTable2,conn, index_label="RowID", if_
exists="replace")


################################################################
print('### Done!! #####################################')
################################################################
```

Now save the Python file. Execute the `Retrieve-IP_DATA_ALL_SQLite.py` with your preferred Python compiler. The data will now be loaded between two SQLite tables.

Date and Time

To generate data and then store the data directly into the SQLite database, do the following. Open your Python editor and create a text file named `Retrieve-Calendar.py` in directory `.\VKHCG\04-Clark\01-Retrieve`.

Here is the Python code you must copy into the file:

```
################################################################
import sys
import os
import sqlite3 as sq
import pandas as pd
import datetime
################################################################
if sys.platform == 'linux':
    Base=os.path.expanduser('~') + 'VKHCG'
else:
    Base='C:/VKHCG'
print('#############################')
print('Working Base :',Base, ' using ', sys.platform)
```

```
print('################################')
################################################################
Company='04-Clark'
################################################################
sDataBaseDir=Base + '/' + Company + '/01-Retrieve/SQLite'
if not os.path.exists(sDataBaseDir):
    os.makedirs(sDataBaseDir)
################################################################
sDatabaseName=sDataBaseDir + '/clark.db'
conn = sq.connect(sDatabaseName)
################################################################
start = pd.datetime(1900, 1, 1)
end = pd.datetime(2018, 1, 1)
date1Data= pd.DataFrame(pd.date_range(start, end))
################################################################
start = pd.datetime(2000, 1, 1)
end = pd.datetime(2020, 1, 1)
date2Data= pd.DataFrame(pd.date_range(start, end))
################################################################
dateData=date1Data.append(date2Data)
dateData.rename(columns={0 : 'FinDate'},inplace=True)
################################################################
print('################')
sTable='Retrieve_Date'
print('Storing :',sDatabaseName,' Table:',sTable)
dateData.to_sql(sTable, conn, if_exists="replace")
print('################')
################################################################
print('################################')
print('Rows : ',dateData.shape[0], ' records')
print('################################')
################################################################
t=0
for h in range(24):
    for m in range(60):
```

```
        for s in range(5):
            nTime=[str(datetime.timedelta(hours=h,minutes=m,seconds=30))]
            if h == 0 and m == 0:
                timeData= pd.DataFrame(nTime)
            else:
                time1Data=  pd.DataFrame(nTime)
                timeData= timeData.append(time1Data)
timeData.rename(columns={0 : 'FinTime'},inplace=True)
################################################################
print('###############')
sTable='Retrieve_Time'
print('Storing :',sDatabaseName,' Table:',sTable)
timeData.to_sql(sTable, conn, if_exists="replace")
print('###############')
################################################################
print('##############################')
print('Rows : ',timeData.shape[0], ' records')
print('##############################')
################################################################
print('### Done!! #########################################')
################################################################
```

Now save the Python file. Execute the `Retrieve-Calendar.py` file with your preferred Python compiler.

The data will now be loaded as two new SQLite tables named `Retrieve_Date` and `Retrieve_Time`.

Other Databases

The important item to understand is that you need only change the library in the code I already proved works for the data science:

```
import sqlite3 as sq
```

The following will work with the same code.

PostgreSQL

PostgreSQL is used in numerous companies, and it enables connections to the database. There are two options: a direct connection using

```
from sqlalchemy import create_engine
engine = create_engine('postgresql://scott:tiger@localhost:5432/vermeulen')
```

and connection via the psycopg2 interface

```
from sqlalchemy import create_engine
engine = create_engine('postgresql+psycopg2://scott:tiger@localhost/
mydatabase')
```

Microsoft SQL Server

Microsoft SQL server is common in companies, and this connector supports your connection to the database.

There are two options. Via the ODBC connection interface, use

```
from sqlalchemy import create_engine
engine = create_engine('mssql+pyodbc://scott:tiger@mydsnvermeulen')
```

Via the direct connection, use

```
from sqlalchemy import create_engine
engine = create_engine('mssql+pymssql://scott:tiger@hostname:port/
vermeulen')
```

MySQL

MySQL is widely used by lots of companies for storing data. This opens that data to your data science with the change of a simple connection string.

There are two options. For direct connect to the database, use

```
from sqlalchemy import create_engine
engine = create_engine('mysql+mysqldb://scott:tiger@localhost/
vermeulen')
```

For connection via the DSN service, use

```
from sqlalchemy import create_engine
engine = create_engine('mssql+pyodbc://mydsnvermeulen')
```

Oracle

Oracle is a common database storage option in bigger companies. It enables you to load data from the following data source with ease:

```
from sqlalchemy import create_engine
engine = create_engine('oracle://andre:vermeulen@127.0.0.1:1521/vermeulen')
```

Microsoft Excel

Excel is common in the data sharing ecosystem, and it enables you to load files using this format with ease. Open your Python editor and create a text file named Retrieve-Country-Currency.py in directory .\VKHCG\04-Clark\01-Retrieve.

Here is the Python code you must copy into the file:

```
################################################################
# -*- coding: utf-8 -*-
################################################################
import os
import pandas as pd
################################################################
Base='C:/VKHCG'
################################################################
sFileDir=Base + '/01-Vermeulen/01-Retrieve/01-EDS/02-Python'
if not os.path.exists(sFileDir):
    os.makedirs(sFileDir)
################################################################
CurrencyRawData = pd.read_excel('C:/VKHCG/01-Vermeulen/00-RawData/Country_
Currency.xlsx')
sColumns = ['Country or territory', 'Currency', 'ISO-4217']
CurrencyData = CurrencyRawData[sColumns]
CurrencyData.rename(columns={'Country or territory': 'Country', 'ISO-4217':
'CurrencyCode'}, inplace=True)
```

```
CurrencyData.dropna(subset=['Currency'],inplace=True)
CurrencyData['Country'] = CurrencyData['Country'].map(lambda x: x.strip())
CurrencyData['Currency'] = CurrencyData['Currency'].map(lambda x:
x.strip())
CurrencyData['CurrencyCode'] = CurrencyData['CurrencyCode'].map(lambda x:
x.strip())
print(CurrencyData)
################################################################
sFileName=sFileDir + '/Retrieve-Country-Currency.csv'
CurrencyData.to_csv(sFileName, index = False)
################################################################
```

Now save the Python file. Execute the `Retrieve-Calendar.py` file with your preferred Python compiler. The code produces `Retrieve-Country-Currency.csv`.

Apache Spark

Apache Spark is now becoming the next standard for distributed data processing. The universal acceptance and support of the processing ecosystem is starting to turn mastery of this technology into a must-have skill.

Use package `pyspark.sql`. The connection is set up by using following code:

```
import pyspark
sc = pyspark.SparkContext()
sql = SQLContext(sc)

df = (sql.read
        .format("IP_DATA_ALL.csv")
        .option("header", "true")
        .load("/path/to_csv.csv"))
```

Apache Cassandra

Cassandra is becoming a widely distributed database engine in the corporate world. The advantage of this technology and ecosystem is that they can scale massively in a great scaling manner. This database can include and exclude nodes into the database ecosystem, without disruption to the data processing.

To access it, use the Python package cassandra.

```
from cassandra.cluster import Cluster

        cluster = Cluster()

        session = cluster.connect('vermeulen')
```

Apache Hive

Access to Hive opens its highly distributed ecosystem for use by data scientists. To achieve this, I suggest using the following:

```
Import pyhs2
with pyhs2.connect(host='localhost',
                port=10000,
                authMechanism="PLAIN",
                user='andre',
                password='vermeulen',
                database='vermeulen') as conn:
    with conn.cursor() as cur:
        #Execute query
       cur.execute("select * from IP_DATA_ALL")
for i in cur.fetch():
            print (i)
```

This enables you to read data from a data lake stored in Hive on top of Hadoop.

Apache Hadoop

Hadoop is one of the most successful data lake ecosystems in highly distributed data science. I suggest that you understand how to access data in this data structure, if you want to perform data science at-scale.

PyDoop

The pydoop package includes a Python MapReduce and HDFS API for Hadoop. Pydoop is a Python interface to Hadoop that allows you to write MapReduce applications and interact with HDFS in pure Python. See https://pypi.python.org/pypi/pydoop.

Luigi

Luigi enables a series of Python features that enable you to build complex pipelines into batch jobs. It handles dependency resolution and workflow management as part of the package.

This will save you from performing complex programming while enabling good-quality processing. See `https://pypi.python.org/pypi/luigi`.

Amazon S3 Storage

S3, or Amazon Simple Storage Service (Amazon S3), creates simple and practical methods to collect, store, and analyze data, irrespective of format, completely at massive scale. I store most of my base data in S3, as it is cheaper than most other methods.

Package s3

Python's s3 module connects to Amazon's S3 REST API. See `https://pypi.python.org/pypi/s3`.

Package Boto

The Boto package is another useful tool that I recommend. It can be accessed from the boto library, as follows:

```
from boto.s3.connection import S3Connection
conn = S3Connection('<aws access key>', '<aws secret key>')
```

Amazon Redshift

Amazon Redshift is a fully managed, petabyte-scale data warehouse service in the cloud. The Python package `redshift-sqlalchemy`, is an Amazon Redshift dialect for sqlalchemy that opens this data source to your data science. See `https://pypi.python.org/pypi/redshift-sqlalchemy`.

This is a typical connection:

```
from sqlalchemy import create_engine
engine = create_engine("redshift+psycopg2://username@host.amazonaws.com:5439/database"
```

You can also connect to Redshift via the package `pandas-redshift`.(See `https://pypi.python.org/pypi/pandas-redshift`.) It combines the capability of packages `psycopg2`, `pandas`, and `boto3` and uses the following connection:

```
engine = create_engine('redshift+psycopg2://username:password@us-east-1.
redshift.amazonaws.com:port/vermeulen')
```

You can effortlessly connect to numerous data sources, and I suggest you investigate how you can use their capabilities for your data science.

Amazon Web Services

When you are working with Amazon Web Services, I suggest you look at the package: `boto3` package. (See: `https://pypi.python.org/pypi/boto3`.)

It is an Amazon Web Services Library Python package that provides interfaces to Amazon Web Services. For example, the following Python code will create an EC2 cloud data processing ecosystem:

```
import boto3
ec2 = boto3.resource('ec2')
instance = ec2.create_instances(
    ImageId='ami-1e299d7e',
    MinCount=1,
    MaxCount=1,
    InstanceType='t2.micro')
print instance[0].id
```

Once that is done, you can download a file from S3's data storage.

```
import boto3
import botocore
BUCKET_NAME = 'vermeulen' # replace with your bucket name
KEY = ' Key_IP_DATA_ALL'# replace with your object key
s3 = boto3.resource('s3')
```

```
try:
    s3.Bucket(BUCKET_NAME).download_file(KEY, 'IP_DATA_ALL.csv')
except botocore.exceptions.ClientError as e:
    if e.response['Error']['Code'] == "404":
        print("The object does not exist.")
    else:
        raise
```

By way of boto3, you have various options with which you can create the services you need in the cloud environment and then load the data you need. Mastering this process is a significant requirement for data scientists.

Summary

Congratulations! You have completed the Retrieve superstep. I have guided you through your first big data processing steps. The results you achieved will form the base data for Chapter 8, on the Assess superstep.

So, what should you know by this stage in the book?

- How to convert the functional requirements from Chapter 4 into a data mapping matrix. Remember: If the data is not loaded via the retrieve layer, it is not part of the process.

- How to convert and update the data mapping matrix as you perform the retrieve processing. Every data source you retrieve is a potential source of secondary hidden information.

- Understand the data profile of your source system. Not all information from sources is in a consistent and/or good data format. Much data will require regular feature development.

- Understand where to get other data sources, even such cloud-based systems as Google, Amazon Web Services, and Microsoft Azure.

Now that you have mastered the Retrieve step, I recommend that you get a beverage, and then we can progress to Chapter 8, in which we will repair more or less all of the data quality concerns in the data you saved.

CHAPTER 8

Assess Superstep

The objectives of this chapter are to show you how to assess your data science data for invalid or erroneous data values.

Caution   I have found that in 90% of my projects, I spend 70% of the processing work on this step, to improve the quality of the data used. This step will definitely improve the quality of your data science.

I urge that you spend the time to "clean up" the data before you progress to the data science, as the incorrect data entries will cause a major impact on the later steps in the process. Perform a data science project on "erroneous" data also, to understand why the upstream processes are producing erroneous data. I have uncovered numerous unknown problems in my customers' ecosystems by investigating this erroneous data.

I will now introduce you to a selection of assessment techniques and algorithms that I scientifically deploy against the data imported during the retrieve phase of the project.

Note   Your specific data sources will have additional requirements. Explore additional quality techniques and algorithms that you need for your particular data.

Assess Superstep

Data quality refers to the condition of a set of qualitative or quantitative variables. Data quality is a multidimensional measurement of the acceptability of specific data sets. In business, data quality is measured to determine whether data can be used as a basis for reliable intelligence extraction for supporting organizational decisions.

© Andreas François Vermeulen 2018
A. F. Vermeulen, *Practical Data Science*, https://doi.org/10.1007/978-1-4842-3054-1_8

Data profiling involves observing in your data sources all the viewpoints that the information offers. The main goal is to determine if individual viewpoints are accurate and complete. The Assess superstep determines what additional processing to apply to the entries that are noncompliant.

I should point out that performing this task is not as straightforward as most customers think. Minor pieces of incorrect data can have major impacts in later data processing steps and can impact the quality of the data science.

Note   You will always have errors. It is the quality techniques and algorithms that resolve these errors, to guide you down the path of successful data science.

Errors

Did you find errors or issues? Typically, I can do one of four things to the data.

Accept the Error

If it falls within an acceptable standard (i.e., West Street instead of West St.), I can decide to accept it and move on to the next data entry.

Take note that if you accept the error, you will affect data science techniques and algorithms that perform classification, such as binning, regression, clustering, and decision trees, because these processes assume that the values in this example are not the same. This option is the easy option, but not always the best option.

Reject the Error

Occasionally, predominantly with first-time data imports, the information is so severely damaged that it is better to simply delete the data entry methodically and not try to correct it. Take note: Removing data is a last resort. I normally add a quality flag and use this flag to avoid this erroneous data being used in data science techniques and algorithms that it will negatively affect. I will discuss specific data science techniques and algorithms in the rest of this book, and at each stage, I will explain how to deal with erroneous data.

Correct the Error

This is the option that a major part of the assess step is dedicated to. Spelling mistakes in customer names, addresses, and locations are a common source of errors, which are methodically corrected. If there are variations on a name, I recommend that you set one data source as the "master" and keep the data consolidated and correct across all the databases using that master as your primary source. I also suggest that you store the original error in a separate value, as it is useful for discovering patterns for data sources that consistently produce errors.

Create a Default Value

This is an option that I commonly see used in my consulting work with companies. Most system developers assume that if the business doesn't enter the value, they should enter a default value. Common values that I have seen are "unknown" or "n/a." Unfortunately, I have also seen many undesirable choices, such as birthdays for dates or pets' names for first name and last name, parents' addresses . . . This address choice goes awry, of course, when more than 300 marketing letters with sample products are sent to parents' addresses by several companies that are using the same service to distribute their marketing work. I suggest that you discuss default values with your customer in detail and agree on an official "missing data" value.

Analysis of Data

I always generate a health report for all data imports. I suggest the following six data quality dimensions.

Completeness

I calculate the number of incorrect entries on each data source's fields as a percentage of the total data. If the data source holds specific importance because of critical data (customer names, phone numbers, e-mail addresses, etc.), I start the analysis of these first, to ensure that the data source is fit to progress to the next phase of analysis for completeness on noncritical data. For example, for personal data to be unique, you need, as a minimum, a first name, last name, and date of birth. If any of this information is not part of the data, it is an incomplete personal data entry. Note that completeness is specific to the business area of the data you are processing.

Uniqueness

I evaluate how unique the specific value is, in comparison to the rest of the data in that field. Also, test the value against other known sources of the same data sets. The last test for uniqueness is to show where the same field is in many data sources. You will report the uniqueness normally, as a histogram across all unique values in each data source.

Timeliness

Record the impact of the date and time on the data source. Are there periods of stability or instability? This check is useful when scheduling extracts from source systems. I have seen countless month-end snapshot extracts performed before the month-end completed. These extracts are of no value. I suggest you work closely with your customer's operational people, to ensure that your data extracts are performed at the correct point in the business cycle.

Validity

Validity is tested against known and approved standards. It is recorded as a percentage of nonconformance against the standard. I have found that most data entries are covered by a standard. For example, country code uses ISO 3166-1; currencies use ISO 4217.

I also suggest that you look at customer-specific standards, for example, International Classification of Diseases (ICD) standards ICD-10. Take note: Standards change over time. For example, ICD-10 is the tenth version of the standard. ICD-7 took effect in 1958, ICD-8A in 1968, ICD-9 in 1979, and ICD-10 in 1999. So, when you validate data, make sure that you apply the correct standard on the correct data period.

Accuracy

Accuracy is a measure of the data against the real-world person or object that is recorded in the data source. There are regulations, such as the European Union's General Data Protection Regulation (GDPR), that require data to be compliant for accuracy.

I recommend that you investigate what standards and regulations you must comply with for accuracy.

Consistency

This measure is recorded as the shift in the patterns in the data. Measure how data changes load after load. I suggest that you measure patterns and checksums for data sources.

In Chapter 9, I will introduce a data structure called a data vault that will assist you with this process. In Chapter 10, I will demonstrate how you use slowly changing dimensions to track these consistencies.

Caution   The study of the errors in the system is just as important as the study of the valid data. Numerous real business issues are due to data errors.

I will now proceed with the practical implementation of the knowledge you have just acquired.

Practical Actions

In Chapter 7, I introduced you to the Python package pandas. The package enables several automatic error-management features.

Missing Values in Pandas

I will guide you first through some basic error-processing concepts, and then we will apply what you have learned to the sample data. I will use Python with the pandas package, but the basic concepts can be used in many other tools.

Following are four basic processing concepts.

Drop the Columns Where All Elements Are Missing Values

Open your Python editor and create this in a file named Assess-Good-Bad-01.py in
directory C:\VKHCG\01-Vermeulen\02-Assess. Now copy the following code into the file:

```
################################################################
# -*- coding: utf-8 -*-
################################################################
Import sys
import os
import pandas as pd
################################################################
if sys.platform == 'linux':
    Base=os.path.expanduser('~') + '/VKHCG'
else:
    Base='C:/VKHCG'
print('##############################')
print('Working Base:',Base,'using',sys.platform)
print('##############################') sInputFileName='Good-or-Bad.csv'
sOutputFileName='Good-or-Bad-01.csv'
Company='01-Vermeulen'
################################################################
################################################################
sFileDir=Base + '/' + Company + '/02-Assess/01-EDS/02-Python'
if not os.path.exists(sFileDir):
    os.makedirs(sFileDir)
################################################################
### Import Warehouse
################################################################
sFileName=Base + '/' + Company + '/00-RawData/' + sInputFileName
print('Loading:',sFileName)
RawData=pd.read_csv(sFileName,header=0)
print('##############################')
print('## Raw Data Values')
print('##############################')
print(RawData)
```

```
print('##############################')
print('## Data Profile')
print('##############################')
print('Rows:',RawData.shape[0])
print('Columns:',RawData.shape[1])
print('##############################')
################################################################
sFileName=sFileDir + '/' + sInputFileName
RawData.to_csv(sFileName, index = False)
################################################################
## This is the important action! The rest of this code snippet
## Only supports this action.

################################################################

TestData=RawData.dropna(axis=1, how='all')
################################################################
print('##############################')
print('## Test Data Values')
print('##############################')
print(TestData)
print('##############################')
print('## Data Profile')
print('##############################')
print('Rows:',TestData.shape[0])
print('Columns:',TestData.shape[1])
print('##############################')
################################################################
sFileName=sFileDir + '/' + sOutputFileName
TestData.to_csv(sFileName, index = False)
################################################################
print('##############################')
print('### Done!! ####################')
print('##############################')
################################################################
```

Save the file Assess-Good-Bad-01.py, then compile and execute with your Python compiler. This will produce a set of displayed values plus two files named Good-or-Bad. csv and Good-or-Bad-01.csv in directory C:/VKHCG/01-Vermeulen/02-Assess/01-EDS/02-Python.

The code loads the following raw data:

ID	FieldA	FieldB	FieldC	FieldD	FieldE	FieldF	FieldG
1	Good	Better	Best	1024		10241	1
2	Good		Best	512		5121	2
3	Good	Better		256		256	3
4	Good	Better	Best			211	4
5	Good	Better		64		6411	5
6	Good		Best	32		32	6
7		Better	Best	16		1611	7
8			Best	8		8111	8
9				4		41	9
10	A	B	C	2		21111	10
							11
10	Good	Better	Best	1024		102411	12
10	Good		Best	512		512	13
10	Good	Better		256		1256	14
10	Good	Better	Best				15
10	Good	Better		64		164	16
10	Good		Best	32		322	17
10		Better	Best	16		163	18
10			Best	8		844	19
10				4		4555	20
10	A	B	C	2		111	21

The `TestData=RawData.dropna(axis=1, how='all')` code results in the following error correction:

ID	FieldA	FieldB	FieldC	FieldD	FieldF	FieldG
1	Good	Better	Best	1024	10241	1
2	Good		Best	512	5121	2
3	Good	Better		256	256	3
4	Good	Better	Best		211	4
5	Good	Better		64	6411	5
6	Good		Best	32	32	6
7		Better	Best	16	1611	7
8			Best	8	8111	8
9				4	41	9
10	A	B	C	2	21111	10
						11
10	Good	Better	Best	1024	102411	12
10	Good		Best	512	512	13
10	Good	Better		256	1256	14
10	Good	Better	Best			15
10	Good	Better		64	164	16
10	Good		Best	32	322	17
10		Better	Best	16	163	18
10			Best	8	844	19
10				4	4555	20
10	A	B	C	2	111	21

All of column E has been deleted, owing to the fact that all values in that column were missing values/errors.

Drop the Columns Where Any of the Elements Is Missing Values

Open your Python editor and create a file named Assess-Good-Bad-02.py in directory
C:\VKHCG\01-Vermeulen\02-Assess. Now copy the following code into the file:

```
################################################################
# -*- coding: utf-8 -*-
################################################################
Import sys
import os
import pandas as pd
################################################################
if sys.platform == 'linux':
    Base=os.path.expanduser('~') + '/VKHCG'
else:
    Base='C:/VKHCG'
print('################################')
print('Working Base:',Base,'using',sys.platform)
print('################################') sInputFileName='Good-or-Bad.csv'
sOutputFileName='Good-or-Bad-02.csv'
Company='01-Vermeulen'
################################################################
Base='C:/VKHCG'
################################################################
sFileDir=Base + '/' + Company + '/02-Assess/01-EDS/02-Python'
if not os.path.exists(sFileDir):
    os.makedirs(sFileDir)
################################################################
### Import Warehouse
################################################################
sFileName=Base + '/' + Company + '/00-RawData/' + sInputFileName
print('Loading:',sFileName)
RawData=pd.read_csv(sFileName,header=0)
print('################################')
print('## Raw Data Values')
print('################################')
```

```
print(RawData)
print('###############################')
print('## Data Profile')
print('###############################')
print('Rows:',RawData.shape[0])
print('Columns:',RawData.shape[1])
print('###############################')
################################################################
sFileName=sFileDir + '/' + sInputFileName
RawData.to_csv(sFileName, index = False)
################################################################
## This is the important action! The rest of this code snippet
## Only supports this action.

################################################################

TestData=RawData.dropna(axis=1, how='any')
################################################################
print('###############################')
print('## Test Data Values')
print('###############################')
print(TestData)
print('###############################')
print('## Data Profile')
print('###############################')
print('Rows:',TestData.shape[0])
print('Columns:',TestData.shape[1])
print('###############################')
################################################################
sFileName=sFileDir + '/' + sOutputFileName
TestData.to_csv(sFileName, index = False)
################################################################
print('###############################')
print('### Done!! ####################')
print('###############################')
################################################################
```

Save the `Assess-Good-Bad-02.py` file, then compile and execute with your Python compiler. This will produce a set of displayed values plus two files named Good-or-Bad. csv and Good-or-Bad-02.csv in directory `C:/VKHCG/01-Vermeulen/02-Assess/01-EDS/02-Python`.

The code loads the following raw data:

ID	FieldA	FieldB	FieldC	FieldD	FieldE	FieldF	FieldG
1	Good	Better	Best	1024		10241	1
2	Good		Best	512		5121	2
3	Good	Better		256		256	3
4	Good	Better	Best			211	4
5	Good	Better		64		6411	5
6	Good		Best	32		32	6
7		Better	Best	16		1611	7
8			Best	8		8111	8
9				4		41	9
10	A	B	C	2		21111	10
							11
10	Good	Better	Best	1024		102411	12
10	Good		Best	512		512	13
10	Good	Better		256		1256	14
10	Good	Better	Best				15
10	Good	Better		64		164	16
10	Good		Best	32		322	17
10		Better	Best	16		163	18
10			Best	8		844	19
10				4		4555	20
10	A	B	C	2		111	21

The `TestData=RawData.dropna(axis=1, how='any')` code results in the following error correction:

FieldG
1
2
3
4
5
6
7
8
9
10
11
12
13
14
15
16
17
18
19
20
21

Columns A, B, C, D, E, and F are removed, owing to the fact that some of their values were missing values/errors. The removal of an entire column is useful when you want to isolate the complete and correct columns.

Keep Only the Rows That Contain a Maximum of Two Missing Values

Open your Python editor and create a file named Assess-Good-Bad-03.py in directory C:\VKHCG\01-Vermeulen\02-Assess. Now copy the following code into the file:

```
################################################################
# -*- coding: utf-8 -*-
################################################################
Import sys
import os
import pandas as pd
################################################################
if sys.platform == 'linux':
    Base=os.path.expanduser('~') + '/VKHCG'
else:
    Base='C:/VKHCG'
print('################################')
print('Working Base:',Base,'using',sys.platform)
print('################################') ###################################
################################
sInputFileName='Good-or-Bad.csv'
sOutputFileName='Good-or-Bad-03.csv'
Company='01-Vermeulen'
################################################################
Base='C:/VKHCG'
################################################################
sFileDir=Base + '/' + Company + '/02-Assess/01-EDS/02-Python'
if not os.path.exists(sFileDir):
    os.makedirs(sFileDir)
################################################################
### Import Warehouse
################################################################
```

288

```python
sFileName=Base + '/' + Company + '/00-RawData/' + sInputFileName
print('Loading:',sFileName)
RawData=pd.read_csv(sFileName,header=0)
print('##############################')
print('## Raw Data Values')
print('##############################')
print(RawData)
print('##############################')
print('## Data Profile')
print('##############################')
print('Rows:',RawData.shape[0])
print('Columns:',RawData.shape[1])
print('##############################')
################################################################
sFileName=sFileDir + '/' + sInputFileName
RawData.to_csv(sFileName, index = False)
################################################################
## This is the important action! The rest of this code snippet
## Only supports this action.

################################################################
TestData=RawData.dropna(thresh=2)
################################################################
print('##############################')
print('## Test Data Values')
print('##############################')
print(TestData)
print('##############################')
print('## Data Profile')
print('##############################')
print('Rows:',TestData.shape[0])
print('Columns:',TestData.shape[1])
print('##############################')
################################################################
sFileName=sFileDir + '/' + sOutputFileName
TestData.to_csv(sFileName, index = False)
```

289

```
################################################################
print('###############################')
print('### Done!! ####################')
print('###############################')
################################################################
```

Save the Assess-Good-Bad-03.py file, then compile and execute with your Python compiler. This will produce a set of displayed values plus two files named Good-or-Bad. csv and Good-or-Bad-03.csv in directory C:/VKHCG/01-Vermeulen/02-Assess/01-EDS/02-Python.

The code loads the following raw data:

ID	FieldA	FieldB	FieldC	FieldD	FieldE	FieldF	FieldG
1	Good	Better	Best	1024		10241	1
2	Good		Best	512		5121	2
3	Good	Better		256		256	3
4	Good	Better	Best			211	4
5	Good	Better		64		6411	5
6	Good		Best	32		32	6
7		Better	Best	16		1611	7
8			Best	8		8111	8
9				4		41	9
10	A	B	C	2		21111	10
							11
10	Good	Better	Best	1024		102411	12
10	Good		Best	512		512	13
10	Good	Better		256		1256	14
10	Good	Better	Best				15
10	Good	Better		64		164	16
10	Good		Best	32		322	17
10		Better	Best	16		163	18

(continued)

ID	FieldA	FieldB	FieldC	FieldD	FieldE	FieldF	FieldG
10			Best	8		844	19
10				4		4555	20
10	A	B	C	2		111	21

The `TestData=RawData.dropna(thresh=2)` results in the following error correction:

ID	FieldA	FieldB	FieldC	FieldD	FieldE	FieldF	FieldG
1	Good	Better	Best	1024		10241	1
2	Good		Best	512		5121	2
3	Good	Better		256		256	3
4	Good	Better	Best			211	4
5	Good	Better		64		6411	5
6	Good		Best	32		32	6
7		Better	Best	16		1611	7
8			Best	8		8111	8
9				4		41	9
10	A	B	C	2		21111	10
10	Good	Better	Best	1024		102411	12
10	Good		Best	512		512	13
10	Good	Better		256		1256	14
10	Good	Better	Best				15
10	Good	Better		64		164	16
10	Good		Best	32		322	17
10		Better	Best	16		163	18
10			Best	8		844	19
10				4		4555	20
10	A	B	C	2		111	21

Row 11 has been removed, owing to the fact that more than two of its values are missing values/errors.

Fill All Missing Values with the Mean, Median, Mode, Minimum, and Maximum of the Particular Numeric Column

Open your Python editor and create a file named Assess-Good-Bad-04.py in directory C:\VKHCG\01-Vermeulen\02-Assess. Now copy the following code into the file:

```
################################################################
# -*- coding: utf-8 -*-
################################################################
Import sys
import os
import pandas as pd
################################################################
Base='C:/VKHCG'
sInputFileName='Good-or-Bad.csv'
sOutputFileNameA='Good-or-Bad-04-A.csv'
sOutputFileNameB='Good-or-Bad-04-B.csv'
sOutputFileNameC='Good-or-Bad-04-C.csv'
sOutputFileNameD='Good-or-Bad-04-D.csv'
sOutputFileNameE='Good-or-Bad-04-E.csv'
Company='01-Vermeulen'
################################################################
if sys.platform == 'linux':
    Base=os.path.expanduser('~') + '/VKHCG'
else:
    Base='C:/VKHCG'
print('################################')
print('Working Base:',Base,'using',sys.platform)
print('################################')
################################################################
sFileDir=Base + '/' + Company + '/02-Assess/01-EDS/02-Python'
if not os.path.exists(sFileDir):
    os.makedirs(sFileDir)
```

```
################################################################
### Import Warehouse
################################################################
sFileName=Base + '/' + Company + '/00-RawData/' + sInputFileName
print('Loading:',sFileName)
RawData=pd.read_csv(sFileName,header=0)

print('###############################')
print('## Raw Data Values')
print('###############################')
print(RawData)
print('###############################')
print('## Data Profile')
print('###############################')
print('Rows:',RawData.shape[0])
print('Columns:',RawData.shape[1])
print('###############################')
################################################################
sFileName=sFileDir + '/' + sInputFileName
RawData.to_csv(sFileName, index = False)
################################################################
TestData=RawData.fillna(RawData.mean())
################################################################
print('###############################')
print('## Test Data Values- Mean')
print('###############################')
print(TestData)
print('###############################')
print('## Data Profile')
print('###############################')
print('Rows:',TestData.shape[0])
print('Columns:',TestData.shape[1])
print('###############################')
################################################################
sFileName=sFileDir + '/' + sOutputFileNameA
TestData.to_csv(sFileName, index = False)
```

```
################################################################
## This is the important action! The rest of this code snippet
## Only supports this action.

################################################################

TestData=RawData.fillna(RawData.median())
################################################################
print('###############################')
print('## Test Data Values - Median')
print('###############################')
print(TestData)
print('###############################')
print('## Data Profile')
print('###############################')
print('Rows:',TestData.shape[0])
print('Columns:',TestData.shape[1])
print('###############################')
################################################################
sFileName=sFileDir + '/' + sOutputFileNameB
TestData.to_csv(sFileName, index = False)
################################################################
################################################################
TestData=RawData.fillna(RawData.mode())
################################################################
print('###############################')
print('## Test Data Values - Mode')
print('###############################')
print(TestData)
print('###############################')
print('## Data Profile')
print('###############################')
print('Rows:',TestData.shape[0])
print('Columns:',TestData.shape[1])
print('###############################')
```

```
################################################################
sFileName=sFileDir + '/' + sOutputFileNameC
TestData.to_csv(sFileName, index = False)
################################################################
################################################################
TestData=RawData.fillna(RawData.min())
################################################################
print('###############################')
print('## Test Data Values - Minumum')
print('###############################')
print(TestData)
print('###############################')
print('## Data Profile')
print('###############################')
print('Rows:',TestData.shape[0])
print('Columns:',TestData.shape[1])
print('###############################')
################################################################
sFileName=sFileDir + '/' + sOutputFileNameD
TestData.to_csv(sFileName, index = False)
################################################################
################################################################
TestData=RawData.fillna(RawData.max())
################################################################
print('###############################')
print('## Test Data Values - Maximum')
print('###############################')
print(TestData)
print('###############################')
print('## Data Profile')
print('###############################')
print('Rows:',TestData.shape[0])
print('Columns:',TestData.shape[1])
print('###############################')
################################################################
```

```
sFileName=sFileDir + '/' + sOutputFileNameE
TestData.to_csv(sFileName, index = False)
################################################################
################################################################
print('###############################')
print('### Done!! ####################')
print('###############################')
################################################################
```

Save the Assess-Good-Bad-04.py file, then compile and execute with your Python compiler. This will produce a set of displayed values plus six files named Good-or-Bad.csv, Good-or-Bad-04-A.csv, Good-or-Bad-04-B.csv, Good-or-Bad-04-C.csv, Good-or-Bad-04-D.csv, and Good-or-Bad-04-E.csv in directory C:/VKHCG/01-Vermeulen/02-Assess/01-EDS/02-Python. I suggest you investigate each result and understand how it was transformed by the code.

Engineering a Practical Assess Superstep

Any source code or other supplementary material referenced by me in this book is available to readers on GitHub, via this book's product page, located at www.apress.com/9781484230534. Please note that this source code assumes that you have completed the source code setup outlined in Chapter 2, as it creates all the directories you will need to complete the examples.

Now that I have explained the various aspects of the Assess superstep, I will explain how to help our company process its data.

Vermeulen PLC

As I guide you through the Assess superstep for the company, the coding will become increasingly multilayered. I will assist you end-to-end through these phases of discovery. Vermeulen PLC has two primary processes that I will lead you through: network routing and job scheduling.

Creating a Network Routing Diagram

The next step along the route is to generate a full network routing solution for the company, to resolve the data issues in the retrieve data.

Note   In the next example, I switch off the `SettingWithCopyWarning` check, as I am using the data manipulation against a copy of the same data: `pd.options.mode.chained_assignment = None`.

Open your Python editor and create a file named `Assess-Network-Routing-Company.py` in directory `C:\VKHCG\01-Vermeulen\02-Assess`.

Now copy the following code into the file:

```
################################################################
import sys
import os
import pandas as pd
################################################################
pd.options.mode.chained_assignment = None
################################################################
################################################################
if sys.platform == 'linux':
    Base=os.path.expanduser('~') + '/VKHCG'
else:
    Base='C:/VKHCG'
print('################################')
print('Working Base:',Base,'using',sys.platform)
print('################################') pri
nt('################################')
print('Working Base:',Base,'using',sys.platform)
print('################################')
################################################################
sInputFileName1='01-Retrieve/01-EDS/01-R/Retrieve_Country_Code.csv'
sInputFileName2='01-Retrieve/01-EDS/02-Python/Retrieve_Router_Location.csv'
################################################################
```

```
sOutputFileName='Assess-Network-Routing-Company.csv'
Company='01-Vermeulen'
################################################################
################################################################
### Import Country Data
################################################################
sFileName=Base + '/' + Company + '/' + sInputFileName1
print('##############################')
print('Loading:',sFileName)
print('##############################')
CountryData=pd.read_csv(sFileName,header=0,low_memory=False,
encoding="latin-1")
print('Loaded Country:',CountryData.columns.values)
print('##############################')
################################################################
## Assess Country Data
################################################################
print('##############################')
print('Changed:',CountryData.columns.values)
```

Tip   I use many of pandas's built-in functions to manipulate data in my data science projects. It will be to your future advantage to investigate the capabilities of this pandas library. See https://pandas.pydata.org/pandas-docs/ stable/api.html, for reference.

```
CountryData.rename(columns={'Country': 'Country_Name'}, inplace=True)
CountryData.rename(columns={'ISO-2-CODE': 'Country_Code'}, inplace=True)
CountryData.drop('ISO-M49', axis=1, inplace=True)
CountryData.drop('ISO-3-Code', axis=1, inplace=True)
CountryData.drop('RowID', axis=1, inplace=True)
print('To:',CountryData.columns.values)
print('##############################')
################################################################
```

```python
################################################################
### Import Company Data
################################################################
sFileName=Base + '/' + Company + '/' + sInputFileName2
print('###############################')
print('Loading:',sFileName)
print('###############################')
CompanyData=pd.read_csv(sFileName,header=0,low_memory=False,
encoding="latin-1")
print('Loaded Company:',CompanyData.columns.values)
print('###############################')
################################################################
## Assess Company Data
################################################################
print('###############################')
print('Changed:',CompanyData.columns.values)
CompanyData.rename(columns={'Country':'Country_Code'}, inplace=True)
print('To:',CompanyData.columns.values)
print('###############################')
################################################################
################################################################
### Import Customer Data
################################################################
sFileName=Base + '/' + Company + '/' + sInputFileName3
print('###############################')
print('Loading:',sFileName)
print('###############################')
CustomerRawData=pd.read_csv(sFileName,header=0,low_memory=False,
encoding="latin-1")
print('###############################')
print('Loaded Customer:',CustomerRawData.columns.values)
print('###############################')
################################################################
CustomerData=CustomerRawData.dropna(axis=0, how='any')
```

```python
print('###############################')
print('Remove Blank Country Code')
print('Reduce Rows from',CustomerRawData.shape[0],'to',CustomerData.
shape[0])
print('###############################')
###################################################################
print('###############################')
print('Changed:',CustomerData.columns.values)
CustomerData.rename(columns={'Country':'Country_Code'},inplace=True)
print('To:',CustomerData.columns.values)
print('###############################')
###################################################################
print('###############################')
print('Merge Company and Country Data')
print('###############################')
CompanyNetworkData=pd.merge(
        CompanyData,
        CountryData,
        how='inner',
        on='Country_Code'
        )
###################################################################
print('###############################')
print('Change',CompanyNetworkData.columns.values)
for i in CompanyNetworkData.columns.values:
    j='Company_'+i
    CompanyNetworkData.rename(columns={i:j},inplace=True)
print('To',CompanyNetworkData.columns.values)
print('###############################')
###################################################################
###################################################################
sFileDir=Base + '/' + Company + '/02-Assess/01-EDS/02-Python'
if not os.path.exists(sFileDir):
    os.makedirs(sFileDir)
```

```
################################################################
sFileName=sFileDir + '/' + sOutputFileName
print('###############################')
print('Storing:',sFileName)
print('###############################')
CompanyNetworkData.to_csv(sFileName, index = False, encoding="latin-1")
################################################################
################################################################
print('###############################')
print('### Done!! ####################')
print('###############################')
################################################################
```

Save the `Assess-Network-Routing-Company.py` file, then compile and execute with your Python compiler. This will produce a set of demonstrated values onscreen, plus a file named `Assess-Network-Routing-Company.csv`.

Company_ Country_Code	Company_ Place_Name	Company_Latitude	Company_ Longitude	Company_Country_ Name
US	New York	40.7528	-73.9725	United States of America
US	New York	40.7214	-74.0052	United States of America
US	New York	40.7662	-73.9862	United States of America

Open your Python editor and create a file called `Assess-Network-Routing-Customer.py` in directory `C:\VKHCG\01-Vermeulen\02-Assess`. Next, create a new file, and move on to the following example.

Here is the code for the example:

```
################################################################
import sys
import os
import pandas as pd
################################################################
pd.options.mode.chained_assignment = None
################################################################
```

```
if sys.platform == 'linux':
    Base=os.path.expanduser('~') + 'VKHCG'
else:
    Base='C:/VKHCG'
print('##############################')
print('Working Base:',Base,'using',sys.platform)
print('##############################')
####################################################################
sInputFileName1='01-Retrieve/01-EDS/01-R/Retrieve_Country_Code.csv'
sInputFileName2='01-Retrieve/01-EDS/02-Python/Retrieve_All_Router_Location.csv'
####################################################################
sOutputFileName='Assess-Network-Routing-Customer.csv'
Company='01-Vermeulen'
####################################################################
####################################################################
### Import Country Data
####################################################################
sFileName=Base + '/' + Company + '/' + sInputFileName1
print('##############################')
print('Loading:',sFileName)
print('##############################')
CountryData=pd.read_csv(sFileName,header=0,low_memory=False,
encoding="latin-1")
print('Loaded Country:',CountryData.columns.values)
print('##############################')
####################################################################
## Assess Country Data
####################################################################
print('##############################')
print('Changed:',CountryData.columns.values)
CountryData.rename(columns={'Country': 'Country_Name'}, inplace=True)
CountryData.rename(columns={'ISO-2-CODE': 'Country_Code'}, inplace=True)
CountryData.drop('ISO-M49', axis=1, inplace=True)
CountryData.drop('ISO-3-Code', axis=1, inplace=True)
CountryData.drop('RowID', axis=1, inplace=True)
print('To:',CountryData.columns.values)
```

```python
print('###############################')
###################################################################
### Import Customer Data
###################################################################
sFileName=Base + '/' + Company + '/' + sInputFileName2
print('###############################')
print('Loading:',sFileName)
print('###############################')
CustomerRawData=pd.read_csv(sFileName,header=0,low_memory=False,
encoding="latin-1")
print('###############################')
print('Loaded Customer:',CustomerRawData.columns.values)
print('###############################')
###################################################################
CustomerData=CustomerRawData.dropna(axis=0, how='any')
print('###############################')
print('Remove Blank Country Code')
print('Reduce Rows from',CustomerRawData.shape[0],'to',CustomerData.
shape[0])
print('###############################')
###################################################################
print('###############################')
print('Changed:',CustomerData.columns.values)
CustomerData.rename(columns={'Country': 'Country_Code'}, inplace=True)
print('To:',CustomerData.columns.values)
print('###############################')
###################################################################
print('###############################')
print('Merge Customer and Country Data')
print('###############################')
CustomerNetworkData=pd.merge(
        CustomerData,
        CountryData,
        how='inner',
        on='Country_Code'
        )
```

```
################################################################
print('##############################')
print('Change',CustomerNetworkData.columns.values)
for i in CustomerNetworkData.columns.values:
    j='Customer_'+i
    CustomerNetworkData.rename(columns={i:j}, inplace=True)
print('To', CustomerNetworkData.columns.values)
print('##############################')
################################################################
sFileDir=Base + '/' + Company + '/02-Assess/01-EDS/02-Python'
if not os.path.exists(sFileDir):
    os.makedirs(sFileDir)
################################################################
sFileName=sFileDir + '/' + sOutputFileName
print('##############################')
print('Storing:', sFileName)
print('##############################')
CustomerNetworkData.to_csv(sFileName, index = False, encoding="latin-1")
################################################################
print('##############################')
print('### Done!! #####################')
print('##############################')
################################################################
```

Save the file `Assess-Network-Routing-Customer.py`, then compile and execute with your Python compiler. This will produce a set of demonstrated values onscreen plus a file named `Assess-Network-Routing-Customer.csv`.

Customer_ Country_Code	Customer_Place_ Name	Customer_ Latitude	Customer_ Longitude	Customer_ Country_Name
BW	Gaborone	-24.6464	25.9119	Botswana
BW	Francistown	-21.1667	27.5167	Botswana
BW	Maun	-19.9833	23.4167	Botswana

Open your Python editor and create a file named `Assess-Network-Routing-Node.py` in directory `C:\VKHCG\01-Vermeulen\02-Assess`. Now copy the following code into the file:

```python
################################################################
import sys
import os
import pandas as pd
################################################################
pd.options.mode.chained_assignment = None
################################################################
if sys.platform == 'linux':
    Base=os.path.expanduser('~') + 'VKHCG'
else:
    Base='C:/VKHCG'
print('################################')
print('Working Base:',Base,'using',sys.platform)
print('################################')
################################################################
sInputFileName='01-Retrieve/01-EDS/02-Python/Retrieve_IP_DATA.csv'
################################################################
sOutputFileName='Assess-Network-Routing-Node.csv'
Company='01-Vermeulen'
################################################################
### Import IP Data
################################################################
sFileName=Base + '/' + Company + '/' + sInputFileName
print('################################')
print('Loading:',sFileName)
print('################################')
IPData=pd.read_csv(sFileName,header=0,low_memory=False, encoding="latin-1")
print('Loaded IP:', IPData.columns.values)
print('################################')
################################################################
print('################################')
print('Changed:',IPData.columns.values)
```

```python
IPData.drop('RowID', axis=1, inplace=True)
IPData.drop('ID', axis=1, inplace=True)
IPData.rename(columns={'Country': 'Country_Code'}, inplace=True)
IPData.rename(columns={'Place.Name': 'Place_Name'}, inplace=True)
IPData.rename(columns={'Post.Code': 'Post_Code'}, inplace=True)
IPData.rename(columns={'First.IP.Number': 'First_IP_Number'}, inplace=True)
IPData.rename(columns={'Last.IP.Number': 'Last_IP_Number'}, inplace=True)
print('To:',IPData.columns.values)
print('##############################')
################################################################
print('##############################')
print('Change',IPData.columns.values)
for i in IPData.columns.values:
    j='Node_'+i
    IPData.rename(columns={i:j}, inplace=True)
print('To', IPData.columns.values)
print('##############################')
################################################################
sFileDir=Base + '/' + Company + '/02-Assess/01-EDS/02-Python'
if not os.path.exists(sFileDir):
    os.makedirs(sFileDir)
################################################################
sFileName=sFileDir + '/' + sOutputFileName
print('##############################')
print('Storing:', sFileName)
print('##############################')
IPData.to_csv(sFileName, index = False, encoding="latin-1")
################################################################
print('##############################')
print('### Done!! ###################')
print('##############################')
################################################################
```

Save the Assess-Network-Routing-Node.py file, then compile and execute with your Python compiler. This will produce a set of demonstrated values onscreen, plus a file named Assess-Network-Routing-Node.csv.

Node_Country_Code	Node_Place_Name	Node_Post_Code	Node_Latitude	Node_Longitude	Node_First_IP_Number	Node_Last_IP_Number
GB	Lerwick	ZE1	60.15	-1.15	523478016	523479039
GB	Lerwick	ZE1	60.15	-1.15	1545369984	1545370111
GB	Selby	YO8	53.7833	-1.0667	35231232	35231359

Graph Theory

I provide a simple introduction to graph theory, before we progress to the examples. Graphs are useful for indicating relationships between entities in the real world. The basic building blocks are two distinct graph components, as follows.

Node

The node is any specific single entity. For example, in "Andre kiss Zhaan," there are two nodes: Andre and Zhaan (Figure 8-1).

Figure 8-1. *Nodes Andre and Zhaan*

Edge

The edge is any specific relationship between two nodes. For example, in "Andre kiss Zhaan," there are two nodes, i.e., Andre and Zhaan. The edge is "kiss." The edge can be recorded as non-directed. This will record "kiss" as "kiss each other" (Figure 8-2).

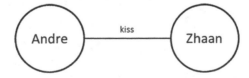

Figure 8-2. *Nodes Andre and Zhaan kiss each other*

The edge can be recorded as directed. This will record the "kiss" as "kiss toward" (Figure 8-3).

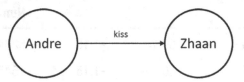

Figure 8-3. *Nodes Andre kiss toward Zhaan*

This enforces the direction of the edge. This concept of direction is useful when dealing with actions that must occur in a specific order, or when you have to follow a specific route. Figure 8-4 shows an example of this.

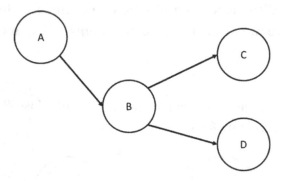

Figure 8-4. *Direction of edges*

You can only travel from A to C via B, A to D via B, A to B, B to C, and B to D. You cannot travel from C to B, D to B, B to A, C to A, and D to A. The directed edge prevents this action.

Directed Acyclic Graph (DAG)

A directed acyclic graph is a specific graph that only has one path through the graph. Figure 8-5 shows a graph that is a DAG, and Figure 8-6 shows a graph that is not a DAG.

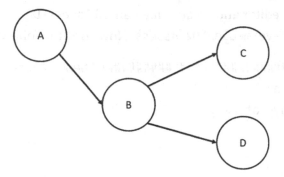

Figure 8-5. *This graph is a DAG*

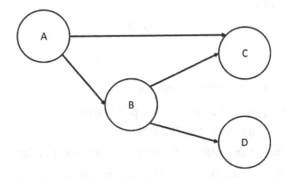

Figure 8-6. *This graph is not a DAG*

The reason Figure 8-6 is not a DAG is because there are two paths between A and C: A to C and A to B to C. Now that you have the basics, let me show you how to use this in your data science.

Building a DAG for Scheduling Jobs

The following example explains how to schedule networking equipment maintenance jobs to perform the required repairs during the month. To create a DAG, I will define what it is first: A DAG is a data structure that enables you to generate a relationship between data entries that can only be performed in a specific order. The DAG is an important structure in the core data science environments, as it is the fundamental structure that enables tools such as Spark, Pig, and Tez in Hadoop to work. It is also used for recording task scheduling and process interdependencies in processing data.

I will introduce a new Python library called networkx.

Open your python editor and create a file named `Assess-DAG-Location.py` in directory `..\VKHCG\01-Vermeulen\02-Assess`. Now copy the following code into the file:

```python
################################################################
import networkx as nx
import matplotlib.pyplot as plt
import sys
import os
import pandas as pd
################################################################
if sys.platform == 'linux':
    Base=os.path.expanduser('~') + 'VKHCG'
else:
    Base='C:/VKHCG'
print('###############################')
print('Working Base:',Base,'using',sys.platform)
print('###############################')
################################################################
sInputFileName='01-Retrieve/01-EDS/02-Python/Retrieve_Router_Location.csv'
sOutputFileName1='Assess-DAG-Company-Country.png'
sOutputFileName2='Assess-DAG-Company-Country-Place.png'
Company='01-Vermeulen'
################################################################
### Import Company Data
################################################################
sFileName=Base + '/' + Company + '/' + sInputFileName
print('###############################')
print('Loading:',sFileName)
print('###############################')
CompanyData=pd.read_csv(sFileName,header=0,low_memory=False,
encoding="latin-1")
print('Loaded Company:',CompanyData.columns.values)
print('###############################')
################################################################
print(CompanyData)
print('###############################')
```

```
print('Rows:',CompanyData.shape[0])
print('############################')
```

We will now create two directed graphs: G1 and G2 using the G=DiGraph(), if you just wanted a graph you would use G=Graph().

```
################################################################
G1=nx.DiGraph()
G2=nx.DiGraph()
################################################################
```

We will now create a node for each of the places in our data. For this, use the G.add_node() function.

Note   If you execute G.add_node("A") and then G.add_node("A") again, you only get one node: "A". The graph automatically resolves any duplicate node requests.

We will now loop through all the country and place-name-country records.

```
for i in range(CompanyData.shape[0]):
    G1.add_node(CompanyData['Country'][i])
    sPlaceName= CompanyData['Place_Name'][i] + '-' + CompanyData['Country'][i]
    G2.add_node(sPlaceName)
```

We will now interconnect all the nodes (G.nodes()), by looping though them in pairs.

```
print('##############################')
for n1 in G1.nodes():
    for n2 in G1.nodes():
        if n1 != n2:
            print('Link:',n1,' to ', n2)
            G1.add_edge(n1,n2)
print('##############################')
```

Now, we can see the nodes and edges you created.

```
print('##############################')
print("Nodes of graph:")
print(G1.nodes())
```

```
print("Edges of graph:")
print(G1.edges())
print('##############################')
```

We can now save our graph, as follows:

```
################################################################
sFileDir=Base + '/' + Company + '/02-Assess/01-EDS/02-Python'
if not os.path.exists(sFileDir):
    os.makedirs(sFileDir)
################################################################
sFileName=sFileDir + '/' + sOutputFileName1
print('##############################')
print('Storing:', sFileName)
print('##############################')
```

Now, you can display the graph as a picture.

```
nx.draw(G1,pos=nx.spectral_layout(G1),
        nodecolor='r',edge_color='b',
        with_labels=True,node_size=8000,
        font_size=12)
plt.savefig(sFileName) # save as png
plt.show() # display
################################################################
```

Now, you can complete the second graph.

```
print('##############################')
for n1 in G2.nodes():
    for n2 in G2.nodes():
        if n1 != n2:
            print('Link:',n1,'to',n2)
            G2.add_edge(n1,n2)
print('##############################')

print('##############################')
print("Nodes of graph:")
print(G2.nodes())
```

```
print("Edges of graph:")
print(G2.edges())
print('###############################')
################################################################
sFileDir=Base + '/' + Company + '/02-Assess/01-EDS/02-Python'
if not os.path.exists(sFileDir):
    os.makedirs(sFileDir)
################################################################
sFileName=sFileDir + '/' + sOutputFileName2
print('###############################')
print('Storing:', sFileName)
print('###############################')
nx.draw(G2,pos=nx.spectral_layout(G2),
        nodecolor='r',edge_color='b',
        with_labels=True,node_size=8000,
        font_size=12)
plt.savefig(sFileName) # save as png
plt.show() # display
################################################################
```

Save the Assess-DAG-Location.py file, then compile and execute with your Python compiler. This will produce a set of demonstrated values onscreen, plus two graphical files named Assess-DAG-Company-Country.png (Figure 8-7) and Assess-DAG-Company-Country-Place.png (Figure 8-8).

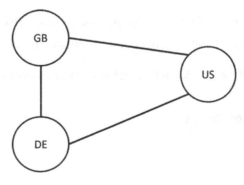

Figure 8-7. *Assess-DAG-Company-Country.png file*

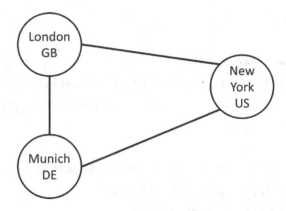

Figure 8-8. `Assess-DAG-Company-Country-Place.png` *file*

Well done. You just completed a simple but valid graph. Let's try a bigger graph.

Remember   There are only two entities. Size has no direct impact on the way to build.

Caution   The `networkx` library works well with fairly large graphs, but it is limited by your memory. My two 32-core AMD CPUs with 1TB DRAM5 hold 100 million nodes and edges easily. But when I cross into a billion nodes and edges, I move onto a distributed graph engine, using my 12-node DataStax Enterprise data platform, with DSE Graph to handle the graphs, via the `dse-graph` library. I can confirm that the examples in this book run on my Quad-Core 4MB RAM laptop easily.

Open your Python editor and create a file named `Assess-DAG-GPS.py` in directory `C:\VKHCG\01-Vermeulen\02-Assess`. Now copy the following code into the file:

```
################################################################
import networkx as nx
import matplotlib.pyplot as plt
import sys
import os
import pandas as pd
################################################################
```

```
if sys.platform == 'linux':
    Base=os.path.expanduser('~') + 'VKHCG'
else:
    Base='C:/VKHCG'
print('##############################')
print('Working Base:',Base,'using',sys.platform)
print('##############################')
##################################################################
sInputFileName='01-Retrieve/01-EDS/02-Python/Retrieve_Router_Location.csv'
sOutputFileName='Assess-DAG-Company-GPS.png'
Company='01-Vermeulen'
##################################################################
### Import Company Data
##################################################################
sFileName=Base + '/' + Company + '/' + sInputFileName
print('##############################')
print('Loading:',sFileName)
print('##############################')
CompanyData=pd.read_csv(sFileName,header=0,low_memory=False,
encoding="latin-1")
print('Loaded Company:',CompanyData.columns.values)
print('##############################')
##################################################################
print(CompanyData)
print('##############################')
print('Rows:',CompanyData.shape[0])
print('##############################')
##################################################################
G=nx.Graph()
##################################################################
```

Tip   I used a simple data science technique in the following code, by adding the round(x,1) on the latitude and longitude values. This reduces the amount of nodes required from 150 to 16, a 90% savings in node processing. Edges reduces the nodes from 11,175 to 120, a 99% savings in edge processing. If you use round(x,2), you get 117 nodes and 6786 edges, a 22% and 39% savings, respectively. This is a useful technique to use when you only want to test your findings at an estimated level.

```
for i in range(CompanyData.shape[0]):
    nLatitude=round(CompanyData['Latitude'][i],1)
    nLongitude=round(CompanyData['Longitude'][i],1)

    if nLatitude < 0:
        sLatitude = str(nLatitude*-1) + 'S'
    else:
        sLatitude = str(nLatitude) + 'N'

    if nLongitude < 0:
        sLongitude = str(nLongitude*-1) + 'W'
    else:
        sLongitude = str(nLongitude) + 'E'

    sGPS= sLatitude + '-' + sLongitude
    G.add_node(sGPS)

print('##############################')
for n1 in G.nodes():
    for n2 in G.nodes():
        if n1 != n2:
            print('Link:',n1,'to',n2)
            G.add_edge(n1,n2)
print('##############################')

print('##############################')
print("Nodes of graph:")
print(G.nodes())
```

```
print("Edges of graph:")
print(G.edges())
print('##############################')
################################################################
sFileDir=Base + '/' + Company + '/02-Assess/01-EDS/02-Python'
if not os.path.exists(sFileDir):
    os.makedirs(sFileDir)
################################################################
sFileName=sFileDir + '/' + sOutputFileName
print('##############################')
print('Storing:', sFileName)
print('##############################')
pos=nx.circular_layout(G,dim=1, scale=2)
nx.draw(G,pos=pos,
        nodecolor='r',edge_color='b',
        with_labels=True,node_size=4000,
        font_size=9)
plt.savefig(sFileName) # save as png
plt.show() # display
################################################################
```

Save the file Assess-DAG-GPS.py, then compile and execute with your Python compiler. This will produce a set of demonstrated values onscreen, plus a graphical file named Assess-DAG-Company-GPS.png (Figure 8-9).

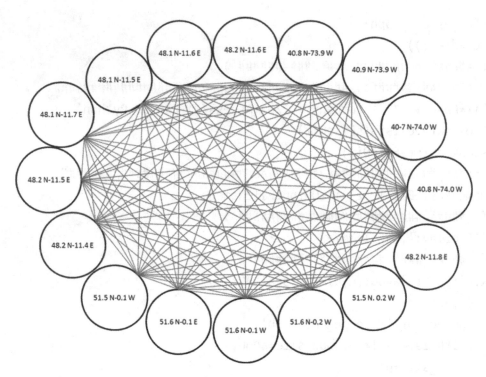

Figure 8-9. `Assess-DAG-Company-GPS.png` *file*

I have guided you through the basic concepts of using graph theory data structures to assess that you have the correct relationship between data entities. I will now demonstrate how to convert the IP data we retrieved into a valid routing network (Figure 8-10).

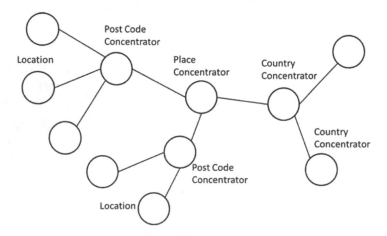

Figure 8-10. *Valid routing network*

The basic validation is as follows:

- Countries interconnect, forming a mesh network.

- Places within countries connect to all in-country nodes, that is, a point in the country where all the network connections are concentrated together, called "place concentrators."

- Post codes can only connect via post code concentrators (devices similar to place concentrators but with one per post code) to place concentrators.

- Locations can only connect to their post code concentrators.

Note There is more than one set of rules to interconnect the nodes. However, the result is still only one graph, when all the rules are applied. If there are overlapping connections between the rules, the graph will resolve only one relationship. This saves you from solving the overlaps. Nice to remember!

Now test your skills against this example. I will guide you through the assess step for Retrieve_IP_DATA_CORE.csv. Open your Python editor and create a file called Assess-DAG-Schedule.py in directory C:\VKHCG\01-Vermeulen\02-Assess.

Now copy this code into the following file:

```
###############################################################
import networkx as nx
import matplotlib.pyplot as plt
import sys
import os
import pandas as pd
###############################################################
if sys.platform == 'linux':
    Base=os.path.expanduser('~') + 'VKHCG'
else:
    Base='C:/VKHCG'
print('###############################')
print('Working Base:',Base,'using',sys.platform)
print('###############################')
```

```
##################################################################
sInputFileName='01-Retrieve/01-EDS/01-R/Retrieve_IP_DATA_CORE.csv'
sOutputFileName='Assess-DAG-Schedule.gml'
Company='01-Vermeulen'
##################################################################
### Import Core Company Data
##################################################################
sFileName=Base + '/' + Company + '/' + sInputFileName
print('###############################')
print('Loading:',sFileName)
print('###############################')
CompanyData=pd.read_csv(sFileName,header=0,low_memory=False,
encoding="latin-1")
print('Loaded Company:',CompanyData.columns.values)
print('###############################')
##################################################################
print(CompanyData)
print('###############################')
print('Rows:',CompanyData.shape[0])
print('###############################')
##################################################################
G=nx.Graph()
##################################################################
for i in range(CompanyData.shape[0]):
    sGroupName0= str(CompanyData['Country'][i])
    sGroupName1= str(CompanyData['Place.Name'][i])
    sGroupName2= str(CompanyData['Post.Code'][i])
    nLatitude=round(CompanyData['Latitude'][i],6)
    nLongitude=round(CompanyData['Longitude'][i],6)
```

Note   You can see here that you can store extra information on the nodes and edges, to enhance the data stored in the graph, while still not causing issues with the basic node-to-edge relationship.

You can add additional information by using the extension of the G.add_node() features.

```
CountryName=sGroupName0
print('Add Node:',sGroupName0)
G.add_node(CountryName,
          routertype='CountryName',
          group0=sGroupName0)

sPlaceName= sGroupName1 + '-' + sGroupName0
G.add_node(sPlaceName,
          routertype='PlaceName',
          group0=sGroupName0,
          group1=sGroupName1)

sPostCodeName= sGroupName1 + '-' + sGroupName2 + '-' + sGroupName0
print('Add Node:',sPostCodeName)
G.add_node(sPostCodeName,
          routertype='PostCode',
          group0=sGroupName0,
          group1=sGroupName1,
          group2=sGroupName2)

if nLatitude < 0:
    sLatitude = str(nLatitude*-1) + 'S'
else:
    sLatitude = str(nLatitude) + 'N'

if nLongitude < 0:
    sLongitude = str(nLongitude*-1) + 'W'
else:
    sLongitude = str(nLongitude) + 'E'

sGPS= sLatitude + '-' + sLongitude
print('Add Node:',sGPS)
G.add_node(sGPS,routertype='GPS',
          group0=sGroupName0,
          group1=sGroupName1,
```

```
                    group2=sGroupName2,
                    sLatitude=sLatitude,
                    sLongitude=sLongitude,
                    nLatitude=nLatitude,
                    nLongitude=nLongitude)
###################################################################
```

You use the extra information by using the extension of the G.node[ ][ ] features. Note, too, nodes_iter(G), which gives you access to all the nodes. We will now add the rules between the nodes.

```
print('###############################')
print('Link Country to Country')
print('###############################')
for n1 in nx.nodes_iter(G):
    if G.node[n1]['routertype'] == 'CountryName':
        for n2 in nx.nodes_iter(G):
            if G.node[n2]['routertype'] == 'CountryName':
                if n1 != n2:
                    print('Link:',n1,'to',n2)
                    G.add_edge(n1,n2)
print('###############################')

print('###############################')
print('Link Country to Place')
print('###############################')
for n1 in nx.nodes_iter(G):
    if G.node[n1]['routertype'] == 'CountryName':
        for n2 in nx.nodes_iter(G):
            if G.node[n2]['routertype'] == 'PlaceName':
                if G.node[n1]['group0'] == G.node[n2]['group0']:
                    if n1 != n2:
                        print('Link:',n1,'to',n2)
                        G.add_edge(n1,n2)
print('###############################')

print('###############################')
```

```
print('Link Place to Post Code')
print('##############################')
for n1 in nx.nodes_iter(G):
    if G.node[n1]['routertype'] == 'PlaceName':
        for n2 in nx.nodes_iter(G):
            if G.node[n2]['routertype'] == 'PostCode':
                if G.node[n1]['group0'] == G.node[n2]['group0']:
                    if G.node[n1]['group1'] == G.node[n2]['group1']:
                        if n1 != n2:
                            print('Link:',n1,'to',n2)
                            G.add_edge(n1,n2)
print('##############################')

print('##############################')
print('Link Post Code to GPS')
print('##############################')
for n1 in nx.nodes_iter(G):
    if G.node[n1]['routertype'] == 'PostCode':
        for n2 in nx.nodes_iter(G):
            if G.node[n2]['routertype'] == 'GPS':
                if G.node[n1]['group0'] == G.node[n2]['group0']:
                    if G.node[n1]['group1'] == G.node[n2]['group1']:
                        if G.node[n1]['group2'] == G.node[n2]['group2']:
                            if n1 != n2:
                                print('Link:',n1,'to',n2)
                                G.add_edge(n1,n2)
print('##############################')

print('##############################')
print("Nodes of graph:",nx.number_of_nodes(G))
print("Edges of graph:",nx.number_of_edges(G))
print('##############################')
#################################################################
sFileDir=Base + '/' + Company + '/02-Assess/01-EDS/02-Python'
if not os.path.exists(sFileDir):
    os.makedirs(sFileDir)
```

```
################################################################
sFileName=sFileDir + '/' + sOutputFileName
print('##############################')
print('Storing:', sFileName)
print('##############################')
nx.write_gml(G,sFileName)
sFileName=sFileName +'.gz'
nx.write_gml(G,sFileName)
################################################################
```

Save the `Assess-DAG-Schedule.py` file, then compile and execute with your Python compiler. This will produce a set of demonstrated values onscreen, plus a graph data file named `Assess-DAG-Schedule.gml`.

Tip   You can use a text editor to view the GML format. I use it many times to fault find graphs.

If you open the file with a text editor, you can see the format it produces. The format is simple but effective.

This is a post code node:

```
node [
    id 327
    label "Munich-80331-DE"
    routertype "PostCode"
    group0 "DE"
    group1 "Munich"
    group2 "80331"
  ]
```

This is a GPS node:

```
node [
    id 328
    label "48.1345 N-11.571 E"
    routertype "GPS"
```

```
    group0 "DE"
    group1 "Munich"
    group2 "80331"
    sLatitude "48.1345 N"
    sLongitude "11.571 E"
    nLatitude 48.134500000000003
    nLongitude 11.571
 ]
```

This is an edge connecting the two nodes:

```
edge [
    source 327
    target 328
]
```

This conversion to a graph data structure has now validated the comprehensive IP structure, by eliminating any duplicate node or edges that were created by the different rules or any duplicate entries in the base data. This natural de-duplication is highly useful when you have more than one system loading requests into the same system.

For example: Two transport agents can request a truck to transport two different boxes from location A to location B. Before the truck leaves, the logistics program will create a shipping graph and then send one truck with the two boxes.

I have proved that this works on the smaller IP core data. Now we can apply it to the larger IP data set called `Retrieve_IP_DATA.csv`.

With the additional scope of the larger data set, I suggest that we process the data, by altering the input and export source design to include the SQLite database, to assist with the resolution of the data processing. As a data scientist, you should constantly explore the accessible tools to develop the effectiveness and efficiency of your solutions. I will guide you through the following processing pattern, to enable you to understand where I am using the strengths of the database engine to resolve the complexity the larger data set requires to complete some of the relationship. We will match the power of the graph data structure with the processing capability of the SQLite database.

> **Warning**   If you have not yet used SQLite on your system, you must install it. See www.sqlite.org for details on how to achieve this installation. You can check the installation by using the sqlite3 version. If you get a result of 3.20.x, you are ready to proceed. Make sure that you also have the python SQLite library installed. Perform an import SQLite in your IPython console. If this does not work, perform a library installation.

Open your Python editor and create a file named Retrieve-IP_2_SQLite_2_DAG.py in directory C:\VKHCG\01-Vermeulen\02-Assess. Now copy the following code into the file:

```
###################################################################
# -*- coding: utf-8 -*-
###################################################################
import networkx as nx
import sys
import os
import sqlite3 as sq
import pandas as pd
###################################################################
if sys.platform == 'linux':
    Base=os.path.expanduser('~') + 'VKHCG'
else:
    Base='C:/VKHCG'
print('#############################')
print('Working Base:',Base,'using',sys.platform)
print('#############################')
###################################################################
sDatabaseName=Base + '/01-Vermeulen/00-RawData/SQLite/vermeulen.db'
conn = sq.connect(sDatabaseName)
###################################################################
sFileName=Base + '/01-Vermeulen/01-Retrieve/01-EDS/02-Python/Retrieve_IP_
DATA.csv'
sOutputFileName='Assess-DAG-Schedule-All.csv'
Company='01-Vermeulen'
```

Now we load the base data into memory as a pandas data structure, and then store it into a SQLite database using the df.tosql() command.

```
################################################################
print('Loading :',sFileName)
IP_DATA_ALL=pd.read_csv(sFileName,header=0,low_memory=False,
encoding="latin-1")
IP_DATA_ALL.index.names = ['RowIDCSV']
IP_DATA_ALL.rename(columns={'Place.Name': 'PlaceName'}, inplace=True)
IP_DATA_ALL.rename(columns={'Post.Code': 'PostCode'}, inplace=True)
#print(IP_DATA_ALL)
print('###############')
sTable='Assess_IP_DATA'
print('Storing:',sDatabaseName,'Table:',sTable)
IP_DATA_ALL.to_sql(sTable, conn, if_exists="replace")
print('###############')
```

We create a non-directed graph, using G=nx.Graph().

```
################################################################
G=nx.Graph()
################################################################
```

We now create a subset of the loaded data from SQLite, to match the formats you require.

Tip   This offloading from the processing to a third-party processing engine enables you to connect to core resources, such as databases to then perform the complex joins and heavier processing. I normally also offload my graph processing in the same way, via DSE Graph, to enhance more complex graphs.

To be able to successfully offload complex data processing is a good skill for a data scientist to have.

Note   I can use a small computer, such as a Raspberry Pi 3, to drive my clusters to perform massive processing tasks with ease.

So, let's restart our example. You must return the distinct list of countries from a data set that has duplicates. I suggest that you use the SQL DISTINCT command to perform the task, via the SQLite data engine.

```
print('###############')
sTable = 'Assess_IP_DATA'
print('Loading:',sDatabaseName,'Table:',sTable)
sSQL="select distinct"
sSQL=sSQL+ "A.Country,"
sSQL=sSQL+ "A.Country AS NodeName,"
sSQL=sSQL+ "A.Country AS GroupName0,"
sSQL=sSQL+ "'Country-Router' AS RouterType"
sSQL=sSQL+ "from"
sSQL=sSQL+ "Assess_IP_DATA as A"
sSQL=sSQL+ "ORDER BY A.Country;"
CompanyData=pd.read_sql_query(sSQL, conn)
print('###############')
```

You now have the distinct list of countries in CompanyData. Well done.

You now need to add these nodes to the graph you already created earlier. You have completed this task before, so just follow this code:

```
for i in range(CompanyData.shape[0]):
    sNode=str(CompanyData['NodeName'][i])
    sRouterType=str(CompanyData['RouterType'][i])
    sGroupName0=str(CompanyData['GroupName0'][i])
    G.add_node(sNode,
               routertype=sRouterType,
               group0=sGroupName0)

print('###############################')
print("Nodes of graph:",nx.number_of_nodes(G))
print("Edges of graph:",nx.number_of_edges(G))
print('###############################')
```

Well done, you are making progress here. You have all the country nodes.

I suggest we store the list of countries for later as table Assess_IP_Country.

```
print('###############')
sTable='Assess_IP_Country'
print('Storing:',sDatabaseName,'Table:',sTable)
CompanyData.to_sql(sTable, conn, if_exists="replace")
```

Now that you have the basic process, you should find this easy to repeat. Next, extract the distinct country and place, add the nodes, and store this as table Assess_IP_ PlaceName.

```
print('###############')
###############################################################
print('###############')
sTable = 'Assess_IP_DATA'
print('Loading:',sDatabaseName,'Table:',sTable)
sSQL="select distinct"
sSQL=sSQL+ "A.Country,"
sSQL=sSQL+ "A.PlaceName,"
sSQL=sSQL+ "A.PlaceName || '-' || A.Country AS NodeName,"
sSQL=sSQL+ "A.Country AS GroupName0,"
sSQL=sSQL+ "A.PlaceName AS GroupName1,"
sSQL=sSQL+ "'Place-Router' AS RouterType"
sSQL=sSQL+ "from"
sSQL=sSQL+ "Assess_IP_DATA as A"
sSQL=sSQL+ "ORDER BY A.Country AND"
sSQL=sSQL+ "A.PlaceName;"
CompanyData=pd.read_sql_query(sSQL, conn)
print('###############')

for i in range(CompanyData.shape[0]):
    sNode=str(CompanyData['NodeName'][i])
    sRouterType=str(CompanyData['RouterType'][i])
    sGroupName0= str(CompanyData['Country'][i])
    sGroupName1= str(CompanyData['PlaceName'][i])

    G.add_node(sNode,
                routertype=sRouterType,
                group0=sGroupName0,
                group1=sGroupName1)
```

```
print('###############################')
print("Nodes of graph:",nx.number_of_nodes(G))
print("Edges of graph:",nx.number_of_edges(G))
print('###############################')

print('###############')
sTable='Assess_IP_PlaceName'
print('Storing:',sDatabaseName,'Table:',sTable)
CompanyData.to_sql(sTable, conn, if_exists="replace")
print('###############')
#################################################################
```

You've now realized more success with the process. Can you now see the advantage of using the database to perform the work?

Next, repeat for country, place-name, and node name data selected. Add nodes and store table as Assess_IP_PostCode.

```
print('###############')
sTable = 'Assess_IP_DATA'
print('Loading:',sDatabaseName,'Table:',sTable)
sSQL="select distinct"
sSQL=sSQL+ "A.Country,"
sSQL=sSQL+ "A.PlaceName,"
sSQL=sSQL+ "A.PostCode,"
sSQL=sSQL+ "A.PlaceName || '-' || A.PostCode || '-' || A.Country AS
NodeName,"
sSQL=sSQL+ "A.Country AS GroupName0,"
sSQL=sSQL+ "A.PlaceName AS GroupName1,"
sSQL=sSQL+ "A.PostCode AS GroupName2,"
sSQL=sSQL+ "'Place-Router' AS RouterType"
sSQL=sSQL+ "from"
sSQL=sSQL+ "Assess_IP_DATA as A"
sSQL=sSQL+ "ORDER BY A.Country AND"
sSQL=sSQL+ "A.PlaceName AND"
sSQL=sSQL+ "A.PostCode;"
CompanyData=pd.read_sql_query(sSQL, conn)
print('###############')
```

```
for i in range(CompanyData.shape[0]):
    sNode=str(CompanyData['NodeName'][i])
    sRouterType=str(CompanyData['RouterType'][i])
    sGroupName0= str(CompanyData['GroupName0'][i])
    sGroupName1= str(CompanyData['GroupName1'][i])
    sGroupName2= str(CompanyData['GroupName2'][i])

    G.add_node(sNode,
                routertype=sRouterType,
                group0=sGroupName0,
                group1=sGroupName1,
                group2=sGroupName2)
print('###############################')
print("Nodes of graph:",nx.number_of_nodes(G))
print("Edges of graph:",nx.number_of_edges(G))
print('###############################')

print('###############')
sTable='Assess_IP_PostCode'
print('Storing:',sDatabaseName,'Table:',sTable)
CompanyData.to_sql(sTable, conn, if_exists="replace")
print('###############')
################################################################
```

Next, repeat for country, place-name, post code, latitude, and longitude data selected. Add nodes and store table as Assess_IP_GPS.

```
print('###############')
sTable = 'Assess_IP_DATA'
print('Loading:',sDatabaseName,'Table:',sTable)
sSQL="select distinct"
sSQL=sSQL+ "A.Country,"
sSQL=sSQL+ "A.PlaceName,"
sSQL=sSQL+ "A.PostCode,"
sSQL=sSQL+ "A.Latitude,"
sSQL=sSQL+ "A.Longitude,"
```

```
sSQL=sSQL+ "(CASE WHEN A.Latitude < O THEN"
sSQL=sSQL+ "'S' || ABS(A.Latitude)"
sSQL=sSQL+ "ELSE "
sSQL=sSQL+ "'N' || ABS(A.Latitude)"
sSQL=sSQL+ "END ) AS sLatitude,"

sSQL=sSQL+ "(CASE WHEN A.Longitude < O THEN"
sSQL=sSQL+ "'W' || ABS(A.Longitude)"
sSQL=sSQL+ "ELSE"
sSQL=sSQL+ "'E' || ABS(A.Longitude)"
sSQL=sSQL+ "END ) AS sLongitude"
sSQL=sSQL+ "from"
sSQL=sSQL+ "Assess_IP_DATA as A;"
CompanyData=pd.read_sql_query(sSQL, conn)
print('################')

sTable='Assess_IP_GPS'
CompanyData.to_sql(sTable, conn, if_exists="replace")
sSQL="select distinct"
sSQL=sSQL+ "A.Country,"
sSQL=sSQL+ "A.PlaceName,"
sSQL=sSQL+ "A.PostCode,"
sSQL=sSQL+ "A.Latitude,"
sSQL=sSQL+ "A.Longitude,"
sSQL=sSQL+ "A.sLatitude,"
sSQL=sSQL+ "A.sLongitude,"
sSQL=sSQL+ "A.sLatitude || '-' || A.sLongitude AS NodeName,"
sSQL=sSQL+ "A.Country AS GroupNameO,"
sSQL=sSQL+ "A.PlaceName AS GroupName1,"
sSQL=sSQL+ "A.PostCode AS GroupName2,"
sSQL=sSQL+ "'GPS-Client' AS RouterType"
sSQL=sSQL+ "from"
sSQL=sSQL+ "Assess_IP_GPS as A"
sSQL=sSQL+ "ORDER BY A.Country AND"
sSQL=sSQL+ "A.PlaceName AND"
sSQL=sSQL+ "A.PostCode AND"
```

```
sSQL=sSQL+ "A.Latitude AND"
sSQL=sSQL+ "A.Longitude;"
CompanyData=pd.read_sql_query(sSQL, conn)

for i in range(CompanyData.shape[0]):
    sNode=str(CompanyData['NodeName'][i])
    sRouterType=str(CompanyData['RouterType'][i])
    sGroupName0= str(CompanyData['GroupName0'][i])
    sGroupName1= str(CompanyData['GroupName1'][i])
    sGroupName2= str(CompanyData['GroupName2'][i])
    nLatitude=round(CompanyData['Latitude'][i],6)
    nLongitude=round(CompanyData['Longitude'][i],6)
    sLatitude= str(CompanyData['sLatitude'][i])
    sLongitude= str(CompanyData['sLongitude'][i])

    G.add_node(sNode,
                routertype=sRouterType,
                group0=sGroupName0,
                group1=sGroupName1,
                group2=sGroupName2,
                sLatitude=sLatitude,
                sLongitude=sLongitude,
                nLatitude=nLatitude,
                nLongitude=nLongitude)

print('################################')
print("Nodes of graph:",nx.number_of_nodes(G))
print("Edges of graph:",nx.number_of_edges(G))
print('################################')

print('################')
sTable='Assess_IP_GPS'
print('Storing:',sDatabaseName,'Table:',sTable)
CompanyData.to_sql(sTable, conn, if_exists="replace")
print('################')
################################################################
```

Well done. You have created all the nodes. Now you will see the true power of using the database.

We must link every country with every other country but never connect the country back to itself. This task would have been complex work without the tables we created while loading the nodes. Now we can perform a join between Assess_IP_Country twice and add an enhancement in the selection, performing only one direction edge creation, by adding the extra requirement of one country being smaller than the other country. The use of the Graph() type graph supplies the add_edge() link in the opposite direction for free. Hence, we have a +/- 50% saving in processing effort. So, proceed with the example to link the different types of nodes.

```
print('##############################')
print('Link Country to Country')
print('##############################')
print('################')
print('Loading:',sDatabaseName,'Table:',sTable)
sSQL="select distinct"
sSQL=sSQL+ "A.NodeName as N1,"
sSQL=sSQL+ "B.NodeName as N2"
sSQL=sSQL+ "from"
sSQL=sSQL+ "Assess_IP_Country as A,"
sSQL=sSQL+ "Assess_IP_Country as B"
sSQL=sSQL+ "WHERE "
sSQL=sSQL+ "A.Country < B.Country AND"
sSQL=sSQL+ "A.NodeName <> B.NodeName;"
CompanyData=pd.read_sql_query(sSQL, conn)
print('################')

    n1= str(CompanyData['N1'][i])
    n2= str(CompanyData['N2'][i])

    print('Link Country:',n1,'to Country:',n2)
    G.add_edge(n1,n2)
print('##############################')

print('##############################')
print("Nodes of graph:",nx.number_of_nodes(G))
```

```python
print("Edges of graph:",nx.number_of_edges(G))
print('###############################')
################################################################
print('###############################')
print('Link Country to Place')
print('###############################')
print('###############')
print('Loading:',sDatabaseName,'Table:',sTable)
sSQL="select distinct"
sSQL=sSQL+ "A.NodeName as N1,"
sSQL=sSQL+ "B.NodeName as N2"
sSQL=sSQL+ "from"
sSQL=sSQL+ "Assess_IP_Country as A,"
sSQL=sSQL+ "Assess_IP_PlaceName as B"
sSQL=sSQL+ "WHERE "
sSQL=sSQL+ "A.Country = B.Country AND"
sSQL=sSQL+ "A.NodeName <> B.NodeName;"
CompanyData=pd.read_sql_query(sSQL, conn)
print('###############')

for i in range(CompanyData.shape[0]):
    n1= str(CompanyData['N2'][i])
    n2= str(CompanyData['N2'][i])

    print('Link Country:',n1,'to Place:',n2)
    G.add_edge(n1,n2)
print('###############################')

print('###############################')
print("Nodes of graph:",nx.number_of_nodes(G))
print("Edges of graph:",nx.number_of_edges(G))
print('###############################')
################################################################
print('###############################')
print('Link Place to Post Code')
print('###############################')
print('###############')
```

```
print('Loading:',sDatabaseName,'Table:',sTable)
sSQL="select distinct"
sSQL=sSQL+ "A.NodeName as N1,"
sSQL=sSQL+ "B.NodeName as N2"
sSQL=sSQL+ "from"
sSQL=sSQL+ "Assess_IP_PlaceName as A,"
sSQL=sSQL+ "Assess_IP_PostCode as B"
sSQL=sSQL+ "WHERE "
sSQL=sSQL+ "A.Country = B.Country AND"
sSQL=sSQL+ "A.PlaceName = B.PlaceName AND"
sSQL=sSQL+ "A.NodeName <> B.NodeName;"
CompanyData=pd.read_sql_query(sSQL, conn)
print('###############')

for i in range(CompanyData.shape[0]):
    n1= str(CompanyData['N1'][i])
    n2= str(CompanyData['N2'][i])

    print('Link Place:',n1,'to Post Code:',n2)
    G.add_edge(n1,n2)
print('################################')

print('################################')
print("Nodes of graph:",nx.number_of_nodes(G))
print("Edges of graph:",nx.number_of_edges(G))
print('################################')
####################################################################
print('################################')
print('Link Post Code to GPS')
print('################################')
print('###############')
print('Loading:',sDatabaseName,'Table:',sTable)
sSQL="select distinct"
sSQL=sSQL+ "A.NodeName as N1,"
sSQL=sSQL+ "B.NodeName as N2"
sSQL=sSQL+ "from"
sSQL=sSQL+ "Assess_IP_PostCode as A,"
```

```
sSQL=sSQL+ "Assess_IP_GPS as B"
sSQL=sSQL+ "WHERE "
sSQL=sSQL+ "A.Country = B.Country AND"
sSQL=sSQL+ "A.PlaceName = B.PlaceName AND"
sSQL=sSQL+ "A.PostCode = B.PostCode AND"
sSQL=sSQL+ "A.NodeName <> B.NodeName;"
CompanyData=pd.read_sql_query(sSQL, conn)
print('###############')

for i in range(CompanyData.shape[0]):
    n1= str(CompanyData['N1'][i])
    n2= str(CompanyData['N2'][i])

    print('Link Post Code:',n1,'to GPS:',n2)
    G.add_edge(n1,n2)
print('#############################')

print('#############################')
print("Nodes of graph:",nx.number_of_nodes(G))
print("Edges of graph:",nx.number_of_edges(G))
print('#############################')
```

Congratulations! You have mastered the graph and database consolidation techniques. Now, save your graph into the Assess-DAG-Schedule-All.gml file.

```
################################################################
sFileDir=Base + '/' + Company + '/02-Assess/01-EDS/02-Python'
if not os.path.exists(sFileDir):
    os.makedirs(sFileDir)
################################################################
sFileName=sFileDir + '/' + sOutputFileName
print('#############################')
print('Storing:',sFileName)
print('#############################')
nx.write_gml(G,sFileName)
sFileName=sFileName +'.gz'
nx.write_gml(G,sFileName)
```

```
##################################################################
##################################################################
print('### Done!! #######################################')
##################################################################
```

Save the file Retrieve-IP_2_SQLite_2_DAG.py, then compile and execute with your Python compiler. This will produce a set of demonstrated values onscreen, plus a graph data file named Assess-DAG-Schedule-All.gml.

Note   You can view the GML file with a text editor.

This produces nodes and edges. This type of analysis is possible in a graph environment, as it is a close-to-reality match with the tangible business problem it is modeling. Graph theory is always a useful tool to use when relationships between business entities require analyzing.

I have now completed the assess step for Vermeulen PLC. At this point, you should be comfortable loading data from files and databases. You can perform SQL requests that offload the work to the database engine. You must, however, understand the format of a graph and how to construct the rules into the graph your customers require.

Tip   If you can draw it on a whiteboard, you can make it a graph.

Debugging Tip   You can open the GML format and use it to draw the nodes on a whiteboard. This is because there are only nodes and edges in the format.

The following two commands will remove nodes (G.remove_node()) and edges (remove_edge(,)) from an existing graph. So, if you want to enforce rules that state you cannot attach two specific nodes, you can simply remove those by performing a remove action.

You can also use graphs to perform what-if scenarios, such as the following: What if we sell the New York office; what will be the impact on our network schedules? The simple action required is to load the graph, remove the node, and save it, as follows:

```
G=nx.read_gml(<location of saved graph>)
G.remove_node('New York')
nx.write_gml(G,(<location of new saved graph>)
```

Any schedules attached to New York are now removed.

If you understand the basic concepts graphs and databases, I suggest that you get some refreshment, and then we can progress to the next set of examples.

Krennwallner AG

Krennwallner AG is the German company that handles all our media business. The company will be used to demonstrate three basic assessment solutions.

Picking Content for Billboards

Krennwallner is responsible for publicizing the customer's advertisements on a set of billboards that the group owns. I will explain the concepts that are required to assess the information relating to billboards that we must process.

You will use the `sqlite3` and `pandas` packages to create a solution. You performed similar actions related to Vermeulen, so the following should be an example that you can handle. The basic process required is to combine two sets of data and then calculate the number of visitors per day from the range of IP addresses that access the billboards in Germany.

Open your Python editor and create a file named `Assess-DE-Billboard.py` in directory `C:\VKHCG\02-Krennwallner\02-Assess`. Now copy the following code into the file:

```
###############################################################
import sys
import os
import sqlite3 as sq
import pandas as pd
###############################################################
```

```
if sys.platform == 'linux':
    Base=os.path.expanduser('~') + 'VKHCG'
else:
    Base='C:/VKHCG'
print('##############################')
print('Working Base:',Base,'using',sys.platform)
print('##############################')
##################################################################
```

The two data sources we must combine are listed following:

```
sInputFileName1='01-Retrieve/01-EDS/02-Python/Retrieve_DE_Billboard_
Locations.csv'
sInputFileName2='01-Retrieve/01-EDS/02-Python/Retrieve_Online_Visitor.csv'
```

The results will be stored in this output file:

```
sOutputFileName='Assess-DE-Billboard-Visitor.csv'
Company='02-Krennwallner'
##################################################################
sDataBaseDir=Base + '/' + Company + '/02-Assess/SQLite'
if not os.path.exists(sDataBaseDir):
    os.makedirs(sDataBaseDir)
##################################################################
```

The database that will support the processing for this example follows:

```
sDatabaseName=sDataBaseDir + '/krennwallner.db'
conn = sq.connect(sDatabaseName)
##################################################################
### Import Billboard Data
##################################################################
sFileName=Base + '/' + Company + '/' + sInputFileName1
print('##############################')
print('Loading:',sFileName)
print('##############################')
```

Load the first file and clean up the duplicates.

```
BillboardRawData=pd.read_csv(sFileName,header=0,low_memory=False,
encoding="latin-1")
BillboardRawData.drop_duplicates(subset=None, keep='first', inplace=True)
BillboardData=BillboardRawData
print('Loaded Company:',BillboardData.columns.values)
print('#############################')
##################################################################
```

Store the cleanly loaded data into SQLite.

```
print('###############')
sTable='Assess_BillboardData'
print('Storing:',sDatabaseName,'Table:',sTable)
BillboardData.to_sql(sTable, conn, if_exists="replace")
print('###############')
##################################################################
print(BillboardData.head())
print('#############################')
print('Rows:',BillboardData.shape[0])
print('#############################')
```

Load the second file and clean up the duplicates.

```
##################################################################
### Import Billboard Data
##################################################################
sFileName=Base + '/' + Company + '/' + sInputFileName2
print('#############################')
print('Loading:',sFileName)
print('#############################')
VisitorRawData=pd.read_csv(sFileName,header=0,low_memory=False,
encoding="latin-1")
VisitorRawData.drop_duplicates(subset=None, keep='first', inplace=True)
```

With the country=DE result, you can now limit the data to be Germany only. This early discarding of superfluous data is a big contributor to effective data science.

Tip   Keep the data set under investigation to the minimum size required. This saves you and your customers time and money.

```
VisitorData=VisitorRawData[VisitorRawData.Country=='DE']
print('Loaded Company:',VisitorData.columns.values)
print('###############################')
##################################################################
```

Store the second set of cleanly loaded data into SQLite.

```
print('################')
sTable='Assess_VisitorData'
print('Storing:',sDatabaseName,'Table:',sTable)
VisitorData.to_sql(sTable, conn, if_exists="replace")
print('################')
##################################################################
print(VisitorData.head())
print('###############################')
print('Rows:',VisitorData.shape[0])
print('###############################')
##################################################################
```

Retrieve an indirectly cleaned data set from SQLite, because you stored the clean data. The reduction in the second data set now also improves the process, with increased processing efficiencies.

```
print('################')
sTable='Assess_BillboardVisitorData'
print('Loading:',sDatabaseName,'Table:',sTable)
sSQL="select distinct"
sSQL=sSQL+ "A.Country AS BillboardCountry,"
sSQL=sSQL+ "A.Place_Name AS BillboardPlaceName,"
sSQL=sSQL+ "A.Latitude AS BillboardLatitude,"
sSQL=sSQL+ "A.Longitude AS BillboardLongitude,"
sSQL=sSQL+ "B.Country AS VisitorCountry,"
sSQL=sSQL+ "B.Place_Name AS VisitorPlaceName,"
```

```
sSQL=sSQL+ "B.Latitude AS VisitorLatitude,"
sSQL=sSQL+ "B.Longitude AS VisitorLongitude,"
sSQL=sSQL+ "(B.Last_IP_Number - B.First_IP_Number) * 365.25 * 24 * 12 AS
VisitorYearRate"
sSQL=sSQL+ "from"
sSQL=sSQL+ "Assess_BillboardData as A"
sSQL=sSQL+ "JOIN"
sSQL=sSQL+ "Assess_VisitorData as B"
sSQL=sSQL+ "ON"
sSQL=sSQL+ "A.Country = B.Country"
sSQL=sSQL+ "AND"
sSQL=sSQL+ "A.Place_Name = B.Place_Name;"
BillboardVistorsData=pd.read_sql_query(sSQL, conn)
print('###############')
################################################################
```

Store the combined set of data back into SQLite.

```
print('###############')
sTable='Assess_BillboardVistorsData'
print('Storing:',sDatabaseName,'Table:',sTable)
BillboardVistorsData.to_sql(sTable, conn, if_exists="replace")
print('###############')
################################################################
print(BillboardVistorsData.head())
print('#############################')
print('Rows:',BillboardVistorsData.shape[0])
print('#############################')
################################################################
sFileDir=Base + '/' + Company + '/02-Assess/01-EDS/02-Python'
if not os.path.exists(sFileDir):
    os.makedirs(sFileDir)
################################################################
print('#############################')
print('Storing:', sFileName)
print('#############################')
```

```
sFileName=sFileDir + '/' + sOutputFileName
BillboardVistorsData.to_csv(sFileName, index = False)
print('###############################')
#####################################################################
print('### Done!! #######################################')
#####################################################################
```

Save the Assess-DE-Billboard.py, then compile and execute with your Python compiler. This will produce a set of demonstrated values onscreen, plus a data file named Assess-DE-Billboard-Visitor.csv.

Billboard Country	Billboard PlaceName	Billboard Latitude	Billboard Longitude	Visitor Country	Visitor PlaceName	Visitor Latitude	Visitor Longitude	Visitor YearRate
DE	Lake	51.7833	8.5667	DE	Lake	51.7833	8.5667	26823960
DE	Horb	48.4333	8.6833	DE	Horb	48.4333	8.6833	26823960
DE	Horb	48.4333	8.6833	DE	Horb	48.4333	8.6833	53753112

You should now understand that you can offload the SQL to the SQLite database, to gain easy access to the data processing power of that the SQL system enables.

We will now investigate the next example.

Understanding Your Online Visitor Data

Online visitors have to be mapped to their closest billboard, to ensure we understand where and what they can access. To achieve this, I will guide you through a graph data processing example, to link the different entities involved in a graph of the visitor activity.

The data that we have, however, is stored with some issues.

- *The billboard names are missing.* With some feature engineering, we can infer the values from the longitude and latitude values.

- *The distance between billboard and visitor is unknown.* With some feature engineering, via the Vincenty's formulae, this can be found.

- *The longitude and latitude requires smoothing, to comply with the billboard naming formatting agreement.*

What are Vincenty's formulae? Thaddeus Vincenty's formulae are two related iterative methods used in geodesy to calculate the distance between two points on the surface of a spheroid. They assume that the true shape of Earth is an oblate spheroid and, therefore, are more accurate than methods that assume a spherical Earth. The distance is called the great-circle distance. (See Figure 8-11)

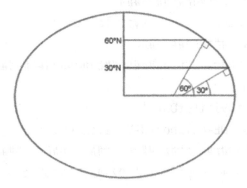

Figure 8-11. *Thaddeus Vincenty's formulae in graphic form*

Tip   In data science, it is always sensible to use proven techniques that have the backing of other experts. Thaddeus Vincenty is an expert in calculating distances on Earth. The geopy library also has a proven Vincenty calculator. The wise data scientist uses these proven formulae, as they give your work additional credibility.

You will use the networkx, sqlite3, pandas, and geopy packages to create a solution. Open your Python editor and create a file called Assess-Billboard_2_Visitor.py in directory C:\VKHCG\ 02-Krennwallner\02-Assess. Now copy the following code into the file:

```
###############################################################
# -*- coding: utf-8 -*-
###############################################################
import networkx as nx
import sys
import os
import sqlite3 as sq
import pandas as pd
from geopy.distance import vincenty
```

```
################################################################
if sys.platform == 'linux':
    Base=os.path.expanduser('~') + 'VKHCG'
else:
    Base='C:/VKHCG'
print('###############################')
print('Working Base:',Base,'using',sys.platform)
print('###############################')
################################################################
Company='02-Krennwallner'
sTable='Assess_BillboardVisitorData'
sOutputFileName='Assess-DE-Billboard-Visitor.gml'
################################################################
sDataBaseDir=Base + '/' + Company + '/02-Assess/SQLite'
if not os.path.exists(sDataBaseDir):
    os.makedirs(sDataBaseDir)
################################################################
sDatabaseName=sDataBaseDir + '/krennwallner.db'
conn = sq.connect(sDatabaseName)
################################################################
print('###############')
print('Loading:',sDatabaseName,'Table:',sTable)
sSQL="select"
sSQL=sSQL+ "A.BillboardCountry,"
sSQL=sSQL+ "A.BillboardPlaceName,"
```

We are smoothing the data using a round(x, 3) formula. This ensures that there is no billboard that has nonstandard numbers.

```
sSQL=sSQL+ "ROUND(A.BillboardLatitude,3) AS BillboardLatitude,"
sSQL=sSQL+ "ROUND(A.BillboardLongitude,3) AS BillboardLongitude,"

sSQL=sSQL+ "(CASE WHEN A.BillboardLatitude < 0 THEN"
sSQL=sSQL+ "'S' || ROUND(ABS(A.BillboardLatitude),3)"
sSQL=sSQL+ "ELSE "
sSQL=sSQL+ "'N' || ROUND(ABS(A.BillboardLatitude),3)"
sSQL=sSQL+ "END ) AS sBillboardLatitude,"
```

```
sSQL=sSQL+ "(CASE WHEN A.BillboardLongitude < 0 THEN"
sSQL=sSQL+ "'W' || ROUND(ABS(A.BillboardLongitude),3)"
sSQL=sSQL+ "ELSE "
sSQL=sSQL+ "'E' || ROUND(ABS(A.BillboardLongitude),3)"
sSQL=sSQL+ "END ) AS sBillboardLongitude,"

sSQL=sSQL+ "A.VisitorCountry,"
sSQL=sSQL+ "A.VisitorPlaceName,"
sSQL=sSQL+ "ROUND(A.VisitorLatitude,3) AS VisitorLatitude,"
sSQL=sSQL+ "ROUND(A.VisitorLongitude,3) AS VisitorLongitude,"

sSQL=sSQL+ "(CASE WHEN A.VisitorLatitude < 0 THEN"
sSQL=sSQL+ "'S' || ROUND(ABS(A.VisitorLatitude),3)"
sSQL=sSQL+ "ELSE "
sSQL=sSQL+ "'N' ||ROUND(ABS(A.VisitorLatitude),3)"
sSQL=sSQL+ "END ) AS sVisitorLatitude,"

sSQL=sSQL+ "(CASE WHEN A.VisitorLongitude < 0 THEN"
sSQL=sSQL+ "'W' || ROUND(ABS(A.VisitorLongitude),3)"
sSQL=sSQL+ "ELSE "
sSQL=sSQL+ "'E' || ROUND(ABS(A.VisitorLongitude),3)"
sSQL=sSQL+ "END ) AS sVisitorLongitude,"

sSQL=sSQL+ "A.VisitorYearRate"
sSQL=sSQL+ "from"
sSQL=sSQL+ "Assess_BillboardVistorsData AS A;"
BillboardVistorsData=pd.read_sql_query(sSQL, conn)
print('###############')


################################################################
```

With this single command, we apply Vincenty's formulae, and we are sure we can support the result in miles.

```
BillboardVistorsData['Distance']=BillboardVistorsData.apply(lambda row:
    round(
    vincenty((row['BillboardLatitude'],row['BillboardLongitude']),
             (row['VisitorLatitude'],row['VisitorLongitude'])).miles
          ,4)
      ,axis=1)
```

347

We now use the newly calculated distance, plus the billboard naming fix, to generate a graph of the billboards and visitors.

```
###############################################################
G=nx.Graph()
###############################################################

for i in range(BillboardVistorsData.shape[0]):
    sNode0='MediaHub-' + BillboardVistorsData['BillboardCountry'][i]

    sNode1='B-'+ BillboardVistorsData['sBillboardLatitude'][i] + '-'
    sNode1=sNode1 + BillboardVistorsData['sBillboardLongitude'][i]
    G.add_node(sNode1,
               Nodetype='Billboard',
               Country=BillboardVistorsData['BillboardCountry'][i],
               PlaceName=BillboardVistorsData['BillboardPlaceName'][i],
               Latitude=round(BillboardVistorsData['BillboardLatitude']
               [i],3),
               Longitude=round(BillboardVistorsData['BillboardLongitude']
               [i],3))

    sNode2='M-'+ BillboardVistorsData['sVisitorLatitude'][i] + '-'
    sNode2=sNode2 + BillboardVistorsData['sVisitorLongitude'][i]
    G.add_node(sNode2,
               Nodetype='Mobile',
               Country=BillboardVistorsData['VisitorCountry'][i],
               PlaceName=BillboardVistorsData['VisitorPlaceName'][i],
               Latitude=round(BillboardVistorsData['VisitorLatitude']
               [i],3),
               Longitude=round(BillboardVistorsData['VisitorLongitude']
               [i],3))

    print('Link Media Hub:',sNode0,'to Billboard:',sNode1)
    G.add_edge(sNode0,sNode1)

    print('Link Post Code:',sNode1,'to GPS:',sNode2)
```

```
G.add_edge(sNode1,sNode2,distance=round(BillboardVistorsData
['Distance'][i]))

################################################################
print('##############################')
print("Nodes of graph:",nx.number_of_nodes(G))
print("Edges of graph:",nx.number_of_edges(G))
print('##############################')
```

Well done. You have a good graph that you can now store for future queries.

```
################################################################
sFileDir=Base + '/' + Company + '/02-Assess/01-EDS/02-Python'
if not os.path.exists(sFileDir):
    os.makedirs(sFileDir)
################################################################
sFileName=sFileDir + '/' + sOutputFileName
print('##############################')
print('Storing:', sFileName)
print('##############################')
nx.write_gml(G,sFileName)
sFileName=sFileName +'.gz'
nx.write_gml(G,sFileName)
################################################################
################################################################
print('### Done!! ##########################################')
################################################################
```

Save the file Assess-Billboard_2_Visitor.py, then compile and execute with your Python compiler. This will produce a set of demonstrated values onscreen, plus a graph data file named Assess-DE-Billboard-Visitor.gml.

Note   You can view the GML file with a text editor.

Summary of Your Achievements

- You can use your own formulae to extract features, such as billboard names.

- You can successfully use other people's formulae to extract features, such as the distance between the billboard and the visitor.

- You can apply a single formula onto a data set in memory, using the `df.apply(lambda row)` technique.

Planning an Event for Top-Ten Customers

It is necessary to generate on a regular basis a preapproved view of the data you are assessing. I will guide you through a process called processing offloading, to find the top-ten customers. I will also show you how to use SQLite to offload processing, using views. The result is that the heavy processing is performed by the database engine. This is also useful when running long-running processes.

Tip   This is the way you would offload processing to systems such as Spark, Hadoop, or Cassandra.

The next examples will demonstrate how you can permanently offload processing to the database engine, by creating views. Once created, you can, in future, use these views for other data science projects.

You will use the `sqlite3` and `pandas` packages to create a solution. Open your Python editor and create a file named `Assess-Visitors.py` in directory `C:\VKHCG\` `02-Krennwallner\02-Assess`.

Now copy the following code into the file:

```
################################################################
import sys
import os
import sqlite3 as sq
import pandas as pd
from pandas.io import sql
################################################################
```

```
if sys.platform == 'linux':
    Base=os.path.expanduser('~') + 'VKHCG'
else:
    Base='C:/VKHCG'
print('################################')
print('Working Base:',Base,'using',sys.platform)
print('################################')
################################################################
Company='02-Krennwallner'
sInputFileName='01-Retrieve/01-EDS/02-Python/Retrieve_Online_Visitor.csv'
################################################################
sDataBaseDir=Base + '/' + Company + '/02-Assess/SQLite'
if not os.path.exists(sDataBaseDir):
    os.makedirs(sDataBaseDir)
################################################################
sDatabaseName=sDataBaseDir + '/krennwallner.db'
conn = sq.connect(sDatabaseName)
################################################################
### Import Country Data
################################################################
```

At this point, we load the data into memory.

```
sFileName=Base + '/' + Company + '/' + sInputFileName
print('################################')
print('Loading:',sFileName)
print('################################')
VisitorRawData=pd.read_csv(sFileName,
                           header=0,
                           low_memory=False,
                           encoding="latin-1",
                           skip_blank_lines=True)
VisitorRawData.drop_duplicates(subset=None, keep='first', inplace=True)
VisitorData=VisitorRawData
print('Loaded Company:',VisitorData.columns.values)
print('################################')
################################################################
```

At this point. we store the data from memory into a database.

```
print('###############')
sTable='Assess_Visitor'
print('Storing:',sDatabaseName,'Table:',sTable)
VisitorData.to_sql(sTable, conn, if_exists="replace")
print('###############')
################################################################
print(VisitorData.head())
print('##############################')
print('Rows:',VisitorData.shape[0])
print('##############################')
################################################################
```

We store the first set of data rules into the database at this point.

```
print('###############')
sView='Assess_Visitor_UseIt'
print('Creating:',sDatabaseName,'View:',sView)
sSQL="DROP VIEW IF EXISTS"+ sView + ";"
sql.execute(sSQL,conn)

sSQL="CREATE VIEW "+ sView + "AS"
sSQL=sSQL+ "SELECT"
sSQL=sSQL+ "A.Country,"
sSQL=sSQL+ "A.Place_Name,"
sSQL=sSQL+ "A.Latitude,"
sSQL=sSQL+ "A.Longitude,"
sSQL=sSQL+ "(A.Last_IP_Number - A.First_IP_Number) AS UsesIt"
sSQL=sSQL+ "FROM"
sSQL=sSQL+ "Assess_Visitor as A"
sSQL=sSQL+ "WHERE"
sSQL=sSQL+ "Country is not null"
sSQL=sSQL+ "AND"
sSQL=sSQL+ "Place_Name is not null;"
sql.execute(sSQL,conn)
################################################################
```

At this point, we store the second set of data rules into the database.

```
print('################')
sView='Assess_Total_Visitors_Location'
print('Creating:',sDatabaseName,'View:',sView)
sSQL="DROP VIEW IF EXISTS"+ sView + ";"
sql.execute(sSQL,conn)

sSQL="CREATE VIEW"+ sView + "AS"
sSQL=sSQL+ "SELECT"
sSQL=sSQL+ "Country,"
sSQL=sSQL+ "Place_Name,"
sSQL=sSQL+ "SUM(UsesIt) AS TotalUsesIt"
sSQL=sSQL+ "FROM"
sSQL=sSQL+ "Assess_Visitor_UseIt"
sSQL=sSQL+ "GROUP BY"
sSQL=sSQL+ "Country,"
sSQL=sSQL+ "Place_Name"
sSQL=sSQL+ "ORDER BY"
sSQL=sSQL+ "TotalUsesIt DESC"
sSQL=sSQL+ "LIMIT 10;"
sql.execute(sSQL,conn)
################################################################
```

At this point, we store the third set of data rules into the database.

```
print('################')
sView='Assess_Total_Visitors_GPS'
print('Creating:',sDatabaseName,'View:',sView)
sSQL="DROP VIEW IF EXISTS "+ sView + ";"
sql.execute(sSQL,conn)

sSQL="CREATE VIEW"+ sView + "AS"
sSQL=sSQL+ "SELECT"
sSQL=sSQL+ "Latitude,"
sSQL=sSQL+ "Longitude,"
sSQL=sSQL+ "SUM(UsesIt) AS TotalUsesIt"
sSQL=sSQL+ "FROM"
```

```
sSQL=sSQL+ "Assess_Visitor_UseIt"
sSQL=sSQL+ "GROUP BY"
sSQL=sSQL+ "Latitude,"
sSQL=sSQL+ "Longitude"
sSQL=sSQL+ "ORDER BY"
sSQL=sSQL+ "TotalUsesIt DESC"
sSQL=sSQL+ "LIMIT 10;"
sql.execute(sSQL,conn)
################################################################
```

We now simply loop through the business rules views and extract the data we need into text files.

```
sTables=['Assess_Total_Visitors_Location', 'Assess_Total_Visitors_GPS']
for sTable in sTables:
    print('################')
    print('Loading:',sDatabaseName,'Table:',sTable)
    sSQL="SELECT"
    sSQL=sSQL+ "*"
    sSQL=sSQL+ "FROM"
    sSQL=sSQL+ " "+ sTable + ";"
    TopData=pd.read_sql_query(sSQL, conn)
    print('################')
    print(TopData)
    print('################')
    print('##############################')
    print('Rows:',TopData.shape[0])
    print('##############################')
################################################################
print('### Done!! ######################################')
################################################################
```

Save the file Assess-Visitors.py, then compile and execute with your Python compiler. You should see two summary results like the following:

Assess_Total_Visitors_Location

	Country	Place_Name	TotalUsesIt
0	CN	Beijing	53139475
1	US	Palo Alto	33682341
2	US	Fort Huachuca	33472427
3	JP	Tokyo	31404799
4	US	Cambridge	25598851
5	US	San Diego	17751367
6	CN	Guangzhou	17563744
7	US	Newark	17270604
8	US	Raleigh	17167484
9	US	Durham	16914033

Assess_Total_Visitors_GPS

	Latitude	Longitude	TotalUsesIt
0	39.9289	116.3883	53139732
1	37.3762	-122.1826	33551404
2	31.5273	-110.3607	33472427
3	35.6427	139.7677	31439772
4	23.1167	113.25	17577053
5	42.3646	-71.1028	16890698
6	40.7355	-74.1741	16813373
7	42.3223	-83.1763	16777212
8	35.7977	-78.6253	16761084
9	32.8072	-117.1649	16747680

This is the concluding assess process requirement for Krennwallner.

Summary of Achievements with Krennwallner AG

- You have successfully used `pandas` to handle the generation of in-memory and on disk data.

- You have successfully used `networkx` to handle graph data.

- You have successfully used `sqlite3` to access SQLite tables to assist with data processing.

- You have successfully used `pandas` to set up SQLite views to persist SQL rules within the database and to offload processing to the SQLite database engine.

Hillman Ltd

Hillman Ltd is our logistics company and works with data related to locations and the distances between these locations.

Planning the Locations of the Warehouses

Planning the location of the warehouses requires the assessment of the GPS locations of these warehouses against the requirements for Hillman's logistics needs.

Warning   The online services provider that supplies the information has a Nominatim API limit of a maximum of one request per second. See `https://operations.osmfoundation.org/policies/nominatim/`. I have limited the requests by limiting the amount of requests per session.

You can install your own services: see `https://wiki.openstreetmap.org/wiki/Nominatim/Installation`. For the purposes of this book, the limited usage is achieving what I intended.

You will use the `geopy` and `pandas` packages to create a solution. Open your Python editor and create a file named `Assess-Warehouse-Address.py` in directory `C:\VKHCG\03-Hillman\02-Assess`.

Now copy the following code into the file:

```
################################################################
################################################################
# -*- coding: utf-8 -*-
################################################################
import os
import pandas as pd
from geopy.geocoders import Nominatim
geolocator = Nominatim()
################################################################
InputDir='01-Retrieve/01-EDS/01-R'
InputFileName='Retrieve_GB_Postcode_Warehouse.csv'
EDSDir='02-Assess/01-EDS'
OutputDir=EDSDir + '/02-Python'
OutputFileName='Assess_GB_Warehouse_Address.csv'
Company='03-Hillman'
################################################################
Base='C:/VKHCG'
################################################################
sFileDir=Base + '/' + Company + '/' + EDSDir
if not os.path.exists(sFileDir):
    os.makedirs(sFileDir)
################################################################
sFileDir=Base + '/' + Company + '/' + OutputDir
if not os.path.exists(sFileDir):
    os.makedirs(sFileDir)
################################################################
sFileName=Base + '/' + Company + '/' + InputDir + '/' + InputFileName
print('##########')
print('Loading:',sFileName)
Warehouse=pd.read_csv(sFileName,header=0,low_memory=False)
Warehouse.sort_values(by='postcode', ascending=1)
################################################################
## Limited to 10 due to service limit on Address Service.
################################################################
```

```
WarehouseGoodHead=Warehouse[Warehouse.latitude != 0].head(5)
WarehouseGoodTail=Warehouse[Warehouse.latitude != 0].tail(5)
```

The next section takes the ten warehouse locations and uses the online service to obtain a full address for the longitude and latitude supplied to the API call. The queries are performed in two cycles of five records each. First, we fix the parameter that is required for each row in memory.

```
################################################################
WarehouseGoodHead['Warehouse_Point']=WarehouseGoodHead.apply(lambda row:
            (str(row['latitude'])+','+str(row['longitude']))
            ,axis=1)
```

Second, we use the new parameter to call the address locator services.

```
WarehouseGoodHead['Warehouse_Address']=WarehouseGoodHead.apply(lambda row:
            geolocator.reverse(row['Warehouse_Point']).address
            ,axis=1)
```

Third, we remove the unwanted columns from the data in memory.

```
WarehouseGoodHead.drop('Warehouse_Point', axis=1, inplace=True)
WarehouseGoodHead.drop('id', axis=1, inplace=True)
WarehouseGoodHead.drop('postcode', axis=1, inplace=True)
```

Here, we simply repeat the previous three steps for a second cycle of five queries.

```
################################################################
WarehouseGoodTail['Warehouse_Point']=WarehouseGoodTail.apply(lambda row:
            (str(row['latitude'])+','+str(row['longitude']))
            ,axis=1)
WarehouseGoodTail['Warehouse_Address']=WarehouseGoodTail.apply(lambda row:
            geolocator.reverse(row['Warehouse_Point']).address
            ,axis=1)
WarehouseGoodTail.drop('Warehouse_Point', axis=1, inplace=True)
WarehouseGoodTail.drop('id', axis=1, inplace=True)
WarehouseGoodTail.drop('postcode', axis=1, inplace=True)
################################################################
```

```
WarehouseGood=WarehouseGoodHead.append(WarehouseGoodTail, ignore_
index=True)
print(WarehouseGood)
###############################################################
```

Here we store our now enhanced data set.

```
sFileName=sFileDir + '/' + OutputFileName
WarehouseGood.to_csv(sFileName, index = False)
###############################################################
print('### Done!! ###################################')
###############################################################
###############################################################
```

Save the Assess-Warehouse-Address.py file, then compile and execute with your Python compiler. This will produce a set of demonstrated values onscreen, plus a graph data file named Assess_GB_Warehouse_Address.csv.

Latitude	Longitude	Warehouse_Address
57.13514	-2.11731	35, Broomhill Road, Kaimhill, Aberdeen, Aberdeen City, Scotland, UK
57.13875	-2.09089	South Esplanade West, Torry, Aberdeen, Aberdeen City, Scotland, UK
57.101	-2.1106	A92, Kincorth, Aberdeen, Aberdeen City, Scotland, UK

Summary of Achievements

- You can now offload processing to online services via APIs.
- You can adapt internal parameters to match external requirements.
- You can extend this model to many other service providers. (See https://geopy.readthedocs.io/en/1.11.0/)

Global New Warehouse

Hillman wants to add extra global warehouses, and you are required to assess where they should be located. We only have to collect the possible locations for warehouses. In Chapters 9 and 10, we will use data science, via clustering, to determine a set of new warehouses for the company.

The following example will show you how to modify the data columns you read in that are totally ambiguous. Open your Python editor and create a file named Assess-Warehouse-Global.py in directory C:\VKHCG\03-Hillman\02-Assess. Now copy the following code into the file:

```python
################################################################
# -*- coding: utf-8 -*-
################################################################
import sys
import os
import pandas as pd
from geopy.geocoders import Nominatim
geolocator = Nominatim()
################################################################
if sys.platform == 'linux':
    Base=os.path.expanduser('~') + 'VKHCG'
else:
    Base='C:/VKHCG'
print('#############################')
print('Working Base:',Base,' using',sys.platform)
print('#############################')
################################################################
Company='03-Hillman'
InputDir='01-Retrieve/01-EDS/01-R'
InputFileName='Retrieve_All_Countries.csv'
EDSDir='02-Assess/01-EDS'
OutputDir=EDSDir + '/02-Python'
OutputFileName='Assess_All_Warehouse.csv'
################################################################
sFileDir=Base + '/' + Company + '/' + EDSDir
if not os.path.exists(sFileDir):
    os.makedirs(sFileDir)
################################################################
sFileDir=Base + '/' + Company + '/' + OutputDir
if not os.path.exists(sFileDir):
    os.makedirs(sFileDir)
```

```
################################################################
sFileName=Base + '/' + Company + '/' + InputDir + '/' + InputFileName
print('###########')
print('Loading:',sFileName)
Warehouse=pd.read_csv(sFileName,header=0,low_memory=False,
encoding="latin-1")
```

The columns are totally ambiguous and require new names, to be useful to any future processing.

```
################################################################
sColumns={'X1': 'Country',
          'X2': 'PostCode',
          'X3': 'PlaceName',
          'X4': 'AreaName',
          'X5': 'AreaCode',
          'X10': 'Latitude',
          'X11': 'Longitude'}
Warehouse.rename(columns=sColumns,inplace=True)
WarehouseGood=Warehouse
################################################################
sFileName=sFileDir + '/' + OutputFileName
WarehouseGood.to_csv(sFileName, index = False)
################################################################
print('### Done!! #####################################')
################################################################
```

Save the file Assess-Warehouse-Global.py file, then compile and execute with your Python compiler. This will produce a set of demonstrated values onscreen, plus a graph data file named Assess_All_Warehouse.csv.

Country	PostCode	PlaceName	AreaName	AreaCode	Latitude	Longitude
HU	8313	Balatongyörök	Zala	ZA	46.75	17.3667
HU	8314	Vonyarcvashegy	Zala	ZA	46.75	17.3167
HU	8315	Gyenesdiás	Zala	ZA	46.7667	17.2833

Planning Shipping Rules for Best-Fit International Logistics

Hillman requires an international logistics solution to support all the required shipping routes.

Tip   I have performed the shipping route solution for about every customer I have worked for over the last ten years. Companies have to move their products according to an effective route and schedule. So, please pay attention to the techniques and the business logic, as they are used by numerous companies.

This example will combine all the skills you mastered in this chapter. Open your Python editor and create a file named `Assess-Best-Fit-Logistics.py` in directory `C:\VKHCG\03-Hillman\02-Assess`.

Now copy the following code into the file:

```
################################################################
# -*- coding: utf-8 -*-
################################################################
```

We will use many libraries in the course of this example, as follows:

```
import sys
import os
import pandas as pd
import networkx as nx
from geopy.distance import vincenty
import sqlite3 as sq
from pandas.io import sql
```

Here we go again—with basic well-known parameters to run the process.

```
################################################################
if sys.platform == 'linux':
    Base=os.path.expanduser('~') + 'VKHCG'
else:
    Base='C:/VKHCG'
```

```
print('##############################')
print('Working Base:',Base,'using',sys.platform)
print('##############################')
################################################################
Company='03-Hillman'
InputDir='01-Retrieve/01-EDS/01-R'
InputFileName='Retrieve_All_Countries.csv'
EDSDir='02-Assess/01-EDS'
OutputDir=EDSDir + '/02-Python'
OutputFileName='Assess_Best_Logistics.gml'
################################################################
sFileDir=Base + '/' + Company + '/' + EDSDir
if not os.path.exists(sFileDir):
    os.makedirs(sFileDir)
################################################################
sFileDir=Base + '/' + Company + '/' + OutputDir
if not os.path.exists(sFileDir):
    os.makedirs(sFileDir)
################################################################
sDataBaseDir=Base + '/' + Company + '/02-Assess/SQLite'
if not os.path.exists(sDataBaseDir):
    os.makedirs(sDataBaseDir)
################################################################
sDatabaseName=sDataBaseDir + '/Hillman.db'
conn = sq.connect(sDatabaseName)
################################################################
```

We must load the data from the data source and fix the columns.

```
sFileName=Base + '/' + Company + '/' + InputDir + '/' + InputFileName
print('###########')
print('Loading:',sFileName)
Warehouse=pd.read_csv(sFileName,header=0,low_memory=False,
encoding="latin-1")
################################################################
```

```
sColumns={'X1': 'Country',
          'X2': 'PostCode',
          'X3': 'PlaceName',
          'X4': 'AreaName',
          'X5': 'AreaCode',
          'X10': 'Latitude',
          'X11': 'Longitude'}
Warehouse.rename(columns=sColumns,inplace=True)
WarehouseGood=Warehouse
#print(WarehouseGood.head())
```

You now have your data in memory. Now we start to manipulate it to fit our requirements. You need the average latitude and longitude for each country in the data source. So, to extract this feature, you must calculate the mean() for each country.

```
################################################################
RoutePointsCountry=pd.DataFrame(WarehouseGood.groupby(['Country'])
[['Latitude','Longitude']].mean())
print(RoutePointsCountry.head())
```

You will now have a list of countries with a single average point for all the known countries. Now store this list as Assess_RoutePointsCountry in the SQLite database.

```
print('###############')
sTable='Assess_RoutePointsCountry'
print('Storing:',sDatabaseName,'Table:',sTable)
RoutePointsCountry.to_sql(sTable, conn, if_exists="replace")
print('###############')
```

You need the average latitude and longitude for each country and post code combined in the data source. So, to extract this feature, you must calculate the mean() for each country and post code and store this list as Assess_RoutePointsPostCode in the SQLite database.

```
################################################################
RoutePointsPostCode=pd.DataFrame(WarehouseGood.groupby(['Country',
'PostCode'])[['Latitude','Longitude']].mean())
#print(RoutePointsPostCode.head())
print('###############')
```

```
sTable='Assess_RoutePointsPostCode'
print('Storing :',sDatabaseName,'Table:',sTable)
RoutePointsPostCode.to_sql(sTable, conn, if_exists="replace")
print('################')
```

You need the average latitude and longitude for each country, post code, and place-name combination in the data source. So, to extract this feature, you must calculate the mean() for each country post code and place-name and store this list as Assess_RoutePointsPlaceName in the SQLite database.

```
############################################################
RoutePointsPlaceName=pd.DataFrame(WarehouseGood.groupby(['Country',
'PostCode','PlaceName'])[['Latitude','Longitude']].mean())
#print(RoutePointsPlaceName.head())
print('################')
sTable='Assess_RoutePointsPlaceName'
print('Storing:',sDatabaseName,'Table:',sTable)
RoutePointsPlaceName.to_sql(sTable, conn, if_exists="replace")
print('################')
```

You will now match the country-to-country routes. I have limited the data to process only country codes GB, DE, BE, AU, US, and IN, to reduce the data processing time. I have processed the complete data set on my cluster in about 20 minutes, but on my not-so-powerful laptop, it took two hours! Hence the limit.

```
############################################################
### Fit Country to Country
############################################################
print('################')
sView='Assess_RouteCountries'
print('Creating:',sDatabaseName,'View:',sView)
sSQL="DROP VIEW IF EXISTS" + sView + ";"
sql.execute(sSQL,conn)

sSQL="CREATE VIEW" + sView + "AS"
sSQL=sSQL+ "SELECT DISTINCT"
sSQL=sSQL+ "S.Country AS SourceCountry,"
sSQL=sSQL+ "S.Latitude AS SourceLatitude,"
```

```
sSQL=sSQL+ "S.Longitude AS SourceLongitude,"
sSQL=sSQL+ "T.Country AS TargetCountry,"
sSQL=sSQL+ "T.Latitude AS TargetLatitude,"
sSQL=sSQL+ "T.Longitude AS TargetLongitude"
sSQL=sSQL+ "FROM"
sSQL=sSQL+ "Assess_RoutePointsCountry AS S"
sSQL=sSQL+ ","
sSQL=sSQL+ "Assess_RoutePointsCountry AS T"
sSQL=sSQL+ "WHERE S.Country <> T.Country"
sSQL=sSQL+ "AND"
sSQL=sSQL+ "S.Country in('GB','DE','BE','AU','US','IN')"
sSQL=sSQL+ "AND"
sSQL=sSQL+ "T.Country in('GB','DE','BE','AU','US','IN');"
sql.execute(sSQL,conn)

print('################')
print('Loading:',sDatabaseName,'Table:',sView)
sSQL="SELECT"
sSQL=sSQL+ "*"
sSQL=sSQL+ "FROM"
sSQL=sSQL+ " " + sView + ";"
RouteCountries=pd.read_sql_query(sSQL, conn)
```

You then apply Vincenty's formulae to get the distance in miles between each combination pair.

```
RouteCountries['Distance']=RouteCountries.apply(lambda row:
    round(
    vincenty((row['SourceLatitude'],row['SourceLongitude']),
              (row['TargetLatitude'],row['TargetLongitude'])).miles
        ,4)
     ,axis=1)

print(RouteCountries.head(5))
```

You will now match the country to post code routes and apply Vincenty's formulae, as previously.

```
###################################################################
### Fit Country to Post Code
###################################################################
print('################')
sView='Assess_RoutePostCode'
print('Creating:',sDatabaseName,'View:',sView)
sSQL="DROP VIEW IF EXISTS" + sView + ";"
sql.execute(sSQL,conn)

sSQL="CREATE VIEW" + sView + "AS"
sSQL=sSQL+ "SELECT DISTINCT"
sSQL=sSQL+ "S.Country AS SourceCountry,"
sSQL=sSQL+ "S.Latitude AS SourceLatitude,"
sSQL=sSQL+ "S.Longitude AS SourceLongitude,"
sSQL=sSQL+ "T.Country AS TargetCountry,"
sSQL=sSQL+ "T.PostCode AS TargetPostCode,"
sSQL=sSQL+ "T.Latitude AS TargetLatitude,"
sSQL=sSQL+ "T.Longitude AS TargetLongitude"
sSQL=sSQL+ "FROM"
sSQL=sSQL+ "Assess_RoutePointsCountry AS S"
sSQL=sSQL+ ","
sSQL=sSQL+ "Assess_RoutePointsPostCode AS T"
sSQL=sSQL+ "WHERE S.Country = T.Country"
sSQL=sSQL+ "AND"
sSQL=sSQL+ "S.Country in ('GB','DE','BE','AU','US','IN')"
sSQL=sSQL+ "AND"
sSQL=sSQL+ "T.Country in ('GB','DE','BE','AU','US','IN');"
sql.execute(sSQL,conn)

print('################')
print('Loading:',sDatabaseName,'Table:',sView)
sSQL="SELECT"
sSQL=sSQL+ "*"
sSQL=sSQL+ "FROM"
sSQL=sSQL+ " " + sView + ";"
RoutePostCode=pd.read_sql_query(sSQL, conn)
```

```
RoutePostCode['Distance']=RoutePostCode.apply(lambda row:
    round(
    vincenty((row['SourceLatitude'],row['SourceLongitude']),
            (row['TargetLatitude'],row['TargetLongitude'])).miles
        ,4)
      ,axis=1)

print(RoutePostCode.head(5))
```

You will now match the post code to place-name routes and apply Vincenty's formulae, as previously.

```
##################################################################
### Fit Post Code to Place Name
##################################################################
print('################')
sView='Assess_RoutePlaceName'
print('Creating:',sDatabaseName,'View:',sView)
sSQL="DROP VIEW IF EXISTS" + sView + ";"
sql.execute(sSQL,conn)

sSQL="CREATE VIEW" + sView + "AS"
sSQL=sSQL+ "SELECT DISTINCT"
sSQL=sSQL+ "S.Country AS SourceCountry,"
sSQL=sSQL+ "S.PostCode AS SourcePostCode,"
sSQL=sSQL+ "S.Latitude AS SourceLatitude,"
sSQL=sSQL+ "S.Longitude AS SourceLongitude,"
sSQL=sSQL+ "T.Country AS TargetCountry,"
sSQL=sSQL+ "T.PostCode AS TargetPostCode,"
sSQL=sSQL+ "T.PlaceName AS TargetPlaceName,"
sSQL=sSQL+ "T.Latitude AS TargetLatitude,"
sSQL=sSQL+ "T.Longitude AS TargetLongitude"
sSQL=sSQL+ "FROM"
sSQL=sSQL+ "Assess_RoutePointsPostCode AS S"
sSQL=sSQL+ ","
sSQL=sSQL+ "Assess_RoutePointsPLaceName AS T"
sSQL=sSQL+ "WHERE"
sSQL=sSQL+ "S.Country = T.Country"
```

```
sSQL=sSQL+ "AND"
sSQL=sSQL+ "S.PostCode = T.PostCode"
sSQL=sSQL+ "AND"
sSQL=sSQL+ "S.Country in('GB','DE','BE','AU','US','IN')"
sSQL=sSQL+ "AND"
sSQL=sSQL+ "T.Country in('GB','DE','BE','AU','US','IN');"
sql.execute(sSQL,conn)

print('################')
print('Loading:',sDatabaseName,'Table:',sView)
sSQL="SELECT"
sSQL=sSQL+ "*"
sSQL=sSQL+ "FROM"
sSQL=sSQL+ " " + sView + ";"
RoutePlaceName=pd.read_sql_query(sSQL, conn)

RoutePlaceName['Distance']=RoutePlaceName.apply(lambda row:
    round(
    vincenty((row['SourceLatitude'],row['SourceLongitude']),
             (row['TargetLatitude'],row['TargetLongitude'])).miles
        ,4)
      ,axis=1)

print(RoutePlaceName.head(5))
```

You will now map your new locations onto a graph. First the countries.

```
###############################################################
G=nx.Graph()
###############################################################
print('Countries:',RouteCountries.shape)
for i in range(RouteCountries.shape[0]):
    sNode0='C-' + RouteCountries['SourceCountry'][i]
    G.add_node(sNode0,
               Nodetype='Country',
               Country=RouteCountries['SourceCountry'][i],
               Latitude=round(RouteCountries['SourceLatitude'][i],4),
               Longitude=round(RouteCountries['SourceLongitude'][i],4))
```

369

```
    sNode1='C-' + RouteCountries['TargetCountry'][i]
    G.add_node(sNode1,
                Nodetype='Country',
                Country=RouteCountries['TargetCountry'][i],
                Latitude=round(RouteCountries['TargetLatitude'][i],4),
                Longitude=round(RouteCountries['TargetLongitude'][i],4))
    G.add_edge(sNode0,sNode1,distance=round(RouteCountries['Distance']
    [i],3))
    #print(sNode0,sNode1)
```

Then the postcodes.

```
###############################################################
print('Post Code:',RoutePostCode.shape)
for i in range(RoutePostCode.shape[0]):
    sNode0='C-' + RoutePostCode['SourceCountry'][i]
    G.add_node(sNode0,
                Nodetype='Country',
                Country=RoutePostCode['SourceCountry'][i],
                Latitude=round(RoutePostCode['SourceLatitude'][i],4),
                Longitude=round(RoutePostCode['SourceLongitude'][i],4))

    sNode1='P-' + RoutePostCode['TargetPostCode'][i]  + '-' +
    RoutePostCode['TargetCountry'][i]
    G.add_node(sNode1,
                Nodetype='PostCode',
                Country=RoutePostCode['TargetCountry'][i],
                PostCode=RoutePostCode['TargetPostCode'][i],
                Latitude=round(RoutePostCode['TargetLatitude'][i],4),
                Longitude=round(RoutePostCode['TargetLongitude'][i],4))
    G.add_edge(sNode0,sNode1,distance=round(RoutePostCode['Distance']
    [i],3))
    #print(sNode0,sNode1)
```

Then the place-names.

```
###############################################################
print('Place Name:',RoutePlaceName.shape)
for i in range(RoutePlaceName.shape[0]):
    sNode0='P-' + RoutePlaceName['TargetPostCode'][i]  + '-'
    sNode0=sNode0 + RoutePlaceName['TargetCountry'][i]
    G.add_node(sNode0,
                Nodetype='PostCode',
                Country=RoutePlaceName['SourceCountry'][i],
                PostCode=RoutePlaceName['TargetPostCode'][i],
                Latitude=round(RoutePlaceName['SourceLatitude'][i],4),
                Longitude=round(RoutePlaceName['SourceLongitude'][i],4))

    sNode1='L-' + RoutePlaceName['TargetPlaceName'][i]  + '-'
    sNode1=sNode1 + RoutePlaceName['TargetPostCode'][i]  + '-'
    sNode1=sNode1 + RoutePlaceName['TargetCountry'][i]
    G.add_node(sNode1,
                Nodetype='PlaceName',
                Country=RoutePlaceName['TargetCountry'][i],
                PostCode=RoutePlaceName['TargetPostCode'][i],
                PlaceName=RoutePlaceName['TargetPlaceName'][i],
                Latitude=round(RoutePlaceName['TargetLatitude'][i],4),
                Longitude=round(RoutePlaceName['TargetLongitude'][i],4))
    G.add_edge(sNode0,sNode1,distance=round(RoutePlaceName['Distance']
    [i],3))
    #print(sNode0,sNode1)
###############################################################
```

You now have a full graph with every possible route. Let's store your hard work.

```
sFileName=sFileDir + '/' + OutputFileName
print('###############################')
print('Storing:', sFileName)
print('###############################')
nx.write_gml(G,sFileName)
sFileName=sFileName +'.gz'
nx.write_gml(G,sFileName)
```

You will now start to extract extra features that are stored in the graph structure. The required feature you are seeking is the shortest path between two locations.

```
###############################################################
print('##############################')
print('Path:', nx.shortest_path(G,source='P-SW1-GB',target='P-01001-
US',weight='distance'))
print('Path length:', nx.shortest_path_length(G,source='P-SW1-
GB',target='P-01001-US',weight='distance'))
print('Path length (1):', nx.shortest_path_length(G,source='P-SW1-
GB',target='C-GB',weight='distance'))
print('Path length (2):', nx.shortest_path_length(G,source='C-
GB',target='C-US',weight='distance'))
print('Path length (3):', nx.shortest_path_length(G,source='C-
US',target='P-01001-US',weight='distance'))
print('##############################')
```

The required feature you are seeking is the shortest path between a given location and any other within one hop of that source.

```
print('Routes from P-SW1-GB < 2:', nx.single_source_shortest_
path(G,source='P-SW1-GB', cutoff=1))
print('Routes from P-01001-US < 2:', nx.single_source_shortest_
path(G,source='P-01001-US', cutoff=1))
print('##############################')
```

There are many features hidden in this graph you've built. If you want to test your new skills, I suggest you look at the online information for networkx, available at https://networkx.github.io/documentation/networkx-1.10/reference/algorithms.html. There are hundreds of possible options and combinations. Enjoy!

But before you start testing your skills, you must perform some database maintenance.

```
###############################################################
print('###############')
print('Vacuum Database')
sSQL="VACUUM;"
sql.execute(sSQL,conn)
```

```
print('################')
##################################################################
print('### Done!! ##############################################')
##################################################################
##################################################################
```

Save the Assess-Best-Fit-Logistics.py file, then compile and execute with your Python compiler. This will produce a set of demonstrated values onscreen, plus a graph data file named Assess_Best_Logistics.gml.

There is a point to the maintenance, as, over the years, I have seen many data scientists complain about their data science running slowly but who neglected simple housekeeping of their systems, which can restore expensive processing power.

Tip   Always clean up after any major data science session. Compress your databases, remove unused files, and clean up the environment. Take care of your tools, and they will take care of you.

Summary of Achievements

- You can now query features out of a graph, such as shortage paths between locations and paths from a given location.

- You can perform housekeeping from your data science to maximize your processing capacity.

Deciding the Best Packing Option for Shipping in Containers

Hillman wants to introduce new shipping containers into its logistics strategy. You need to model a data set that will support this project.

I will guide you through a process of assessing the possible container sizes.

Tip   This example introduces features with ranges or tolerances.

These not-so-perfect data values can cause major issues in the data science ecosystem. I will guide you through how to take precise measurements and add the given tolerances and create the ranges of values you need for the data science in Chapters 9 and 10.

Open your Python editor and create a file named Assess-Shipping-Containers.py in directory C:\VKHCG\03-Hillman\02-Assess.

Now copy the following code into the file:

```python
###############################################################
# -*- coding: utf-8 -*-
###############################################################
import sys
import os
import pandas as pd
import sqlite3 as sq
from pandas.io import sql
###############################################################
if sys.platform == 'linux':
    Base=os.path.expanduser('~') + 'VKHCG'
else:
    Base='C:/VKHCG'
print('###############################')
print('Working Base:',Base,'using',sys.platform)
print('###############################')
###############################################################
Company='03-Hillman'
InputDir='01-Retrieve/01-EDS/02-Python'
InputFileName1='Retrieve_Product.csv'
InputFileName2='Retrieve_Box.csv'
InputFileName3='Retrieve_Container.csv'
EDSDir='02-Assess/01-EDS'
OutputDir=EDSDir + '/02-Python'
OutputFileName='Assess_Shipping_Containers.csv'
###############################################################
sFileDir=Base + '/' + Company + '/' + EDSDir
if not os.path.exists(sFileDir):
    os.makedirs(sFileDir)
###############################################################
sFileDir=Base + '/' + Company + '/' + OutputDir
if not os.path.exists(sFileDir):
```

```
    os.makedirs(sFileDir)
################################################################
sDataBaseDir=Base + '/' + Company + '/02-Assess/SQLite'
if not os.path.exists(sDataBaseDir):
    os.makedirs(sDataBaseDir)
################################################################
sDatabaseName=sDataBaseDir + '/hillman.db'
conn = sq.connect(sDatabaseName)
################################################################
################################################################
### Import Product Data
################################################################
sFileName=Base + '/' + Company + '/' + InputDir + '/' + InputFileName1
print('###########')
print('Loading:',sFileName)
ProductRawData=pd.read_csv(sFileName,
                    header=0,
                    low_memory=False,
                    encoding="latin-1"
                    )
ProductRawData.drop_duplicates(subset=None, keep='first', inplace=True)
ProductRawData.index.name = 'IDNumber'
```

I have limited the data set to speed up the processing. This code refers only to the current top-ten products having a length value of at least a half-meter.

Note   Feel free to experiment with different parameters, if you want to. Think what if you only use products between 25 centimeters and 50 centimeters, or an entire product range with a width greater than 1 meter? This is data science. Have fun with it.

```
ProductData=ProductRawData[ProductRawData.Length <= 0.5].head(10)

print('Loaded Product:',ProductData.columns.values)
print('#############################')
```

```
################################################################
print('################')
sTable='Assess_Product'
print('Storing:',sDatabaseName,'Table:',sTable)
ProductData.to_sql(sTable, conn, if_exists="replace")
print('################')
################################################################
print(ProductData.head())
print('##############################')
print('Rows:',ProductData.shape[0])
print('##############################')
################################################################
################################################################
### Import Box Data
################################################################
sFileName=Base + '/' + Company + '/' + InputDir + '/' + InputFileName2
print('###########')
print('Loading:',sFileName)
BoxRawData=pd.read_csv(sFileName,
                       header=0,
                       low_memory=False,
                       encoding="latin-1"
                       )
BoxRawData.drop_duplicates(subset=None, keep='first', inplace=True)
BoxRawData.index.name = 'IDNumber'
```

I have limited the data set to speed up the processing. This code refers only to the current top-1000 packing boxes, with a length value of less than one meter. You may alter these limitations, if you want to.

```
BoxData=BoxRawData[BoxRawData.Length <= 1].head(1000)

print('Loaded Product:',BoxData.columns.values)
print('##############################')
################################################################
print('################')
```

```
sTable='Assess_Box'
print('Storing:',sDatabaseName,'Table:',sTable)
BoxData.to_sql(sTable, conn, if_exists="replace")
print('################')
################################################################
print(BoxData.head())
print('##############################')
print('Rows:',BoxData.shape[0])
print('##############################')
################################################################
################################################################
### Import Container Data
################################################################
sFileName=Base + '/' + Company + '/' + InputDir + '/' + InputFileName3
print('###########')
print('Loading:',sFileName)
ContainerRawData=pd.read_csv(sFileName,
                    header=0,
                    low_memory=False,
                    encoding="latin-1"
                    )
ContainerRawData.drop_duplicates(subset=None, keep='first', inplace=True)
ContainerRawData.index.name = 'IDNumber'
```

I have limited the data set to speed up the processing. This code is only for current top-ten shipping containers, with a length value of less than two meters. You may alter these limitations, as you wish.

```
ContainerData=ContainerRawData[ContainerRawData.Length <= 2].head(10)

print('Loaded Product:',ContainerData.columns.values)
print('##############################')
################################################################
print('################')
sTable='Assess_Container'
print('Storing:',sDatabaseName,'Table:',sTable)
```

```
BoxData.to_sql(sTable, conn, if_exists="replace")
print('################')
####################################################################
print(ContainerData.head())
print('##############################')
print('Rows:',ContainerData.shape[0])
print('##############################')
####################################################################
```

You are tasked with finding the correct packing box for each of your selected products. The product must fit in the box, with an amount of packing foam to protect the product.

```
####################################################################
### Fit Product in Box
####################################################################
print('################')
sView='Assess_Product_in_Box'
print('Creating:',sDatabaseName,'View:',sView)
sSQL="DROP VIEW IF EXISTS " + sView + ";"
sql.execute(sSQL,conn)

sSQL="CREATE VIEW" + sView + "AS"
sSQL=sSQL+ "SELECT"
sSQL=sSQL+ "P.UnitNumber AS ProductNumber,"
sSQL=sSQL+ "B.UnitNumber AS BoxNumber,"
sSQL=sSQL+ "(B.Thickness * 1000) AS PackSafeCode,"
sSQL=sSQL+ "(B.BoxVolume - P.ProductVolume) AS PackFoamVolume,"
```

I have given you a formula to calculate the air freight volumetric (chargeable weight), because when you ship by air, you pay for the shipment per kilogram, against the higher value of either the air freight volumetric (chargeable weight) or your item's true weight. The general value is 167 kilograms per cubic centimeter.

```
sSQL=sSQL+ "((B.Length*10) * (B.Width*10) * (B.Height*10)) * 167 AS Air_
Dimensional_Weight,"
```

I have given you a formula here to calculate the road freight volumetric (chargeable weight), because when you ship by road, you pay for the shipment per kilogram, against the higher value of either the road freight volumetric (chargeable weight) or your item's true weight. The general value is 333 kilograms per cubic centimeter.

```
sSQL=sSQL+ "((B.Length*10) * (B.Width*10) * (B.Height*10)) * 333 AS Road_
Dimensional_Weight,"
```

I have given you a formula here to calculate the sea freight volumetric (chargeable weight), because when you ship by sea, you pay for the shipment per kilogram, against the higher value of either the sea freight volumetric (chargeable weight) or your item's true weight. The general value is 1000 kilograms per cubic centimeter.

```
sSQL=sSQL+ "((B.Length*10) * (B.Width*10) * (B.Height*10)) * 1000 AS Sea_
Dimensional_Weight,"
```

```
sSQL=sSQL+ "P.Length AS Product_Length,"
sSQL=sSQL+ "P.Width AS Product_Width,"
sSQL=sSQL+ "P.Height AS Product_Height,"
sSQL=sSQL+ "P.ProductVolume AS Product_cm_Volume,"
sSQL=sSQL+ "((P.Length*10) * (P.Width*10) * (P.Height*10)) AS Product_ccm_
Volume,"
```

Remember I said earlier that you will require ranges or options. Here we have a tolerance set up for packing the product in the prescribed box for a range of 5% below and 5% above the true size. Hence, a product will fit in more than one box, with variances in packing material thickness.

Note   I saved one of my customers nearly £2000 per day in shipping by using smaller boxes for packing. Remember to consider the volumetric (chargeable weight) you pay for to take up extra shipping space.

```
sSQL=sSQL+ "(B.Thickness * 0.95) AS Minimum_Pack_Foam,"
sSQL=sSQL+ "(B.Thickness * 1.05) AS Maximum_Pack_Foam,"
sSQL=sSQL+ "B.Length - (B.Thickness * 1.10) AS Minimum_Product_Box_Length,"
```

```
sSQL=sSQL+ "B.Length - (B.Thickness * 0.95) AS Maximum_Product_Box_Length,"
sSQL=sSQL+ "B.Width - (B.Thickness * 1.10) AS Minimum_Product_Box_Width,"
sSQL=sSQL+ "B.Width - (B.Thickness * 0.95) AS Maximum_Product_Box_Width,"
sSQL=sSQL+ "B.Height - (B.Thickness * 1.10) AS Minimum_Product_Box_Height,"
sSQL=sSQL+ "B.Height - (B.Thickness * 0.95) AS Maximum_Product_Box_Height,"

sSQL=sSQL+ "B.Length AS Box_Length,"
sSQL=sSQL+ "B.Width AS Box_Width,"
sSQL=sSQL+ "B.Height AS Box_Height,"
sSQL=sSQL+ "B.BoxVolume AS Box_cm_Volume,"
sSQL=sSQL+ "((B.Length*10) * (B.Width*10) * (B.Height*10)) AS Box_ccm_
Volume,"
sSQL=sSQL+ "(2 * B.Length * B.Width) + (2 * B.Length * B.Height) + (2 *
B.Width * B.Height) AS Box_sqm_Area,"
```

The next limitation is the boxes' packing-strength rating. Any box has a maximum approved weight rating.

Note   In logistics, if you overload a box, and it gets damaged, your insurance does not cover the cost. That small miscalculation is the most common cause of damage in logistics. Do not overload the boxes!

Our example uses three different strength of boxes ratings: 3.5 kilograms, 7.7 kilograms, and 10 kilograms per cubic centimeter.

```
sSQL=sSQL+ "((B.Length*10) * (B.Width*10) * (B.Height*10)) *  3.5 AS Box_A_
Max_Kg_Weight,"
sSQL=sSQL+ "((B.Length*10) * (B.Width*10) * (B.Height*10)) *  7.7 AS Box_B_
Max_Kg_Weight,"
sSQL=sSQL+ "((B.Length*10) * (B.Width*10) * (B.Height*10)) * 10.0 AS Box_C_
Max_Kg_Weight"

sSQL=sSQL+ "FROM"
sSQL=sSQL+ "Assess_Product as P"
sSQL=sSQL+ ","
sSQL=sSQL+ "Assess_Box as B"
sSQL=sSQL+ "WHERE"
```

Here is the main range mapping process for the volume of the product against the box:

```
sSQL=sSQL+ "P.Length >= (B.Length - (B.Thickness * 1.10))"
sSQL=sSQL+ "AND"
sSQL=sSQL+ "P.Width >= (B.Width - (B.Thickness * 1.10))"
sSQL=sSQL+ "AND"
sSQL=sSQL+ "P.Height >= (B.Height - (B.Thickness * 1.10))"
sSQL=sSQL+ "AND"
sSQL=sSQL+ "P.Length <= (B.Length - (B.Thickness * 0.95))"
sSQL=sSQL+ "AND"
sSQL=sSQL+ "P.Width <= (B.Width - (B.Thickness * 0.95))"
sSQL=sSQL+ "AND"
sSQL=sSQL+ "P.Height <= (B.Height - (B.Thickness * 0.95))"
```

It is also advisable that the box be bigger than the product. But you probably know that. One of my customers did not, however, and ordered 20,000 boxes per year that were too small.

```
sSQL=sSQL+ "AND"
sSQL=sSQL+ "(B.Height - B.Thickness) >= 0"
sSQL=sSQL+ "AND"
sSQL=sSQL+ "(B.Width - B.Thickness) >= 0"
sSQL=sSQL+ "AND"
sSQL=sSQL+ "(B.Height - B.Thickness) >= 0"
sSQL=sSQL+ "AND"
sSQL=sSQL+ "B.BoxVolume >= P.ProductVolume;"
sql.execute(sSQL,conn)
```

I will now show you how to fit the boxes on a shipping pallet.

```
################################################################
### Fit Box in Pallet
################################################################
t=0
for l in range(2,8):
    for w in range(2,8):
        for h in range(4):
            t += 1
```

```
            PalletLine=[('IDNumber',[t]),
                        ('ShipType',['Pallet']),
                        ('UnitNumber',('L-'+format(t,"06d"))),
                        ('Box_per_Length',(format(2**l,"4d"))),
                        ('Box_per_Width',(format(2**w,"4d"))),
                        ('Box_per_Height',(format(2**h,"4d")))]
        if t==1:
            PalletFrame = pd.DataFrame.from_items(PalletLine)
        else:
            PalletRow = pd.DataFrame.from_items(PalletLine)
            PalletFrame = PalletFrame.append(PalletRow)
PalletFrame.set_index(['IDNumber'],inplace=True)
################################################################
PalletFrame.head()
print('##############################')
print('Rows:',PalletFrame.shape[0])
print('##############################')
################################################################
### Fit Box on Pallet
################################################################
print('###############')
sView='Assess_Box_on_Pallet'
print('Creating:',sDatabaseName,'View:',sView)
sSQL="DROP VIEW IF EXISTS" + sView + ";"
sql.execute(sSQL,conn)

sSQL="CREATE VIEW" + sView + "AS"
sSQL=sSQL+ "SELECT DISTINCT"
sSQL=sSQL+ "P.UnitNumber AS PalletNumber,"
sSQL=sSQL+ "B.UnitNumber AS BoxNumber,"
```

The size of the pallet is determined by your stacking plan. Do you stack three boxes wide by four boxes long by three boxes high? Or do you stack them five by five by five? That is the question. Experiment once you have the data.

```
sSQL=sSQL+ "round(B.Length*P.Box_per_Length,3) AS Pallet_Length,"
sSQL=sSQL+ "round(B.Width*P.Box_per_Width,3) AS Pallet_Width,"
sSQL=sSQL+ "round(B.Height*P.Box_per_Height,3) AS Pallet_Height,"
sSQL=sSQL+ "P.Box_per_Length * P.Box_per_Width * P.Box_per_Height AS
Pallet_Boxes"
sSQL=sSQL+ "FROM"
sSQL=sSQL+ "Assess_Box as B"
sSQL=sSQL+ ","
sSQL=sSQL+ "Assess_Pallet as P"
```

The only limitation I suggest is that the pallet not be bigger than 20 meters × 9 meters × 5 meters, the biggest item that can be picked up with the crane available. But, once again, feel free to change this restriction, if you want to experiment.

```
sSQL=sSQL+ "WHERE"
sSQL=sSQL+ "round(B.Length*P.Box_per_Length,3) <= 20"
sSQL=sSQL+ "AND"
sSQL=sSQL+ "round(B.Width*P.Box_per_Width,3) <= 9"
sSQL=sSQL+ "AND"
sSQL=sSQL+ "round(B.Height*P.Box_per_Height,3) <= 5;"
sql.execute(sSQL,conn)
```

Let's now calculate the product in box and box on pallet options.

```
#################################################################
sTables=['Assess_Product_in_Box','Assess_Box_on_Pallet']
for sTable in sTables:
    print('################')
    print('Loading:',sDatabaseName,'Table:',sTable)
    sSQL="SELECT "
    sSQL=sSQL+ "*"
    sSQL=sSQL+ "FROM"
    sSQL=sSQL+ " " + sTable + ";"
    SnapShotData=pd.read_sql_query(sSQL, conn)
    print('################')
    sTableOut=sTable + '_SnapShot'
    print('Storing:',sDatabaseName,'Table:',sTable)
```

```
SnapShotData.to_sql(sTableOut, conn, if_exists="replace")
print('###############')
```

I will now show you how to calculate the options for loading pallets into a correct shipping container. The rules are simple: if the pallet fits the container opening, it is OK to ship via that container.

```
################################################################
### Fit Pallet in Container
################################################################
sTables=['Length','Width','Height']
for sTable in sTables:

    sView='Assess_Pallet_in_Container_' + sTable
    print('Creating:',sDatabaseName,'View:',sView)
    sSQL="DROP VIEW IF EXISTS" + sView + ";"
    sql.execute(sSQL,conn)

    sSQL="CREATE VIEW" + sView + "AS"
    sSQL=sSQL+ "SELECT DISTINCT"
    sSQL=sSQL+ "C.UnitNumber AS ContainerNumber,"
    sSQL=sSQL+ "P.PalletNumber,"
    sSQL=sSQL+ "P.BoxNumber,"
    sSQL=sSQL+ "round(C." + sTable + "/P.Pallet_" + sTable + ",0)"
    sSQL=sSQL+ "AS Pallet_per_" + sTable + ","
    sSQL=sSQL+ "round(C." + sTable + "/P.Pallet_" + sTable + ",0)"
    sSQL=sSQL+ "* P.Pallet_Boxes AS Pallet_" + sTable + "_Boxes,"
    sSQL=sSQL+ "P.Pallet_Boxes"
    sSQL=sSQL+ "FROM"
    sSQL=sSQL+ "Assess_Container as C"
    sSQL=sSQL+ ","
    sSQL=sSQL+ "Assess_Box_on_Pallet_SnapShot as P"
    sSQL=sSQL+ "WHERE"
```

Here is the check: if it fits, it goes. You can load many pallets on a container, but it has to fit.

```
    sSQL=sSQL+ "round(C.Length/P.Pallet_Length,0) > 0"
    sSQL=sSQL+ "AND"
```

```
sSQL=sSQL+ "round(C.Width/P.Pallet_Width,0) > 0"
sSQL=sSQL+ "AND"
sSQL=sSQL+ "round(C.Height/P.Pallet_Height,0) > 0;"
sql.execute(sSQL,conn)

print('################')
print('Loading:',sDatabaseName,'Table:',sView)
sSQL="SELECT"
sSQL=sSQL+ "*"
sSQL=sSQL+ "FROM"
sSQL=sSQL+ " " + sView + ";"
SnapShotData=pd.read_sql_query(sSQL, conn)
print('################')
sTableOut= sView + '_SnapShot'
print('Storing:',sDatabaseName,'Table:',sTableOut)
SnapShotData.to_sql(sTableOut, conn, if_exists="replace")
print('################')
```

Now we simply extract the pallet-in-container option for later use.

```
################################################################
print('################')
sView='Assess_Pallet_in_Container'
print('Creating:',sDatabaseName,'View:',sView)
sSQL="DROP VIEW IF EXISTS" + sView + ";"
sql.execute(sSQL,conn)

sSQL="CREATE VIEW" + sView + "AS"
sSQL=sSQL+ "SELECT"
sSQL=sSQL+ "CL.ContainerNumber,"
sSQL=sSQL+ "CL.PalletNumber,"
sSQL=sSQL+ "CL.BoxNumber,"
sSQL=sSQL+ "CL.Pallet_Boxes AS Boxes_per_Pallet,"
sSQL=sSQL+ "CL.Pallet_per_Length,"
sSQL=sSQL+ "CW.Pallet_per_Width,"
sSQL=sSQL+ "CH.Pallet_per_Height,"
sSQL=sSQL+ "CL.Pallet_Length_Boxes * CW.Pallet_Width_Boxes * CH.Pallet_
Height_Boxes AS Container_Boxes"
```

385

```
sSQL=sSQL+ "FROM"
sSQL=sSQL+ "Assess_Pallet_in_Container_Length_SnapShot as CL"
sSQL=sSQL+ "JOIN"
sSQL=sSQL+ "Assess_Pallet_in_Container_Width_SnapShot as CW"
sSQL=sSQL+ "ON"
sSQL=sSQL+ "CL.ContainerNumber = CW.ContainerNumber"
sSQL=sSQL+ "AND"
sSQL=sSQL+ "CL.PalletNumber = CW.PalletNumber"
sSQL=sSQL+ "AND"
sSQL=sSQL+ "CL.BoxNumber = CW.BoxNumber"
sSQL=sSQL+ "JOIN"
sSQL=sSQL+ "Assess_Pallet_in_Container_Height_SnapShot as CH"
sSQL=sSQL+ "ON"
sSQL=sSQL+ "CL.ContainerNumber = CH.ContainerNumber"
sSQL=sSQL+ "AND"
sSQL=sSQL+ "CL.PalletNumber = CH.PalletNumber"
sSQL=sSQL+ "AND"
sSQL=sSQL+ "CL.BoxNumber = CH.BoxNumber;"
sql.execute(sSQL,conn)
################################################################
sTables=['Assess_Product_in_Box','Assess_Pallet_in_Container']
for sTable in sTables:
    print('################')
    print('Loading:',sDatabaseName,'Table:',sTable)
    sSQL="SELECT "
    sSQL=sSQL+ "*"
    sSQL=sSQL+ "FROM"
    sSQL=sSQL+ " " + sTable + ";"
    PackData=pd.read_sql_query(sSQL, conn)
    print('################')
    print(PackData)
    print('################')
    print('################################')
    print('Rows:',PackData.shape[0])
    print('################################')
```

```
    sFileName=sFileDir + '/' + sTable + '.csv'
    print(sFileName)
    PackData.to_csv(sFileName, index = False)
################################################################
print('### Done!! #####################################')
################################################################
```

Save the file Assess-Shipping-Containers.py file, then compile and execute with your Python compiler. This will produce a set of demonstrated values onscreen, plus a data file named Assess_Shipping_Containers.csv.

Creating a Delivery Route

Hillman requires the complete grid plan of the delivery routes for the company, to ensure the suppliers, warehouses, shops, and customers can be reached by its new strategy. This new plan will enable the optimum routes between suppliers, warehouses, shops, and customers.

Open your Python editor and create a file named Assess-Shipping-Routes.py in directory C:\VKHCG\03-Hillman\02-Assess.

Now copy the following code into the file:

```
################################################################
# -*- coding: utf-8 -*-
################################################################
import sys
import os
import pandas as pd
import sqlite3 as sq
from pandas.io import sql
import networkx as nx
from geopy.distance import vincenty
################################################################
nMax=100
nMaxPath=100
nSet=True
nVSet=False
################################################################
```

```python
if sys.platform == 'linux':
    Base=os.path.expanduser('~') + 'VKHCG'
else:
    Base='C:/VKHCG'
print('##############################')
print('Working Base:',Base,'using',sys.platform)
print('##############################')
###############################################################
Company='03-Hillman'
InputDir1='01-Retrieve/01-EDS/01-R'
InputDir2='01-Retrieve/01-EDS/02-Python'
InputFileName1='Retrieve_GB_Postcode_Warehouse.csv'
InputFileName2='Retrieve_GB_Postcodes_Shops.csv'
EDSDir='02-Assess/01-EDS'
OutputDir=EDSDir + '/02-Python'
OutputFileName1='Assess_Shipping_Routes.gml'
OutputFileName2='Assess_Shipping_Routes.txt'
###############################################################
sFileDir=Base + '/' + Company + '/' + EDSDir
if not os.path.exists(sFileDir):
    os.makedirs(sFileDir)
###############################################################
sFileDir=Base + '/' + Company + '/' + OutputDir
if not os.path.exists(sFileDir):
    os.makedirs(sFileDir)
###############################################################
sDataBaseDir=Base + '/' + Company + '/02-Assess/SQLite'
if not os.path.exists(sDataBaseDir):
    os.makedirs(sDataBaseDir)
###############################################################
sDatabaseName=sDataBaseDir + '/hillman.db'
conn = sq.connect(sDatabaseName)
###############################################################
###############################################################
### Import Warehouse Data
```

```
################################################################
sFileName=Base + '/' + Company + '/' + InputDir1 + '/' + InputFileName1
print('###########')
print('Loading:',sFileName)
WarehouseRawData=pd.read_csv(sFileName,
                header=0,
                low_memory=False,
                encoding="latin-1"
                )
WarehouseRawData.drop_duplicates(subset=None, keep='first', inplace=True)
WarehouseRawData.index.name = 'IDNumber'
```

> **Note**   I have taken a subset of warehouse locations here, to enable quicker processing. Feel free to change the parameters if you want more or fewer locations.

```
WarehouseData=WarehouseRawData.head(nMax)
WarehouseData=WarehouseData.append(WarehouseRawData.tail(nMax))
WarehouseData=WarehouseData.append(WarehouseRawData[WarehouseRawData.
postcode=='KA13'])
WarehouseData=WarehouseData.append(WarehouseRawData[WarehouseRawData.
postcode=='SW1W'])
WarehouseData.drop_duplicates(subset=None, keep='first', inplace=True)

print('Loaded Warehouses:',WarehouseData.columns.values)
print('##############################')
################################################################
print('###############')
sTable='Assess_Warehouse_UK'
print('Storing:',sDatabaseName,'Table:',sTable)
WarehouseData.to_sql(sTable, conn, if_exists="replace")
print('###############')
################################################################
print(WarehouseData.head())
print('##############################')
```

```
print('Rows:',WarehouseData.shape[0])
print('#############################')
##################################################################
### Import Shop Data
##################################################################
sFileName=Base + '/' + Company + '/' + InputDir1 + '/' + InputFileName2
print('###########')
print('Loading:',sFileName)
ShopRawData=pd.read_csv(sFileName,
                    header=0,
                    low_memory=False,
                    encoding="latin-1"
                    )
ShopRawData.drop_duplicates(subset=None, keep='first', inplace=True)
ShopRawData.index.name = 'IDNumber'
ShopData=ShopRawData
print('Loaded Shops:',ShopData.columns.values)
print('##############################')
```

Note   I have taken a subset of shop locations here, to enable quicker processing.
Feel free to change the parameters, if you want more or fewer locations.

```
##################################################################
print('################')
sTable='Assess_Shop_UK'
print('Storing:',sDatabaseName,'Table:',sTable)
ShopData.to_sql(sTable, conn, if_exists="replace")
print('################')
##################################################################
print(ShopData.head())
print('##############################')
print('Rows:',ShopData.shape[0])
print('##############################')
##################################################################
```

```
### Connect HQ
################################################################
print('###############')
sView='Assess_HQ'
print('Creating:',sDatabaseName,'View:',sView)
sSQL="DROP VIEW IF EXISTS" + sView + ";"
sql.execute(sSQL,conn)

sSQL="CREATE VIEW" + sView + "AS"
sSQL=sSQL+ "SELECT"
sSQL=sSQL+ "W.postcode AS HQ_PostCode,"
sSQL=sSQL+ "'HQ-' || W.postcode AS HQ_Name,"
sSQL=sSQL+ "round(W.latitude,6) AS HQ_Latitude,"
sSQL=sSQL+ "round(W.longitude,6) AS HQ_Longitude"
sSQL=sSQL+ "FROM"
sSQL=sSQL+ "Assess_Warehouse_UK as W"
sSQL=sSQL+ "WHERE"
sSQL=sSQL+ "TRIM(W.postcode) in ('KA13','SW1W');"
sql.execute(sSQL,conn)
```

To calculate the warehouse-to-warehouse routes, I suggest a simple process for all warehouses to ship to all warehouses.

```
################################################################
### Connect Warehouses
################################################################
print('###############')
sView='Assess_Warehouse'
print('Creating:',sDatabaseName,'View:',sView)
sSQL="DROP VIEW IF EXISTS" + sView + ";"
sql.execute(sSQL,conn)

sSQL="CREATE VIEW" + sView + "AS"
sSQL=sSQL+ "SELECT"
sSQL=sSQL+ "W.postcode AS Warehouse_PostCode,"
sSQL=sSQL+ "'WH-' || W.postcode AS Warehouse_Name,"
sSQL=sSQL+ "round(W.latitude,6) AS Warehouse_Latitude,"
```

```
sSQL=sSQL+ "round(W.longitude,6) AS Warehouse_Longitude"
sSQL=sSQL+ "FROM"
sSQL=sSQL+ "Assess_Warehouse_UK as W;"
sql.execute(sSQL,conn)
```

To calculate the warehouse-to-shop routes, I suggest a simple process for all warehouses to ship only to shops in the same post code.

```
################################################################
### Connect Warehouse to Shops by PostCode
################################################################
print('###############')
sView='Assess_Shop'
print('Creating:',sDatabaseName,'View:',sView)
sSQL="DROP VIEW IF EXISTS" + sView + ";"
sql.execute(sSQL,conn)

sSQL="CREATE VIEW" + sView + "AS"
sSQL=sSQL+ "SELECT"
sSQL=sSQL+ "TRIM(S.postcode) AS Shop_PostCode,"
sSQL=sSQL+ "'SP-' || TRIM(S.FirstCode) || '-' || TRIM(S.SecondCode) AS
Shop_Name,"
sSQL=sSQL+ "TRIM(S.FirstCode) AS Warehouse_PostCode,"
sSQL=sSQL+ "round(S.latitude,6) AS Shop_Latitude,"
sSQL=sSQL+ "round(S.longitude,6) AS Shop_Longitude"
sSQL=sSQL+ "FROM"
sSQL=sSQL+ "Assess_Warehouse_UK as W"
sSQL=sSQL+ "JOIN"
sSQL=sSQL+ "Assess_Shop_UK as S"
sSQL=sSQL+ "ON"
sSQL=sSQL+ "TRIM(W.postcode) = TRIM(S.FirstCode);"
sql.execute(sSQL,conn)
################################################################
```

I suggest you now use your knowledge in graph theory to calculate the routes. Make a graph, as follows:

```
################################################################
G=nx.Graph()
################################################################
print('###############')
sTable = 'Assess_HQ'
print('Loading:',sDatabaseName,'Table:',sTable)
sSQL="SELECT DISTINCT"
sSQL=sSQL+ "*"
sSQL=sSQL+ "FROM"
sSQL=sSQL+ " " + sTable + ";"
RouteData=pd.read_sql_query(sSQL, conn)
print('###############')
```

Add the HQ nodes to the graph.

```
################################################################
print(RouteData.head())
print('#############################')
print('HQ Rows:',RouteData.shape[0])
print('#############################')
################################################################
for i in range(RouteData.shape[0]):
    sNode0=RouteData['HQ_Name'][i]
    G.add_node(sNode0,
            Nodetype='HQ',
            PostCode=RouteData['HQ_PostCode'][i],
            Latitude=round(RouteData['HQ_Latitude'][i],6),
            Longitude=round(RouteData['HQ_Longitude'][i],6))
```

Add the warehouse nodes.

```
################################################################
print('###############')
sTable = 'Assess_Warehouse'
print('Loading:',sDatabaseName,'Table:',sTable)
```

```
sSQL="SELECT DISTINCT"
sSQL=sSQL+ "*"
sSQL=sSQL+ "FROM"
sSQL=sSQL+ " " + sTable + ";"
RouteData=pd.read_sql_query(sSQL, conn)
print('###############')
##############################################################
print(RouteData.head())
print('###############################')
print('Warehouse Rows:',RouteData.shape[0])
print('###############################')
##############################################################
for i in range(RouteData.shape[0]):
    sNode0=RouteData['Warehouse_Name'][i]
    G.add_node(sNode0,
                Nodetype='Warehouse',
                PostCode=RouteData['Warehouse_PostCode'][i],
                Latitude=round(RouteData['Warehouse_Latitude'][i],6),
                Longitude=round(RouteData['Warehouse_Longitude'][i],6))
```

Add the shops' nodes.

```
##############################################################
print('###############')
sTable = 'Assess_Shop'
print('Loading:',sDatabaseName,'Table:',sTable)
sSQL=" SELECT DISTINCT"
sSQL=sSQL+ "*"
sSQL=sSQL+ "FROM"
sSQL=sSQL+ " " + sTable + ";"
RouteData=pd.read_sql_query(sSQL, conn)
print('###############')
##############################################################
print(RouteData.head())
print('###############################')
print('Shop Rows:',RouteData.shape[0])
```

```
print('###############################')
###############################################################
for i in range(RouteData.shape[0]):
    sNode0=RouteData['Shop_Name'][i]
    G.add_node(sNode0,
                Nodetype='Shop',
                PostCode=RouteData['Shop_PostCode'][i],
                WarehousePostCode=RouteData['Warehouse_PostCode'][i],
                Latitude=round(RouteData['Shop_Latitude'][i],6),
                Longitude=round(RouteData['Shop_Longitude'][i],6))
```

Now we create the shipping routes. We will calculate a Vincenty's distance for each route, plus a transport method, in the form of a correct-size truck or a forklift.

```
###############################################################
## Create Edges
###############################################################
print('###############################')
print('Loading Edges')
print('###############################')
```

Let's loop through all the nodes and set up the routes that are valid.

```
for sNode0 in nx.nodes_iter(G):
    for sNode1 in nx.nodes_iter(G):
```

Let's create the HQ routes with Vincenty's distance between the HQs:

```
    if G.node[sNode0]['Nodetype']=='HQ' and \
       G.node[sNode1]['Nodetype']=='HQ' and \
       sNode0 != sNode1:
            distancemeters=round(\
                                vincenty(\
                                        (\
                                        G.node[sNode0]['Latitude'],\
                                        G.node[sNode0]['Longitude']\
                                        ),\
                                        (\
                                        G.node[sNode1]['Latitude']\
```

```
                                                      ,\
                                                      G.node[sNode1]['Longitude']\
                                                      )\
                                                      ).meters\
                                  ,0)
          distancemiles=round(\
                              vincenty(\
                                      (\
                                      G.node[sNode0]['Latitude'],\
                                      G.node[sNode0]['Longitude']\
                                      ),\
                                      (\
                                      G.node[sNode1]['Latitude']\
                                      ,\
                                      G.node[sNode1]['Longitude']\
                                      )\
                                      ).miles\
                                  ,3)

      if distancemiles >= 0.05:
          cost = round(150+(distancemiles * 2.5),6)
          vehicle='V001'
      else:
          cost = round(2+(distancemiles * 0.10),6)
          vehicle='ForkLift'

      G.add_edge(sNode0,sNode1,DistanceMeters=distancemeters,\
              DistanceMiles=distancemiles,\
              Cost=cost,Vehicle=vehicle)
    if nVSet==True:
      print('Edge-H-H:',sNode0,' to ', sNode1,\
          ' Distance:',distancemeters,'meters',\
        distancemiles,'miles','Cost', cost,'Vehicle',vehicle)
```

Let's create the headquarters (HQ)-to-warehouse routes with the Vincenty's distance between the HQ's and warehouses.'

```
if G.node[sNode0]['Nodetype']=='HQ' and\
    G.node[sNode1]['Nodetype']=='Warehouse' and\
    sNode0 != sNode1:
        distancemeters=round(\
                            vincenty(\
                                (\
                                G.node[sNode0]['Latitude'],\
                                G.node[sNode0]['Longitude']\
                                ),\
                                (\
                                G.node[sNode1]['Latitude']\
                                ,\
                                G.node[sNode1]['Longitude']\
                                )\
                                ).meters\
                        ,0)
        distancemiles=round(\
                            vincenty(\
                                (\
                                G.node[sNode0]['Latitude'],\
                                G.node[sNode0]['Longitude']\
                                ),\
                                (\
                                G.node[sNode1]['Latitude']\
                                ,\
                                G.node[sNode1]['Longitude']\
                                )\
                                ).miles\
                        ,3)
        if distancemiles >= 10:
            cost = round(50+(distancemiles * 2),6)
            vehicle='V002'
        else:
            cost = round(5+(distancemiles * 1.5),6)
            vehicle='V003'
        if distancemiles <= 50:
```

```
            G.add_edge(sNode0,sNode1,DistanceMeters=distancemeters,\
                    DistanceMiles=distancemiles,\
                    Cost=cost,Vehicle=vehicle)
        if nVSet==True:
            print('Edge-H-W:',sNode0,'to',sNode1,\
                'Distance:',distancemeters,'meters',\
                distancemiles,'miles','Cost',cost,'Vehicle',vehicle)
```

Let's create the intra-warehouse routes with the Vincenty's distance between the warehouses.

```
if nSet==True and \
    G.node[sNode0]['Nodetype']=='Warehouse' and\
    G.node[sNode1]['Nodetype']=='Warehouse' and\
    sNode0 != sNode1:
        distancemeters=round(\
                            vincenty(\
                                (\
                                G.node[sNode0]['Latitude'],\
                                G.node[sNode0]['Longitude']\
                                ),\
                                (\
                                G.node[sNode1]['Latitude']\
                                ,\
                                G.node[sNode1]['Longitude']\
                                )\
                                ).meters\
                        ,0)
        distancemiles=round(\
                            vincenty(\
                                (\
                                G.node[sNode0]['Latitude'],\
                                G.node[sNode0]['Longitude']\
                                ),\
                                (\
                                G.node[sNode1]['Latitude']\
```

```
                                    ,\
                                    G.node[sNode1]['Longitude']\
                                    )\
                                    ).miles\
                        ,3)
    if distancemiles >= 10:
        cost = round(50+(distancemiles * 1.10),6)
        vehicle='V004'
    else:
        cost = round(5+(distancemiles * 1.05),6)
        vehicle='V005'

    if distancemiles <= 20:
        G.add_edge(sNode0,sNode1,DistanceMeters=distancemeters,\
                    DistanceMiles=distancemiles,\
                    Cost=cost,Vehicle=vehicle)
        if nVSet==True:
            print('Edge-W-W:',sNode0,'to',sNode1,\
                'Distance:',distancemeters,'meters',\
                distancemiles,'miles','Cost',cost,'Vehicle',vehicle)
```

Let's create the warehouse-to-shop routes with the Vincenty's distance.

```
if G.node[sNode0]['Nodetype']=='Warehouse' and \
    G.node[sNode1]['Nodetype']=='Shop' and \
    G.node[sNode0]['PostCode']==G.node[sNode1]['WarehousePostCode'] and\
    sNode0 != sNode1:

        distancemeters=round(\
                        vincenty(\
                                (\
                                G.node[sNode0]['Latitude'],\
                                G.node[sNode0]['Longitude']\
                                ),\
                                (\
                                G.node[sNode1]['Latitude']\
                                ,\
```

```
                                          G.node[sNode1]['Longitude']\
                                          )\
                                          ).meters\
                             ,0)
        distancemiles=round(\
                          vincenty(\
                                 (\
                                 G.node[sNode0]['Latitude'],\
                                 G.node[sNode0]['Longitude']\
                                 ),\
                                 (\
                                  G.node[sNode1]['Latitude']\
                                  ,\
                                  G.node[sNode1]['Longitude']\
                                 )\
                                 ).miles\
                         ,3)
        if distancemiles >= 10:
            cost = round(50+(distancemiles * 1.50),6)
            vehicle='V006'
        else:
            cost = round(5+(distancemiles * 0.75),6)
            vehicle='V007'
        if distancemiles <= 10:
            G.add_edge(sNode0,sNode1,DistanceMeters=distancemeters,\
                      DistanceMiles=distancemiles,\
                      Cost=cost,Vehicle=vehicle)
            if nVSet==True:
                print('Edge-W-S:',sNode0,'to',sNode1,\
                 'Distance:',distancemeters,'meters',\
                  distancemiles,'miles','Cost', cost,'Vehicle',vehicle)
```

Let's create the intra-shop routes with the Vincenty's distance.

```
    if nSet==True and\
        G.node[sNode0]['Nodetype']=='Shop' and\
        G.node[sNode1]['Nodetype']=='Shop' and\
```

```
G.node[sNode0]['WarehousePostCode']==G.node[sNode1]
['WarehousePostCode'] and\
sNode0 != sNode1:

    distancemeters=round(\
                        vincenty(\
                            (\
                            G.node[sNode0]['Latitude'],\
                            G.node[sNode0]['Longitude']\
                            ),\
                            (\
                            G.node[sNode1]['Latitude']\
                            ,\
                             G.node[sNode1]['Longitude']\
                             )\
                             ).meters\
                        ,0)
    distancemiles=round(\
                        vincenty(\
                            (\
                            G.node[sNode0]['Latitude'],\
                            G.node[sNode0]['Longitude']\
                            ),\
                            (\
                             G.node[sNode1]['Latitude']\
                             ,\
                             G.node[sNode1]['Longitude']\
                             )\
                             ).miles\
                        ,3)

    if distancemiles >= 0.05:
        cost = round(5+(distancemiles * 0.5),6)
        vehicle='V008'
    else:
        cost = round(1+(distancemiles * 0.1),6)
        vehicle='V009'
```

```
            if distancemiles <= 0.075:
                G.add_edge(sNode0,sNode1,DistanceMeters=distancemeters,\
                        DistanceMiles=distancemiles, \
                        Cost=cost,Vehicle=vehicle)
            if nVSet==True:
                print('Edge-S-S:',sNode0,'to',sNode1,\
                    'Distance:',distancemeters,'meters',\

    if nSet==True and\
        G.node[sNode0]['Nodetype']=='Shop' and\
        G.node[sNode1]['Nodetype']=='Shop' and\
        G.node[sNode0]['WarehousePostCode']!=G.node[sNode1]
        ['WarehousePostCode'] and \
        sNode0 != sNode1:

            distancemeters=round(\
                            vincenty(\
                                (\
                                G.node[sNode0]['Latitude'],\
                                G.node[sNode0]['Longitude']\
                                ),\
                                (\
                                G.node[sNode1]['Latitude']\
                                 ,\
                                 G.node[sNode1]['Longitude']\
                                 )\
                                 ).meters\
                        ,0)
            distancemiles=round(\
                            vincenty(\
                                (\
                                G.node[sNode0]['Latitude'],\
                                G.node[sNode0]['Longitude']\
                                ),\
                                (\
                                G.node[sNode1]['Latitude']\
```

```
                                                ,\
                                        G.node[sNode1]['Longitude']\
                                        )\
                                        ).miles\
                        ,3)

            cost = round(1+(distancemiles * 0.1),6)
            vehicle='V010'

             if distancemiles <= 0.025:
                G.add_edge(sNode0,sNode1,DistanceMeters=distancemeters,\
                        DistanceMiles=distancemiles,\
                        Cost=cost,Vehicle=vehicle)
                if nVSet==True:
                    print('Edge-S-S:',sNode0,'to',sNode1,\
                      'Distance:',distancemeters,'meters',\
                      distancemiles,'miles','Cost', cost,'Vehicle',vehicle)
################################################################
sFileName=sFileDir + '/' + OutputFileName1
print('##############################')
print('Storing:',sFileName)
print('##############################')
nx.write_gml(G,sFileName)
sFileName=sFileName +'.gz'
nx.write_gml(G,sFileName)
################################################################
print('Nodes:',nx.number_of_nodes(G))
print('Edges:',nx.number_of_edges(G))
################################################################
sFileName=sFileDir + '/' + OutputFileName2
print('##############################')
print('Storing:',sFileName)
print('##############################')
```

I will now guide you through the process of extracting the routes or path from the shipping plan in the graph. We will use nx.shortest_path(G,x,y) to extract the shortest path between nodes x and y on graph G.

```
################################################################
## Create Paths
################################################################
print('###############################')
print('Loading Paths')
print('###############################')
f = open(sFileName,'w')
l=0
sline = 'ID|Cost|StartAt|EndAt|Path|Measure'
if nVSet==True: print ('0', sline)
f.write(sline+ '\n')
for sNode0 in nx.nodes_iter(G):
    for sNode1 in nx.nodes_iter(G):
        if sNode0 != sNode1 and\
            nx.has_path(G, sNode0, sNode1)==True and\
            nx.shortest_path_length(G, \
              source=sNode0, \
              target=sNode1, \
              weight='DistanceMiles') < nMaxPath:
                l+=1
                sID='{:.0f}'.format(l)
                spath = ','.join(nx.shortest_path(G,\
                  source=sNode0,\
                  target=sNode1,\
                  weight='DistanceMiles'))
                slength= '{:.6f}'.format(\
                  nx.shortest_path_length(G,\
                  source=sNode0,\
                  target=sNode1,\
                  weight='DistanceMiles'))
                sline = sID + '|"DistanceMiles"|"' + sNode0 + '"|"'\
                + sNode1 + '"|"' + spath + '"|' + slength
                if nVSet==True: print (sline)
                f.write(sline + '\n')
                l+=1
```

```python
sID='{:.0f}'.format(l)
spath = ','.join(nx.shortest_path(G,\
  source=sNode0,\
  target=sNode1,\
  weight='DistanceMeters'))
slength= '{:.6f}'.format(\
  nx.shortest_path_length(G,\
  source=sNode0,\
  target=sNode1,\
  weight='DistanceMeters'))
sline = sID + '|"DistanceMeters"|"' + sNode0 + '"|"'\
+ sNode1 + '"|"' + spath + '"|' + slength
if nVSet==True: print(sline)
f.write(sline + '\n')
l+=1
sID='{:.0f}'.format(l)
spath = ','.join(nx.shortest_path(G,\
  source=sNode0,\
  target=sNode1,\
  weight='Cost'))
slength= '{:.6f}'.format(\
  nx.shortest_path_length(G,\
  source=sNode0,\
  target=sNode1,\
  weight='Cost'))
sline = sID + '|"Cost"|"' + sNode0 + '"|"'\
+ sNode1 + '"|"' + spath + '"|' + slength
if nVSet==True: print (sline)
f.write(sline + '\n')
f.close()
################################################################
print('Nodes:',nx.number_of_nodes(G))
print('Edges:',nx.number_of_edges(G))
print('Paths:',sID)
################################################################
```

405

```
###############################################################
print('###############')
print('Vacuum Database')
sSQL="VACUUM;"
sql.execute(sSQL,conn)
print('###############')
###############################################################
print('### Done!! ###########################################')
###############################################################
```

Save the Assess-Shipping-Routes.py file, then compile and execute with your Python compiler. This will produce a set of demonstrated values onscreen, plus a graph data file named Assess_Shipping_Routes.gml.

Note   You can view the GML file with a text editor.

The path finder generates a file with shortest paths, named Assess_Shipping_Routes.txt. You now have the basis of a working shipping route graph that you can use to generate any queries you may have against the solution. This final Python script completes the access step for Hillman Ltd.

Clark Ltd

Clark Ltd is the accountancy company that handles everything related to the VKHCG's finances and personnel. I will let you investigate Clark with your new knowledge. You can do this!

Simple Forex Trading Planner

Clark requires the assessment of the group's forex data, for processing and data quality issues. I will guide you through an example of a forex solution.

Open your Python editor and create a file named Assess-Forex.py in directory C:\ VKHCG\04-Clark\02-Assess.

Now copy the following code into the file:

```python
###############################################################
import sys
import os
import sqlite3 as sq
import pandas as pd
###############################################################
if sys.platform == 'linux':
    Base=os.path.expanduser('~') + 'VKHCG'
else:
    Base='C:/VKHCG'
print('###############################')
print('Working Base:',Base,'using',sys.platform)
print('###############################')
###############################################################
Company='04-Clark'
sInputFileName1='01-Vermeulen/01-Retrieve/01-EDS/02-Python/Retrieve-
Country-Currency.csv'
sInputFileName2='04-Clark/01-Retrieve/01-EDS/01-R/Retrieve_Euro_
ExchangeRates.csv'
###############################################################
sDataBaseDir=Base + '/' + Company + '/02-Assess/SQLite'
if not os.path.exists(sDataBaseDir):
    os.makedirs(sDataBaseDir)
###############################################################
sDatabaseName=sDataBaseDir + '/clark.db'
conn = sq.connect(sDatabaseName)
###############################################################
### Import Country Data
###############################################################
sFileName1=Base + '/' + sInputFileName1
print('###############################')
print('Loading:',sFileName1)
print('###############################')
```

```
CountryRawData=pd.read_csv(sFileName1,header=0,low_memory=False,
encoding="latin-1")
CountryRawData.drop_duplicates(subset=None, keep='first', inplace=True)
CountryData=CountryRawData
print('Loaded Company:',CountryData.columns.values)
print('###############################')
################################################################
print('###############')
sTable='Assess_Country'
print('Storing:',sDatabaseName,' Table:',sTable)
CountryData.to_sql(sTable, conn, if_exists="replace")
print('###############')
################################################################
print(CountryData.head())
print('###############################')
print('Rows: ',CountryData.shape[0])
print('###############################')
################################################################
### Import Forex Data
################################################################
sFileName2=Base + '/' + sInputFileName2
print('###############################')
print('Loading:',sFileName2)
print('###############################')
ForexRawData=pd.read_csv(sFileName2,header=0,low_memory=False,
encoding="latin-1")
ForexRawData.drop_duplicates(subset=None, keep='first', inplace=True)
ForexData=ForexRawData.head(5)
print('Loaded Company:',ForexData.columns.values)
print('###############################')
################################################################
print('###############')
sTable='Assess_Forex'
print('Storing:',sDatabaseName,'Table:',sTable)
ForexData.to_sql(sTable, conn, if_exists="replace")
print('###############')
```

```
################################################################
print(ForexData.head())
print('##############################')
print('Rows:',ForexData.shape[0])
print('##############################')
################################################################
print('###############')
sTable='Assess_Forex'
print('Loading:',sDatabaseName,'Table:',sTable)
sSQL="select distinct"
sSQL=sSQL+ "A.CodeIn"
sSQL=sSQL+ "from"
sSQL=sSQL+ "Assess_Forex as A;"
CodeData=pd.read_sql_query(sSQL, conn)
print('###############')
################################################################

for c in range(CodeData.shape[0]):
    print('###############')
    sTable='Assess_Forex & 2x Country > ' + CodeData['CodeIn'][c]
    print('Loading:',sDatabaseName,'Table:',sTable)
    sSQL="select distinct"
    sSQL=sSQL+ "A.Date,"
    sSQL=sSQL+ "A.CodeIn,"
    sSQL=sSQL+ "B.Country as CountryIn,"
    sSQL=sSQL+ "B.Currency as CurrencyNameIn,"
    sSQL=sSQL+ "A.CodeOut,"
    sSQL=sSQL+ "C.Country as CountryOut,"
    sSQL=sSQL+ "C.Currency as CurrencyNameOut,"
    sSQL=sSQL+ "A.Rate"
    sSQL=sSQL+ "from"
    sSQL=sSQL+ "Assess_Forex as A"
    sSQL=sSQL+ "JOIN"
    sSQL=sSQL+ "Assess_Country as B"
    sSQL=sSQL+ "ON A.CodeIn = B.CurrencyCode"
    sSQL=sSQL+ "JOIN"
```

```
    sSQL=sSQL+ "Assess_Country as C"
    sSQL=sSQL+ "ON A.CodeOut = C.CurrencyCode"
    sSQL=sSQL+ "WHERE"
    sSQL=sSQL+ "A.CodeIn ='" + CodeData['CodeIn'][c] + "';"
    ForexData=pd.read_sql_query(sSQL, conn).head(1000)
    print('###############')
    print(ForexData)
    print('###############')
    sTable='Assess_Forex_' + CodeData['CodeIn'][c]
    print('Storing:',sDatabaseName,'Table:',sTable)
    ForexData.to_sql(sTable, conn, if_exists="replace")
    print('###############')
    print('#############################')
    print('Rows:',ForexData.shape[0])
    print('#############################')
################################################################
print('### Done!! #####################################')
################################################################
```

Save the Assess-Forex.py file, then compile and execute with your Python compiler. This will produce a set of demonstrated values onscreen

Financials

Clark requires you to process the balance sheet for the VKHCG group companies. I will guide you through a sample balance sheet data assessment, to ensure that only the good data is processed.

Open your Python editor and create a file named Assess-Financials.py in directory C:\VKHCG\04-Clark\02-Assess. Now copy the following code into the file:

```
################################################################
import sys
import os
import sqlite3 as sq
import pandas as pd
################################################################
if sys.platform == 'linux':
```

```
    Base=os.path.expanduser('~') + 'VKHCG'
else:
    Base='C:/VKHCG'
print('##############################')
print('Working Base:',Base,'using',sys.platform)
print('##############################')
################################################################
Company='04-Clark'
sInputFileName='01-Retrieve/01-EDS/01-R/Retrieve_Profit_And_Loss.csv'
################################################################
sDataBaseDir=Base + '/' + Company + '/02-Assess/SQLite'
if not os.path.exists(sDataBaseDir):
    os.makedirs(sDataBaseDir)
################################################################
sDatabaseName=sDataBaseDir + '/clark.db'
conn = sq.connect(sDatabaseName)
################################################################
### Import Financial Data
################################################################
sFileName=Base + '/' + Company + '/' + sInputFileName
print('##############################')
print('Loading:',sFileName)
print('##############################')
FinancialRawData=pd.read_csv(sFileName,header=0,low_memory=False,
encoding="latin-1")
FinancialData=FinancialRawData
print('Loaded Company:',FinancialData.columns.values)
print('##############################')
################################################################
print('################')
sTable='Assess-Financials'
print('Storing:',sDatabaseName,'Table:',sTable)
FinancialData.to_sql(sTable, conn, if_exists="replace")
print('################')
################################################################
print(FinancialData.head())
```

411

```
print('###############################')
print('Rows:',FinancialData.shape[0])
print('###############################')
################################################################
################################################################
print('### Done!! #####################################')
################################################################
```

Save the Assess-Financials.py file, then compile and execute with your Python compiler. This will produce a set of demonstrated values onscreen

Financial Calendar

Clark stores all the master records for the financial calendar. So, I suggest we import the calendar from the retrieve step's data storage.

Open your Python editor and create a file named Assess-Calendar.py in directory C:\VKHCG\04-Clark\02-Assess. Now copy the following code into the file:

```
################################################################
import sys
import os
import sqlite3 as sq
import pandas as pd
################################################################
if sys.platform == 'linux':
    Base=os.path.expanduser('~') + 'VKHCG'
else:
    Base='C:/VKHCG'
print('###############################')
print('Working Base:',Base,'using',sys.platform)
print('###############################')
################################################################
Company='04-Clark'
################################################################
sDataBaseDirIn=Base + '/' + Company + '/01-Retrieve/SQLite'
if not os.path.exists(sDataBaseDirIn):
    os.makedirs(sDataBaseDirIn)
```

```python
sDatabaseNameIn=sDataBaseDirIn + '/clark.db'
connIn = sq.connect(sDatabaseNameIn)
################################################################
sDataBaseDirOut=Base + '/' + Company + '/01-Retrieve/SQLite'
if not os.path.exists(sDataBaseDirOut):
    os.makedirs(sDataBaseDirOut)
sDatabaseNameOut=sDataBaseDirOut + '/clark.db'
connOut = sq.connect(sDatabaseNameOut)
################################################################
sTableIn='Retrieve_Date'
sSQL='select * FROM' + sTableIn + ';'
print('###############')
sTableOut='Assess_Time'
print('Loading:',sDatabaseNameIn,'Table:',sTableIn)
dateRawData=pd.read_sql_query(sSQL, connIn)
dateData=dateRawData
################################################################
print('###############################')
print('Load Rows:',dateRawData.shape[0],'records')
print('###############################')
dateData.drop_duplicates(subset='FinDate', keep='first', inplace=True)
################################################################
print('###############')
sTableOut='Assess_Date'
print('Storing:',sDatabaseNameOut,'Table:',sTableOut)
dateData.to_sql(sTableOut, connOut, if_exists="replace")
print('###############')
################################################################
print('###############################')
print('Store Rows:',dateData.shape[0],' records')
print('###############################')
################################################################
################################################################
sTableIn='Retrieve_Time'
sSQL='select * FROM' + sTableIn + ';'
```

413

```
print('################')
sTableOut='Assess_Time'
print('Loading:',sDatabaseNameIn,'Table:',sTableIn)
timeRawData=pd.read_sql_query(sSQL, connIn)
timeData=timeRawData
################################################################
print('#################################')
print('Load Rows: ',timeData.shape[0],'records')
print('#################################')
timeData.drop_duplicates(subset=None, keep='first', inplace=True)
################################################################
print('################')
sTableOut='Assess_Time'
print('Storing:',sDatabaseNameOut,'Table:',sTableOut)
timeData.to_sql(sTableOut, connOut, if_exists="replace")
print('################')
################################################################
print('#################################')
print('Store Rows:',timeData.shape[0],'records')
print('#################################')
################################################################
print('### Done!! #########################################')
################################################################
```

Save the Assess-Calendar.py file, then compile and execute with your Python compiler. This will produce database tables named Assess_Date and Assess_Time.

People

Clark Ltd generates the payroll, so it holds all the staff records. Clark also handles all payments to suppliers and receives payments from customers' details on all companies.

Open your Python editor and create a file named Assess-People.py in directory C:\ VKHCG\04-Clark\02-Assess. Now copy the following code into the file:

```
################################################################
import sys
import os
```

```python
import sqlite3 as sq
import pandas as pd
################################################################
if sys.platform == 'linux':
    Base=os.path.expanduser('~') + 'VKHCG'
else:
    Base='C:/VKHCG'
print('#############################')
print('Working Base:',Base,'using',sys.platform)
print('#############################')
################################################################
Company='04-Clark'
sInputFileName1='01-Retrieve/01-EDS/02-Python/Retrieve-Data_female-names.
csv'
sInputFileName2='01-Retrieve/01-EDS/02-Python/Retrieve-Data_male-names.csv'
sInputFileName3='01-Retrieve/01-EDS/02-Python/Retrieve-Data_last-names.csv'
sOutputFileName1='Assess-Staff.csv'
sOutputFileName2='Assess-Customers.csv'
################################################################
sDataBaseDir=Base + '/' + Company + '/02-Assess/SQLite'
if not os.path.exists(sDataBaseDir):
    os.makedirs(sDataBaseDir)
################################################################
sDatabaseName=sDataBaseDir + '/clark.db'
conn = sq.connect(sDatabaseName)
################################################################
### Import Female Data
################################################################
sFileName=Base + '/' + Company + '/' + sInputFileName1
print('#############################')
print('Loading :',sFileName)
print('#############################')
print(sFileName)
FemaleRawData=pd.read_csv(sFileName,header=0,low_memory=False,
encoding="latin-1")
FemaleRawData.rename(columns={'NameValues':'FirstName'},inplace=True)
```

415

```python
FemaleRawData.drop_duplicates(subset=None, keep='first', inplace=True)
FemaleData=FemaleRawData.sample(100)
print('##############################')
################################################################
print('###############')
sTable='Assess_FemaleName'
print('Storing:',sDatabaseName,'Table:',sTable)
FemaleData.to_sql(sTable, conn, if_exists="replace")
print('###############')
################################################################
print('##############################')
print('Rows:',FemaleData.shape[0],'records')
print('##############################')
################################################################
### Import Male Data
################################################################
sFileName=Base + '/' + Company + '/' + sInputFileName2
print('##############################')
print('Loading:',sFileName)
print('##############################')
MaleRawData=pd.read_csv(sFileName,header=0,low_memory=False,
encoding="latin-1")
MaleRawData.rename(columns={'NameValues':'FirstName'},inplace=True)
MaleRawData.drop_duplicates(subset=None, keep='first', inplace=True)
MaleData=MaleRawData.sample(100)
print('##############################')
################################################################
print('###############')
sTable='Assess_MaleName'
print('Storing:',sDatabaseName,'Table:',sTable)
MaleData.to_sql(sTable, conn, if_exists="replace")
print('###############')
################################################################
print('##############################')
print('Rows:',MaleData.shape[0],'records')
print('##############################')
```

```
################################################################
### Import Surname Data
################################################################
sFileName=Base + '/' + Company + '/' + sInputFileName3
print('###############################')
print('Loading:',sFileName)
print('###############################')
SurnameRawData=pd.read_csv(sFileName,header=0,low_memory=False,
encoding="latin-1")
SurnameRawData.rename(columns={'NameValues':'LastName'},inplace=True)
SurnameRawData.drop_duplicates(subset=None, keep='first', inplace=True)
SurnameData=SurnameRawData.sample(200)
print('###############################')
################################################################
print('###############')
sTable='Assess_Surname'
print('Storing:',sDatabaseName,'Table:',sTable)
SurnameData.to_sql(sTable, conn, if_exists="replace")
print('###############')
################################################################
print('###############################')
print('Rows:',SurnameData.shape[0],'records')
print('###############################')
################################################################
################################################################
print('###############')
sTable='Assess_FemaleName & Assess_MaleName'
print('Loading:',sDatabaseName,'Table:',sTable)
sSQL="select distinct"
sSQL=sSQL+ "A.FirstName,"
sSQL=sSQL+ "'Female' as Gender"
sSQL=sSQL+ "from"
sSQL=sSQL+ "Assess_FemaleName as A"
sSQL=sSQL+ "UNION"
sSQL=sSQL+ "select distinct"
sSQL=sSQL+ "A.FirstName,"
```

417

```
sSQL=sSQL+ "'Male' as Gender"
sSQL=sSQL+ "from"
sSQL=sSQL+ "Assess_MaleName as A;"
FirstNameData=pd.read_sql_query(sSQL, conn)
print('################')
####################################################################
#print('################')
sTable='Assess_FirstName'
print('Storing:',sDatabaseName,'Table:',sTable)
FirstNameData.to_sql(sTable, conn, if_exists="replace")
print('################')
####################################################################
####################################################################
print('################')
sTable='Assess_FirstName x2 & Assess_Surname'
print('Loading:',sDatabaseName,'Table:',sTable)
sSQL="select distinct"
sSQL=sSQL+ "A.FirstName,"
sSQL=sSQL+ "B.FirstName AS SecondName,"
sSQL=sSQL+ "C.LastName,"
sSQL=sSQL+ "A.Gender"
sSQL=sSQL+ "from"
sSQL=sSQL+ "Assess_FirstName as A"
sSQL=sSQL+ ","
sSQL=sSQL+ "Assess_FirstName as B"
sSQL=sSQL+ ","
sSQL=sSQL+ "Assess_Surname as C"
sSQL=sSQL+ "WHERE"
sSQL=sSQL+ "A.Gender = B.Gender"
sSQL=sSQL+ "AND"
sSQL=sSQL+ "A.FirstName <> B.FirstName;"
PeopleRawData=pd.read_sql_query(sSQL, conn)
People1Data=PeopleRawData.sample(10000)

sTable='Assess_FirstName & Assess_Surname'
print('Loading:',sDatabaseName,'Table:',sTable)
```

```
sSQL="select distinct"
sSQL=sSQL+ "A.FirstName,"
sSQL=sSQL+ "'' AS SecondName,"
sSQL=sSQL+ "B.LastName,"
sSQL=sSQL+ "A.Gender"
sSQL=sSQL+ "from"
sSQL=sSQL+ "Assess_FirstName as A"
sSQL=sSQL+ ","
sSQL=sSQL+ "Assess_Surname as B;"
PeopleRawData=pd.read_sql_query(sSQL, conn)
People2Data=PeopleRawData.sample(10000)
PeopleData=People1Data.append(People2Data)
print(PeopleData)
print('###############')
###################################################################
#print('###############')
sTable='Assess_People'
print('Storing:',sDatabaseName,'Table:',sTable)
PeopleData.to_sql(sTable, conn, if_exists="replace")
print('###############')
###################################################################
sFileDir=Base + '/' + Company + '/02-Assess/01-EDS/02-Python'
if not os.path.exists(sFileDir):
    os.makedirs(sFileDir)
###################################################################
sOutputFileName = sTable+'.csv'
sFileName=sFileDir + '/' + sOutputFileName
print('##############################')
print('Storing:', sFileName)
print('##############################')
PeopleData.to_csv(sFileName, index = False)
print('##############################')
###################################################################
print('### Done!! #######################################')
###################################################################
```

Save the `Assess-People.py` file, then compile and execute with your Python compiler. This will produce a set of demonstrated values onscreen

Summary

Congratulations! You have completed the Assess superstep. You now have a stable base for blending data sources into a data vault.

I have guided you through your first data-quality processing steps. The results you achieved will serve as the base data for Chapter 9.

So, what should you know at this stage in the book? You should know how to

- Effectively solve different data-quality concerns.

- Convert and update the data mapping matrix as you perform the assess processing. Every data source you assess is a clean source of data science information.

- Successfully perform feature engineering.

- Introduce standards, via the assessments, against other data sources, and look-ups against standards such as the currency code and country codes.

- Transform column-format data into graph relationships. This is an important data science proficiency to master.

Note   In real-life data science projects, you would by now have consumed as high as 50% of the project resources to reach the stable data set you accomplished today. Well done! Data preparation is the biggest drain on your labor, budget, and computational resources.

Now, you have mastered the Assess superstep and become skilled at handling data quality. The next steps are more structured in nature, as you will now create a common data structure to form the basis of the full-scale data science investigation that begins in Chapter 9.

I recommend that you get a good assortment of snacks, and then we can advance to the next chapter, in which we will construct a data vault for your data science.

CHAPTER 9

Process Superstep

The Process superstep adapts the assess results of the retrieve versions of the data sources into a highly structured data vault that will form the basic data structure for the rest of the data science steps. This data vault involves the formulation of a standard data amalgamation format across a range of projects.

Tip   If you follow the rules of the data vault, it results in a clean and stable structure for your future data science.

The Process superstep is the amalgamation process that pipes your data sources into five main categories of data (Figure 9-1).

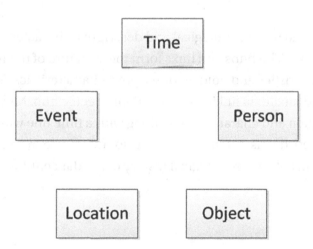

Figure 9-1. *Five categories of data*

Using only these five hubs in your data vault, and with good modeling, you can describe most activities of your customers. This enable you to then fine-tune your data science algorithms, to simply understand the five hubs' purpose and relationships that enable good data science.

© Andreas François Vermeulen 2018
A. F. Vermeulen, *Practical Data Science*, https://doi.org/10.1007/978-1-4842-3054-1_9

Data Vault

Data vault modeling is a database modeling method designed by Dan Linstedt. The data structure is designed to be responsible for long-term historical storage of data from multiple operational systems. It supports chronological historical data tracking for full auditing and enables parallel loading of the structures.

Hubs

Data vault hubs contain a set of unique business keys that normally do not change over time and, therefore, are stable data entities to store in hubs. Hubs hold a surrogate key for each hub data entry and metadata labeling the source of the business key.

Links

Data vault links are associations between business keys. These links are essentially many-to-many joins, with additional metadata to enhance the particular link.

Satellites

Data vault satellites hold the chronological and descriptive characteristics for a specific section of business data. The hubs and links form the structure of the model but have no chronological characteristics and hold no descriptive characteristics. Satellites consist of characteristics and metadata linking them to their specific hub. Metadata labeling the origin of the association and characteristics, along with a time line with start and end dates for the characteristics, is put in safekeeping, for future use from the data section. Each satellite holds an entire chronological history of the data entities within the specific satellite.

Reference Satellites

Reference satellites are referenced from satellites but under no circumstances bound with metadata for hub keys. They prevent redundant storage of reference characteristics that are used regularly by other satellites.

Typical reference satellites are

- *Standard codes*: These are codes such as ISO 3166 for country codes, ISO 4217 for currencies, and ISO 8601 for time zones.

- *Fixed lists for specific characteristics*: These can be standard lists that reduce other standard lists. For example, the list of countries your business has offices in may be a reduced fixed list from the ISO 3166 list. You can also generate your own list of, say, business regions, per your own reporting structures.

- *Conversion lookups*: Look at Global Positioning System (GPS) transformations, such as the Molodensky transformation, Helmert transformation, Molodensky-Badekas transformation, or Gauss-Krüger coordinate system.

   The most common is WGS84, the standard U.S. Department of Defense definition of a global location system for geospatial information, and is the reference system for the GPS.

Time-Person-Object-Location-Event Data Vault

The data vault we use is based on the Time-Person-Object-Location-Event (T-P-O-L-E) design principle (Figure 9-2).

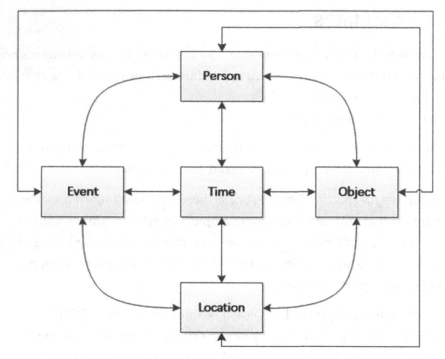

Figure 9-2. *T-P-O-L-E—high-level design*

Note   The entire data structures discussed over the next pages can be found in the `datavault.db` file, which can be found in directory `\VKHCG\88-DV`.

Time Section

The time section contains the complete data structure for all data entities related to recording the time at which everything occurred.

Time Hub

The time hub consists of the following fields:

```
CREATE TABLE [Hub-Time] (
    IDNumber       VARCHAR (100) PRIMARY KEY,
    IDTimeNumber   Integer,
    ZoneBaseKey    VARCHAR (100),
```

```
DateTimeKey    VARCHAR (100),
DateTimeValue DATETIME
);
```

This time hub acts as the core connector between time zones and other date-time associated values.

Time Links

The time links link the time hub to the other hubs (Figure 9-3).

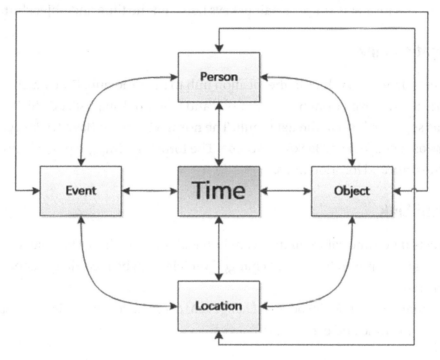

Figure 9-3. *Time links*

The following links are supported.

Time-Person Link

This connects date-time values within the person hub to the time hub. The physical link structure is stored as a many-to-many relationship time within the data vault. Later in this chapter, I will supply details on how to construct the tables.

Dates such as birthdays, marriage anniversaries, and the date of reading this book can be recorded as separate links in the data vault. The normal format is `BirthdayOn`, `MarriedOn`, or `ReadBookOn`. The format is simply a pair of keys between the time and person hubs.

Time-Object Link

This connects date-time values within the object hub to the time hub. Dates such as those on which you bought a car, sold a car, and read this book can be recorded as separate links in the data vault. The normal format is `BoughtCarOn`, `SoldCarOn`, or `ReadBookOn`. The format is simply a pair of keys between the time and object hubs.

Time-Location Link

This connects date-time values in the location hub to the time hub. Dates such as moved to post code SW1, moved from post code SW1, and read book at post code SW1 can be recorded as separate links in the data vault. The normal format is `MovedToPostCode`, `MovedFromPostCode`, or `ReadBookAtPostCode`. The format is simply a pair of keys between the time and location hubs.

Time-Event Link

This connects date-time values in the event hub with the time hub. Dates such as those on which you have moved house and changed vehicles can be recorded as separate links in the data vault.

The normal format is `MoveHouse` or `ChangeVehicle`. The format is simply a pair of keys between the time and event hubs.

Time Satellites

Time satellites are the part of the vault that stores the following fields.

Note   The `<Time Zone>` part of the table name is from a list of time zones that I will discuss later in the chapter.

```
CREATE TABLE [Satellite-Time-<Time Zone>] (
    IDZoneNumber       VARCHAR (100) PRIMARY KEY,
    IDTimeNumber       INTEGER,
    ZoneBaseKey        VARCHAR (100),
    DateTimeKey        VARCHAR (100),
    UTCDateTimeValue DATETIME,
    Zone               VARCHAR (100),
    DateTimeValue      DATETIME
);
```

Time satellites enable you to work more easily with international business patterns. You can move between time zones to look at such patterns as "In the morning . . . " or "At lunchtime . . . " These capabilities will be used during the Transform superstep, to discover patterns and behaviors around the world.

Person Section

The person section contains the complete data structure for all data entities related to recording the person involved.

Person Hub

The person hub consists of a series of fields that supports a "real" person. The person hub consists of the following fields:

```
CREATE TABLE [Hub-Person] (
    IDPersonNumber     INTEGER,
    FirstName          VARCHAR (200),
    SecondName         VARCHAR (200),
    LastName           VARCHAR (200),
    Gender             VARCHAR (20),
    TimeZone           VARCHAR (100),
    BirthDateKey       VARCHAR (100),
    BirthDate          DATETIME
);
```

Person Links

This links the person hub to the other hubs (Figure 9-4).

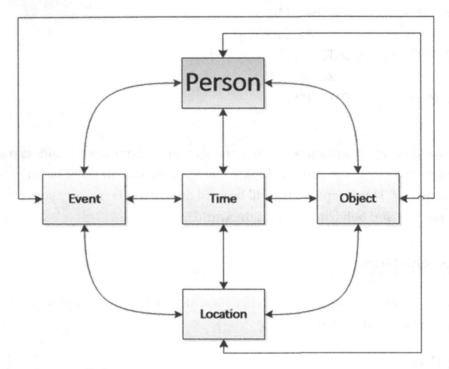

Figure 9-4. *Person links*

The following links are supported in person links.

Person-Time Link

This link joins the person to the time hub, to describe the relationships between the two hubs. The link consists of the following fields:

```
CREATE TABLE [Link-Person-Time] (
    IDPersonNumber    INTEGER,
    IDTimeNumber      INTEGER,
    ValidDate         DATETIME
);
```

Person-Object Link

This link joins the person to the object hub to describe the relationships between the two hubs. The link consists of the following fields:

```
CREATE TABLE [Link-Person-Object] (
    IDPersonNumber    INTEGER,
    IDObjectNumber    INTEGER,
    ValidDate         DATETIME
);
```

Person-Location Link

This link joins the person to the location hub, to describe the relationships between the two hubs. The link consists of the following fields:

```
CREATE TABLE [Link-Person-Time] (
    IDPersonNumber      INTEGER,
    IDLocationNumber    INTEGER,
    ValidDate           DATETIME
);
```

Person-Event Link

This link joins the person to the event hub, to describe the relationships between the two hubs. The link consists of the following fields:

```
CREATE TABLE [Link-Person-Time] (
    IDPersonNumber    INTEGER,
    IDEventNumber     INTEGER,
    ValidDate         DATETIME
);
```

Person Satellites

The person satellites are the part of the vault that stores the temporal attributes and descriptive attributes of the data. The satellite is of the following format:

```
CREATE TABLE [Satellite-Person-Gender] (
PersonSatelliteID VARCHAR (100),
```

```
IDPersonNumber INTEGER,
FirstName VARCHAR (200),
SecondName VARCHAR (200),
LastName VARCHAR (200),
BirthDateKey VARCHAR (20),
Gender VARCHAR (10),
);
```

Object Section

The object section contains the complete data structure for all data entities related to recording the object involved.

Object Hub

The object hub consists of a series of fields that supports a "real" object. The object hub consists of the following fields:

```
CREATE TABLE [Hub-Object-Species] (
IDObjectNumber INTEGER,
ObjectBaseKey VARCHAR (100),
ObjectNumber VARCHAR (100),
ObjectValue VARCHAR (200),
);
```

This structure enables you as a data scientist to categorize the objects in the business environment.

Object Links

These link the object hub to the other hubs (Figure 9-5).

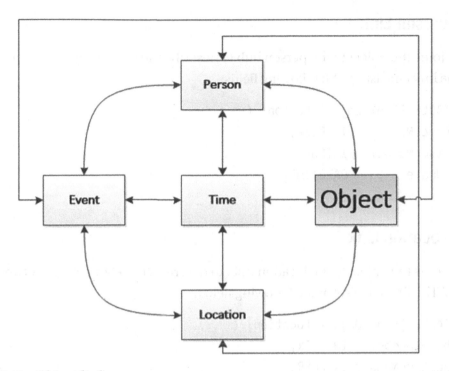

Figure 9-5. *Object links*

The following links are supported:

Object-Time Link

This link joins the object to the time hub, to describe the relationships between the two hubs. The link consists of the following fields:

```
CREATE TABLE [Link-Object-Time] (
    IDObjectNumber    INTEGER,
    IDTimeNumber      INTEGER,
    ValidDate         DATETIME
);
```

Object-Person Link

This link joins the object to the person hub to describe the relationships between the two hubs. The link consists of the following fields:

```
CREATE TABLE [Link-Object-Person] (
    IDObjectNumber     INTEGER,
    IDPersonNumber     INTEGER,
    ValidDate          DATETIME
);
```

Object-Location Link

This link joins the object to the location hub, to describe the relationships between the two hubs. The link consists of the following fields:

```
CREATE TABLE [Link-Object-Location] (
    IDObjectNumber     INTEGER,
    IDLocationNumber INTEGER,
    ValidDate          DATETIME
);
```

Object-Event Link

This link joins the object to the event hub to describe the relationships between the two hubs.

Object Satellites

Object satellites are the part of the vault that stores and provisions the detailed characteristics of objects. The typical object satellite has the following data fields:

```
CREATE TABLE [Satellite-Object-Make-Model] (
IDObjectNumber INTEGER,
ObjectSatelliteID VARCHAR (200),
ObjectType VARCHAR (200),
ObjectKey VARCHAR (200),
ObjectUUID VARCHAR (200),
```

```
Make VARCHAR (200),
Model VARCHAR (200)
);
```

The object satellites will hold additional characteristics, as each object requires additional information to describe the object. I keep each set separately, to ensure future expansion, as the objects are the one hub that evolves at a high rate of change, as new characteristics are discovered by your data science.

Location Section

The location section contains the complete data structure for all data entities related to recording the location involved.

Location Hub

The location hub consists of a series of fields that supports a GPS location. The location hub consists of the following fields:

```
CREATE TABLE [Hub-Location] (
    IDLocationNumber INTEGER,
    ObjectBaseKey  VARCHAR (200),
    LocationNumber INTEGER,
    LocationName   VARCHAR (200),
    Longitude      DECIMAL (9, 6),
    Latitude       DECIMAL (9, 6)
);
```

The location hub enables you to link any location, address, or geospatial information to the rest of the data vault.

Location Links

The location links join the location hub to the other hubs (Figure 9-6).

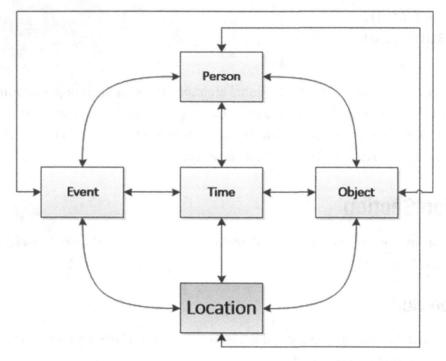

Figure 9-6. *Location links*

The following links are supported.

Location-Time Link

The link joins the location to the time hub, to describe the relationships between the two hubs. The link consists of the following fields:

```
CREATE TABLE [Link-Location-Time] (
    IDLocationNumber    INTEGER,
    IDTimeNumber        INTEGER,
    ValidDate           DATETIME
);
```

These links support business actions such as `ArrivedAtShopAtDateTime` or `ShopOpensAtTime`.

Location-Person Link

This link joins the location to the person hub, to describe the relationships between the two hubs. The link consists of the following fields:

```
CREATE TABLE [Link-Location-Person] (
    IDLocationNumber    INTEGER,
    IDPersonNumber      INTEGER,
    ValidDate           DATETIME
);
```

These links support such business actions as ManagerAtShop or SecurityAtShop.

Location-Object Link

This link joins the location to the object hub, to describe the relationships between the two hubs. The link consists of the following fields:

```
CREATE TABLE [Link-Location-Object] (
    IDLocationNumber    INTEGER,
    IDObjectNumber      INTEGER,
    ValidDate           DATETIME
);
```

These links support such business actions as ShopDeliveryVan or RackAtShop.

Location-Event Link

This link joins the location to the event hub, to describe the relationships between the two hubs. The link consists of the following fields:

```
CREATE TABLE [Link-Location-Event] (
    IDLocationNumber    INTEGER,
    IDEventNumber       INTEGER,
    ValidDate           DATETIME
);
```

These links support such business actions as ShopOpened or PostCodeDeliveryStarted.

Location Satellites

The location satellites are the part of the vault that stores and provisions the detailed characteristics of where entities are located. The typical location satellite has the following data fields:

```
CREATE TABLE [Satellite-Location-PostCode] (
IDLocationNumber INTEGER,
LocationSatelliteID VARCHAR (200),
LocationType VARCHAR (200),
LocationKey VARCHAR (200),
LocationUUID VARCHAR (200),
CountryCode VARCHAR (20),
PostCode VARCHAR (200)
);
```

The location satellites will also hold additional characteristics that are related only to your specific customer. They may split their business areas into their own regions, e.g., Europe, Middle-East, and China. These can be added as a separate location satellite.

Event Section

The event section contains the complete data structure for all data entities related to recording the event that occurred.

Event Hub

The event hub consists of a series of fields that supports events that happens in the real world. The event hub consists of the following fields:

```
CREATE TABLE [Hub-Event] (
    IDEventNumber    INTEGER,
    EventType        VARCHAR (200),
    EventDescription VARCHAR (200)
);
```

Event Links

Event links join the event hub to the other hubs (see Figure 9-7).

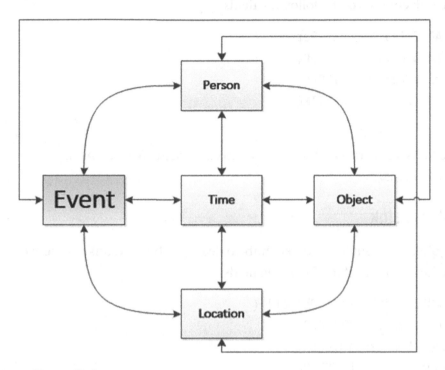

Figure 9-7. *Event links*

The following links are supported.

Event-Time Link

This link joins the event to the time hub, to describe the relationships between the two hubs. The link consists of the following fields:

```
CREATE TABLE [Link-Event-Time] (
    IDEventNumber    INTEGER,
    IDTimeNumber     INTEGER,
    ValidDate        DATETIME
);
```

These links support such business actions as DeliveryDueAt or DeliveredAt.

Event-Person Link

This link joins the event to the person hub, to describe the relationships between the two hubs. The link consists of the following fields:

```
CREATE TABLE [Link-Event-Person] (
    IDEventNumber     INTEGER,
    IDPersonNumber    INTEGER,
    ValidDate         DATETIME
);
```

These links support such business actions as ManagerAppointAs or StaffMemberJoins.

Event-Object Link

This link joins the event to the object hub, to describe the relationships between the two hubs. The link consists of the following fields:

```
CREATE TABLE [Link-Event-Object] (
    IDEventNumber     INTEGER,
    IDObjectNumber    INTEGER,
    ValidDate         DATETIME
);
```

These links support such business actions as VehicleBuy, VehicleSell, or ItemInStock.

Event-Location Link

The link joins the event to the location hub to describe the relationships between the two hubs. The link consists of the following fields:

```
CREATE TABLE [Link-Event-Location] (
    IDEventNumber     INTEGER,
    IDTimeNumber      INTEGER,
    ValidDate         DATETIME
);
```

These links support such business actions as DeliveredAtPostCode or PickupFromGPS.

Event Satellites

The event satellites are the part of the vault that stores the details related to all the events that occur within the systems you will analyze with your data science. I suggest that you keep to one type of event per satellite. This enables future expansion and easier long-term maintenance of the data vault.

Engineering a Practical Process Superstep

I will now begin with a series of informational and practical sections with the end goal of having you use the Process step to generate a data vault using the T-P-O-L-E design model previously discussed.

Note   I use the T-P-O-L-E model, but the "pure" data vault can use various other hubs, links, or satellites. The model is a simplified one that enables me to reuse numerous utilities and processes I have developed over years of working with data science.

Time

Time is the characteristic of the data that is directly associated with time recording. ISO 8601-2004 describes an international standard for data elements and interchange formats for dates and times. Following is a basic set of rules to handle the data interchange.

The calendar is based on the Gregorian calendar. The following data entities (year, month, day, hour, minute, second, and fraction of a second) are officially part of the ISO 8601-2004 standard. The data is recorded from largest to smallest. Values must have a pre-agreed fixed number of digits that are padded with leading zeros. I will now introduce you to these different data entities.

Year

The standard year must use four-digit values within the range 0000 to 9999, with the optional agreement that any date denoted by AC/BC requires a conversion whereby AD 1 = year 1, 1 BC = year 0, 2 BC = year -1. That means that 2017AC becomes +2017, but 2017BC becomes -2016.

Valid formats are: YYYY and +/-YYYY. The rule for a valid year is 20 October 2017 becomes 2017-10-20 or +2017-10-20 or a basic 20171020. The rule for missing data on a year is 20 October becomes --10-20 or --1020.

The use of YYYYMMDD is common in source systems. I advise you to translate to YYYY-MM-DD during the Assess step, to generate a consistency with the date in later stages of processing.

Warning   Discuss with your customer the dates before October 1582 (the official start of the Gregorian calendar). The Julian calendar was used from January 1, 45 BC. If the earlier dates are vital to your customer's data processing, I suggest that you perform an in-depth study into the other, more complex differences in dates. This is very important when you work with text mining on older documents.

Open your IPython editor and investigate the following date format:

```
from datetime import datetime
from pytz import timezone, all_timezones

now_date = datetime(2001,2,3,4,5,6,7)
now_utc=now_date.replace(tzinfo=timezone('UTC'))
print('Date:',str(now_utc.strftime("%Y-%m-%d %H:%M:%S (%Z) (%z)")))
print('Year:',str(now_utc.strftime("%Y")))
```

You should see

```
Date: 2001-02-03 04:05:06 (UTC) (+0000)
Year: 2001
```

This enables you to easily extract a year value from a given date.

Month

The standard month must use two-digit values within the range of 01 through 12. The rule for a valid month is 20 October 2017 becomes 2017-10-20 or +2017-10-20. The rule for valid month without a day is October 2017 becomes 2017-10 or +2017-10.

Warning   You cannot use YYYYMM, so 201710 is not valid. This standard avoids confusion with the truncated representation YYMMDD, which is still often used in source systems. 201011 could be 2010-11 or 1920-10-11.

Open your IPython editor and investigate the following date format:

```
print('Date:',str(now_utc.strftime("%Y-%m-%d %H:%M:%S (%Z) (%z)")))
print('Month:',str(now_utc.strftime("%m")))
```

You should get back

```
Date: 2001-02-03 04:05:06 (UTC) (+0000)
Month: 02
```

Now you can get the month.

This follows the standard English names for months.

Number	Name
1	January
2	February
3	March
4	April
5	May
6	June
7	July
8	August
9	September
10	October
11	November
12	December

Open your IPython editor and investigate the following date format:

```
print('Date:',str(now_utc.strftime("%Y-%m-%d %H:%M:%S (%Z) (%z)")))
print('MonthName:', str(now_utc.strftime("%B")))
```

441

This gives you the months in words.

```
Date: 2001-02-03 04:05:06 (UTC) (+0000)
MonthName: February
```

Day

The standard day of the month must use a two-digit value with possible values within the range of 01 through 31, as per an agreed format for the specific month and year. The rule for a valid month is 20 October 2017 becomes 2017-10-20 or +2017-10-20.

Open your IPython editor and investigate the following date format:

```
print('Date:',str(now_utc.strftime("%Y-%m-%d %H:%M:%S (%Z) (%z)")))
print('Day:',str(now_utc.strftime("%d")))
```

This returns the day of the date.

```
Date: 2001-02-03 04:05:06 (UTC) (+0000)
Day: 03
```

There are two standards that apply to validate dates: non-leap year and leap year. Owing to the length of the solar year being marginally less than 365¼ days—by about 11 minutes—every 4 years, the calendar indicates a leap year, unless the year is divisible by 400. This fixes the 11 minutes gained over 400 years.

Open your IPython editor and investigate the following date format. I will show you a way to find the leap years.

```
import datetime
for year in range(1960,2025):
    month=2
    day=29
    hour=0
    correctDate = None
    try:
        newDate = datetime.datetime(year=year,month=month,day=day,hour=hour)
        correctDate = True
    except ValueError:
        correctDate = False
```

```
    if correctDate == True:
        if year%400 == 0:
            print(year, 'Leap Year (400)')
        else:
            print(year, 'Leap Year')
    else:
        print(year,'Non Leap Year')
```

The result shows that the leap years are as follows:

```
1960,1964,1968,1972,1976,1980,1984,1988,1992,1996,2000(This was a 400)
2004,2008,2012,2016,2020,2024.
```

Now, we can look at the time component of the date time.

Hour

The standard hour must use a two-digit value within the range of 00 through 24. The valid format is hhmmss or hh:mm:ss. The shortened format hhmm or hh:mm is accepted but not advised. The format hh is not supported. The use of 00:00:00 is the beginning of the calendar day. The use of 24:00:00 is only to indicate the end of the calendar day.

Open your IPython editor and investigate the following date format:

```
print('Date:',str(now_utc.strftime("%Y-%m-%d %H:%M:%S (%Z) (%z)")))
print('Hour:',str(now_utc.strftime("%H")))
```

The results are:

```
Date: 2001-02-03 04:05:06 (UTC) (+0000)
Hour: 04
```

Minute

The standard minute must use two-digit values within the range of 00 through 59. The valid format is hhmmss or hh:mm:ss. The shortened format hhmm or hh:mm is accepted but not advised.

Open your IPython editor and investigate the following date format:

```
print('Date:',str(now_utc.strftime("%Y-%m-%d %H:%M:%S (%Z) (%z)")))
print('Minute:',str(now_utc.strftime("%M")))
```

The results are

```
Date: 2001-02-03 04:05:06 (UTC) (+0000)
Minute: 05
```

Second

The standard second must use two-digit values within the range of 00 through 59. The valid format is hhmmss or hh:mm:ss.

Open your IPython editor and investigate the following date format:

```
print('Date:',str(now_utc.strftime("%Y-%m-%d %H:%M:%S (%Z) (%z)")))
print('Second:',str(now_utc.strftime("%S")))
```

This returns the seconds of the date time.

```
Date: 2001-02-03 04:05:06 (UTC) (+0000)
Second: 06
```

Note This is normally where most systems stop recording time, but I have a few clients that require a more precise time-recording level.

So, now we venture into sub-second times. This is a fast world with interesting dynamics.

Fraction of a Second

The fraction of a second is only defined as a format: hhmmss,sss or hh:mm:ss,sss or hhmmss.sss or hh:mm:ss.sss. I prefer the hh:mm:ss.sss format, as the use of the comma causes issues with exports to CSV file formats, which are common in the data science ecosystem.

The current commonly used formats are the following:

- hh:mm:ss.s: Tenth of a second

- hh:mm:ss.ss: Hundredth of a second

- hh:mm:ss.sss: Thousandth of a second

Data science works best if you can record the data at the lowest possible levels. I suggest that you record your time at the hh:mm:ss.sss level.

Open your IPython editor and investigate the following date format:

```
print('Date:',str(now_utc.strftime("%Y-%m-%d %H:%M:%S (%Z) (%z)")))
print('Millionth of Second:',str(now_utc.strftime("%f")))
```

This returns the fraction of a second in one-millionth of a second, or vicrosecond, as follows:

Date: 2001-02-03 04:05:06 (UTC) (+0000)

Millionth of Second: 000007

There are some smaller units, such as the following:

- *Nanosecond*: One thousand-millionth of a second

  - Scale comparison: One nanosecond is to one second as one second is to 31.71 years.

- *Picosecond*: One trillionth or one millionth of one millionth of a second

  - Scale comparison: A picosecond is to one second as one second is to 31,710 years.

I have a customer who owns a scientific device called a streak, or intensified CCD (ICCD), camera. These are able to picture the motion of light, wherein +/-3.3 picoseconds is the time it takes for light to travel 1 millimeter.

Note   In 2011, MIT researchers announced they had a trillion-frame-per-second video camera. This camera's ability is exceptional, as the standard for film is 24 frames per second (fps), and 30 fps for video, with some 48 fps models.

In the measurement of high-performance computing, it takes +/-330 picoseconds for a common 3.0 GHz computer CPU to add two integers. The world of a data scientists is a world of extreme measures.

Remember the movie-to-frame example you performed earlier in Chapter 5? This camera would produce enough frames in a second for 1,056,269,593.8 years of video playback at 30 fps.

Local Time and Coordinated Universal Time (UTC)

The date and time we use is bound to a specific location on the earth and the specific time of the calendar year. The agreed local time is approved by the specific country.

I am discussing time zones because, as a data scientist, you will have to amalgamate data from multiple data sources across the globe. The issue is the impact of time zones and the irregular application of rules by customers.

Local Time

The approved local time is agreed by the specific country, by approving the use of a specific time zone for that country. Daylight Saving Time can also be used, which gives a specific country a shift in its local time during specific periods of the year, if they use it. The general format for local time is hh:mm:ss.

Open your IPython editor and investigate the following date format:

```
from datetime import datetime
now_date = datetime.now()
print('Date:',str(now_date.strftime("%Y-%m-%d %H:%M:%S (%Z) (%z)")))
```

The results is

```
2017-10-01 10:45:58 () ()
```

The reason for the two empty brackets is that the date time is in a local time setting.

Warning   Most data systems record the local time in the local time setting for the specific server, PC, or tablet.

As a data scientist, it is required that you understand that not all systems in the ecosystem record time in a reliable manner.

Tip   There is a Simple Network Time Protocol (SNTP) that you can use automatically to synchronize your system's time with that of a remote server. The SNTP can be used to update the clock on a machine with a remote server. This keeps your machine's time accurate, by synchronizing with servers that are known to have accurate times. I normally synchronize my computers that record data at a customer with the same NTP server the customer is using, to ensure that I get the correct date and time, as per their system.

Now you can start investigating the world of international date and time settings.

Coordinated Universal Time (UTC)

Coordinated Universal Time is a measure of time within about 1 second of mean solar time at 0° longitude. The valid format is hh:mm:ss±hh:mm or hh:mm:ss±hh.

For example, on September 11, 2017, nine a.m. in London, UK, is 09:00:00 local time and UTC 10:00:00+01:00. In New York, US, it would be 06:00:00 local time and UTC 10:00:00-04:00. In New Delhi, India, it would be 15:30:00 local time and UTC 10:00:00+05:30.

The more precise method would be to record it as 2017-09-11T10:00:00Z. I will explain over the next part of this chapter, why the date is important when you work with multi-time zones as a data scientist.

Open your IPython editor and investigate the following date format:

```
from datetime import datetime
now_date = datetime.now()
print('Local Date Time:',str(now_date.strftime("%Y-%m-%d %H:%M:%S (%Z)
(%z)")))
now_utc=now_date.replace(tzinfo=timezone('UTC'))
print('UTC Date Time:',str(now_utc.strftime("%Y-%m-%d %H:%M:%S (%Z)
(%z)")))
```

Warning   I am in working in the London, Edinburgh time zone, so my results will be based on that fact. My time zone is Europe/London.

```
from datetime import datetime
now_date = datetime.now()
print('Local Date Time:',str(now_date.strftime("%Y-%m-%d %H:%M:%S (%Z)
(%z)")))
now_utc=now_date.replace(tzinfo=timezone('UTC'))
print('UTC Date Time:',str(now_utc.strftime("%Y-%m-%d %H:%M:%S (%Z)
(%z)")))
now_london=now_date.replace(tzinfo=timezone('Europe/London'))
print('London Date Time:',str(now_london.strftime("%Y-%m-%d %H:%M:%S (%Z)
(%z)")))
```

So, my reference results are

```
Local Date Time: 2017-10-03 16:14:43 () ()
UTC Date Time: 2017-10-03 16:14:43 (UTC) (+0000)
London Date Time: 2017-10-03 16:14:43 (LMT) (-0001)
```

Now, with that knowledge noted, let's look at your results again.

```
from datetime import datetime
now_date = datetime.now()
print('Local Date Time:',str(now_date.strftime("%Y-%m-%d %H:%M:%S (%Z)
(%z)")))
now_utc=now_date.replace(tzinfo=timezone('UTC'))
print('UTC Date Time:',str(now_utc.strftime("%Y-%m-%d %H:%M:%S (%Z)
(%z)")))
```

My results are

```
Local Date Time: 2017-10-03 16:14:43 () ()
UTC Date Time: 2017-10-03 16:14:43 (UTC) (+0000)
```

Must companies now have different time zones? Yes! VKHCG is no different.

Open your Python editor, and let's investigate VKHCG's time zones.

```
from datetime import datetime
from pytz import timezone, all_timezones
```

Get the current time.

```
now_date_local=datetime.now()
```

Change the local time to 'Europe/London' local time by tagging it as 'Europe/London' time.

```
now_date=now_date_local.replace(tzinfo=timezone('Europe/London'))
```

```
print('Local Date Time:',str(now_date.strftime("%Y-%m-%d %H:%M:%S (%Z)
(%z)")))
```

Now you have a time-zone-enabled data time value that you can use to calculate other time zones. The first conversion you perform is to the UTC time zone, as follows:

```
now_utc=now_date.astimezone(timezone('UTC'))
print('UTC Date Time:',str(now_utc.strftime("%Y-%m-%d %H:%M:%S (%Z)
(%z)")))
```

What is the time in London, UK?

```
now_eu_london=now_date.astimezone(timezone('Europe/London'))
print('London Date Time:',str(now_eu_london.strftime("%Y-%m-%d %H:%M:%S
(%Z) (%z)")))
```

What is the time in Berlin, Germany?

```
now_eu_berlin=now_date.astimezone(timezone('Europe/Berlin'))
print('Berlin Date Time:',str(now_eu_berlin.strftime("%Y-%m-%d %H:%M:%S
(%Z) (%z)")))
```

What is the time in the Jersey Islands, UK?

```
now_eu_jersey=now_date.astimezone(timezone('Europe/Jersey'))
print('Jersey Date Time:',str(now_eu_jersey.strftime("%Y-%m-%d %H:%M:%S
(%Z) (%z)")))
```

What is the time in New York, US?

```
now_us_eastern=now_date.astimezone(timezone('US/Eastern'))
print('USA Easten Date Time:',str(now_us_eastern.strftime("%Y-%m-%d
%H:%M:%S (%Z) (%z)")))
```

What is the time in Arizona, US?

```
now_arizona=now_date.astimezone(timezone('US/Arizona'))
print('USA Arizona Date Time:',str(now_arizona.strftime("%Y-%m-%d %H:%M:%S
(%Z) (%z)")))
```

What is the time in Auckland, Australia?

```
now_auckland=now_date.astimezone(timezone('Pacific/Auckland'))
print('Auckland Date Time:',str(now_auckland.strftime("%Y-%m-%d %H:%M:%S
(%Z) (%z)")))
```

What is the time in Yukon, Canada?

```
now_yukon=now_date.astimezone(timezone('Canada/Yukon'))
print('Canada Yukon Date Time:',str(now_yukon.strftime("%Y-%m-%d %H:%M:%S
(%Z) (%z)")))
```

What is the time in Reykjavik, Iceland?

```
now_reyk=now_date.astimezone(timezone('Atlantic/Reykjavik'))
print('Reykjavik Date Time:',str(now_reyk.strftime("%Y-%m-%d %H:%M:%S (%Z)
(%z)")))
```

What is the time in Mumbai, India?

```
now_india=now_date.astimezone(timezone('Etc/GMT-7'))
print('India Date Time:',str(now_india.strftime("%Y-%m-%d %H:%M:%S (%Z)
(%z)")))
```

What time zones does Vermeulen use?

```
print('Vermeulen Companies')
print('Local Date Time:',str(now_date_local.strftime("%Y-%m-%d %H:%M:%S
(%Z) (%z)")))
print('HQ Edinburgh:',str(now_utc.strftime("%Y-%m-%d %H:%M:%S (%Z) (%z)")))
print('Iceland Thor Computers:',str(now_reyk.strftime("%Y-%m-%d %H:%M:%S
(%Z) (%z)")))
print('USA Arizona Computers:',str(now_arizona.strftime("%Y-%m-%d %H:%M:%S
(%Z) (%z)")))
```

Here are the Vermeulen results:

```
Vermeulen Companies
Local Date Time: 2017-10-03 16:14:43 () ()
HQ Edinburgh: 2017-10-03 16:15:43 (UTC) (+0000)
Iceland Thor Computers: 2017-10-03 16:15:43 (GMT) (+0000)
USA Arizona Computers: 2017-10-03 09:15:43 (MST) (-0700)
```

What about Krennwallner?

```
print('Krennwallner Companies')
print('Local Date Time:',str(now_date_local.strftime("%Y-%m-%d %H:%M:%S
(%Z) (%z)")))
print('HQ Berlin:',str(now_eu_berlin.strftime("%Y-%m-%d %H:%M:%S (%Z) (%z)")))
print('HQ USA:',str(now_us_eastern.strftime("%Y-%m-%d %H:%M:%S (%Z) (%z)")))
```

Following are Krennwallner's results:

```
Krennwallner Companies
Local Date Time: 2017-10-03 16:14:43 () ()
HQ Berlin: 2017-10-03 18:15:43 (CEST) (+0200)
HQ USA: 2017-10-03 12:15:43 (EDT) (-0400)
```

What about Hillman?

```
print('Hillman Companies')
print('HQ London:',str(now_london.strftime("%Y-%m-%d %H:%M:%S (%Z) (%z)")))
print('HQ USA:',str(now_arizona.strftime("%Y-%m-%d %H:%M:%S (%Z) (%z)")))
print('HQ Canada:',str(now_yukon.strftime("%Y-%m-%d %H:%M:%S (%Z) (%z)")))
print('HQ Australia:',str(now_auckland.strftime("%Y-%m-%d %H:%M:%S (%Z)
(%z)")))
print('HQ India:',str(now_india.strftime("%Y-%m-%d %H:%M:%S (%Z) (%z)")))
```

Hillman's results are only for headquarters (HQs).

```
Hillman Companies
HQ London: 2017-10-03 16:14:43 (LMT) (-0001)
HQ USA: 2017-10-03 09:15:43 (MST) (-0700)
HQ Canada: 2017-10-03 09:15:43 (PDT) (-0700)
HQ Australia: 2017-10-04 05:15:43 (NZDT) (+1300)
HQ India: 2017-10-03 23:15:43 (+07) (+0700)
```

What about Clark?

```
print('Clark Companies')
print('HQ Jersey:',str(now_eu_jersey.strftime("%Y-%m-%d %H:%M:%S (%Z)
(%z)")))
print('HQ Berlin:',str(now_eu_berlin.strftime("%Y-%m-%d %H:%M:%S (%Z)
(%z)")))
print('HQ USA:',str(now_us_eastern.strftime("%Y-%m-%d %H:%M:%S (%Z)
(%z)")))
```

What about Hillman?

```
HQ Jersey: 2017-10-03 17:15:43 (BST) (+0100)
HQ Berlin: 2017-10-03 18:15:43 (CEST) (+0200)
HQ USA: 2017-10-03 12:15:43 (EDT) (-0400)
```

This shows how our date and time is investigated by data scientists. As humans, we have separated the concepts of date and time for thousands of years. The date was calculated from the stars. Time was once determined by an hourglass. Later, the date was on a calendar, and time is now on a pocket watch. But in the modern international world there is another date-time data item. Mine is a smartphone—period!

Combining Date and Time

When using date and time in one field, ISO supports the format YYYY-MM-DDThh:mm:ss. sss for local time and YYYY-MM-DDThh:mm:ss.sssZ for Coordinated Universal Time.

These date and time combinations are regularly found in international companies' financials or logistics shipping and on my smartphone.

Time Zones

Time zones enable the real world to create a calendar of days and hours that people can use to run their lives. The date and time is normally expressed, for example, as October 20, 2017, and 9:00. In the world of the data scientist, this is inadequate information. The data scientist requires a date, time, and time zone. YYYY-MM-DDThh:mm:ss.sssZ is the format for a special time zone for the data scientist's data processing needs.

Note   This format was discussed earlier in this chapter, as an alternative to UTC.

In international data, I have also seen time zones set in other formats and must discuss with you how to handle these formats in a practical manner.

Open your IPython editor and investigate the following date format:

```
from pytz import all_timezones
for zone in all_timezones:
    print(zone)
```

There are 593 zones. Yes, read correctly! The issue is that companies do not even adhere to these standard zones, due to lack of understanding. For example one of my customers uses YYYY-DDThh:mm:ss.sssY for UTC-12:00, or Yankee Time Zone.

Warning   These zones are not valid formats of ISO! But they are generally used by source systems around the world.

I will offer you the list that I have seen before, if you have to convert back to UTC format or local time.

Tip   I normally add a UTC time zone with YYYY-DDThh:mm:ss.sssZ to all my data science, as it ensures that you always have a common variable for date and time processing between data from multi-time zones.

The point of data source recording is normally the time zone you use.

Warning   The time zone is dependent on your source system's longitude and latitude or the clock settings on the hardware. The data below is generalized. I suggest you investigate the source of your data first, before picking the correct time zone.

Once you have found the correct original time zone, I suggest you do the following:

```
now_date_local=datetime.now()
now_date=now_date_local.replace(tzinfo=timezone('Europe/London'))
print('Local Date Time:',str(now_date_local.strftime("%Y-%m-%d %H:%M:%S
(%Z) (%z)")))
print('Local Date Time:',str(now_date.strftime("%Y-%m-%d %H:%M:%S (%Z)
(%z)")))
```

The result is

```
Local Date Time: 2017-10-01 12:54:03 () ()
Local Date Time: 2017-10-01 12:54:03 (LMT) (-0001)
```

'Europe/London' applies the correct time zone from the list of 593 zones. Now convert to UTC.

```
now_utc=now_date.astimezone(timezone('UTC'))
print('UTC Date Time:',str(now_utc.strftime("%Y-%m-%d %H:%M:%S (%Z)
(%z)")))
```

The result is

```
UTC Date Time: 2017-10-01 12:55:03 (UTC) (+0000)
```

I advise that you only use this with caution, as your preferred conversion. You will note that there is a more or less inaccurate meaning to YYYY-DDDThh:mmM, which converts to UTC+12:00, UTC+12:45, UTC+13:00, and UTC+14:00.

UTC+14:00 is UTC+14 and stretches as far as 30° east of the 180° longitude line, creating a large fold in the International Date Line and interesting data issues for the data scientist.

Interesting Facts

The Republic of Kiribati introduced a change of date and time for its eastern half on January 1, 1995, from time zones -11 and -10 to +13 and +14. Before this, the time zones UTC+13 and UTC+14 did not exist. The reason was that Kiritimati Island started the year 2000 before any other country on Earth, a feature the Kiribati government capitalized on as a prospective tourist asset. It was successful, but it also caused major issues with the credit card transactions of the tourists who took up the offer. The local times for the Phoenix and Line Islands government offices are on opposite sides of the International Date Line and can only communicate by radio or telephone on the four days of the week when both sides experience weekdays concurrently. The problems with changes always appear in the data system. The GPS and time zone validation checks on encryption engines do not support this nonstandard change.

Tonga used UTC+14 for daylight saving time from 1999 to 2002 and from 2016 to the present, to enable the celebration of New Year 2000 at the same time as the Line Islands in Kiribati and returned to the changed time zone in 2016 to get the New Year tourist business back. It became active on November 5, 2017, and inactive on January 21, 2018.

On December 29, 2011 (UTC-10), Samoa changed its standard time from UTC-11 to UTC+13 and its daylight-saving time from UTC-10 to UTC+14, to move the International Date Line to the other side of the country. Wonder why? Eureka, New Year's tourist business.

Alaska (formerly Russian America) used local times from GMT+11:30 to GMT+15:10 until 1867 (UTC was introduced in 1960).

The Mexican state of Quintana Roo changed permanently on February 1, 2015, to UTC-5, to boost the tourism sector by creating longer, lighter evenings.

Namibia changed on September 3, 2017, from UTC+1 and daylight saving of UTC+2 to permanent UTC+2.

UTC+14 was used as a daylight-saving time before 1982 in parts of eastern Russia (Chukotka).

UTC-12:00 is the last local time zone to start a new day.

The 180° meridian was nominated as the International Date Line, because it generally passes through the sparsely populated central Pacific Ocean, and approved in 1884. Now, we have many research probes, cruise liners, and airplanes regularly crossing this line and recording data via the source systems.

The data scientist must take note of these exceptions when dealing with data from different time zones, as a minor shift in the data from one year to another may hide a seasonal trend. I was paid by a hotel group in Cancún, Quintana Roo, Mexico, to investigate if a change in time zone yielded growth in their own income and what potential additional income would accrue if they actively promoted the additional open times of their hotel bars.

Intervals Identified by Start and End Dates and Time

There is a standard format for supporting a date and time interval in a single field: YYYYMMDDThhmmss/YYYYMMDDThhmmss or YYYY-MM-DDThh:mm:ss/YYYY-MM-DDThh:mm:ss. So, October 20, 2017, can be recorded as 2017-10-20T000000/2017-10-20T235959 and Holiday December 2017 as 2017-12-22T120001/2017-12-27T085959, in the same data table.

Note   The preceding example was found in one of my source systems. Complex data issues are common in data science. Spot them quickly, and deal with them quickly!

As a general note of caution, I advise that when you discover these complex structures, convert them to more simple structures as soon as you can. I suggest that you add an assess step, to split these into four fields: Start Date Time, End Date Time, Original Date Time, and Business Term.

Using the previously mentioned format, October 20, 2017, recorded as 2017-10-20T000000/2017-10-20T235959, becomes the following:

Start Date Time	End Date Time	Original Date Time	Business Term
2017-10-20T000000	2017-10-20T235959	2017-10-20T000000/2017-10-20T235959	20 October 2017

Holiday December 2017 can be recorded as 2017-12-22T120001/2017-12-27T085959 in the same data table, as follows:

Start Date Time	End Date Time	Original Date Time	Business Term
2017-12-22T120001	2017-12-22T235959	2017-12-22T120001/2017-12-27T085959	Holiday December 2017
2017-12-23T000000	2017-12-23T235959	2017-12-22T120001/2017-12-27T085959	Holiday December 2017
2017-12-24T000000	2017-12-24T235959	2017-12-22T120001/2017-12-27T085959	Holiday December 2017
2017-12-25T000000	2017-12-25T235959	2017-12-22T120001/2017-12-27T085959	Holiday December 2017
2017-12-26T000000	2017-12-26T235959	2017-12-22T120001/2017-12-27T085959	Holiday December 2017
2017-12-27T000000	2017-12-27T085959	2017-12-22T120001/2017-12-27T085959	Holiday December 2017

Now you have same day time periods that are easier to convert into date time formats.

Special Date Formats

The following are special date formats.

Day of the Week

ISO 8601-2004 sets the following standard:

Number	Name
1	Monday
2	Tuesday
3	Wednesday
4	Thursday
5	Friday
6	Saturday
7	Sunday

Open your IPython editor and investigate the following date format:

```
now_date_local=datetime.now()
now_date=now_date_local.replace(tzinfo=timezone('Europe/London'))
print('Weekday:',str(now_date.strftime("%w")))
```

The result is

```
(It is Sunday)
Weekday: 0
```

Caution   I have seen variations! Sunday = 0, Sunday = 1, and Monday = 0. Make sure you have used the correct format before joining data.

Week Dates

The use of week dates in many source systems causes some complications when working with dates. Week dates are denoted by the use of a capital *W* and *W* with *D* in the format.

The following are all valid formats: YYYY-Www or YYYYWww or YYYY-Www-D or YYYYWwwD. For examples, October 22, 2017, is 2017-W42, if you use ISO 8601(European), but 2017-W43, if you use ISO 8601(US).

457

> **Warning**   ISO 8601(European) is an official format. The standard is that the
> first week of a year is a week that includes the first Thursday of the year or that
> contains January 4th. By this method, January 1st, 2nd, and 3rd can be included
> in the last week of the previous year, or December 29th, 30th, and 31st can be
> included in the first week of next year.
>
> ISO 8601(US) is widely used, and according to this standard, the first week begins
> on January1st. The next week begins on the next Sunday.

Other options are that the first week of the year begins on January 1st, and the next week begins on the following Monday, and the first week of the year begins on January 1st, and the next week begins on January 8th.

```
now_date_local=datetime.now()
now_date=now_date_local.replace(tzinfo=timezone('Europe/London'))
print('Week of Year:',str(now_date.strftime("%YW%WD%w")))
```

The result is

```
Week of Year: 2017W39D0
```

Day of the Year

The day of the year is calculated simply by counting the amount of days from January 1st of the given year. For example: October 20, 2017 is 2017/293.

> **Caution**   Change this to the ISO standard as soon as possible!

I found this format stored as 2017293 in a birth date field. The reason was to hide a staff member's age. This successfully broke the pension planner's projection, which read the value as the US date for March 29, 2017, hence, 2017/29/3. This staff member would potentially have received two birthday cards and cake from his management in 2017. I fixed the problem in September 2017.

So, let's look at this day of the year as it was intended to be used by the format. The day of the year is useful to determine seasonality in the data when comparing data between years. Check if the same thing happens on the same day every year. Not really that simple, but close!

Here is the format for the day of year:

```
now_date_local=datetime.now()
now_date=now_date_local.replace(tzinfo=timezone('Europe/London'))
print('Day of year:',str(now_date.strftime("%Y/%j")))
```

The result is

```
Day of year: 2017/274
```

Well done. You have now completed the basic introduction to date and time in the Process step. In the practical part of this chapter, I will show how important the YYYY-MM-DDThh:mm:ss.sssZ value for data and time is to the data science.

This completes the basic information on date and time. Next, you will start to build the time hub, links, and satellites.

Open your Python editor and create a file named Process_Time.py. Save it into directory .. \VKHCG\01-Vermeulen\03-Process. The first part consists of standard imports and environment setups.

```
###############################################################
# -*- coding: utf-8 -*-
###############################################################
import sys
import os
from datetime import datetime
from datetime import timedelta
from pytz import timezone, all_timezones
import pandas as pd
import sqlite3 as sq
from pandas.io import sql
import uuid
pd.options.mode.chained_assignment = None
###############################################################
if sys.platform == 'linux':
    Base=os.path.expanduser('~') + '/VKHCG'
else:
    Base='C:/VKHCG'
print('###############################')
```

```
print('Working Base :',Base, ' using ', sys.platform)
print('##############################')
###################################################################
Company='03-Hillman'
InputDir='00-RawData'
InputFileName='VehicleData.csv'
###################################################################
sDataBaseDir=Base + '/' + Company + '/03-Process/SQLite'
if not os.path.exists(sDataBaseDir):
    os.makedirs(sDataBaseDir)
###################################################################
sDatabaseName=sDataBaseDir + '/Hillman.db'
conn1 = sq.connect(sDatabaseName)
###################################################################
sDataVaultDir=Base + '/88-DV'
if not os.path.exists(sDataBaseDir):
    os.makedirs(sDataBaseDir)
###################################################################
sDatabaseName=sDataVaultDir + '/datavault.db'
conn2 = sq.connect(sDatabaseName)
```

You will set up a time hub for a period of ten years before January 1, 2018. If you want to experiment with different periods, simply change the parameters.

```
###################################################################
base = datetime(2018,1,1,0,0,0)
numUnits=10*365*24
###################################################################
date_list = [base - timedelta(hours=x) for x in range(0, numUnits)]
t=0
for i in date_list:
    now_utc=i.replace(tzinfo=timezone('UTC'))
    sDateTime=now_utc.strftime("%Y-%m-%d %H:%M:%S")
    sDateTimeKey=sDateTime.replace(' ','-').replace(':','-')
    t+=1
    IDNumber=str(uuid.uuid4())
    TimeLine=[('ZoneBaseKey', ['UTC']),
```

```
                ('IDNumber', [IDNumber]),
                ('nDateTimeValue', [now_utc]),
                ('DateTimeValue', [sDateTime]),
                ('DateTimeKey', [sDateTimeKey])]
    if t==1:
        TimeFrame = pd.DataFrame.from_items(TimeLine)
    else:
        TimeRow = pd.DataFrame.from_items(TimeLine)
        TimeFrame = TimeFrame.append(TimeRow)
################################################################
TimeHub=TimeFrame[['IDNumber','ZoneBaseKey','DateTimeKey','DateTimeValue']]
TimeHubIndex=TimeHub.set_index(['IDNumber'],inplace=False)
################################################################
TimeFrame.set_index(['IDNumber'],inplace=True)
################################################################
sTable = 'Process-Time'
print('Storing :',sDatabaseName,' Table:',sTable)
TimeHubIndex.to_sql(sTable, conn1, if_exists="replace")
################################################################
sTable = 'Hub-Time'
print('Storing :',sDatabaseName,' Table:',sTable)
TimeHubIndex.to_sql(sTable, conn2, if_exists="replace")
################################################################
```

You have successfully built the hub for time in the data vault. Now, we will build satellites for each of the time zones. So, let's create a loop to perform these tasks:

```
active_timezones=all_timezones
z=0
for zone in active_timezones:
    t=0
    for j in range(TimeFrame.shape[0]):
        now_date=TimeFrame['nDateTimeValue'][j]
        DateTimeKey=TimeFrame['DateTimeKey'][j]
        now_utc=now_date.replace(tzinfo=timezone('UTC'))
        sDateTime=now_utc.strftime("%Y-%m-%d %H:%M:%S")
        now_zone = now_utc.astimezone(timezone(zone))
```

461

```
        sZoneDateTime=now_zone.strftime("%Y-%m-%d %H:%M:%S")
        t+=1
        z+=1
        IDZoneNumber=str(uuid.uuid4())
        TimeZoneLine=[('ZoneBaseKey', ['UTC']),
                      ('IDZoneNumber', [IDZoneNumber]),
                      ('DateTimeKey', [DateTimeKey]),
                      ('UTCDateTimeValue', [sDateTime]),
                      ('Zone', [zone]),
                      ('DateTimeValue', [sZoneDateTime])]
        if t==1:
            TimeZoneFrame = pd.DataFrame.from_items(TimeZoneLine)
        else:
            TimeZoneRow = pd.DataFrame.from_items(TimeZoneLine)
            TimeZoneFrame = TimeZoneFrame.append(TimeZoneRow)

    TimeZoneFrameIndex=TimeZoneFrame.set_index(['IDZoneNumber'],
    inplace=False)
    sZone=zone.replace('/','-').replace(' ','')
    ############################################################
    sTable = 'Process-Time-'+sZone
    print('Storing :',sDatabaseName,' Table:',sTable)
    TimeZoneFrameIndex.to_sql(sTable, conn1, if_exists="replace")
############################################################
    ############################################################
    sTable = 'Satellite-Time-'+sZone
    print('Storing :',sDatabaseName,' Table:',sTable)
    TimeZoneFrameIndex.to_sql(sTable, conn2, if_exists="replace")
############################################################
```

You will note that your databases are growing at an alarming rate. I suggest that we introduce a minor database maintenance step that compacts the database to take up less space on the disk. The required command is vacuum. It performs an action similar to that of a vacuum-packing system, by compressing the data into the smallest possible disk footprint.

```
print('################')
print('Vacuum Databases')
sSQL="VACUUM;"
sql.execute(sSQL,conn1)
sql.execute(sSQL,conn2)
print('################')
####################################################################
print('### Done!! #######################################')
####################################################################
```

Congratulations! You have built your first hub and satellites for time in the data vault. The data vault has been built in directory ..\ VKHCG\88-DV\datavault.db. You can access it with your SQLite tools, if you want to investigate what you achieved. In Figure 9-8, you should see a structure that contains these data items.

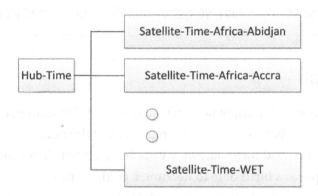

Figure 9-8. *Structure with three data items.*

Person

This structure records the person(s) involved within the data sources. A person is determined as a set of data items that, combined, describes a single "real" person in the world.

Remember   Angus MacGyver, the baker down the street, is a person.

Angus, my dog, is an object.

Angus MacGyver, a fictional character on TV played by Richard Dean Anderson, is an object. Richard Dean Anderson, the actor, is a person.

You would be surprised how many fictional characters data scientists find in databases around the world.

Warning   Beware the use of unwanted superfluous data entries in your data science!

Twice in the last year, a public safety system dispatched numerous emergency vehicles and staff because a data scientist assumed it would be tolerable to use "Angus MacGyver" to name a terror suspect who plants bombs in high-rise buildings at random. The data scientist defended his actions by reason that he tests his data science code by hard-coding it into his data science algorithms. Not the hallmark of a good data scientist!

Tip   Remove self-generated data errors from production systems immediately. They represent bad science! And, yes, they can damage property and hurt people.

Golden Nominal

A golden nominal record is a single person's record, with distinctive references for use by all systems. This gives the system a single view of the person. I use first name, other names, last name, and birth date as my golden nominal. The data we have in the assess directory requires a birth date to become a golden nominal. I will explain how to generate a golden nominal using our sample data set.

Open your Python editor and create a file called `Process-People.py` in the `..\VKHCG\04-Clark\03-Process` directory.

```
###############################################################
import sys
import os
import sqlite3 as sq
import pandas as pd
from datetime import datetime, timedelta
from pytz import timezone, all_timezones
from random import randint
###############################################################
```

```
if sys.platform == 'linux':
    Base=os.path.expanduser('~') + '/VKHCG'
else:
    Base='C:/VKHCG'
print('################################')
print('Working Base :',Base, ' using ', sys.platform)
print('################################')
#################################################################
Company='04-Clark'
sInputFileName='02-Assess/01-EDS/02-Python/Assess_People.csv'
#################################################################
sDataBaseDir=Base + '/' + Company + '/03-Process/SQLite'
if not os.path.exists(sDataBaseDir):
    os.makedirs(sDataBaseDir)
#################################################################
sDatabaseName=sDataBaseDir + '/clark.db'
conn = sq.connect(sDatabaseName)
#################################################################
### Import Female Data
#################################################################
sFileName=Base + '/' + Company + '/' + sInputFileName
print('################################')
print('Loading :',sFileName)
print('################################')
print(sFileName)
```

Now load the previous assess data into memory to process.

```
RawData=pd.read_csv(sFileName,header=0,low_memory=False,
encoding="latin-1")
RawData.drop_duplicates(subset=None, keep='first', inplace=True)
```

Let's assume nobody was born before January 1, 1900.

```
start_date = datetime(1900,1,1,0,0,0)
start_date_utc=start_date.replace(tzinfo=timezone('UTC'))
```

465

Let's assume, too, that the people who were born spread by random allocation over a period of 100 years.

```
HoursBirth=100*365*24
RawData['BirthDateUTC']=RawData.apply(lambda row:
                (start_date_utc + timedelta(hours=randint(0, HoursBirth)))
                ,axis=1)
zonemax=len(all_timezones)-1
RawData['TimeZone']=RawData.apply(lambda row:
                (all_timezones[randint(0, zonemax)])
                ,axis=1)
RawData['BirthDateISO']=RawData.apply(lambda row:
                row["BirthDateUTC"].astimezone(timezone(row['TimeZone']))
                ,axis=1)
RawData['BirthDateKey']=RawData.apply(lambda row:
                row["BirthDateUTC"].strftime("%Y-%m-%d %H:%M:%S")
                ,axis=1)
RawData['BirthDate']=RawData.apply(lambda row:
                row["BirthDateISO"].strftime("%Y-%m-%d %H:%M:%S")
                ,axis=1)

################################################################

Data=RawData.copy()
```

You have to drop to two ISO complex date time structures, as the code does not translate into SQLite's data types.

```
Data.drop('BirthDateUTC', axis=1,inplace=True)
Data.drop('BirthDateISO', axis=1,inplace=True)

print('###############################')
```

You now save your new golden nominal to SQLite.

```
################################################################
print('###############')
sTable='Process_Person'
print('Storing :',sDatabaseName,' Table:',sTable)
```

```
Data.to_sql(sTable, conn, if_exists="replace")
print('###############')
```

Now save your new golden nominal to a CSV file.

```
###################################################################
sFileDir=Base + '/' + Company + '/03-Process/01-EDS/02-Python'
if not os.path.exists(sFileDir):
    os.makedirs(sFileDir)
###################################################################
sOutputFileName = sTable + '.csv'
sFileName=sFileDir + '/' + sOutputFileName
print('##############################')
print('Storing :', sFileName)
print('##############################')
RawData.to_csv(sFileName, index = False)
print('##############################')
###################################################################
print('### Done!! ##########################################')
###################################################################
```

A real-world example is Ada, countess of Lovelace, who was born Augusta Ada King-Noel. She was an English mathematician and writer, known principally for her work on Charles Babbage's proposed mechanical general-purpose computer, the Analytical Engine. She was born on December 10, 1815.

First Name = Augusta

Other Names = Ada King-Noel

Last Name = Lovelace

Birth Date = 1815-10-10 00:00:00

You use an official name, as on her birth record, not her common name. You could add a satellite called CommonKnownAs and put "Ada Lovelace" in that table.

Note   The next section covers some interesting concepts that I discovered while working all over the world. It gives you some insight into the diversity of requirements data scientists can be presented with in executing their work.

The Meaning of a Person's Name

We all have names, and our names supply a larger grouping, or even hidden meanings, to data scientists. If people share the same last name, there is a probability they may be related. I will discuss this again later.

Around the world, there are also many other relationships connected to people's names.

Let's look at a few to which I was exposed over the years that I processed data.

Given Name, Matronymic, and Patronymic Meanings

There can be a specific relationship or meaning embedded in people's names. The Icelandic name Björk Guðmundsdóttir expresses a complex relationship between the individual and her family members.

Björk is the person's name. Guðmundsdóttir is a relationship: daughter (-dóttir) of Guðmundur. This makes the person female.

Björk's father is Guðmundur Gunnarsson. This is the most common practice.

Björk could also have the following matronymic name: Hildardottir, or even matro- and patronymic names: Hildar- og Guðmundsdóttir, because Björk's mother is Hildur R'una Hauksdottir.

Gunnarsson denotes the relationship "son of Gunnar." Therefore, Ólafur Guðmundsson is Björk's brother. Ólafur's son is Jón Ólafurson.

I've just contradicted my own earlier assumption. Different last names can equally denote members of the same family!

Caution   Do not think that algorithms and assumptions that work in your country work in all other countries with similar success.

The Malay name Isa bin Osman also indicates a hidden relationship. Isa is the given name. bin Osman denotes the relationship "son of Osman." Isa's sister would be Mira binti Osman, as *binti* means "daughter of."

Encik Isa would mean "Mr. Isa bin Osman," but the person writing knows Isa personally. So, do not address Isa as Encik Isa, if you do not know him personally.

Here are a few more naming conventions to take note of.

Language	Suffix "son"	Suffix "daughter"
Old Norse	-son	-dóttir
Danish	-sen	-datter
Faroese	-son	-dóttir
Finnish	-poika	-tytär
Icelandic	-son -sson	-dóttir
Norwegian	-sen/-son/-søn	-datter/-dotter
Swedish	-son	-dotter

Different Order of Parts of a Name

A full name is not always in the order first name, middle name, and last name. In the Chinese name 毛泽东 (Mao Zedong), for example, the family name (surname) is 毛 (Mao). The middle character, 泽 (Ze), is a generational name and is common to all siblings of the same family. According to the pīnyīn system governing the transliteration of Chinese into English, the last two characters, or elements of a Chinese name are generally combined as one word, e.g., *Zedong*, rather than *Ze Dong*. Mao's brothers and sister are 毛泽民 (Mao Zemin), 毛泽覃 (Mao Zetan), and 毛泽红 (Mao Zehong).

Note   "Ze Dong" is rarely, if ever, used in transliteration.

Among people with whom he was not on familiar terms, Mao may have been referred to as 毛泽东先生 (Mao Zedong xiānshēng) or 毛先生 (Mao xiānshēng) (*xiānshēng* being the equivalent of "Mr."). If you are on familiar terms with someone called 毛泽东, you would normally refer to them using 泽东 (Zedong), not just 东 (Dong).

In Japan, Korea, and Hungary, the order of names is family name, followed by given name(s).

Note   Chinese people who deal with Westerners will often adopt an additional given name that is easier for Westerners to use. Yao Ming (family name Yao, given name Ming) also renders his as Fred Yao Ming or Fred Ming Yao, for Westerners.

With Spanish-speaking people, it is common to have two family names. For example, María-Jose Carreño Quiñones refers to the daughter of Antonio Carreño Rodríguez and María Quiñones Marqués. You would address her as Señorita Carreño, not Señorita Quiñones.

Brazilians have similar customs and may have three or four family names, chaining the names of ancestors, such as, for example, José Eduardo Santos Tavares Melo Silva. Typically, two Spanish family names have the order paternal+maternal, whereas Portuguese names in Brazil have the order maternal+paternal. However, this order may change, just to confuse the data scientist.

Note   Some names consist of particles, such as de or e, between family names, for example, Carreño de Quiñones or Tavares e Silva.

Russians use patronymics as their middle name but also use family names, in the order given name+patronymic+family name. The endings of the patronymic and family names indicate whether the person is male or female. For example, the wife of Борис Николаевич Ельцин (Boris Nikolayevich Yeltsin) is Наина Иосифовна Ельцина (Naina Iosifovna Yeltsina).

Note   According to Russian naming conventions, a male's name ends in a consonant, while a female's name (even the patronymic) ends in *a*. This russian naming convention applies to all males and females people, not only married individuals.

Inheritance of Names

It would be wrong to assume that members of the same immediate family share the same last name. There is a growing trend in the West for married women to keep their maiden names after marriage. There are, however, other cultures, such as the Chinese, in which this is the standard practice. In some countries, a married woman may or may not take the name of her husband. If, for example, the Malay girl Zaiton married Isa, mentioned previously, she might remain Mrs. Zaiton, or she might choose to become Zaiton Isa, in which case you could refer to her as Mrs. Isa.

These data characteristics can make grouping data complex for the data scientist.

Other Relationships

The Dutch culture has its own naming conventions. The rules governing it are the following:

- First-born sons are named after paternal grandfathers.

- First-born daughters are named after maternal grandmothers.

- Second sons are named after maternal grandfathers.

- Second daughters are named after paternal grandmothers.

- Subsequent children are often named after uncles and aunts.

However, if a son has died before his next brother is born, the younger brother is usually given the same name as his deceased brother. The same goes for a daughter. When a father dies before the birth of a son, the son is named after him. When a mother dies upon the birth of a daughter, the daughter is named after the mother.

I found out the hard way, from a data set, that there is this structure to Dutch names, when we flagged a person deceased after finding a death certificate. It is not common in non-dutch families to re-use a dead person's name in the same generation and our data science always marked it as an error. So we had to remove this error rule when we moved system from america into Europe to resolve these cultural nonconformity.

Dutch names also have a historical and cultural relationship to places of birth or family occupations.

For example, Vermeulen is a Belgian variant of the Dutch surname ver Meulen, which indicates that my family was "from the far mills" or were "traveling millers" by trade. Hence the windmill on my family crest.

In line with Dutch tradition, marriage used to require a woman to precede her maiden name with her husband's name, linked by a hyphen. An Anna Pietersen who married a Jan Jansen became Anna Jansen-Pietersen. The current law in the Netherlands gives people more freedom. Upon marriage within the Netherlands, both partners default to keeping their own surnames, but both are given the choice of legally using their partner's surname or a combination of the two. In simple terms, Anna Pietersen, Jan Pietersen, Anna Jansen, Jan Jansen, Anna Jansen-Pietersen, and Jan Jansen-Pietersen could be the names of the same two people over a period of time.

These variances and additional meanings to names can enable the data scientist to train his or her algorithms to find relationships between people, by simply using their names. I have shared with you some of the interesting discoveries I have made during my years of working in many countries. For the purposes of this book, I will keep the person names simple and follow the Western convention of first name and last name.

Now, we load the person into the data vault. Open your Python editor and create a new file named Process-People.py in directory ..\VKHCG\04-Clark\03-Process. Now copy the following Python code into the file:

```
################################################################
import sys
import os
import sqlite3 as sq
import pandas as pd
from pandas.io import sql
from datetime import datetime, timedelta
from pytz import timezone, all_timezones
from random import randint
################################################################
if sys.platform == 'linux':
    Base=os.path.expanduser('~') + '/VKHCG'
else:
    Base='C:/VKHCG'
print('############################')
print('Working Base :',Base, ' using ', sys.platform)
print('############################')
################################################################
Company='04-Clark'
sInputFileName='02-Assess/01-EDS/02-Python/Assess_People.csv'
################################################################
sDataBaseDir=Base + '/' + Company + '/03-Process/SQLite'
if not os.path.exists(sDataBaseDir):
    os.makedirs(sDataBaseDir)
################################################################
sDatabaseName=sDataBaseDir + '/clark.db'
conn1 = sq.connect(sDatabaseName)
################################################################
```

```
sDataVaultDir=Base + '/88-DV'
if not os.path.exists(sDataBaseDir):
    os.makedirs(sDataBaseDir)
################################################################
sDatabaseName=sDataVaultDir + '/datavault.db'
conn2 = sq.connect(sDatabaseName)
################################################################
### Import Female Data
################################################################
sFileName=Base + '/' + Company + '/' + sInputFileName
print('###############################')
print('Loading :',sFileName)
print('###############################')
print(sFileName)
RawData=pd.read_csv(sFileName,header=0,low_memory=False,
encoding="latin-1")
RawData.drop_duplicates(subset=None, keep='first', inplace=True)

start_date = datetime(1900,1,1,0,0,0)
start_date_utc=start_date.replace(tzinfo=timezone('UTC'))

HoursBirth=100*365*24
RawData['BirthDateUTC']=RawData.apply(lambda row:
                (start_date_utc + timedelta(hours=randint(0, HoursBirth)))
                ,axis=1)
zonemax=len(all_timezones)-1
RawData['TimeZone']=RawData.apply(lambda row:
                (all_timezones[randint(0, zonemax)])
                ,axis=1)
RawData['BirthDateISO']=RawData.apply(lambda row:
                row["BirthDateUTC"].astimezone(timezone(row['TimeZone']))
                ,axis=1)
RawData['BirthDateKey']=RawData.apply(lambda row:
                row["BirthDateUTC"].strftime("%Y-%m-%d %H:%M:%S")
                ,axis=1)
```

473

```python
RawData['BirthDate']=RawData.apply(lambda row:
                row["BirthDateISO"].strftime("%Y-%m-%d %H:%M:%S")
                ,axis=1)

###############################################################

Data=RawData.copy()
Data.drop('BirthDateUTC', axis=1,inplace=True)
Data.drop('BirthDateISO', axis=1,inplace=True)
indexed_data = Data.set_index(['PersonID'])

print('###############################')
###############################################################
print('################')
sTable='Process_Person'
print('Storing :',sDatabaseName,' Table:',sTable)
indexed_data.to_sql(sTable, conn1, if_exists="replace")
print('################')
###############################################################
PersonHubRaw=Data[['PersonID','FirstName','SecondName','LastName',
'BirthDateKey']]
PersonHubRaw['PersonHubID']=RawData.apply(lambda row:
                str(uuid.uuid4())
                ,axis=1)
PersonHub=PersonHubRaw.drop_duplicates(subset=None, \
                                                keep='first',\
                                                inplace=False)
indexed_PersonHub = PersonHub.set_index(['PersonHubID'])
sTable = 'Hub-Person'
print('Storing :',sDatabaseName,' Table:',sTable)
indexed_PersonHub.to_sql(sTable, conn2, if_exists="replace")
###############################################################
PersonSatelliteGenderRaw=Data[['PersonID','FirstName','SecondName',
'LastName'\
                        ,'BirthDateKey','Gender']]
PersonSatelliteGenderRaw['PersonSatelliteID']=RawData.apply(lambda row:
                str(uuid.uuid4())
                ,axis=1)
```

474

```
PersonSatelliteGender=PersonSatelliteGenderRaw.drop_duplicates
(subset=None, \
                                                    keep='first', \
                                                    inplace=False)
indexed_PersonSatelliteGender = PersonSatelliteGender.set_index(
['PersonSatelliteID'])
sTable = 'Satellite-Person-Gender'
print('Storing :',sDatabaseName,' Table:',sTable)
indexed_PersonSatelliteGender.to_sql(sTable, conn2, if_exists="replace")
################################################################
PersonSatelliteBirthdayRaw=Data[['PersonID','FirstName','SecondName',
'LastName',\
                        'BirthDateKey','TimeZone','BirthDate']]
PersonSatelliteBirthdayRaw['PersonSatelliteID']=RawData.apply(lambda row:
            str(uuid.uuid4())
            ,axis=1)
PersonSatelliteBirthday=PersonSatelliteBirthdayRaw.drop_duplicates
(subset=None, \
                                                    keep='first',\
                                                    inplace=False)
indexed_PersonSatelliteBirthday = PersonSatelliteBirthday.set_index(
['PersonSatelliteID'])
sTable = 'Satellite-Person-Names'
print('Storing :',sDatabaseName,' Table:',sTable)
indexed_PersonSatelliteBirthday.to_sql(sTable, conn2, if_exists="replace")
################################################################
sFileDir=Base + '/' + Company + '/03-Process/01-EDS/02-Python'
if not os.path.exists(sFileDir):
    os.makedirs(sFileDir)
################################################################
sOutputFileName = sTable + '.csv'
sFileName=sFileDir + '/' + sOutputFileName
print('###############################')
print('Storing :', sFileName)
print('###############################')
```

```
RawData.to_csv(sFileName, index = False)
print('##############################')
##################################################################
print('###############')
print('Vacuum Databases')
sSQL="VACUUM;"
sql.execute(sSQL,conn1)
sql.execute(sSQL,conn2)
print('###############')
##################################################################
print('### Done!! #######################################')
##################################################################
```

Object

This structure records the other objects and nonhuman entities that are involved in the data sources. The object must have three basic fields: Object Type, Object Group, and Object Code to be valid. I also add a short and long description for information purposes.

The data science allows for the data vault's object hub to be a simple combination of Object Type, Object Group, and Object Code, to formulate a full description of any nonhuman entities in the real world.

I must note that as you start working with objects, you will discover that most Object Types have standards to support the specific object's processing. Let's look at a few common objects I've had to handle by data science.

Animals

Animals belong to many different groups, starting with the Animal Kingdom.

Kingdom

All living organisms are first placed into different kingdoms. There are five different kingdoms that classify life on Earth: Animals, Plants, Fungi, Bacteria, and Protists (single-celled organisms).

Phylum

The Animal Kingdom is divided into 40 smaller groups, known as phyla. Accordingly, animals are grouped by their main features. Animals usually fall into one of five different phyla: Cnidaria (invertebrates), Chordata (vertebrates), Arthropods, Molluscs, and Echinoderms.

Class

A phylum is divided into even smaller groups, known as classes. The Chordata (vertebrates) phylum splits into Mammalia (mammals), Actinopterygii (bony fish), Chondrichthyes (cartilaginous fish), Aves (birds), Amphibia (amphibians), and Reptilia (reptiles).

Order

Each class is divided into even smaller groups, known as orders. The class Mammalia (mammals) splits into different groups, including Carnivora, Primate, Artiodactyla, and Rodentia.

Family

In every order, there are different families of animals, which all have very similar features. The Carnivora order breaks into families that include Felidae (cats), Canidae (dogs), Ursidae (bears), and Mustelidae (weasels).

Genus

Every animal family is then divided into groups known as genera. Each genus contains animals that have very similar features and are closely related. For example, the Felidae (cat) family contains the genera *Felis* (small cats and domestic cats), *Panthera* (tigers, leopards, jaguars, and lions), and *Puma* (panthers and cougars).

Species

Each individual species within a genus is named after its individual features and characteristics. The names of animals are in Latin, as are genera, so that they can be understood worldwide, and consist of two words. The first part in the name of an animal will be the genus, and the second part indicates the specific species.

Let's classify Tigger, my toy tiger.

Kingdom	Animalia (Animal)
Phylum	Chordata (Vertebrate)
Class	Mammalia (Mammal)
Order	Carnivora (Carnivore)
Family	Felidae (Cat)
Genus	*Panthera*
Species	*tigris* (Tiger)

Now let's classify my dog, Angus.

Kingdom	Animalia
Class	Mammalia
Order	Carnivora
Family	Canidae
Genus	*Canis*
Specie	*Lupus familiaris*

He is classified as *Canis Lupus familiaris.*

You must look at an object as revealed by recently discovered knowledge. Now, I will guide you through a Python code session, to convert the flat file into a graph with knowledge. The only information you have are in file `Animals.csv` in directory `..\VKHCG\03-Hillman\00-RawData`.

The format is

ItemLevel	ParentID	ItemID	ItemName
0	0	50	Bacteria
0	0	202422	Plantae
1	50	956096	Negibacteria
1	50	956097	Posibacteria

The field has the following meanings:

- `ItemLevel` is how far the specific item is from the top node in the classification.

- `ParentID` is the ItemID for the parent of the Item listed.

- `ItemID` is the unique identifier for the item.

- `ItemName` is the full name of the item.

The data fits together as a considerable tree of classifications. You must create a graph that gives you the following:

`Bacteria-> Negibacteria and Bacteria-> Posibacteria`

Following is the code to transform it. You will perform a few sections of data preparation, data storage for the retrieve, Assess supersteps, and then we will complete the Process step into the data vault. You start with the standard framework, so please transfer the code to your Python editor. First, let's set up the data:

```
################################################################
# -*- coding: utf-8 -*-
################################################################
import sys
import os
import pandas as pd
import networkx as nx
```

```python
import sqlite3 as sq
import numpy as np
################################################################
if sys.platform == 'linux':
    Base=os.path.expanduser('~') + '/VKHCG'
else:
    Base='C:/VKHCG'
print('################################')
print('Working Base :',Base, ' using ', sys.platform)
print('################################')
################################################################
ReaderCode='SuperDataScientist'
```

Please replace the 'Practical Data Scientist' in the next line with your name.

```python
ReaderName='Practical Data Scientist'
```

You now set up the locations of all the deliverables of the code.

```python
################################################################
Company='03-Hillman'
InputRawFileName='Animals.csv'

EDSRetrieveDir='01-Retrieve/01-EDS'
InputRetrieveDir=EDSRetrieveDir + '/02-Python'
InputRetrieveFileName='Retrieve_All_Animals.csv'

EDSAssessDir='02-Assess/01-EDS'
InputAssessDir=EDSAssessDir + '/02-Python'
InputAssessFileName='Assess_All_Animals.csv'
InputAssessGraphName='Assess_All_Animals.gml'
```

You now create the locations of all the deliverables of the code.

```python
################################################################
sFileRetrieveDir=Base + '/' + Company + '/' + InputRetrieveDir
if not os.path.exists(sFileRetrieveDir):
    os.makedirs(sFileRetrieveDir)
################################################################
```

```
sFileAssessDir=Base + '/' + Company + '/' + InputAssessDir
if not os.path.exists(sFileAssessDir):
    os.makedirs(sFileAssessDir)
################################################################
sDataBaseDir=Base + '/' + Company + '/03-Process/SQLite'
if not os.path.exists(sDataBaseDir):
    os.makedirs(sDataBaseDir)
################################################################
sDatabaseName=sDataBaseDir + '/Hillman.db'
conn = sq.connect(sDatabaseName)
################################################################
# Raw to Retrieve
################################################################
```

You upload the CSV file with the flat structure.

```
sFileName=Base + '/' + Company + '/00-RawData/' + InputRawFileName
print('###########')
print('Loading :',sFileName)
AnimalRaw=pd.read_csv(sFileName,header=0,low_memory=False, encoding =
"ISO-8859-1")
AnimalRetrieve=AnimalRaw.copy()
print(AnimalRetrieve.shape)
################################################################
```

You store the Retrieve steps data now.

```
sFileName=sFileRetrieveDir + '/' + InputRetrieveFileName
print('###########')
print('Storing Retrieve :',sFileName)
AnimalRetrieve.to_csv(sFileName, index = False)
```

You store the Assess steps data now.

```
################################################################
# Retrieve to Assess
################################################################
AnimalGood1 = AnimalRetrieve.fillna('0', inplace=False)
AnimalGood2=AnimalGood1[AnimalGood1.ItemName!=0]
```

481

```
AnimalGood2[['ItemID','ParentID']]=AnimalGood2[['ItemID','ParentID']].
astype(np.int32)

AnimalAssess=AnimalGood2
print(AnimalAssess.shape)
################################################################
sFileName=sFileAssessDir + '/' + InputAssessFileName
print('###########')
print('Storing Assess :',sFileName)
AnimalAssess.to_csv(sFileName, index = False)
################################################################
print('###############')
sTable='All_Animals'
print('Storing :',sDatabaseName,' Table:',sTable)
AnimalAssess.to_sql(sTable, conn, if_exists="replace")
print('###############')
```

You start with the Process steps, to process the flat data into a graph. You can now extract the nodes, as follows:

```
################################################################

print('###############')
sTable='All_Animals'
print('Loading Nodes :',sDatabaseName,' Table:',sTable)
sSQL=" SELECT DISTINCT"
sSQL=sSQL+ " CAST(ItemName AS VARCHAR(200)) AS NodeName,"
sSQL=sSQL+ " CAST(ItemLevel AS INT) AS NodeLevel"
sSQL=sSQL+ " FROM"
sSQL=sSQL+ " " + sTable + ";"

AnimalNodeData=pd.read_sql_query(sSQL, conn)

print(AnimalNodeData.shape)
```

You have now successfully extracted the nodes. Well done. You can now extract the edges. You will start with the Process step, to convert the data into an appropriate graph structure.

```
################################################################
print('###############')
sTable='All_Animals'
print('Loading Edges :',sDatabaseName,' Table:',sTable)
sSQL=" SELECT DISTINCT"
sSQL=sSQL+ " CAST(A1.ItemName AS VARCHAR(200)) AS Node1,"
sSQL=sSQL+ " CAST(A2.ItemName AS VARCHAR(200)) AS Node2"
sSQL=sSQL+ " FROM"
sSQL=sSQL+ " " + sTable + " AS A1"
sSQL=sSQL+ " JOIN"
sSQL=sSQL+ " " + sTable + " AS A2"
sSQL=sSQL+ " ON"
sSQL=sSQL+ " A1.ItemID=A2.ParentID;"

AnimalEdgeData=pd.read_sql_query(sSQL, conn)
print(AnimalEdgeData.shape)
```

You have now extracted the edges. So, let's build a graph.

```
################################################################
G=nx.Graph()
t=0
G.add_node('world', NodeName='World')
################################################################
```

You add the nodes first.

```
GraphData=AnimalNodeData
print(GraphData)
################################################################
m=GraphData.shape[0]
for i in range(m):
    t+=1
    sNode0Name=str(GraphData['NodeName'][i]).strip()
    print('Node :',t,' of ',m,sNode0Name)
    sNode0=sNode0Name.replace(' ', '-').lower()
```

```
    G.add_node(sNode0, NodeName=sNode0Name)
    if GraphData['NodeLevel'][i] == 0:
        G.add_edge(sNode0,'world')
###############################################################
```

You add the edges second.

```
GraphData=AnimalEdgeData
t=0
###############################################################
m=GraphData.shape[0]

for i in range(m):
    t+=1
    sNode0Name=str(GraphData['Node1'][i]).strip()
    sNode1Name=str(GraphData['Node2'][i]).strip()
    print('Link :',t,' of ',m,sNode0Name,' to ',sNode1Name)
    sNode0=sNode0Name.replace(' ', '-').lower()
    sNode1=sNode1Name.replace(' ', '-').lower()
    G.add_edge(sNode0,sNode1)
```

You have nodes and edges. Now add yourself as a node.

```
###############################################################
RCode=ReaderCode.replace(' ', '-').lower()
G.add_node(RCode,NodeName=ReaderName)
G.add_edge('homo-sapiens',RCode)

###############################################################
```

Now store the graph.

```
sFileName= sFileAssessDir + '/' + InputAssessGraphName
print('###############################')
print('Storing :', sFileName)
print('###############################')
nx.write_gml(G,sFileName)
sFileName=sFileName +'.gz'
nx.write_gml(G,sFileName)
```

Let's consider some of the knowledge we already can extract. Look at how the following objects relate to each other:

Yourself?

The author?

Angus, my dog?

Tigger, my toy tiger?

Chris Hillman, my technical advisor and friend?

Enter the following Python file into your editor, and you will discover new knowledge.

```python
###############################################################
# Find Lists of Objects
###############################################################
TargetNodes=[ReaderCode,'Andre Vermeulen','Angus', 'Tigger', 'Chris Hillman']
for j in range(len(TargetNodes))  :
    TargetNodes[j]=TargetNodes[j].replace(' ', '-').lower()
###############################################################
for TargetNode in TargetNodes:
    if TargetNode in nx.nodes(G):
        print('=============================')
        print('Path:','World',' to ',G.node[TargetNode]['NodeName'])
        print('=============================')
        for nodecode in nx.shortest_path(G,source='world',target=TargetNode):
            print(G.node[nodecode]['NodeName'])
        print('=============================')
    else:
        print('=============================')
        print('No data - ', TargetNode, ' is missing!')
        print('=============================')
```

If you are classed as a *Homo sapiens*, well done!

Now, let's look at what it would take via DNA for Angus to become Tigger (other than an 87 kilogram weight loss and allowing him on the bed)?

```
################################################################
print('=============================')
print(' How do we turn Angus into Tigger?')
print('=============================')
for nodecode in nx.shortest_path(G,source='angus',target='tigger'):
    print(G.node[nodecode]['NodeName'])
################################################################
```

Let's look at what it would take for Chris to become Dr. Chris.

```
print('=============================')
print('How do you make Chris a Doctor?')
print('=============================')
for nodecode in nx.shortest_path(G,source='chris-hillman',target='dr-
chris'):
    print(G.node[nodecode]['NodeName'])
print('=============================')

################################################################
print('### Done!! #####################################')
################################################################
```

You have now successfully created a new graph, which you will use to populate the animal object part of the data vault. Create a new file in your Python editor named Process-Animal-Graph.py in directory ..\VKHCG\03-Hillman\03-Process. Copy the following Python code into the file. First, we will load some of the libraries you need and create a set of basic file names and directories, to enable the ecosystem.

```
################################################################
# -*- coding: utf-8 -*-
################################################################
import sys
import os
import pandas as pd
import networkx as nx
import sqlite3 as sq
from pandas.io import sql
import uuid
```

```python
################################################################
if sys.platform == 'linux':
    Base=os.path.expanduser('~') + '/VKHCG'
else:
    Base='C:/VKHCG'
print('################################')
print('Working Base :',Base, ' using ', sys.platform)
print('################################')
################################################################
Company='03-Hillman'
```

Here is the path to the graph you created earlier in this chapter:

```python
InputAssessGraphName='Assess_All_Animals.gml'
EDSAssessDir='02-Assess/01-EDS'
InputAssessDir=EDSAssessDir + '/02-Python'
################################################################
sFileAssessDir=Base + '/' + Company + '/' + InputAssessDir
if not os.path.exists(sFileAssessDir):
    os.makedirs(sFileAssessDir)
################################################################
sDataBaseDir=Base + '/' + Company + '/03-Process/SQLite'
if not os.path.exists(sDataBaseDir):
    os.makedirs(sDataBaseDir)
################################################################
sDatabaseName=sDataBaseDir + '/Hillman.db'
conn1 = sq.connect(sDatabaseName)
################################################################
sDataVaultDir=Base + '/88-DV'
if not os.path.exists(sDataBaseDir):
    os.makedirs(sDataBaseDir)
################################################################
sDatabaseName=sDataVaultDir + '/datavault.db'
conn2 = sq.connect(sDatabaseName)
################################################################
```

You now load the graph you created earlier in this chapter.

```
sFileName= sFileAssessDir + '/' + InputAssessGraphName
print('###############################')
print('Loading Graph :', sFileName)
print('###############################')
G=nx.read_gml(sFileName)
print('Nodes: ', G.number_of_nodes())
print('Edges: ', G.number_of_edges())
```

You will now read the nodes of the graph and add them to the data vault as objects.

```
################################################################
t=0
tMax=G.number_of_nodes()
for node in nx.nodes_iter(G):
    t+=1
    IDNumber=str(uuid.uuid4())
    NodeName=G.node[node]['NodeName']
    print('Extract:',t,' of ',tMax,':',NodeName)
    ObjectLine=[('ObjectBaseKey', ['Species']),
                ('IDNumber', [IDNumber]),
                ('ObjectNumber', [str(t)]),
                ('ObjectValue', [NodeName])]
    if t==1:
        ObjectFrame = pd.DataFrame.from_items(ObjectLine)
    else:
        ObjectRow = pd.DataFrame.from_items(ObjectLine)
        ObjectFrame = ObjectFrame.append(ObjectRow)

################################################################
ObjectHubIndex=ObjectFrame.set_index(['IDNumber'],inplace=False)
################################################################
sTable = 'Process-Object-Species'
print('Storing :',sDatabaseName,' Table:',sTable)
ObjectHubIndex.to_sql(sTable, conn1, if_exists="replace")
```

```
################################################################
sTable = 'Hub-Object-Species'
print('Storing :',sDatabaseName,' Table:',sTable)
ObjectHubIndex.to_sql(sTable, conn2, if_exists="replace")
################################################################
print('###############')
print('Vacuum Databases')
sSQL="VACUUM;"
sql.execute(sSQL,conn1)
sql.execute(sSQL,conn2)
print('###############')
################################################################
print('### Done!! ######################################')
################################################################
```

Well done. You now have an additional object hub in the data vault for species. Have a look at these two pieces of code. Loop through all nodes and return the shortest path from 'World' node.

```
for node1 in nx.nodes_iter(G):
    p=nx.shortest_path(G, source='world', target=node1, weight=None)
    print('Path from World to ',node1,':',p)
```

Loop through all nodes and return the shortest path from the entire node set.

```
for node1 in nx.nodes_iter(G):
    for node2 in nx.nodes_iter(G):
        p=nx.shortest_path(G, source=node1, target=node1, weight=None)
        print('Path:',p)
```

Let's quickly discuss a new algorithm. Dijkstra's algorithm is an algorithm for discovery of the shortest paths between nodes in a graph, designed by computer scientist Edsger Wybe Dijkstra in 1956. Figure 9-9 shows an example.

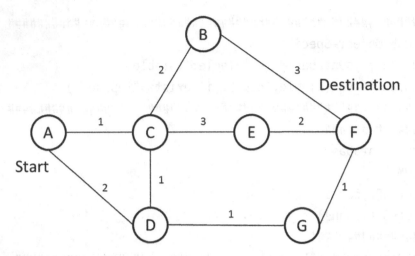

Figure 9-9. *Example of Dijkstra's algorithm*

```
import networkx as nx
G = nx.Graph()
G.add_edge('a', 'c', weight=1)
G.add_edge('a', 'd', weight=2)
G.add_edge('b', 'c', weight=2)
G.add_edge('c', 'd', weight=1)
G.add_edge('b', 'f', weight=3)
G.add_edge('c', 'e', weight=3)
G.add_edge('e', 'f', weight=2)
G.add_edge('d', 'g', weight=1)
G.add_edge('g', 'f', weight=1)
d=nx.dijkstra_path(G, source='a', target='f', weight=None)
print(d)
```

I calculate that ['a', 'c', 'b', 'f'] is the shortest path.

Yes, this algorithm was conceived in 1956 and published three years later. Data science is not as new as many people think. We deploy countless well-established algorithms in our work.

Loop through all nodes and return the shortest path from the entire node set, using Dijkstra's algorithm:

```
for node1 in nx.nodes_iter(G):
    for node2 in nx.nodes_iter(G):
        d=nx.dijkstra_path(G, source=node1, target=node1, weight=None)
        print('Path:',d)
```

Question: Could you create a link table that uses this path as the link?

Here's the table you must populate:

```
CREATE TABLE [Link-Object-Species] (
    LinkID      VARCHAR (100),
    NodeFrom    VARCHAR (200),
    NodeTo      VARCHAR (200),
    ParentNode  VARCHAR (200),
    Node        VARCHAR (200),
    Step        INTEGER
);
```

Try your new skills against the graph. Can you get it to work?

Let's look at the next object type.

Vehicles

The international classification of vehicles is a complex process. There are standards, but these are not universally applied or similar between groups or countries. For example, my Cadillac Escalade ESV 4WD PLATINUM is a

- "Full-size SUV pickup truck" in market segment (American English)

- "Large 4×4" in market segment (British English)

- "Upper Large SUV" in market segment (Australian English)

- "Standard Sport Utility Vehicle" in US EPA size class

- "Off-roader" in European NCAP structural category

- "Large Off-Road 4×4" in European NCAP class (1997–2009)

- "J-segment sport utility car" in European market segment

- "Luxury Station Wagon" in US Highway Loss Data Institute classification

491

To specifically categorize my vehicle, I require a vehicle identification number (VIN). A single VIN supplies a data scientist with massive amounts of classification details. There are two standards for vehicles: ISO 3779:2009 (road vehicles—VIN—content and structure) and ISO 3780:2009 (road vehicles—world manufacturer identifier [WMI] code). I have found that these two combined can categorize the vehicle in good detail, i.e., I can decode the VIN.

VIN: 1HGBH41JXMN109186

- *Model*: Prelude

- *Type*: Passenger Car

- *Make*: Honda

- *Model year*: 1991

- *Manufacturer*: Honda of America Mfg. Inc.

- *Manufactured in*: United States (North America)

- *Sequential number*: 109186

- *Body type*: Two-door coupe

- *Vehicle type*: Passenger car

Warning   A vehicle's number plate or tags is not unique to that vehicle. There are private plate numbers that belong to a person, not a vehicle. This results in several different vehicles having the same plate number over time. When a car is resold, the new owner can also receive new plate numbers. So, do not use the plate number as a unique key.

As a data scientist, you must understand the data you can infer from simple codes and classifications of the object been recorded.

Tip   Vehicles are systematically coded, and you will appreciate the longer-term rewards of this, if you take the time to understand the coding standards first.

Let's load the vehicle data for Hillman Ltd into the data vault, as we will need it later. Create a new file named `Process-Vehicle-Logistics.py` in the Python editor in directory ..\VKHCG\03-Hillman\03-Process. Copy the code into your new file, as I guide you through it. Start with the standard structure that creates the ecosystem.

```
###############################################################
# -*- coding: utf-8 -*-
###############################################################
import sys
import os
import pandas as pd
import sqlite3 as sq
from pandas.io import sql
import uuid

pd.options.mode.chained_assignment = None
###############################################################
if sys.platform == 'linux':
    Base=os.path.expanduser('~') + '/VKHCG'
else:
    Base='C:/VKHCG'
print('###############################')
print('Working Base :',Base, ' using ', sys.platform)
print('###############################')
###############################################################
Company='03-Hillman'
InputDir='00-RawData'
InputFileName='VehicleData.csv'
###############################################################
sDataBaseDir=Base + '/' + Company + '/03-Process/SQLite'
if not os.path.exists(sDataBaseDir):
    os.makedirs(sDataBaseDir)
###############################################################
sDatabaseName=sDataBaseDir + '/Hillman.db'
conn1 = sq.connect(sDatabaseName)
###############################################################
sDataVaultDir=Base + '/88-DV'
if not os.path.exists(sDataBaseDir):
    os.makedirs(sDataBaseDir)
```

```
#############################################################
sDatabaseName=sDataVaultDir + '/datavault.db'
conn2 = sq.connect(sDatabaseName)
#############################################################
```

Now load the vehicle to be added to the data vault.

```
sFileName=Base + '/' + Company + '/' + InputDir + '/' + InputFileName
print('###########')
print('Loading :',sFileName)
VehicleRaw=pd.read_csv(sFileName,header=0,low_memory=False,
encoding="latin-1")
#############################################################
sTable='Process_Vehicles'
print('Storing :',sDatabaseName,' Table:',sTable)
VehicleRaw.to_sql(sTable, conn1, if_exists="replace")
#############################################################
```

Now subset the vehicle characteristics that you need to ready for the data vault to Make and Model only.

```
VehicleRawKey=VehicleRaw[['Make','Model']].copy()
```

Remove any duplicates.

```
VehicleKey=VehicleRawKey.drop_duplicates()
```

Create a vehicle key for the hub. You can clean up the VehicleKey by removing unwanted characters.

```
VehicleKey['ObjectKey']=VehicleKey.apply(lambda row:
    str('('+ str(row['Make']).strip().replace(' ', '-').replace('/', '-').
    lower() +
     ')-(' + (str(row['Model']).strip().replace(' ', '-').replace(' ',
     '-').lower())
    +')')
        ,axis=1)
Set all records a been 'vehicle' as a objectType.
```

```
VehicleKey['ObjectType']=VehicleKey.apply(lambda row:
    'vehicle'
        ,axis=1)
```

Set every record with unique id.

```
VehicleKey['ObjectUUID']=VehicleKey.apply(lambda row:
    str(uuid.uuid4())
        ,axis=1)
VehicleHub=VehicleKey[['ObjectType','ObjectKey','ObjectUUID']].copy()
VehicleHub.index.name='ObjectHubID'
sTable = 'Hub-Object-Vehicle'
print('Storing :',sDatabaseName,' Table:',sTable)
VehicleHub.to_sql(sTable, conn2, if_exists="replace")
```

Create a vehicle satellite for the hub, to store any extra vehicle characteristics.

```
################################################################
### Vehicle Satellite
################################################################
#
VehicleSatellite=VehicleKey[['ObjectType','ObjectKey','ObjectUUID','Make','
Model']].copy()
VehicleSatellite.index.name='ObjectSatelliteID'
sTable = 'Satellite-Object-Make-Model'
print('Storing :',sDatabaseName,' Table:',sTable)
VehicleSatellite.to_sql(sTable, conn2, if_exists="replace")
```

Create a vehicle dimension that joins the hub to the satellite.

```
################################################################
### Vehicle Dimension
################################################################
sView='Dim-Object'
print('Storing :',sDatabaseName,' View:',sView)
sSQL="CREATE VIEW IF NOT EXISTS [" + sView + "] AS"
sSQL=sSQL+ " SELECT DISTINCT"
sSQL=sSQL+ "    H.ObjectType,"
sSQL=sSQL+ "    H.ObjectKey AS VehicleKey,"
```

```
sSQL=sSQL+ "    TRIM(S.Make) AS VehicleMake,"
sSQL=sSQL+ "    TRIM(S.Model) AS VehicleModel"
sSQL=sSQL+ " FROM"
sSQL=sSQL+ "    [Hub-Object-Vehicle] AS H"
sSQL=sSQL+ " JOIN"
sSQL=sSQL+ "    [Satellite-Object-Make-Model] AS S"
sSQL=sSQL+ " ON"
sSQL=sSQL+ "        H.ObjectType=S.ObjectType"
sSQL=sSQL+ " AND"
sSQL=sSQL+ " H.ObjectUUID=S.ObjectUUID;"
sql.execute(sSQL,conn2)
```

Let's test the vehicle dimension that joins the hub to the satellite.

```
print('###############')
print('Loading :',sDatabaseName,' Table:',sView)
sSQL=" SELECT DISTINCT"
sSQL=sSQL+ " VehicleMake,"
sSQL=sSQL+ " VehicleModel"
sSQL=sSQL+ " FROM"
sSQL=sSQL+ " [" + sView + "]"
sSQL=sSQL+ " ORDER BY"
sSQL=sSQL+ " VehicleMake"
sSQL=sSQL+ " AND"
sSQL=sSQL+ " VehicleMake;"
DimObjectData=pd.read_sql_query(sSQL, conn2)

DimObjectData.index.name='ObjectDimID'
DimObjectData.sort_values(['VehicleMake','VehicleModel'],inplace=True,
ascending=True)
print('###############')
print(DimObjectData)
################################################################
print('###############')
print('Vacuum Databases')
sSQL="VACUUM;"
sql.execute(sSQL,conn1)
```

```
sql.execute(sSQL,conn2)
print('###############')
################################################################
conn1.close()
conn2.close()
################################################################
#print('### Done!! #####################################')
################################################################
```

Congratulations! You now have two object hubs created. I will cover several other objects that you can investigate.

Chemical Compounds

Objects consist of numerous chemicals, and these chemicals can inform you about the categorization of any object. There is a related library named pyEQL (`pip3 install pyEQL`).

Open your Python editor and try the following:

```
from pyEQL import chemical_formula as cf
print(cf.hill_order('CH2(CH3)4COOH'))
```

This will help you resolve chemical formulas. Let's look at a few examples.

- Table salt (sodium chloride) is NaCl. It contains one part sodium and one part chloride.

```
print(cf.get_element_mole_ratio('NaCl','Na')) print(cf.get_element_mole_
ratio('NaCl','Cl'))
```

- Drain cleaner (sulfuric acid) is H_2SO_4. It contains two parts hydrogen, one part sulfur, and four parts oxygen.

- Baking Soda (sodium bicarbonate) is $NaHCO_3$. What are the ratios?

- Glycerol is $C_3H_8O_3$. What are the ratios?

- What are combined to make the substance?

```
print(cf.get_element_names('C3H8O3'))
```

Check if a formula is valid.

```
print(cf.is_valid_formula('Fe2(SO4)3'))
print(cf.is_valid_formula('Fe2(SO4)3+Bad'))
```

You can also investigate other options. See library `periodictable` at `https://pypi.python.org/pypi/periodictable`.

```
from periodictable import *
for el in elements:
        print("%s %s"%(el.symbol, el.name))
```

You should now have a list of elements.

You can also investigate `biopython` (see `http://biopython.org`). You can access this by installing `conda install -c anaconda biopython`. This provides insights into the world of computational molecular biology. Can you combine it with your knowledge, to create extra object hubs?

Tip   You can extract various objects from the world around you and then start analyzing how they change the art of data science.

The object data structure will change as your customers' data changes from business to business. I advise that you start with the analysis for any nonperson entities' classifications as soon as possible in your project. Talk to your customers' subject-matter experts, get their classification model, and understand them 100%.

Also include all angles of the object's description. A retail store might sell table salt, but the manufacturer produces sodium chloride. By connecting these within your data vault, your algorithms will understand these relationships. Algorithms such as deep learning, logistic regression, and random forests, which I will cover in Chapter 10, work with improved net results, if you have already created the basic relationships, by linking different areas of insight.

Tip   Start by Object Type and Object Group first when identifying your business requirements. If you have the two fields sorted, you are 80+% of the way to completing the object structure.

This is just a basic start to the processing of objects in data science. This part of the data vault will always be the largest portion of the data science work in the Process superstep.

I have also enriched my customers' data by getting external experts to categorize objects for which particular customers have never identified the common characteristics.

Warning   It is better to load in large amounts of overlapping details than to have inadequate and incomplete details. Do not over-summarize your Object Type and Object Group groupings.

For the purpose of the practical examples, I will cover the algorithms and techniques that will assist you to perform the categorization of the objects in data sources.

Location

This structure records any location where an interaction between data entities occurred in the data sources. The location of a data action is a highly valuable piece of data for any business. It enables geospatial insight of the complete data sets. *Geospatial* refers to information that categorizes where defined structures are on the earth's surface.

Let's discuss a few basics concerning location.

Latitude, Longitude, and Altitude

Latitude, longitude, and altitude are some of the oldest official methods of describing the location on the earth, elaborated by Hipparchus, a Greek astronomer (190–120 BC). Lines of longitude are perpendicular to and lines of latitude are parallel to the equator. Longitude is a range of values between -° to 180°, with the Greenwich Prime Meridian set at 0°. Latitude is a range of values between -90° to 90°, with the equator set at 0°. Altitude is the distance above sea level at any given latitude and longitude. The range is -11,033 meters (Mariana Trench) to +8,850 meters (Mt. Everest, in Nepal). The three deepest trenches are the Mariana Trench (11,033 meters), Tonga Trench (10,882 meters), and Kuril-Kamchatka Trench (10,500 meters). The three highest mountains are Mt. Everest (+8,850 meters), K2/Qogir (+8,611 meters), and Kangchenjunga (+8,586 meters). Locations on Earth have the potential to be anywhere on a 510.1 trillion square meter grid.

> **Tip**   I always reduce any location to a latitude, longitude, and altitude, with a short location description.

When allocating locations of equipment, I use latitude, longitude, and altitude, plus a cabinet and rack location. If I have servers at Thor Data Center, I record their location as latitude=64.0496287 and longitude=-21.9958472, with a description field of "Thor Data Center," plus a cabinet number and rack number.

Global Positioning System (GPS) and Global Navigation Satellite System (GLONASS)

The Global Positioning System and Global Navigation Satellite System are space-based radio navigation systems that supply data to any GPS/GLONASS receiver, to enable the receiver to calculate its own position on the earth's surface. This system still uses latitude, longitude, and altitude but has an advanced technique that uses a series of overlapping distant mappings from a minimum of four satellites, to map the receiver's location. Hence, this system is highly accurate.

Location Characteristics

Any location has more than one type of geographic characteristic. I will discuss some of the more commonly used location classifications and data entries.

Absolute Location

Absolute location provides a definite reference to locate a specific place. The absolute reference latitude, longitude, and attitude in the data I process are the core data I record. Using a GPS data source is always advisable.

Examples: Big Ben, Westminster, London, SW1A 0AA, UK is at latitude: 51.510357 and longitude: -0.116773. The London Eye, London, is at latitude: 51.503399 and longitude: -0.119519. The Statue of Liberty National Monument, New York, NY 10004, USA, is at latitude: 40.689247 and longitude: -74.044502.

Relative Location

Relative location describes a place with respect to its environment and its connection to other places. The data is dependent on other data being correct. I have seen a few relative location data sources.

I will use Big Ben, Westminster, London, SW1A 0AA, UK, at latitude: 51.510357 and longitude: -0.116773, and the Statue of Liberty National Monument, New York, NY 10004, USA, at latitude: 40.689247 and longitude: -74.044502, to show the process you can expect.

Descriptive Route

These data sources are useful for getting an approximate relative location between absolute locations.

Example: From Big Ben to the London Eye is a nine-minute (0.5-mile) walk, via A302 and The Queen's Walk, or six minutes, via A302 with a bicycle, or five minutes via taxi. From Big Ben to the Statue of Liberty National Monument is 7.5 hours by plane.

Distance, Bearing Between Latitude/Longitude Points

The distance, bearing between latitude/longitude points is a calculation used in the Assess step with the logistics calculations. It gives you a more accurate relative relationship between absolute locations.

Big Ben to London Eye results:

> *Distance*: 0.7737 km
>
> *Initial bearing*: 179° 45′ 55″
>
> *Final bearing*: 179° 45′ 56″
>
> *Midpoint*: 51° 30′ 25″ N, 000° 00′ 07″ E

Big Ben to Statue of Liberty results:

> *Distance*: 5583 km
>
> *Initial bearing*: 071° 35′ 41″
>
> *Final bearing*: 128° 50′ 54″
>
> *Midpoint*: 52° 23′ 15″ N, 041° 16′ 05″ E

This method supplies data scientists with more precise information between two absolute locations.

Warning   The method only works if the absolute locations are 100% correct!

Vincenty's Formula for Direct Azimuth and Distance

If you start at Big Ben and travel at bearing 289 for a distance of 5583 kilometers using the WGS84 ellipsoid, you end at

> *Latitude*: 41°06′19″N 41.10539784
>
> *Longitude*: 74°22′37″W -74.37694307
>
> *Final bearing*: 231°23′36″ 231.39333234
>
> *Back bearing*: 51°23′36″ 51.39333234

We are in New York, but not at the Statue of Liberty.

Tip   With more precise bearing and distance values, it is possible to get to the Statue of Liberty.

The relative location of a location is useful when you want to find the closest shop or calculate the best method of transport between locations.

Natural Characteristics

A topographical map is a good format for the physical characteristics of a location. I suggest that you collect data on a specific location, to enhance your customer's data. I also suggest that you look at data sources for weather measurements, temperature, land types, soil types, underground minerals, water levels, plant life, and animal life for specific locations.

Note   Any natural information that can enhance the location data is good to collect.

Tip   Collect the information over a period. That way, you get a time series that you can use to check for changes in the natural characteristics of the location.

These natural characteristics are a good source of data for testing the correlations between business behavior and the natural location. Such natural characteristics as a forest, desert, or lake region can indicate to businesses like a chain store when it will or will not sell, say, a motorboat or walking boots. An area covered in red mud 90% of the year may not be selling white trousers. The lack or excess of wind in an area may determine the success of energy-saving solutions, such as wind power generation. Or sinkholes will increase the cost of homeowner's insurance.

Human Characteristics

I suggest that you collect data on any human-designed or cultural features of a location. These features include the type of government, land use, architectural styles, forms of livelihood, religious practices, political systems, common foods, local folklore, transportation, languages spoken, money availability, and methods of communication.

These natural characteristics are a good source of data to test for correlations between business behavior and the natural location. Human characteristics can also assist with categorization of the location.

Human-Environment Interaction

The interaction of humans with their environment is a major relationship that guides people's behavior and the characteristics of the location. Activities such as mining and other industries, roads, and landscaping at a location create both positive and negative effects on the environment, but also on humans. A location earmarked as a green belt, to assist in reducing the carbon footprint, or a new interstate change its current and future characteristics. The location is a main data source for the data science, and, normally, we find unknown or unexpected effects on the data insights.

In the Python editor, open a new file named `Process_Location.py` in directory `..\VKHCG\01-Vermeulen\03-Process`. Start with the standard imports and setup of the ecosystem.

```
################################################################
# -*- coding: utf-8 -*-
################################################################
import sys
import os
import pandas as pd
import sqlite3 as sq
from pandas.io import sql
import uuid
################################################################
if sys.platform == 'linux':
    Base=os.path.expanduser('~') + '/VKHCG'
else:
    Base='C:/VKHCG'
print('################################')
print('Working Base :',Base, ' using ', sys.platform)
print('################################')
################################################################
Company='01-Vermeulen'
InputAssessGraphName='Assess_All_Animals.gml'
EDSAssessDir='02-Assess/01-EDS'
InputAssessDir=EDSAssessDir + '/02-Python'
################################################################
sFileAssessDir=Base + '/' + Company + '/' + InputAssessDir
if not os.path.exists(sFileAssessDir):
    os.makedirs(sFileAssessDir)
################################################################
sDataBaseDir=Base + '/' + Company + '/03-Process/SQLite'
if not os.path.exists(sDataBaseDir):
    os.makedirs(sDataBaseDir)
################################################################
sDatabaseName=sDataBaseDir + '/Vermeulen.db'
conn1 = sq.connect(sDatabaseName)
################################################################
```

```
sDataVaultDir=Base + '/88-DV'
if not os.path.exists(sDataBaseDir):
    os.makedirs(sDataBaseDir)
################################################################
sDatabaseName=sDataVaultDir + '/datavault.db'
conn2 = sq.connect(sDatabaseName)
################################################################
tMax=360*180
################################################################
```

You will now loop through longitude and latitude.

Warning   I am using a step value of 1 in the following sample code. In real-life systems, I use a step value as small as 0.000001. This will result in a long-running loop. To do this, generate the pandas data frame.

```
for Longitude in range(-180,180,1):
    for Latitude in range(-90,90,1):
        t+=1
        IDNumber=str(uuid.uuid4())
        LocationName='L'+format(round(Longitude,3)*1000, '+07d') +\
                        '-'+format(round(Latitude,3)*1000, '+07d')
        print('Create:',t,' of ',tMax,':',LocationName)
        LocationLine=[('ObjectBaseKey', ['GPS']),
                    ('IDNumber', [IDNumber]),
                    ('LocationNumber', [str(t)]),
                    ('LocationName', [LocationName]),
                    ('Longitude', [Longitude]),
                    ('Latitude', [Latitude])]
        if t==1:
            LocationFrame = pd.DataFrame.from_items(LocationLine)
        else:
            LocationRow = pd.DataFrame.from_items(LocationLine)
            LocationFrame = LocationFrame.append(LocationRow)
```

505

```
################################################################
LocationHubIndex=LocationFrame.set_index(['IDNumber'],inplace=False)
################################################################
sTable = 'Process-Location'
print('Storing :',sDatabaseName,' Table:',sTable)
LocationHubIndex.to_sql(sTable, conn1, if_exists="replace")
################################################################
sTable = 'Hub-Location'
print('Storing :',sDatabaseName,' Table:',sTable)
LocationHubIndex.to_sql(sTable, conn2, if_exists="replace")
################################################################
print('###############')
print('Vacuum Databases')
sSQL="VACUUM;"
sql.execute(sSQL,conn1)
sql.execute(sSQL,conn2)
print('###############')
################################################################
print('### Done!! #########################################')
################################################################
```

Congratulations! You have created the hub for location. You can expand the knowledge by adding other satellites, for example, post code, common names, regional information, and other business-specific descriptions. You should investigate libraries such as geopandas (see http://geopandas.org), to enhance your location with extra knowledge. You can install this by using install -c conda-forge geopandas.

Example:

```
import geopandas as gpd
from matplotlib import pyplot as pp
world = gpd.read_file(gpd.datasets.get_path('naturalearth_lowres'))
southern_world = world.cx[:, :0]
southern_world.plot(figsize=(10, 3));
world.plot(figsize=(10, 6));
pp.show()import geopandas as gpd
from matplotlib import pyplot as pp
world = gpd.read_file(gpd.datasets.get_path('naturalearth_lowres'))
```

```
southern_world = world.cx[:, :0]
southern_world.plot(figsize=(10, 3));
world.plot(figsize=(10, 6));
pp.show()
```

The library supports an easy technique to handle locations. Now that we have completed location structure successfully, we have only to handle events, to complete the data vault.

Event

This structure records any specific event or action that is discovered in the data sources. An event is any action that occurs within the data sources. Events are recorded using three main data entities: Event Type, Event Group, and Event Code. The details of each event are recorded as a set of details against the event code. There are two main types of events.

Explicit Event

This type of event is stated in the data source clearly and with full details. There is clear data to show that the specific action was performed.

Following are examples of explicit events:

- A security card with number 1234 was used to open door A.

- You are reading Chapter 9 of *Practical Data Science*.

- I bought ten cans of beef curry.

Explicit events are the events that the source systems supply, as these have direct data that proves that the specific action was performed.

Implicit Event

This type of event is formulated from characteristics of the data in the source systems plus a series of insights on the data relationships.

The following are examples of implicit events:

- A security card with number 8887 was used to open door X.

- A security card with number 8887 was issued to Mr. Vermeulen.

- Room 302 is fitted with a security reader marked door X.

These three events would imply that Mr. Vermeulen entered room 302 as an event. Not true!

Warning   Beware the natural nature of humans to assume that a set of actions always causes a specific known action. Record as events only proven fact!

If I discover that all visitor passes are issued at random and not recovered on exit by the visitor's desk, there is a probability that there is more than one security card with the number 1234. As a data scientist, you must ensure that you do not create implicit events from assumptions that can impact the decisions of the business. You can only record facts as explicit events.

The event is the last of the core structures of the Process step. I will now assist you to convert the basic knowledge of the time, person, object, location, and event structures into a data source that will assist the data scientist to perform all the algorithms and techniques required to convert the data into business insights.

These five main pillars (time, person, object, location, and event), as supported by the data extracted from the data sources, will be used by the data scientist to construct the data vault. I will return to the data vault later in this chapter. I want to introduce a few data science concepts that will assist you in converting the assess data into a data vault. I suggest you open your Python editor and see if you can create the event hub on your own.

```
###############################################################
# -*- coding: utf-8 -*-
###############################################################
import sys
import os
import pandas as pd
import sqlite3 as sq
from pandas.io import sql
import uuid
###############################################################
if sys.platform == 'linux':
    Base=os.path.expanduser('~') + '/VKHCG'
else:
    Base='C:/VKHCG'
```

```
print('###############################')
print('Working Base :',Base, ' using ', sys.platform)
print('###############################')
################################################################
Company='01-Vermeulen'
InputFileName='Action_Plan.csv'
################################################################
sDataBaseDir=Base + '/' + Company + '/03-Process/SQLite'
if not os.path.exists(sDataBaseDir):
    os.makedirs(sDataBaseDir)
################################################################
sDatabaseName=sDataBaseDir + '/Vermeulen.db'
conn1 = sq.connect(sDatabaseName)
################################################################
sDataVaultDir=Base + '/88-DV'
if not os.path.exists(sDataBaseDir):
    os.makedirs(sDataBaseDir)
################################################################
sDatabaseName=sDataVaultDir + '/datavault.db'
conn2 = sq.connect(sDatabaseName)
################################################################
sFileName=Base + '/' + Company + '/00-RawData/' + InputFileName
print('Loading :',sFileName)
EventRawData=pd.read_csv(sFileName,header=0,low_memory=False,
encoding="latin-1")
EventRawData.index.names=['EventID']
EventHubIndex=EventRawData
################################################################
sTable = 'Process-Event'
print('Storing :',sDatabaseName,' Table:',sTable)
EventHubIndex.to_sql(sTable, conn1, if_exists="replace")
################################################################
sTable = 'Hub-Event'
print('Storing :',sDatabaseName,' Table:',sTable)
EventHubIndex.to_sql(sTable, conn2, if_exists="replace")
```

```
####################################################################
print('###############')
print('Vacuum Databases')
sSQL="VACUUM;"
sql.execute(sSQL,conn1)
sql.execute(sSQL,conn2)
print('###############')
####################################################################
print('### Done!! #######################################')
####################################################################
```

Well done. You just completed the next hub.

Data Science Process

I want to discuss a basic data science process at this point, before I progress to the rest of the Process step. This basic process will guide the process forward and formulate the basic investigation techniques that I will teach you as we progress through the further data discovery.

Roots of Data Science

Data science is at its core about curiosity and inquisitiveness. This core is rooted in the 5 Whys. The 5 Whys is a technique used in the analysis phase of data science.

Benefits of the 5 Whys

The 5 Whys assist the data scientist to identify the root cause of a problem and determine the relationship between different root causes of the same problem. It is one of the simplest investigative tools—easy to complete without intense statistical analysis.

When Are the 5 Whys Most Useful?

The 5 Whys are most useful for finding solutions to problems that involve human factors or interactions that generate multilayered data problems.

In day-to-day business life, they can be used in real-world businesses to find the root causes of issues.

510

How to Complete the 5 Whys

Write down the specific problem. This will help you to formalize the problem and describe it completely. It also helps the data science team to focus on the same problem.

Ask why the problem occurred and write the answer below the problem. If the answer you provided doesn't identify the root cause of the problem that you wrote down first, ask why again, and write down that answer.

Loop back to the preceding step until you and your customer are in agreement that the problem's root cause is identified. Again, this may require fewer or more than the 5 Whys.

Tip   I have found that most questions take only four whys, but if you require seven, I recommend that you go back over the process, as it is not solving the root cause.

As a data scientist, I have found that it helps to perform the 5 Whys when dealing with data and processing the data into the core data vault. Seek explicit reasons for loading that specific data entry into the specific part of the data vault.

Caution   The data-mapping decisions you take will affect the quality of the data science and the value of the insights you derive from the data vault.

Fishbone Diagrams

I have found this simple diagram is a useful tool to guide the questions that I need to resolve related to where each data source's data fits into the data vault. The diagram (Figure 9-10) is drawn up as you complete the 5 Whys process, as you will discover that there are normally many causes for why specific facts have been recorded.

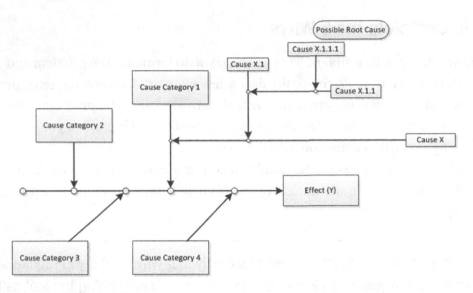

Figure 9-10. *Fishbone diagram*

Let's return to my earlier event, the one in which I bought ten cans of beef curry. The ten cans are the effect (Y), but the four root causes of the purchase are

- I was hungry, so I bought ten tins. I did not like the brand of curry that I bought 10 cans of the previous week.

- My neighbor needed five cans, as she was no longer able to walk, and she requested the brand that I purchased, as it is the cheapest, at £1.10.

- I fed two cans to the dog, because I feel dog food is not nutritious, but I was not prepared to buy a more expensive brand of canned beef curry for the dog.

- I put three cans in the charity bin outside the local school, as it wanted that brand of curry for its soup kitchen and the tokens (vouchers) for its "Fund your School's new laptops" at £2.50 per can campaign.

I had a customer's data scientist (an ex-employee now) categorize my ten cans plus the 9,000 others sold that week as a major success for their new brand of curry. The customer's buyer signed a prepaid six-month purchase order as a result of this success. Some 100,000 cans showed up at the retailer and were sold within eight days. The sad part is that the reason the product sold so well was the £2.50-per-tin token (voucher) that obscured a 25-cents (US) cash-back offer that the local wholesaler erroneously covered with a £2.50 "Fund your School's new laptops" sticker that was meant for the more exclusive £25.00 schoolbags he was also supplying. The wholesaler's buyer was illegally shipping black market expired products into the country. The £2.50 sticker also covered the expiry date. The data system advised by the wholesaler's data scientist used an implicit event, whereby the system marked the unknown expiry date as the scan date into the wholesaler's warehouses.

Caution   If it sounds too good to be true, it is probably wrong!

This trivial sales disaster cost two data scientists and a buyer their jobs; a wholesaler was declared bankrupt; six people contracted food poisoning at the soup kitchen; and my dog had to be taken to the vet. The formula is simple: Dog + Curry = Sick puppy! I still place the liability on the shop; my wife disagrees.

The "Fund your School's new laptops" campaign was a big success, with nearly 90% of the tokens now redeemed, sadly, without an increase in sales of the £25.00 schoolbags.

My neighbor received five tins of our better-brand curry, because the dog got sick. I got more work. All was good.

I'm still not sure why only 2,000 unopened cans were returned by customers to the retailer, and only 1 in 200 still had the token attached.

Tip   Always ask why, and never assume the answer is correct until you've proven it.

So, let's look at another 5 Whys example.

5 Whys Example

So, let's look at the data science I performed with the retailer after the trouble with the cans of curry.

Problem Statement: Customers are unhappy because they are being shipped products that don't meet their specifications.

1. Why are customers being shipped bad products?

   - Because manufacturing built the products to a specification that is different from what the customer and the salesperson agreed to.

2. Why did manufacturing build the products to a different specification than that of sales?

- Because the salesperson accelerates work on the shop floor by calling the head of manufacturing directly to begin work. An error occurred when the specifications were being communicated or written down.

3. Why does the salesperson call the head of manufacturing directly to start work instead of following the procedure established by the company?

- Because the "start work" form requires the sales director's approval before work can begin and slows the manufacturing process (or stops it when the director is out of the office).

4. Why does the form contain an approval for the sales director?

- Because the sales director must be continually updated on sales for discussions with the CEO, as my retailer customer was a top-ten key account.

In this case, only four whys were required to determine that a non-value-added signature authority helped to cause a process breakdown in the quality assurance for a key account! The rest was just criminal.

The external buyer at the wholesaler knew this process was regularly bypassed and started buying the bad tins to act as an unofficial backfill for the failing process in the quality-assurance process in manufacturing, to make up the shortfalls in sales demand. The wholesaler simply relabeled the product and did not change how it was manufactured. The reason? Big savings lead to big bonuses. A key client's orders had to be filled. . . . Sales are important!

Tip   It takes just a quick why—times five—to verify!

The next technique I employ is useful in 90% of my data science. It usually offers multifaceted solutions but requires a further level of analysis.

Monte Carlo Simulation

I want to introduce you to a powerful data science tool called a Monte Carlo simulation. This technique performs analysis by building models of possible results, by substituting a range of values—a probability distribution—for parameters that have inherent uncertainty. It then calculates results over and over, each time using a different set of random values from the probability functions. Depending on the number of uncertainties and the ranges specified for them, a Monte Carlo simulation can involve thousands or tens of thousands of recalculations before it is complete. Monte Carlo simulation produces distributions of possible outcome values. As a data scientist, this gives you an indication of how your model will react under real-life situations. It also gives the data scientist a tool to check complex systems, wherein the input parameters are high-volume or complex. I will show you a practical use of this tool in the next section.

Causal Loop Diagrams

A causal loop diagram (CLD) is a causal diagram that aids in visualizing how a number of variables in a system are interrelated and drive cause-and-effect processes. The diagram consists of a set of nodes and edges. Nodes represent the variables, and edges are the links that represent a connection or a relation between the two variables. I normally use my graph theory, covered in Chapter 8, to model the paths through the system. I simply create a directed graph and then step through it one step at a time.

Example: The challenge is to keep the "Number of Employees Available to Work and Productivity" as high as possible.

I modeled the following as a basic diagram, while I investigated the tricky process (Figure 9-11).

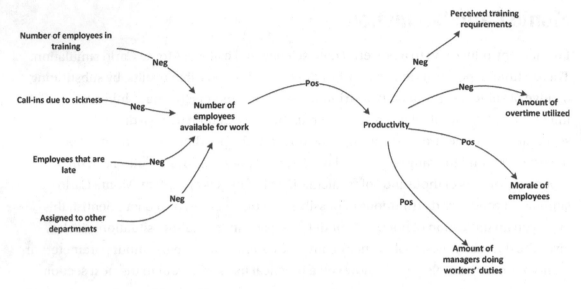

Figure 9-11. *Causal loop diagram*

Our first conclusion was "We need more staff." I then simulated the process, using a hybrid data science technique formulated from two other techniques—Monte Carlo simulation and Deep Reinforcement Learning—against a graph data model. The result was the discovery of several other additional pieces of the process that affected the outcome I was investigating. This was the result (Figure 9-12).

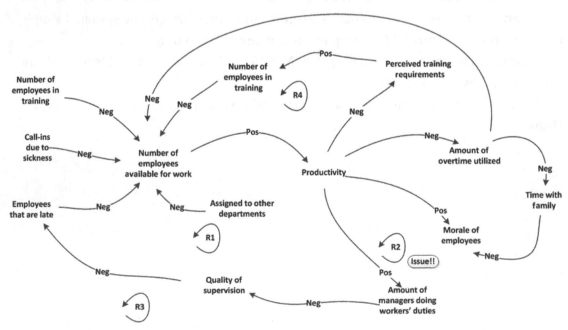

Figure 9-12. *Monte Carlo result*

The Monte Carlo simulation cycled through a cause-and-effect model by changing the values at points R1 through R4 for three hours, as a training set. The simulation was then run for three days with a reinforced deep learning model that adapted the key drivers in the system.

The result was "Managers need to manage not work." The R2—percentage of manage doing employees' duties—was the biggest cause and impact driver in the system.

Tip   Do not deliver results early, if you are not sure of the true data science results. Work effectively and efficiently not quickly!

I routinely have numerous projects going through analysis in parallel. This way, I can process parallel data science tasks while still supporting excellent, effective, and efficient data science.

Pareto Chart

Pareto charts are a technique I use to perform a rapid processing plan for the data science. Pareto charts can be constructed by segmenting the range of the data into groups (also called segments, bins, or categories). (See Figure 9-13.)

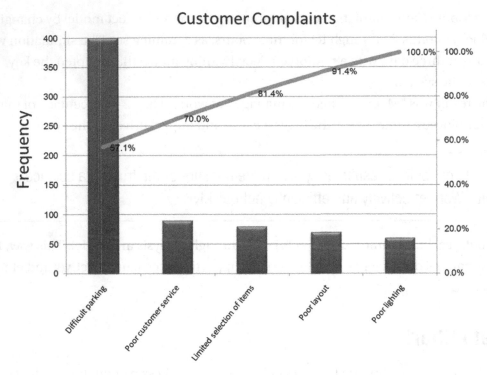

Figure 9-13. *Pareto chart*

Questions the Pareto chart answers:

- What are the largest issues facing our team or my customer's business?

- What 20% of sources are causing 80% of the problems (80/20 Rule)?

- Where should we focus our efforts to achieve the greatest improvements?

I perform a rapid assessment of the data science processes and determine what it will take to do 80% of the data science effectively and efficiently in the most rapid time frame. It is a maximum-gain technique.

Tip   If you are saving your customer 80% of its proven loses, it is stress-free to identify resources to complete the remaining 20% savings.

Correlation Analysis

The most common analysis I perform at this step is the correlation analysis of all the data in the data vault. Feature development is performed between data items, to find relationships between data values.

```
import pandas as pd
a = [ [1, 2, 3], [5, 6, 9], [7, 3, 13], [5, 3, 19], [5, 3, 12], [5, 6, 11],
[5, 6, 13], [5, 3, 4]]
df = pd.DataFrame(data=a)
cr=df.corr()
print(cr)
```

Forecasting

Forecasting is the ability to project a possible future, by looking at historical data. The data vault enables these types of investigations, owing to the complete history it collects as it processes the source's systems data. You will perform many forecasting projects during your career as a data scientist and supply answers to such questions as the following:

- What should we buy?

- What should we sell?

- Where will our next business come from?

People want to know what you calculate to determine what is about to happen. I will cover an example on forecasting now.

VKHCG financed a number of other companies to generate additional capital. Check in \VKHCG\04-Clark\ 00-RawData for a file named VKHCG_Shares.csv. Here is our investment portfolio:

Shares	sTable	Units
WIKI/GOOGL	WIKI_Google	1000000
WIKI/MSFT	WIKI_Microsoft	1000000
WIKI/UPS	WIKI_UPS	1000000
WIKI/AMZN	WIKI_Amazon	1000000

519

Shares	sTable	Units
LOCALBTC/USD	LOCALBTC_USD	1000000
PERTH/AUD_USD_M	PERTH_AUD_USD_M	1000
PERTH/AUD_USD_D	PERTH_AUD_USD_D	1000
PERTH/AUD_SLVR_M	PERTH_SLVR_USD_M	1000
PERTH/AUD_SLVR_D	PERTH_SLVR_USD_D	1000
BTER/AURBTC	BTER_AURBTC	1000000
FED/RXI_US_N_A_UK	FED_RXI_US_N_A_UK	1000
FED/RXI_N_A_CA	FED_RXI_N_A_CA	1000

I will guide you through a technique to determine if we can forecast what our shares will do in the near-future. Open a new file in your Python editor and save it as Process-Shares-Data.py in directory \VKHCG\04-Clark\03-Process. I will guide you through this process. You will require a library called quandl (conda install -c anaconda quandl), which enables you to use online share numbers for the examples. As before, start the ecosystem by setting up the libraries and the basic directories of the environment.

```
################################################################
import sys
import os
import sqlite3 as sq
import quandl
import pandas as pd
################################################################
if sys.platform == 'linux':
    Base=os.path.expanduser('~') + '/VKHCG'
else:
    Base='C:/VKHCG'
print('################################')
print('Working Base :',Base, ' using ', sys.platform)
print('################################')
################################################################
```

```
Company='04-Clark'
sInputFileName='00-RawData/VKHCG_Shares.csv'
sOutputFileName='Shares.csv'
################################################################
sDataBaseDir=Base + '/' + Company + '/03-Process/SQLite'
if not os.path.exists(sDataBaseDir):
    os.makedirs(sDataBaseDir)
################################################################
sFileDir1=Base + '/' + Company + '/01-Retrieve/01-EDS/02-Python'
if not os.path.exists(sFileDir1):
    os.makedirs(sFileDir1)
################################################################
sFileDir2=Base + '/' + Company + '/02-Assess/01-EDS/02-Python'
if not os.path.exists(sFileDir2):
    os.makedirs(sFileDir2)
################################################################
sFileDir3=Base + '/' + Company + '/03-Process/01-EDS/02-Python'
if not os.path.exists(sFileDir3):
    os.makedirs(sFileDir3)
################################################################
sDatabaseName=sDataBaseDir + '/clark.db'
conn = sq.connect(sDatabaseName)
################################################################
```

You will now import the list of shares we own to create the forecasts.

```
################################################################
### Import Share Names Data
################################################################
sFileName=Base + '/' + Company + '/' + sInputFileName
print('###############################')
print('Loading :',sFileName)
print('###############################')
RawData=pd.read_csv(sFileName,header=0,low_memory=False,
encoding="latin-1")
```

```
RawData.drop_duplicates(subset=None, keep='first', inplace=True)
print('Rows    :',RawData.shape[0])
print('Columns:',RawData.shape[1])
print('###############')
###################################################################
sFileName=sFileDir1 + '/Retrieve_' + sOutputFileName
print('###############################')
print('Storing :', sFileName)
print('###############################')
RawData.to_csv(sFileName, index = False)
print('###############################')
###################################################################
sFileName=sFileDir2 + '/Assess_' + sOutputFileName
print('###############################')
print('Storing :', sFileName)
print('###############################')
RawData.to_csv(sFileName, index = False)
print('###############################')
###################################################################
sFileName=sFileDir3 + '/Process_' + sOutputFileName
print('###############################')
print('Storing :', sFileName)
print('###############################')
RawData.to_csv(sFileName, index = False)
print('###############################')
```

You will now loop through each of the shares and perform a forecast for each share.

```
###################################################################
### Import Shares Data Details
###################################################################
for sShare in range(RawData.shape[0]):
    sShareName=str(RawData['Shares'][sShare])
    ShareData = quandl.get(sShareName)
    UnitsOwn=RawData['Shares'][sShare]
    ShareData['UnitsOwn']=ShareData.apply(lambda row:(UnitsOwn),axis=1)
```

```
    print('################')
    print('Share  :',sShareName)
    print('Rows   :',ShareData.shape[0])
    print('Columns:',ShareData.shape[1])
    print('################')
    ####################################################################
    print('################')
    sTable=str(RawData['sTable'][sShare])
    print('Storing :',sDatabaseName,' Table:',sTable)
    ShareData.to_sql(sTable, conn, if_exists="replace")
    print('################')
    ####################################################################
    sOutputFileName = sTable + '.csv'
    sFileName=sFileDir1 + '/Retrieve_' + sOutputFileName
    print('##############################')
    print('Storing :', sFileName)
    print('##############################')
    RawData.to_csv(sFileName, index = False)
    print('##############################')
    ####################################################################
    sOutputFileName = sTable + '.csv'
    sFileName=sFileDir2 + '/Assess_' + sOutputFileName
    print('##############################')
    print('Storing :', sFileName)
    print('##############################')
    RawData.to_csv(sFileName, index = False)
    print('##############################')
    ####################################################################
    sOutputFileName = sTable + '.csv'
    sFileName=sFileDir3 + '/Process_' + sOutputFileName
    print('##############################')
    print('Storing :', sFileName)
    print('##############################')
    RawData.to_csv(sFileName, index = False)
    print('##############################')
####################################################################
```

```
###################################################################
print('### Done!! #######################################')
###################################################################
```

What do you think? Are we doing well, or are we heading for bankruptcy? Would you advise us to buy more shares?

The shares will be stored in the data vault as objects, and the specific shares, i.e., share codes, will be stored in the object hub. Store the real daily share prices in the object-share-price satellite, to track the daily changes. Store your daily projections or forecasts in an object-share-price-projected hub. That way, you can compare your projections with real pricing, and you can also spot trends in the share price over time.

To track share price over time, you will also require a link to the time hub, to enable investigation of the share price against any other activities in the data vault. So, as I stated before, you can model any data you get from your data lake or process from the insights into the data within the data vault.

Data Science

You must understand that data science works best when you follow approved algorithms and techniques. You can experiment and wrangle the data, but in the end, you must verify and support your results. Applying good data science ensures that you can support your work with acceptable proofs and statistical tests. I believe that data science that works follows these basic steps:

1. Start with a question. Make sure you have fully addressed the 5 Whys.

2. Follow a good pattern to formulate a model. Formulate a model, guess a prototype for the data, and start a virtual simulation of the real-world parameters. If you have operational research or econometrics skills, this is where you will find use for them.

   Mix some mathematics and statistics into the solution, and you have the start of a data science model.

   Remember: All questions have to be related to the customer's business and must provide insights into the question that results in an actionable outcome and, normally, a quantifiable return-on-investment of the application of the data science processes you plan to deploy.

3. Gather observations and use them to generate a hypothesis. Start the investigation by collecting the required observations, as per your model. Process your model against the observations and prove your hypothesis to be true or false.

4. Use real-world evidence to judge the hypothesis. Relate the findings back to the real world and, through storytelling, convert the results into real-life business advice and insights.

Caution   You can have the best results, but if you cannot explain why they are important, you have not achieved your goal.

---

5. Collaborate early and often with customers and with subject matter experts along the way.

You also must communicate early and often with your relevant experts to ensure that you take them with you along the journey of discovery. Businesspeople want to be part of solutions to their problems. Your responsibility is to supply good scientific results to support the business.

Summary

You have successfully completed the Process step of this chapter. You should now have a solid data vault with processed data ready for extraction by the Transform step, to turn your new knowledge into wisdom and understanding.

At this point, the Retrieve, Assess and Process steps would have enabled you to change the unknown data in the data lake into a classified series of hub, i.e., T-P-O-L-E structures. You should have the following in the data vault:

- Time hub with satellites and links

- Person hub with satellites and links

- Object hub with satellites and links

- Location hub with satellites and links

- Event hub with satellites and links

The rest of the data science is now performed against the standard structure. This will help you to develop solid and insightful data science techniques and methods against the T-P-O-L-E structure.

Note   If you have processed the features from the data lake into the data vault, you should now have completed most of the data engineering or data transformation tasks.

You have achieved a major milestone by building the data vault.

Note   The data vault creates a natural history, as it stores all of the preceding processing runs as snapshots in the data vault. You can reprocess the complete Transform superstep and recover a full history of every step in the processing.

The next step is for you to start performing more pure data science tasks, to investigate what data you have found and how you are going to formulate the data from the data vault into a query-able solution for the business.

CHAPTER 10

Transform Superstep

The Transform superstep allows you, as a data scientist, to take data from the data vault and formulate answers to questions raised by your investigations. The transformation step is the data science process that converts results into insights.

It takes standard data science techniques and methods to attain insight and knowledge about the data that then can be transformed into actionable decisions, which, through storytelling, you can explain to non-data scientists what you have discovered in the data lake.

Any source code or other supplementary material referenced by me in this book is available to readers on GitHub, via this book's product page, located at www.apress.com/9781484230534. Please note that this source code assumes that you have completed the source code setup outlined in Chapter 2.

Transform Superstep

The Transform superstep uses the data vault from the process step as its source data. The transformations are tuned to work with the five dimensions of the data vault. As a reminder of what the structure looks like, see Figure 10-1.

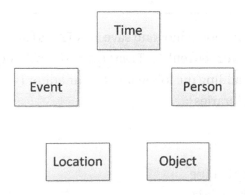

Figure 10-1. *Five categories of data*

Dimension Consolidation

The data vault consists of five categories of data, with linked relationships and additional characteristics in satellite hubs.

To perform dimension consolidation, you start with a given relationship in the data vault and construct a sun model for that relationship, as shown in Figure 10-2.

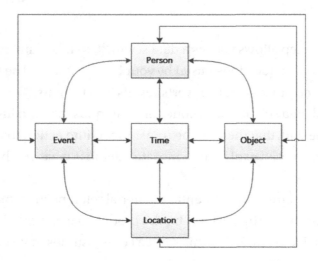

Figure 10-2. *T-P-O-L-E hHigh-level design*

I will cover the example of a person being born, to illustrate the consolidation process.

Note   You will need a Python editor to complete this chapter, so please start it, then we can proceed to the data science required.

Open a new file in the Python editor and save it as `Transform-Gunnarsson_is_Born.py` in directory `..\VKHCG\01-Vermeulen\04-Transform`. You will require the Python ecosystem, so set it up by adding the following to your editor (you must set up the ecosystem by adding the libraries):

```
import sys
import os
from datetime import datetime
from datetime import timedelta
```

```
from pytz import timezone, all_timezones
import pandas as pd
import sqlite3 as sq
from pandas.io import sql
import uuid

pd.options.mode.chained_assignment = None
```

Find the working directory of the examples.

```
################################################################
if sys.platform == 'linux':
    Base=os.path.expanduser('~') + '/VKHCG'
else:
    Base='C:/VKHCG'
print('################################')
print('Working Base :',Base, ' using ', sys.platform)
print('################################')
```

Set up the company you are processing.

```
Company='01-Vermeulen'
Add the company work space:
sDataBaseDir=Base + '/' + Company + '/04-Transform/SQLite'
if not os.path.exists(sDataBaseDir):
    os.makedirs(sDataBaseDir)
sDatabaseName=sDataBaseDir + '/Vermeulen.db'
conn1 = sq.connect(sDatabaseName)
```

Add the data vault.

```
sDataVaultDir=Base + '/88-DV'
if not os.path.exists(sDataVaultDir):
    os.makedirs(sDataVaultDir)
sDatabaseName=sDataVaultDir + '/datavault.db'
conn2 = sq.connect(sDatabaseName)
```

Add the data warehouse.

```
sDataWarehousetDir=Base + '/99-DW'
if not os.path.exists(sDataWarehousetDir):
    os.makedirs(sDataWarehousetDir)
sDatabaseName=sDataVaultDir + '/datawarehouse.db'
conn3 = sq.connect(sDatabaseName)
```

Execute the Python code now, to set up the basic ecosystem.

Note   The new data structure, called a data warehouse, is in directory ../99-DW.

The data warehouse is the only data structure delivered from the Transform step. Let's look at a real-world scenario.

Guðmundur Gunnarsson was born on December 20, 1960, at 9:15 in Landspítali, Hringbraut 101, 101 Reykjavík, Iceland. Following is what I would expect to find in the data vault.

Time

You need a date and time of December 20, 1960, at 9:15 in Reykjavík, Iceland. Enter the following code into your editor (you start with a UTC time):

```
print('Time Category')
print('UTC Time')
BirthDateUTC = datetime(1960,12,20,10,15,0)
BirthDateZoneUTC=BirthDateUTC.replace(tzinfo=timezone('UTC'))
BirthDateZoneUTCStr=BirthDateZoneUTC.strftime("%Y-%m-%d %H:%M:%S (%Z) (%z)")
BirthDateLocal=BirthDate.strftime("%Y-%m-%d %H:%M:%S")
print(BirthDateZoneUTCStr)
```

Formulate a Reykjavík local time.

```
print('Birth Date in Reykjavik :')
BirthZone = 'Atlantic/Reykjavik'
BirthDate = BirthDateZoneUTC.astimezone(timezone(BirthZone))
BirthDateStr=BirthDate.strftime("%Y-%m-%d %H:%M:%S (%Z) (%z)")
print(BirthDateStr)
```

You have successfully discovered the time key for the time hub and the time zone satellite for Atlantic/Reykjavik.

- *Time Hub*: You have a UTC date and time of December 20, 1960, at 9:15 in Reykjavík, Iceland, as follows:

  1960-12-20 10:15:00 (UTC) (+0000)

- *Time Satellite*: Birth date in Reykjavík:

  1960-12-20 09:15:00 (-01) (-0100)

Now you can save your work, by adding the following to your code.

Build a data frame, as follows:

```
IDZoneNumber=str(uuid.uuid4())
sDateTimeKey=BirthDateZoneStr.replace(' ','-').replace(':','-')
TimeLine=[('ZoneBaseKey', ['UTC']),
            ('IDNumber', [IDZoneNumber]),
            ('DateTimeKey', [sDateTimeKey]),
            ('UTCDateTimeValue', [BirthDateZoneUTC]),
            ('Zone', [BirthZone]),
            ('DateTimeValue', [BirthDateStr])]
TimeFrame = pd.DataFrame.from_items(TimeLine)
```

Create the time hub.

```
TimeHub=TimeFrame[['IDNumber','ZoneBaseKey','DateTimeKey','DateTimeValue']]
TimeHubIndex=TimeHub.set_index(['IDNumber'],inplace=False)
```

```
sTable = 'Hub-Time-Gunnarsson'
print('\n###############################')
print('Storing :',sDatabaseName,'\n Table:',sTable)
print('\n###############################')
TimeHubIndex.to_sql(sTable, conn2, if_exists="replace")
sTable = 'Dim-Time-Gunnarsson'
TimeHubIndex.to_sql(sTable, conn3, if_exists="replace")
```

Create the time satellite.

```
TimeSatellite=TimeFrame[['IDNumber','DateTimeKey','Zone','DateTimeValue']]
TimeSatelliteIndex=TimeSatellite.set_index(['IDNumber'],inplace=False)
```

531

```
BirthZoneFix=BirthZone.replace(' ','-').replace('/','-')
sTable = 'Satellite-Time-' + BirthZoneFix + '-Gunnarsson'
print('\n###############################')
print('Storing :',sDatabaseName,'\n Table:',sTable)
print('\n###############################')
TimeSatelliteIndex.to_sql(sTable, conn2, if_exists="replace")
sTable = 'Dim-Time-' + BirthZoneFix + '-Gunnarsson'
TimeSatelliteIndex.to_sql(sTable, conn3, if_exists="replace")
```

Well done. You now have a Time category. The next category is Person.

Person

You must record that Guðmundur Gunnarsson was born on December 20, 1960, at 9:15 in Iceland.

Add the following to your existing code:

```
print('Person Category')
FirstName = 'Guðmundur'
LastName = 'Gunnarsson'
print('Name:',FirstName,LastName)
print('Birth Date:',BirthDateLocal)
print('Birth Zone:',BirthZone)
print('UTC Birth Date:',BirthDateZoneStr)
```

You just created the person in the person hub.

```
IDPersonNumber=str(uuid.uuid4())
PersonLine=[('IDNumber', [IDPersonNumber]),
            ('FirstName', [FirstName]),
            ('LastName', [LastName]),
            ('Zone', ['UTC']),
            ('DateTimeValue', [BirthDateZoneStr])]
PersonFrame = pd.DataFrame.from_items(PersonLine)
###############################################################
TimeHub=PersonFrame
TimeHubIndex=TimeHub.set_index(['IDNumber'],inplace=False)
###############################################################
```

```
sTable = 'Hub-Person-Gunnarsson'
print('\n###############################')
print('Storing :',sDatabaseName,'\n Table:',sTable)
print('\n###############################')
TimeHubIndex.to_sql(sTable, conn2, if_exists="replace")
sTable = 'Dim-Person-Gunnarsson'
TimeHubIndex.to_sql(sTable, conn3, if_exists="replace")
```

Well done. You now have a person hub.

Can you create a location hub and satellite for National University Hospital of Iceland?

> Latitude: 64° 08' 10.80" N or 64.136332788 Longitude: -21° 55' 22.79" W or -21.922996308

What else can you process from the address data?

> Landspítali: Local Name
>
> Hringbraut 101: Street Address
>
> 101: Post Code
>
> Reykjavík: City Name
>
> Iceland: Country

You can add to the information an entry in the location hub, if the specific latitude and longitude for Reykjavík does not exist yet. You can then add to a satellite named Hospitals the key from the location hub, plus all the extra information. Alternatively, you can simply add it as a satellite called BuildingAt with the same extra information and a type_of_building indicator.

Whatever way you perform it, I advise discussing it with your customer.

Note   You can data-mine many characteristics from a simple connection to a location or building.

Can you now create an event hub for "Birth"? Yes, you can—easily—but you have an event called "Born." Use that.

Note   I have found that various people have diverse names for similar events, for example, *born* and *birth*, *death* and *deceased*.

I have a library of events that we model for the same event. I keep these in an additional table, and when we create an event for one of the words in a similar list, we also create it for the other lists.

For example, if we create an event for born and birth, that facilitates queries from the Transform step. If you asked "Who was born?" or "Whose birth was it?" you would get results retrieved from the data vault into the data warehouse.

So, what about Object? Yes, you could use the genus/species data. Guðmundur is a *Homo sapiens* (human). Try to find the correct record to link to.

Tip   You should now understand that even a simple business action, such as "Guðmundur Gunnarsson's Birth," can generate massive amounts of extra data points.

I suggest you try and look at major events in your own life: your birth, first day at school. What if you look at the actions in the event hub, which apply to you?

If you understand that with enhancement of the data vault you can achieve major enhancement for the Transform step's dimension consolidation, you have achieved a major milestone in your capability as a data scientist.

Sun Model

The use of sun models is a technique that enables the data scientist to perform consistent dimension consolidation, by explaining the intended data relationship with the business, without exposing it to the technical details required to complete the transformation processing. So, let's revisit our business statement: Guðmundur Gunnarsson was born on December 20, 1960, at 9:15 in Landspítali, Hringbraut 101, 101 Reykjavík, Iceland.

Person-to-Time Sun Model

The following sun model in Figure 10-3 explains the relationship between the Time and Person categories in the data vault.

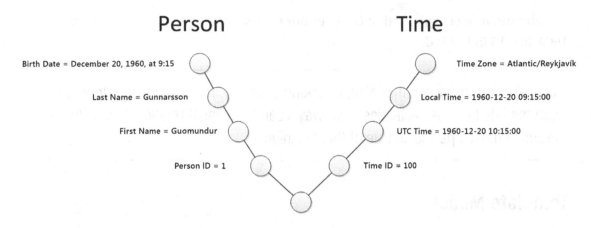

Figure 10-3. *Person-to-Time sun model*

The sun model is constructed to show all the characteristics from the two data vault hub categories you are planning to extract. It explains how you will create two dimensions and a fact via the Transform step from Figure 10-3. You will create two dimensions (Person and Time) with one fact (PersonBornAtTime), as shown in Figure 10-4.

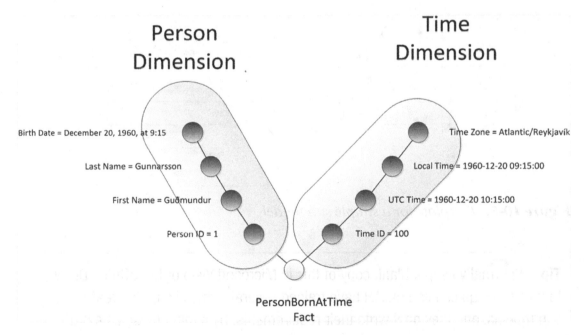

Figure 10-4. *Person-to-Time sun model (explained)*

The sun model explains that Guðmundur Gunnarsson was born on December 20, 1960, at 9:15 in Iceland.

Tip I practice my sun modeling by taking everyday interactions and drawing sun models for those activities. That way, I can draw most relationships with ease. Practice makes perfect is true of this technique.

Template Model

I also have several printed copies of sun models with blank entries (Figure 10-5), to quickly record relationships while I am meeting with clients.

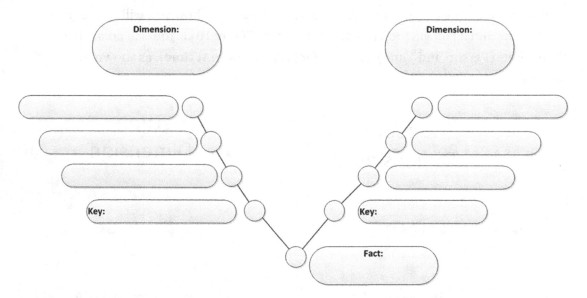

Figure 10-5. *Template for a simple sun model*

Tip I normally keep a blank copy of this in Microsoft Visio or LibreOffice Draw, to give me a quick head start. I have trained several of my clients to create sun models, and they now write their requirements, by adding these as extra information. It gives them a sense of ownership.

Person-to-Object Sun Model

Can you find the dimensions and facts on these already completed sun models?
(See Figure 10-6.)

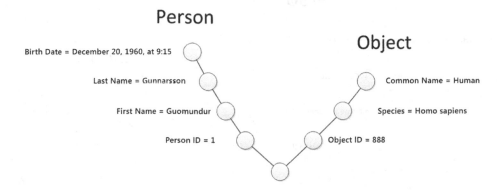

Figure 10-6. *Sun model for the PersonIsSpecies fact*

How did you progress? In Figure 10-6, dimensions are Person and Object. Fact is
PersonIsSpecies. This describes Guðmundur Gunnarsson as a *Homo sapiens*.

Person-to-Location Sun Model

If you have dimensions Person and Location and fact PersonAtLocation, you have
successfully read the sun model. (Figure 10-7).

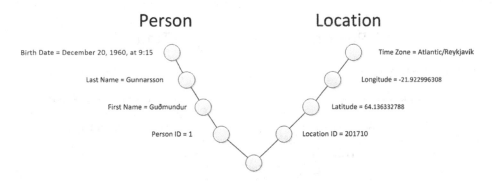

Figure 10-7. *Sun model for PersonAtLocation fact*

This describes that Guðmundur Gunnarsson was born at:

Latitude: 64° 08' 10.80" N or 64.136332788

Longitude: -21° 55' 22.79" W or -21.922996308

Person-to-Event Sun Model

And this event sun model? Can you extract the information in Figure 10-8?

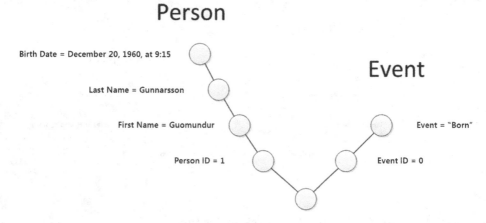

Figure 10-8. *Sun model for PersonBorn fact*

If you have dimensions Person and Event and fact PersonBorn, you have successfully read the sun model (Figure 10-8).

Sun Model to Transform Step

I will guide you through the creation of the Transform step, as shown by the sun model in Figure 10-9.

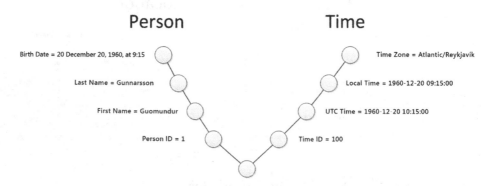

Figure 10-9. *Sun model for PersonBornAtTime fact*

You must build three items: dimension Person, dimension Time, and fact PersonBornAtTime. Open your Python editor and create a file named `Transform-Gunnarsson-Sun-Model.py` in directory `..\VKHCG\01-Vermeulen\04-Transform`. Here is your basic ecosystem:

```
################################################################
# -*- coding: utf-8 -*-
################################################################
import sys
import os
from datetime import datetime
from pytz import timezone
import pandas as pd
import sqlite3 as sq
import uuid
pd.options.mode.chained_assignment = None
################################################################
if sys.platform == 'linux' or sys.platform == ' Darwin':
    Base=os.path.expanduser('~') + '/VKHCG'
else:
    Base='C:/VKHCG'
print('##############################')
print('Working Base :',Base, ' using ', sys.platform)
print('##############################')
################################################################
Company='01-Vermeulen'
################################################################
sDataBaseDir=Base + '/' + Company + '/04-Transform/SQLite'
if not os.path.exists(sDataBaseDir):
    os.makedirs(sDataBaseDir)
################################################################
```

```
sDatabaseName=sDataBaseDir + '/Vermeulen.db'
conn1 = sq.connect(sDatabaseName)
################################################################
sDataWarehousetDir=Base + '/99-DW'
if not os.path.exists(sDataWarehousetDir):
    os.makedirs(sDataWarehousetDir)
################################################################
sDatabaseName=sDataWarehousetDir + '/datawarehouse.db'
conn2 = sq.connect(sDatabaseName)
################################################################
```

Here is your Time dimension:

```
print('\n#################################')
print('Time Dimension')
BirthZone = 'Atlantic/Reykjavik'
BirthDateUTC = datetime(1960,12,20,10,15,0)
BirthDateZoneUTC=BirthDateUTC.replace(tzinfo=timezone('UTC'))
BirthDateZoneStr=BirthDateZoneUTC.strftime("%Y-%m-%d %H:%M:%S")
BirthDateZoneUTCStr=BirthDateZoneUTC.strftime("%Y-%m-%d %H:%M:%S (%Z)
(%z)")
BirthDate = BirthDateZoneUTC.astimezone(timezone(BirthZone))
BirthDateStr=BirthDate.strftime("%Y-%m-%d %H:%M:%S (%Z) (%z)")
BirthDateLocal=BirthDate.strftime("%Y-%m-%d %H:%M:%S")
################################################################
IDTimeNumber=str(uuid.uuid4())
TimeLine=[('TimeID', [IDTimeNumber]),
         ('UTCDate', [BirthDateZoneStr]),
         ('LocalTime', [BirthDateLocal]),
         ('TimeZone', [BirthZone])]
TimeFrame = pd.DataFrame.from_items(TimeLine)
################################################################
DimTime=TimeFrame
DimTimeIndex=DimTime.set_index(['TimeID'],inplace=False)
################################################################
```

```
sTable = 'Dim-Time'
print('\n###############################')
print('Storing :',sDatabaseName,'\n Table:',sTable)
print('\n###############################')
DimTimeIndex.to_sql(sTable, conn1, if_exists="replace")
DimTimeIndex.to_sql(sTable, conn2, if_exists="replace")
```

Well done. you have a Time dimension. Let's build the Person dimension.

```
print('\n###############################')
print('Dimension Person')
print('\n###############################')
FirstName = 'Guðmundur'
LastName = 'Gunnarsson'
##################################################################
IDPersonNumber=str(uuid.uuid4())
PersonLine=[('PersonID', [IDPersonNumber]),
            ('FirstName', [FirstName]),
            ('LastName', [LastName]),
            ('Zone', ['UTC']),
            ('DateTimeValue', [BirthDateZoneStr])]
PersonFrame = pd.DataFrame.from_items(PersonLine)
#################################################################
DimPerson=PersonFrame
DimPersonIndex=DimPerson.set_index(['PersonID'],inplace=False)
#################################################################
sTable = 'Dim-Person'
print('\n###############################')
print('Storing :',sDatabaseName,'\n Table:',sTable)
print('\n###############################')
DimPersonIndex.to_sql(sTable, conn1, if_exists="replace")
DimPersonIndex.to_sql(sTable, conn2, if_exists="replace")
```

Finally, we add the fact, as follows:

```
print('\n#############################')
print('Fact - Person - time')
print('\n#############################')
IDFactNumber=str(uuid.uuid4())
PersonTimeLine=[('IDNumber', [IDFactNumber]),
                ('IDPersonNumber', [IDPersonNumber]),
                ('IDTimeNumber', [IDTimeNumber])]
PersonTimeFrame = pd.DataFrame.from_items(PersonTimeLine)
################################################################
FctPersonTime=PersonTimeFrame
FctPersonTimeIndex=FctPersonTime.set_index(['IDNumber'],inplace=False)
################################################################
sTable = 'Fact-Person-Time'
print('\n#############################')
print('Storing:',sDatabaseName,'\n Table:',sTable)
print('\n#############################')
FctPersonTimeIndex.to_sql(sTable, conn1, if_exists="replace")
FctPersonTimeIndex.to_sql(sTable, conn2, if_exists="replace")
```

Can you now understand how to formulate a sun model and build the required dimensions and facts?

Building a Data Warehouse

As you have performed so well up to now, I will ask you to open the Transform-Sun-Models.py file from directory ..\VKHCG\01-Vermeulen\04-Transform.

Note   The Python program will build you a good and solid warehouse with which to try new data science techniques. Please be patient; it will supply you with a big push forward.

Execute the program and have some coffee once it runs. You can either code it for yourself or simply upload the code from the samples directory. The code follows, if you want to understand the process. Can you understand the transformation from data vault to data warehouse?

```python
################################################################
# -*- coding: utf-8 -*-
################################################################
import sys
import os
from datetime import datetime
from pytz import timezone
import pandas as pd
import sqlite3 as sq
import uuid
pd.options.mode.chained_assignment = None
################################################################
if sys.platform == 'linux':
    Base=os.path.expanduser('~') + '/VKHCG'
else:
    Base='C:/VKHCG'
print('###############################')
print('Working Base :',Base, ' using ', sys.platform)
print('###############################')
################################################################
Company='01-Vermeulen'
################################################################
sDataBaseDir=Base + '/' + Company + '/04-Transform/SQLite'
if not os.path.exists(sDataBaseDir):
    os.makedirs(sDataBaseDir)
################################################################
sDatabaseName=sDataBaseDir + '/Vermeulen.db'
conn1 = sq.connect(sDatabaseName)
################################################################
```

```
sDataVaultDir=Base + '/88-DV'
if not os.path.exists(sDataVaultDir):
    os.makedirs(sDataVaultDir)
################################################################
sDatabaseName=sDataVaultDir + '/datavault.db'
conn2 = sq.connect(sDatabaseName)
################################################################
sDataWarehouseDir=Base + '/99-DW'
if not os.path.exists(sDataWarehouseDir):
    os.makedirs(sDataWarehouseDir)
################################################################
sDatabaseName=sDataWarehouseDir + '/datawarehouse.db'
conn3 = sq.connect(sDatabaseName)
################################################################
sSQL=" SELECT DateTimeValue FROM [Hub-Time];"
DateDataRaw=pd.read_sql_query(sSQL, conn2)
DateData=DateDataRaw.head(1000)
print(DateData)
################################################################
print('\n#################################')
print('Time Dimension')
print('\n#################################')
t=0
mt=DateData.shape[0]
for i in range(mt):
    BirthZone = ('Atlantic/Reykjavik','Europe/London','UCT')
    for j in range(len(BirthZone)):
        t+=1
        print(t,mt*3)
        BirthDateUTC = datetime.strptime(DateData['DateTimeValue'][i],
        "%Y-%m-%d %H:%M:%S")
        BirthDateZoneUTC=BirthDateUTC.replace(tzinfo=timezone('UTC'))
        BirthDateZoneStr=BirthDateZoneUTC.strftime("%Y-%m-%d %H:%M:%S")
        BirthDateZoneUTCStr=BirthDateZoneUTC.strftime("%Y-%m-%d %H:%M:%S
        (%Z) (%z)")
```

```
            BirthDate = BirthDateZoneUTC.astimezone(timezone(BirthZone[j]))
            BirthDateStr=BirthDate.strftime("%Y-%m-%d %H:%M:%S (%Z) (%z)")
            BirthDateLocal=BirthDate.strftime("%Y-%m-%d %H:%M:%S")
            ################################################################
            IDTimeNumber=str(uuid.uuid4())
            TimeLine=[('TimeID', [str(IDTimeNumber)]),
                      ('UTCDate', [str(BirthDateZoneStr)]),
                      ('LocalTime', [str(BirthDateLocal)]),
                      ('TimeZone', [str(BirthZone)])]
        if t==1:
            TimeFrame = pd.DataFrame.from_items(TimeLine)
        else:
            TimeRow = pd.DataFrame.from_items(TimeLine)
            TimeFrame=TimeFrame.append(TimeRow)
################################################################
DimTime=TimeFrame
DimTimeIndex=DimTime.set_index(['TimeID'],inplace=False)
################################################################
sTable = 'Dim-Time'
print('\n###############################')
print('Storing :',sDatabaseName,'\n Table:',sTable)
print('\n###############################')
DimTimeIndex.to_sql(sTable, conn1, if_exists="replace")
DimTimeIndex.to_sql(sTable, conn3, if_exists="replace")
################################################################
sSQL=" SELECT " + \
     " FirstName," + \
     " SecondName," + \
     " LastName," + \
     " BirthDateKey " + \
     " FROM [Hub-Person];"
PersonDataRaw=pd.read_sql_query(sSQL, conn2)
PersonData=PersonDataRaw.head(1000)
################################################################
print('\n###############################')
```

545

```
print('Dimension Person')
print('\n###############################')
t=0
mt=DateData.shape[0]
for i in range(mt):
    t+=1
    print(t,mt)
    FirstName = str(PersonData["FirstName"])
    SecondName = str(PersonData["SecondName"])
    if len(SecondName) > 0:
        SecondName=""
    LastName = str(PersonData["LastName"])
    BirthDateKey = str(PersonData["BirthDateKey"])
    ################################################################
    IDPersonNumber=str(uuid.uuid4())
    PersonLine=[('PersonID', [str(IDPersonNumber)]),
                ('FirstName', [FirstName]),
                ('SecondName', [SecondName]),
                ('LastName', [LastName]),
                ('Zone', [str('UTC')]),
                ('BirthDate', [BirthDateKey])]
    if t==1:
        PersonFrame = pd.DataFrame.from_items(PersonLine)
    else:
        PersonRow = pd.DataFrame.from_items(PersonLine)
        PersonFrame = PersonFrame.append(PersonRow)
################################################################
DimPerson=PersonFrame
print(DimPerson)
DimPersonIndex=DimPerson.set_index(['PersonID'],inplace=False)
################################################################
sTable = 'Dim-Person'
print('\n###############################')
print('Storing :',sDatabaseName,'\n Table:',sTable)
print('\n###############################')
```

```
DimPersonIndex.to_sql(sTable, conn1, if_exists="replace")
DimPersonIndex.to_sql(sTable, conn3, if_exists="replace")
################################################################
```

You should now have a good example of a data vault to data warehouse transformation. Congratulations on your progress!

Transforming with Data Science

You now have a good basis for data exploration and preparation from the data lake into data vault and from the data vault to the data warehouse. I will now introduce you to the basic data science to transform your data into insights. You must understand a selected set of basic investigation practices, to gain insights from your data.

Steps of Data Exploration and Preparation

You must keep detailed notes of what techniques you employed to prepare the data. Make sure you keep your data traceability matrix up to date after each data engineering step has completed. Update your Data Lineage and Data Providence, to ensure that you have both the technical and business details for the entire process. Now, I will take you through a small number of the standard transform checkpoints, to ensure that your data science is complete.

Missing Value Treatment

You must describe in detail what the missing value treatments are for the data lake transformation. Make sure you take your business community with you along the journey. At the end of the process, they must trust your techniques and results. If they trust the process, they will implement the business decisions that you, as a data scientist, aspire to achieve.

Why Missing Value Treatment Is Required

Explain with notes on the data traceability matrix why there is missing data in the data lake. Remember: Every inconsistency in the data lake is conceivably the missing insight your customer is seeking from you as a data scientist. So, find them and explain them. Your customer will exploit them for business value.

Why Data Has Missing Values

The 5 Whys is the technique that helps you to get to the root cause of your analysis. The use of cause-and-effect fishbone diagrams will assist you to resolve those questions.

I have found the following common reasons for missing data:

- Data fields renamed during upgrades

- Migration processes from old systems to new systems where mappings were incomplete

- Incorrect tables supplied in loading specifications by subject-matter expert

- Data simply not recorded, as it was not available

- Legal reasons, owing to data protection legislation, such as the General Data Protection Regulation (GDPR), resulting in a not-to-process tag on the data entry

- Someone else's "bad" data science. People and projects make mistakes, and you will have to fix their errors in your own data science.

Warning   Ensure that your data science processing is not the reason you are missing data. That is the quickest way to lose your customer's trust.

What Methods Treat Missing Values?

During your progress through the supersteps, you have used many techniques to resolve missing data. Record them in your lineage, but also make sure you collect precisely how each technique applies to the processing flow.

Techniques of Outlier Detection and Treatment

During the processing, you will have detected several outliers that are not complying with your expected ranges, e.g., you expected "Yes" or "No" but found some "N/A"s, or you expected number ranges between 1 and 10 but got 11, 12, and 13 also. These out-of-order items are the outliers.

I suggest you treat them as you treat the missing data. Make sure that your customer agrees with the process, as it will affect the insights you will process and their decisions.

Elliptic Envelope

I will introduce a function called `EllipticEnvelope`. The basic idea is to assume that a data set is from a known distribution and then evaluate any entries not complying to that assumption. Fitting an elliptic envelope is one of the more common techniques used to detect outliers in a Gaussian distributed data set.

The `scikit-learn` package provides an object `covariance.EllipticEnvelope` that fits a robust covariance estimate to the data, and thus fits an ellipse to the central data points, ignoring points outside the central mode. For instance, if the inlier data are Gaussian distributed, it will estimate the inlier location and covariance in a robust way (i.e., without being influenced by outliers). The Mahalanobis distances obtained from this estimate are used to derive a measure of outlyingness. If you want more in-depth details on the function, visit `http://scikit-learn.org/stable/modules/generated/sklearn.covariance.EllipticEnvelope.html`.

Example:

```
# -*- coding: utf-8 -*-
import numpy as np
from sklearn.covariance import EllipticEnvelope
from sklearn.svm import OneClassSVM
import matplotlib.pyplot as plt
import matplotlib.font_manager
from sklearn.datasets import load_boston

# Get data
X1 = load_boston()['data'][:, [8, 10]]  # two clusters
X2 = load_boston()['data'][:, [5, 12]]  # "banana"-shaped

# Define "classifiers" to be used
classifiers = {
    "Empirical Covariance": EllipticEnvelope(support_fraction=1.,
                                             contamination=0.261),
    "Robust Covariance (Minimum Covariance Determinant)":
    EllipticEnvelope(contamination=0.261),
    "OCSVM": OneClassSVM(nu=0.261, gamma=0.05)}
```

The classifiers assume you are testing the normal data entries. Let's look at `EllipticEnvelope(support_fraction=1., contamination=0.261)`. The `support_fraction` is the portion of the complete population you want to use to determine the border between inliers and outliers. In this case, we use 1, which means 100%. The contamination is the indication of what portion of the population could be outliers, hence, the amount of contamination of the data set, i.e., the proportion of outliers in the data set. In your case, this is set to 0.261 against a possible 0.5, more generally described as 26.1% contamination.

The `EllipticEnvelope(contamination=0.261)` is only a change of the included population, by using the defaults for all the settings, except for contamination that is set to 26.1%.

Third is another type of detection called `sklearn.svm.OneClassSVM`, which is discussed later in this chapter.

```
colors = ['m', 'g', 'b']
legend1 = {}
legend2 = {}

# Learn a frontier for outlier detection with several classifiers
xx1, yy1 = np.meshgrid(np.linspace(-8, 28, 500), np.linspace(3, 40, 500))
xx2, yy2 = np.meshgrid(np.linspace(3, 10, 500), np.linspace(-5, 45, 500))
for i, (clf_name, clf) in enumerate(classifiers.items()):
    fig1a=plt.figure(1)
    fig1a.set_size_inches(10, 10)
    clf.fit(X1)
    Z1 = clf.decision_function(np.c_[xx1.ravel(), yy1.ravel()])
    Z1 = Z1.reshape(xx1.shape)
    legend1[clf_name] = plt.contour(
        xx1, yy1, Z1, levels=[0], linewidths=2, colors=colors[i])
    plt.figure(2)
    clf.fit(X2)
    Z2 = clf.decision_function(np.c_[xx2.ravel(), yy2.ravel()])
    Z2 = Z2.reshape(xx2.shape)
    legend2[clf_name] = plt.contour(
        xx2, yy2, Z2, levels=[0], linewidths=2, colors=colors[i])
```

```
legend1_values_list = list(legend1.values())
legend1_keys_list = list(legend1.keys())

# Plot the results (= shape of the data points cloud)
fig1b=plt.figure(1)  # two clusters
fig1b.set_size_inches(10, 10)
plt.title("Outlier detection on a real data set (boston housing)")
plt.scatter(X1[:, 0], X1[:, 1], color='black')
bbox_args = dict(boxstyle="round", fc="0.8")
arrow_args = dict(arrowstyle="->")
plt.annotate("several confounded points", xy=(24, 19),
             xycoords="data", textcoords="data",
             xytext=(13, 10), bbox=bbox_args, arrowprops=arrow_args)
plt.xlim((xx1.min(), xx1.max()))
plt.ylim((yy1.min(), yy1.max()))
plt.legend((legend1_values_list[0].collections[0],
            legend1_values_list[1].collections[0],
            legend1_values_list[2].collections[0]),
           (legend1_keys_list[0], legend1_keys_list[1], legend1_keys_
           list[2]),
           loc="upper center",
           prop=matplotlib.font_manager.FontProperties(size=12))
plt.ylabel("accessibility to radial highways")
plt.xlabel("pupil-teacher ratio by town")

legend2_values_list = list(legend2.values())
legend2_keys_list = list(legend2.keys())

fig2a=plt.figure(2)  # "banana" shape
fig2a.set_size_inches(10, 10)
plt.title("Outlier detection on a real data set (boston housing)")
plt.scatter(X2[:, 0], X2[:, 1], color='black')
plt.xlim((xx2.min(), xx2.max()))
plt.ylim((yy2.min(), yy2.max()))
plt.legend((legend2_values_list[0].collections[0],
            legend2_values_list[1].collections[0],
            legend2_values_list[2].collections[0]),
```

```
            (legend2_keys_list[0], legend2_keys_list[1], legend2_keys_
            list[2]),
            loc="upper center",
            prop=matplotlib.font_manager.FontProperties(size=12))
plt.ylabel("% lower status of the population")
plt.xlabel("average number of rooms per dwelling")

plt.show()
```

There are a few other outlier detection techniques you can investigate.

Isolation Forest

One efficient way of performing outlier detection in high-dimensional data sets is to use random forests. The ensemble.IsolationForest tool "isolates" observations by randomly selecting a feature and then randomly selecting a split value between the maximum and minimum values of the selected feature.

Because recursive partitioning can be represented by a tree structure, the number of splittings required to isolate a sample is equivalent to the path length from the root node to the terminating node. This path length, averaged over a forest of such random trees, is a measure of normality and our decision function. Random partitioning produces a noticeably shorter path for anomalies. Hence, when a forest of random trees collectively produces shorter path lengths for particular samples, they are highly likely to be anomalies. See http://scikit-learn.org/stable/modules/generated/sklearn. ensemble.IsolationForest.html#sklearn.ensemble.IsolationForest.

Novelty Detection

Novelty detection simply performs an evaluation in which we add one more observation to a data set. Is the new observation so different from the others that we can doubt that it is regular? (I.e., does it come from the same distribution?) Or, on the contrary, is it so similar to the other that we cannot distinguish it from the original observations? This is the question addressed by the novelty detection tools and methods.

The sklearn.svm.OneClassSVM tool is a good example of this unsupervised outlier detection technique. For more information, see http://scikit-learn.org/stable/ modules/generated/sklearn.svm.OneClassSVM.html#sklearn.svm.OneClassSVM.

Local Outlier Factor

An efficient way to perform outlier detection on moderately high-dimensional data sets is to use the local outlier factor (LOF) algorithm. The `neighbors.LocalOutlierFactor` algorithm computes a score (called a local outlier factor) reflecting the degree of abnormality of the observations. It measures the local density deviation of a given data point with respect to its neighbors. The idea is to detect the samples that have a substantially lower density than their neighbors.

In practice, the local density is obtained from the k-nearest neighbors. The LOF score of an observation is equal to the ratio of the average local density of its k-nearest neighbors and its own local density. A normal instance is expected to have a local density like that of its neighbors, while abnormal data are expected to have a much smaller local density.

See `http://scikit-learn.org/stable/modules/generated/sklearn.neighbors.LocalOutlierFactor.html#sklearn.neighbors.LocalOutlierFactor` for additional information.

When the amount of contamination is known, this algorithm illustrates three different ways of performing—based on a robust estimator of covariance, which assumes that the data are Gaussian distributed—and performs better than the one-class SVM, in that case. The first is the one-class SVM, which has the ability to capture the shape of the data set and, hence, perform better when the data is strongly non-Gaussian, i.e., with two well-separated clusters. The second is the isolation forest algorithm, which is based on random forests and, hence, better adapted to large-dimensional settings, even if it performs quite well in the example you will perform next. Third is the local outlier factor to measure the local deviation of a given data point with respect to its neighbors, by comparing their local density.

Example:

Here, the underlying truth about inliers and outliers is given by the points' colors. The orange-filled area indicates which points are reported as inliers by each method. You assume to know the fraction of outliers in the data sets, and the following provides an example of what you can achieve.

Open your Python editor, set up this ecosystem, and investigate the preceding techniques.

```python
import numpy as np
from scipy import stats
import matplotlib.pyplot as plt
import matplotlib.font_manager

from sklearn import svm
from sklearn.covariance import EllipticEnvelope
from sklearn.ensemble import IsolationForest
from sklearn.neighbors import LocalOutlierFactor

rng = np.random.RandomState(42)

# Your example settings

n_samples = 200
outliers_fraction = 0.25
clusters_separation = [0, 1, 2]

# define two outlier detection tools to be compared
classifiers = {
    "One-Class SVM": svm.OneClassSVM(nu=0.95 * outliers_fraction + 0.05,
                                     kernel="rbf", gamma=0.1),
    "Robust covariance": EllipticEnvelope(contamination=outliers_fraction),
    "Isolation Forest": IsolationForest(max_samples=n_samples,
                                        contamination=outliers_fraction,
                                        random_state=rng),
    "Local Outlier Factor": LocalOutlierFactor(
        n_neighbors=35,
        contamination=outliers_fraction)}

# Compare given classifiers under given settings
xx, yy = np.meshgrid(np.linspace(-7, 7, 100), np.linspace(-7, 7, 100))
n_inliers = int((1. - outliers_fraction) * n_samples)
n_outliers = int(outliers_fraction * n_samples)
ground_truth = np.ones(n_samples, dtype=int)
ground_truth[-n_outliers:] = -1

# Fit the problem with varying cluster separation
for i, offset in enumerate(clusters_separation):
```

```
np.random.seed(42)
# Data generation
X1 = 0.3 * np.random.randn(n_inliers // 2, 2) - offset
X2 = 0.3 * np.random.randn(n_inliers // 2, 2) + offset
X = np.r_[X1, X2]
# Add outliers
X = np.r_[X, np.random.uniform(low=-6, high=6, size=(n_outliers, 2))]

# Fit the model
plt.figure(figsize=(9, 7))
for i, (clf_name, clf) in enumerate(classifiers.items()):
    # fit the data and tag outliers
    if clf_name == "Local Outlier Factor":
        y_pred = clf.fit_predict(X)
        scores_pred = clf.negative_outlier_factor_
    else:
        clf.fit(X)
        scores_pred = clf.decision_function(X)
        y_pred = clf.predict(X)
    threshold = stats.scoreatpercentile(scores_pred,
                                        100 * outliers_fraction)
    n_errors = (y_pred != ground_truth).sum()
    # plot the levels lines and the points
    if clf_name == "Local Outlier Factor":
        # decision_function is private for LOF
        Z = clf._decision_function(np.c_[xx.ravel(), yy.ravel()])
    else:
        Z = clf.decision_function(np.c_[xx.ravel(), yy.ravel()])
    Z = Z.reshape(xx.shape)
    subplot = plt.subplot(2, 2, i + 1)
    subplot.contourf(xx, yy, Z, levels=np.linspace(Z.min(), threshold, 7),
                     cmap=plt.cm.Blues_r)
    a = subplot.contour(xx, yy, Z, levels=[threshold],
                        linewidths=2, colors='red')
    subplot.contourf(xx, yy, Z, levels=[threshold, Z.max()],
                     colors='orange')
```

```
        b = subplot.scatter(X[:-n_outliers, 0], X[:-n_outliers, 1], c='white',
                        s=20, edgecolor='k')
        c = subplot.scatter(X[-n_outliers:, 0], X[-n_outliers:, 1], c='black',
                        s=20, edgecolor='k')
        subplot.axis('tight')
        subplot.legend(
            [a.collections[0], b, c],
            ['learned decision function', 'true inliers', 'true outliers'],
            prop=matplotlib.font_manager.FontProperties(size=10),
            loc='lower right')
        subplot.set_xlabel("%d. %s (errors: %d)" % (i + 1, clf_name, n_
        errors))
        subplot.set_xlim((-7, 7))
        subplot.set_ylim((-7, 7))
    plt.subplots_adjust(0.04, 0.1, 0.96, 0.94, 0.1, 0.26)
    plt.suptitle("Outlier detection")
plt.show()
```

You can now detect outliers in a data set. That is a major achievement, as the most interesting features of data science work are found in these outliers and why they exist. This supports you in performing the feature engineering of the data sets.

What Is Feature Engineering?

Feature engineering is your core technique to determine the important data characteristics in the data lake and ensure they get the correct treatment through the steps of processing. Make sure that any featuring extraction process technique is documented in the data transformation matrix and the data lineage.

Common Feature Extraction Techniques

I will introduce you to several common feature extraction techniques that will help you to enhance any existing data warehouse, by applying data science to the data in the warehouse.

Binning

Binning is a technique that is used to reduce the complexity of data sets, to enable the data scientist to evaluate the data with an organized grouping technique. Binning is a good way for you to turn continuous data into a data set that has specific features that you can evaluate for patterns. A simple example is the cost of candy in your local store, which might range anywhere from a penny to ten dollars, but if you subgroup the price into, say, a rounded-up value that then gives you a range of five values against five hundred, you have just reduced your processing complexity to 1/500th of what it was before. There are several good techniques, which I will discuss next.

I have two binning techniques that you can use against the data sets. Open your Python editor and try these examples. The first technique is to use the `digitize` function.

```
import numpy
data = numpy.random.random(100)
bins = numpy.linspace(0, 1, 10)
digitized = numpy.digitize(data, bins)
bin_means = [data[digitized == i].mean() for i in range(1, len(bins))]
print(bin_means)
```

The second is to use the `histogram` function.

```
bin_means2 = (numpy.histogram(data, bins, weights=data)[0] /
              numpy.histogram(data, bins)[0])
print(bin_means2)
```

This transform technique can be used to reduce the location dimension into three latitude bins and four longitude bins.

You will require the NumPy library.

```
import numpy as np
```

Set up the latitude and longitude data sets.

```
LatitudeData = np.array(range(-90,90,1))
LongitudeData = np.array(range(-180,180,1))
```

Set up the latitude and longitude data bins.

```
LatitudeBins = np.array(range(-90,90,45))
LongitudeBins = np.array(range(-180,180,60))
```

Digitize the data sets with the data bins.

```
LatitudeDigitized = np.digitize(LatitudeData, LatitudeBins)
LongitudeDigitized = np.digitize(LongitudeData, LongitudeBins)
Calculate the mean against the bins:
LatitudeBinMeans = [LatitudeData[LatitudeDigitized == i].mean() for i in
range(1, len(LatitudeBins))]
LongitudeBinMeans = [LongitudeData[LongitudeDigitized == i].mean() for i in
range(1, len(LongitudeBins))]
```

Well done. You have the three latitude bins and four longitude bins.

```
print(LatitudeBinMeans)
print(LongitudeBinMeans)
```

You can also use the histogram function to achieve similar results.

```
LatitudeBinMeans2 = (np.histogram(LatitudeData, LatitudeBins,\
            weights=LatitudeData)[0] /
            np.histogram(LatitudeData, LatitudeBins)[0])
LongitudeBinMeans2 = (np.histogram(LongitudeData, LongitudeBins,\
            weights=LongitudeData)[0] /
            np.histogram(LongitudeData, LongitudeBins)[0])
print(LatitudeBinMeans2)
print(LongitudeBinMeans2)
```

Now you can apply two different techniques for binning.

Averaging

The use of averaging enables you to reduce the amount of records you require to report any activity that demands a more indicative, rather than a precise, total.

Example:

Create a model that enables you to calculate the average position for ten sample points. First, set up the ecosystem.

```
import numpy as np
import pandas as pd
```

Create two series to model the latitude and longitude ranges.

```
LatitudeData = pd.Series(np.array(range(-90,91,1)))
LongitudeData = pd.Series(np.array(range(-180,181,1)))
You then select 10 samples for each range:
LatitudeSet=LatitudeData.sample(10)
LongitudeSet=LongitudeData.sample(10)
```

Calculate the average of each.

```
LatitudeAverage = np.average(LatitudeSet)
LongitudeAverage = np.average(LongitudeSet)
```

See your results.

```
print('Latitude')
print(LatitudeSet)
print('Latitude (Avg):',LatitudeAverage)
print('##############')
print('Longitude')
print(LongitudeSet)
print('Longitude (Avg):', LongitudeAverage)
```

You can now calculate the average of any range of numbers. If you run the code several times, you should get different samples. (See Transform-Average-Location.py in ..\VKHCG\01-Vermeulen\04-Transform for code.)

Challenge Question One:

In directory ..\VKHCG\01-Vermeulen\00-RawData, there is a file called IP_DATA_CORE.csv. Try to import the latitude and longitude columns and calculate the average values.

Tip   Look at pandas and numpy and have some fun!

Challenge Question Two:

Try to calculate the average amount of IP addresses:

> Per country

> Per place-name

> Per post code

Tip   Amount of IP address = (-1*First IP Number + Last IP Number)

Latent Dirichlet Allocation (LDA)

A latent Dirichlet allocation (LDA) is a statistical model that allows sets of observations to be explained by unobserved groups that elucidates why they match or belong together within text documents. This technique is useful when investigating text from a collection of documents that are common in the data lake, as companies store all their correspondence in a data lake. This model is also useful for Twitter or e-mail analysis.

Note   To run the example, you will require `pip install lda`.

In your Python editor, create a new file named `Transform_Latent_Dirichlet_allocation.py` in directory `.. \VKHCG\01-Vermeulen\04-Transform`. Following is an example of what you can achieve:

```
import numpy as np
import lda
import lda.datasets
X = lda.datasets.load_reuters()
vocab = lda.datasets.load_reuters_vocab()
titles = lda.datasets.load_reuters_titles()
X.shape
X.sum()
```

You can experiment with ranges of `n_topics` and `n_iter` values to observe the impact on the process.

```
model = lda.LDA(n_topics=50, n_iter=1500, random_state=1)
model.fit(X)
topic_word = model.topic_word_
n_top_words = 10
for i, topic_dist in enumerate(topic_word):
    topic_words = np.array(vocab)[np.argsort(topic_dist)][:-(n_top_words+1):-1]
    print('Topic {}: {}'.format(i, ' '.join(topic_words)))
```

Investigate the top-ten topics.

```
doc_topic = model.doc_topic_
for i in range(10):
    print("{} (top topic: {})".format(titles[i], doc_topic[i].argmax()))
```

Well done. You can now analyze text documents.

Do you think if you had Hillman's logistics shipping document you could use this technique? Indeed, you could, as this technique will work on any text note fields or e-mail, even Twitter entries.

Note   If you want to read your Twitter account feed, I suggest using the library `twitter`. Install it using `pip install -i https://pypi.anaconda.org/pypi/simple twitter`.

Now, you can read your Twitter accounts.

Tip   For e-mail, I suggest the standard Python `email` library. See `https://docs.python.org/3.4/library/email.html`.

Now, you can read e-mail.

The complete process is about getting data to the data lake and then guiding it through the steps: retrieve, assess, process, and transform.

Tip   I have found, on average, that it is only after the third recheck that 90% of the data science is complete. The process is an iterative design process. The methodology is based on a cyclic process of prototyping, testing, analyzing, and refining. Success will be achieved as you close out the prototypes.

I will now explain a set of common data science terminology that you will encounter in the field of data science.

Hypothesis Testing

Hypothesis testing is not precisely an algorithm, but it's a must-know for any data scientist. You cannot progress until you have thoroughly mastered this technique.

Hypothesis testing is the process by which statistical tests are used to check if a hypothesis is true, by using data. Based on hypothetical testing, data scientists choose to accept or reject the hypothesis. When an event occurs, it can be a trend or happen by chance. To check whether the event is an important occurrence or just happenstance, hypothesis testing is necessary.

There are many tests for hypothesis testing, but the following two are most popular.

T-Test

A t-test is a popular statistical test to make inferences about single means or inferences about two means or variances, to check if the two groups' means are statistically different from each other, where n < 30 and standard deviation is unknown. For more information on the t-test, see https://docs.scipy.org/doc/scipy/reference/generated/scipy.stats.t.html#scipy.stats.t and https://docs.scipy.org/doc/scipy/reference/generated/scipy.stats.ttest_ind.html.

Example:

First you set up the ecosystem, as follows:

```
import numpy as np
from scipy.stats import ttest_ind, ttest_ind_from_stats
from scipy.special import stdtr
```

Create a set of "unknown" data. (This can be a set of data you want to analyze.) In the following example, there are five random data sets. You can select them by having nSet equal 1, 2, 3, 4, or 5.

```
nSet=1
if nSet==1:
    a = np.random.randn(40)
```

```
    b = 4*np.random.randn(50)

if nSet==2:
    a=np.array([27.1,22.0,20.8,23.4,23.4,23.5,25.8,22.0,24.8,20.2,21.9,
            22.1,22.9,20.5,24.4])
    b=np.array([27.1,22.0,20.8,23.4,23.4,23.5,25.8,22.0,24.8,20.2,21.9,
            22.1,22.9,20.5,24.41])

if nSet==3:
    a=np.array([17.2,20.9,22.6,18.1,21.7,21.4,23.5,24.2,14.7,21.8])
    b=np.array([21.5,22.8,21.0,23.0,21.6,23.6,22.5,20.7,23.4,21.8,20.7,
            21.7,21.5,22.5,23.6,21.5,22.5,23.5,21.5,21.8])

if nSet==4:
    a=np.array([19.8,20.4,19.6,17.8,18.5,18.9,18.3,18.9,19.5,22.0])
    b=np.array([28.2,26.6,20.1,23.3,25.2,22.1,17.7,27.6,20.6,13.7,23.2,
            17.5,20.6,18.0,23.9,21.6,24.3,20.4,24.0,13.2])

if nSet==5:
    a = np.array([55.0, 55.0, 47.0, 47.0, 55.0, 55.0, 55.0, 63.0])
    b = np.array([55.0, 56.0, 47.0, 47.0, 55.0, 55.0, 55.0, 63.0])
```

First, you will use scipy's t-test.

```
# Use scipy.stats.ttest_ind.
t, p = ttest_ind(a, b, equal_var=False)
print("t-Test_ind:            t = %g  p = %g" % (t, p))
```

Second, you will get the descriptive statistics.

```
# Compute the descriptive statistics of a and b.
abar = a.mean()
avar = a.var(ddof=1)
na = a.size
adof = na - 1

bbar = b.mean()
bvar = b.var(ddof=1)
nb = b.size
bdof = nb - 1
```

```
# Use scipy.stats.ttest_ind_from_stats.
t2, p2 = ttest_ind_from_stats(abar, np.sqrt(avar), na,
                              bbar, np.sqrt(bvar), nb,
                              equal_var=False)
print("t-Test_ind_from_stats: t = %g  p = %g" % (t2, p2))
```

Look at Welch's t-test formula.

Third, you can use the formula to calculate the test.

```
# Use the formulas directly.
tf = (abar - bbar) / np.sqrt(avar/na + bvar/nb)
dof = (avar/na + bvar/nb)**2 / (avar**2/(na**2*adof) + bvar**2/
(nb**2*bdof))
pf = 2*stdtr(dof, -np.abs(tf))

print("Formula:              t = %g  p = %g" % (tf, pf))
P=1-p
if P < 0.001:
    print('Statistically highly significant:',P)
else:
    if P < 0.05:
        print('Statistically significant:',P)
    else:
        print('No conclusion')
```

You should see results like this:

```
t-Test_ind:             t = -1.5827  p = 0.118873
t-Test_ind_from_stats: t = -1.5827  p = 0.118873
Formula:                t = -1.5827  p = 0.118873
No conclusion
```

Your results are as follows. The p means the probability, or how likely your results are occurring by chance. In this case, it's 11%, or p-value = 0.11.

The p-value results can be statistically significant when $P < 0.05$ and statistically highly significant if $P < 0.001$ (a less than one-in-a-thousand chance of being wrong). So, in this case, it cannot be noted as either statistically significant or statistically highly significant, as it is 0.11.

Go back and change nSet at the beginning of the code you just entered. Remember: I mentioned that you can select them by nSet = 1, 2, 3, 4, or 5.

Retest the data sets. You should now see that the p-value changes, and you should also understand that the test gives you a good indicator of whether the two results sets are similar.

Can you find the 99.99%?

Chi-Square Test

A chi-square (or squared [χ^2]) test is used to examine if two distributions of categorical variables are significantly different from each other.

Example:

Try these examples that are generated with five different data sets. First, set up the ecosystem.

```
import numpy as np
import scipy.stats as st
```

Create data sets.

```
np.random.seed(1)
# Create sample data sets.
nSet=1
if nSet==1:
    a = abs(np.random.randn(50))
    b = abs(50*np.random.randn(50))

if nSet==2:
    a=np.array([27.1,22.0,20.8,23.4,23.4,23.5,25.8,22.0,24.8,20.2,21.9,
                22.1,22.9,20.5,24.4])
    b=np.array([27.1,22.0,20.8,23.4,23.4,23.5,25.8,22.0,24.8,20.2,21.9,
                22.1,22.9,20.5,24.41])

if nSet==3:
    a=np.array([17.2,20.9,22.6,18.1,21.7,21.4,23.5,24.2,14.7,21.8])
    b=np.array([21.5,22.8,21.0,23.0,21.6,23.6,22.5,20.7,23.4,21.8])
```

```
if nSet==4:
    a=np.array([19.8,20.4,19.6,17.8,18.5,18.9,18.3,18.9,19.5,22.0])
    b=np.array([28.2,26.6,20.1,23.3,25.2,22.1,17.7,27.6,20.6,13.7])

if nSet==5:
    a = np.array([55.0, 55.0, 47.0, 47.0, 55.0, 55.0, 55.0, 63.0])
    b = np.array([55.0, 56.0, 47.0, 47.0, 55.0, 55.0, 55.0, 63.0])

obs = np.array([a,b])
```

Perform the test.

```
chi2, p, dof, expected = st.chi2_contingency(obs)
```

Display the results.

```
msg = "Test Statistic : {}\np-value: {}\ndof: {}\n"
print( msg.format( chi2, p , dof,expected) )

P=1-p
if P < 0.001:
    print('Statistically highly significant:',P)
else:
    if P < 0.05:
        print('Statistically significant:',P)
    else:
        print('No conclusion')
```

Can you understand what the test indicates as you cycle the nSet through samples 1–5?

Overfitting and Underfitting

Overfitting and underfitting are major problems when data scientists retrieve data insights from the data sets they are investigating. Overfitting is when the data scientist generates a model to fit a training set perfectly, but it does not generalize well against an unknown future real-world data set, as the data science is so tightly modeled against the known data set, the most minor outlier simply does not get classified correctly. The solution only works for the specific data set and no other data set. For example, if a person earns more than $150,000, that person is rich; otherwise, the person is poor. A binary classification of rich or poor will not work, as can a person earning about $145,000 be poor?

Underfitting the data scientist's results into the data insights has been so nonspecific that to some extent predictive models are inappropriately applied or questionable as regards to insights. For example, your person classifier has a 48% success rate to determine the sex of a person. That will never work, as with a binary guess, you could achieve a 50% rating by simply guessing.

Your data science must offer a significant level of insight for you to secure the trust of your customers, so they can confidently take business decisions, based on the insights you provide them.

Polynomial Features

The polynomic formula is the following: $(a_1x+b_1)(a_2x+b_2)=a_1a_2x^2+(a_1b_2+a_2b_1)x+b_1b_2$. The polynomial feature extraction can use a chain of polynomic formulas to create a hyperplane that will subdivide any data sets into the correct cluster groups. The higher the polynomic complexity, the more precise the result that can be achieved.

Example:

```
import numpy as np
import matplotlib.pyplot as plt

from sklearn.linear_model import Ridge
from sklearn.preprocessing import PolynomialFeatures
from sklearn.pipeline import make_pipeline

def f(x):
    """ function to approximate by polynomial interpolation"""
    return x * np.sin(x)

# generate points used to plot
x_plot = np.linspace(0, 10, 100)

# generate points and keep a subset of them
x = np.linspace(0, 10, 100)
rng = np.random.RandomState(0)
rng.shuffle(x)
```

```
x = np.sort(x[:20])
y = f(x)

# create matrix versions of these arrays
X = x[:, np.newaxis]
X_plot = x_plot[:, np.newaxis]

colors = ['teal', 'yellowgreen', 'gold']
lw = 2
plt.plot(x_plot, f(x_plot), color='cornflowerblue', linewidth=lw,
         label="Ground Truth")
plt.scatter(x, y, color='navy', s=30, marker='o', label="training points")

for count, degree in enumerate([3, 4, 5]):
    model = make_pipeline(PolynomialFeatures(degree), Ridge())
    model.fit(X, y)
    y_plot = model.predict(X_plot)
    plt.plot(x_plot, y_plot, color=colors[count], linewidth=lw,
             label="Degree %d" % degree)

plt.legend(loc='lower left')

plt.show()
```

Now that you know how to generate a polynomic formula to match any curve, I will show you a practical application using a real-life data set.

Common Data-Fitting Issue

These higher order polynomic formulas are, however, more prone to overfitting, while lower order formulas are more likely to underfit. It is a delicate balance between two extremes that support good data science.

Example:

```
import numpy as np
import matplotlib.pyplot as plt
from sklearn.pipeline import Pipeline
from sklearn.preprocessing import PolynomialFeatures
```

```python
from sklearn.linear_model import LinearRegression
from sklearn.model_selection import cross_val_score

def true_fun(X):
    return np.cos(1.5 * np.pi * X)

np.random.seed(0)

n_samples = 30
degrees = [1, 4, 15]

X = np.sort(np.random.rand(n_samples))
y = true_fun(X) + np.random.randn(n_samples) * 0.1

plt.figure(figsize=(14, 5))
for i in range(len(degrees)):
    ax = plt.subplot(1, len(degrees), i + 1)
    plt.setp(ax, xticks=(), yticks=())

    polynomial_features = PolynomialFeatures(degree=degrees[i],
                                             include_bias=False)
    linear_regression = LinearRegression()
    pipeline = Pipeline([("polynomial_features", polynomial_features),
                         ("linear_regression", linear_regression)])
    pipeline.fit(X[:, np.newaxis], y)

    # Evaluate the models using crossvalidation
    scores = cross_val_score(pipeline, X[:, np.newaxis], y,
                             scoring="neg_mean_squared_error", cv=10)

    X_test = np.linspace(0, 1, 100)
    plt.plot(X_test, pipeline.predict(X_test[:, np.newaxis]), label="Model")
    plt.plot(X_test, true_fun(X_test), label="True function")
    plt.scatter(X, y, edgecolor='b', s=20, label="Samples")
    plt.xlabel("x")
    plt.ylabel("y")
    plt.xlim((0, 1))
    plt.ylim((-2, 2))
    plt.legend(loc="best")
```

```
    plt.title("Degree {}\nMSE = {:.2e}(+/- {:.2e})".format(
        degrees[i], -scores.mean(), scores.std()))
plt.show()
```

Precision-Recall

Precision-recall is a useful measure for successfully predicting when classes are extremely imbalanced. In information retrieval,

- Precision is a measure of result relevancy.

- Recall is a measure of how many truly relevant results are returned.

Precision-Recall Curve

The precision-recall curve shows the trade-off between precision and recall for different thresholds. A high area under the curve represents both high recall and high precision, where high precision relates to a low false positive rate, and high recall relates to a low false negative rate. High scores for both shows that the classifier is returning accurate results (high precision), as well as returning a majority of all positive results (high recall).

A system with high recall but low precision returns many results, but most of its predicted labels are incorrect when compared to the training labels. A system with high precision but low recall is just the opposite, returning very few results, but most of its predicted labels are correct when compared to the training labels. An ideal system with high precision and high recall will return many results, with all results labeled correctly.

Precision (P) is defined as the number of true positives (Tp) over the number of true positives (Tp) plus the number of false positives (Fp).

$$P = \frac{Tp}{Tp + Fp}$$

Recall (R) is defined as the number of true positives (Tp) over the number of true positives (Tp) plus the number of false negatives (Fn).

$$R = \frac{Tp}{Tp + Fn}$$

The true negative rate (TNR) is the rate that indicates the recall of the negative items.

$$TNR = \frac{Tn}{Tn+Fp}$$

Accuracy (A) is defined as

$$A = \frac{Tp+Tn}{Tp+Fp+Tn+Fn}$$

Sensitivity and Specificity

Sensitivity and specificity are statistical measures of the performance of a binary classification test, also known in statistics as a classification function. Sensitivity (also called the true positive rate, the recall, or probability of detection) measures the proportion of positives that are correctly identified as such (e.g., the percentage of sick people who are correctly identified as having the condition). Specificity (also called the true negative rate) measures the proportion of negatives that are correctly identified as such (e.g., the percentage of healthy people who are correctly identified as not having the condition).

F1-Measure

The F1-score is a measure that combines precision and recall in the harmonic mean of precision and recall.

$$F1 = 2*\frac{P*R}{P+R}$$

Note   The precision may not decrease with recall.

I have found the following `sklearn` functions useful when calculation these measures:

- `sklearn.metrics.average_precision_score`
- `sklearn.metrics.recall_score`
- `sklearn.metrics.precision_score`
- `sklearn.metrics.f1_score`

I suggest you practice with sets of data and understand these parameters and what affects them positively and negatively. They are the success indicators of your data science.

Example:

```
###########################################################################
# In binary classification settings
# ---------------------------------------------------------
#
# Create simple data
# .................
#
# Try to differentiate the two first classes of the iris data
from sklearn import svm, datasets
from sklearn.model_selection import train_test_split
import numpy as np

iris = datasets.load_iris()
X = iris.data
y = iris.target

# Add noisy features
random_state = np.random.RandomState(0)
n_samples, n_features = X.shape
X = np.c_[X, random_state.randn(n_samples, 200 * n_features)]

# Limit to the two first classes, and split into training and test
X_train, X_test, y_train, y_test = train_test_split(X[y < 2], y[y < 2],
                                                    test_size=.5,
                                                    random_state=random_
                                                    state)

# Create a simple classifier
classifier = svm.LinearSVC(random_state=random_state)
classifier.fit(X_train, y_train)
y_score = classifier.decision_function(X_test)
```

```
###########################################################################
# Compute the average precision score
# ...............................
from sklearn.metrics import average_precision_score
average_precision = average_precision_score(y_test, y_score)

print('Average precision-recall score: {0:0.2f}'.format(
        average_precision))

###########################################################################
# Plot the Precision-Recall curve
# ..............................
from sklearn.metrics import precision_recall_curve
import matplotlib.pyplot as plt

precision, recall, _ = precision_recall_curve(y_test, y_score)

plt.step(recall, precision, color='b', alpha=0.2,
         where='post')
plt.fill_between(recall, precision, step='post', alpha=0.2,
                 color='b')

plt.xlabel('Recall')
plt.ylabel('Precision')
plt.ylim([0.0, 1.05])
plt.xlim([0.0, 1.0])
plt.title('2-class Precision-Recall curve: AUC={0:0.2f}'.format(
          average_precision))

###########################################################################
# In multi-label settings
# -----------------------
#
# Create multi-label data, fit, and predict
# .........................................
```

573

```
#
# We create a multi-label dataset, to illustrate the precision-recall in
# multi-label settings

from sklearn.preprocessing import label_binarize

# Use label_binarize to be multi-label like settings
Y = label_binarize(y, classes=[0, 1, 2])
n_classes = Y.shape[1]

# Split into training and test
X_train, X_test, Y_train, Y_test = train_test_split(X, Y, test_size=.5,
                                    random_state=random_state)

# We use OneVsRestClassifier for multi-label prediction
from sklearn.multiclass import OneVsRestClassifier

# Run classifier
classifier = OneVsRestClassifier(svm.LinearSVC(random_state=random_state))
classifier.fit(X_train, Y_train)
y_score = classifier.decision_function(X_test)

###########################################################################
# The average precision score in multi-label settings
# .................................................
from sklearn.metrics import precision_recall_curve
from sklearn.metrics import average_precision_score

# For each class
precision = dict()
recall = dict()
average_precision = dict()
for i in range(n_classes):
    precision[i], recall[i], _ = precision_recall_curve(Y_test[:, i],
                                                y_score[:, i])
```

```
    average_precision[i] = average_precision_score(Y_test[:, i], y_score
[:, i])

# A "micro-average": quantifying score on all classes jointly
precision["micro"], recall["micro"], _ = precision_recall_curve(Y_test.
ravel(),
    y_score.ravel())
average_precision["micro"] = average_precision_score(Y_test, y_score,
                                            average="micro")
print('Average precision score, micro-averaged over all classes: {0:0.2f}'
    .format(average_precision["micro"]))

##########################################################################
# Plot the micro-averaged Precision-Recall curve
# ...............................................
#

plt.figure()
plt.step(recall['micro'], precision['micro'], color='b', alpha=0.2,
        where='post')
plt.fill_between(recall["micro"], precision["micro"], step='post',
alpha=0.2,
                color='b')

plt.xlabel('Recall')
plt.ylabel('Precision')
plt.ylim([0.0, 1.05])
plt.xlim([0.0, 1.0])
plt.title(
    'Average precision score, micro-averaged over all classes:
AUC={0:0.2f}'
    .format(average_precision["micro"]))

##########################################################################
# Plot Precision-Recall curve for each class and iso-f1 curves
# ...............................................................
```

```
#
from itertools import cycle
# setup plot details
colors = cycle(['navy', 'turquoise', 'darkorange', 'cornflowerblue', 'teal'])

plt.figure(figsize=(12, 8))
f_scores = np.linspace(0.2, 0.8, num=4)
lines = []
labels = []
for f_score in f_scores:
    x = np.linspace(0.01, 1)
    y = f_score * x / (2 * x - f_score)
    l, = plt.plot(x[y >= 0], y[y >= 0], color='gray', alpha=0.2)
    plt.annotate('f1={0:0.1f}'.format(f_score), xy=(0.9, y[45] + 0.02))

lines.append(l)
labels.append('iso-f1 curves')
l, = plt.plot(recall["micro"], precision["micro"], color='gold', lw=2)
lines.append(l)
labels.append('micro-average Precision-recall (area = {0:0.2f})'
            ''.format(average_precision["micro"]))

for i, color in zip(range(n_classes), colors):
    l, = plt.plot(recall[i], precision[i], color=color, lw=2)
    lines.append(l)
    labels.append('Precision-recall for class {0} (area = {1:0.2f})'
                ''.format(i, average_precision[i]))

fig = plt.gcf()
fig.subplots_adjust(bottom=0.25)
plt.xlim([0.0, 1.0])
plt.ylim([0.0, 1.05])
plt.xlabel('Recall')
plt.ylabel('Precision')
plt.title('Extension of Precision-Recall curve to multi-class')
plt.legend(lines, labels, loc=(0, -.38), prop=dict(size=14))

plt.show()
```

Receiver Operating Characteristic (ROC) Analysis Curves

A receiver operating characteristic (ROC) analysis curve is a graphical plot that illustrates the diagnostic ability of a binary classifier system as its discrimination threshold is varied. The ROC curve plots the true positive rate (TPR) against the false positive rate (FPR) at various threshold settings. The true positive rate is also known as sensitivity, recall, or probability of detection.

You will find the ROC analysis curves useful for evaluating whether your classification or feature engineering is good enough to determine the value of the insights you are finding. This helps with repeatable results against a real-world data set. So, if you suggest that your customers should take a specific action as a result of your findings, ROC analysis curves will support your advice and insights but also relay the quality of the insights at given parameters.

You should now open your Python editor and create the following ecosystem.

Example:

```
import numpy as np
from scipy import interp
import matplotlib.pyplot as plt

from sklearn import svm, datasets
from sklearn.metrics import roc_curve, auc
from sklearn.model_selection import StratifiedKFold

# ###############################################################################
# Data IO and generation

# Import some data to play with
iris = datasets.load_iris()
X = iris.data
y = iris.target
X, y = X[y != 2], y[y != 2]
n_samples, n_features = X.shape

# Add noisy features
random_state = np.random.RandomState(0)
X = np.c_[X, random_state.randn(n_samples, 200 * n_features)]
```

```
# ##########################################################################
# Classification and ROC analysis

# Run classifier with cross-validation and plot ROC curves
cv = StratifiedKFold(n_splits=6)
classifier = svm.SVC(kernel='linear', probability=True,
                     random_state=random_state)

tprs = []
aucs = []
mean_fpr = np.linspace(0, 1, 100)

i = 0
for train, test in cv.split(X, y):
    probas_ = classifier.fit(X[train], y[train]).predict_proba(X[test])
    # Compute ROC curve and area the curve
    fpr, tpr, thresholds = roc_curve(y[test], probas_[:, 1])
    tprs.append(interp(mean_fpr, fpr, tpr))
    tprs[-1][0] = 0.0
    roc_auc = auc(fpr, tpr)
    aucs.append(roc_auc)
    plt.plot(fpr, tpr, lw=1, alpha=0.3,
             label='ROC fold %d (AUC = %0.2f)' % (i, roc_auc))

    i += 1
plt.plot([0, 1], [0, 1], linestyle='--', lw=2, color='r',
         label='Luck', alpha=.8)

mean_tpr = np.mean(tprs, axis=0)
mean_tpr[-1] = 1.0
mean_auc = auc(mean_fpr, mean_tpr)
std_auc = np.std(aucs)
plt.plot(mean_fpr, mean_tpr, color='b',
         label=r'Mean ROC (AUC = %0.2f $\pm$ %0.2f)' % (mean_auc, std_auc),
         lw=2, alpha=.8)

std_tpr = np.std(tprs, axis=0)
tprs_upper = np.minimum(mean_tpr + std_tpr, 1)
```

```
tprs_lower = np.maximum(mean_tpr - std_tpr, 0)
plt.fill_between(mean_fpr, tprs_lower, tprs_upper, color='gray', alpha=.2,
                 label=r'$\pm$ 1 std. dev.')

plt.xlim([-0.05, 1.05])
plt.ylim([-0.05, 1.05])
plt.xlabel('False Positive Rate')
plt.ylabel('True Positive Rate')
plt.title('Receiver operating characteristic example')
plt.legend(loc="lower right")
plt.show()
```

Cross-Validation Test

Cross-validation is a model validation technique for evaluating how the results of a statistical analysis will generalize to an independent data set. It is mostly used in settings where the goal is the prediction. Knowing how to calculate a test such as this enables you to validate the application of your model on real-world, i.e., independent data sets.

I will guide you through a test. Open your Python editor and create the following ecosystem:

```
import numpy as np
from sklearn.model_selection import cross_val_score
from sklearn import datasets, svm
import matplotlib.pyplot as plt

digits = datasets.load_digits()
X = digits.data
y = digits.target
```

Let's pick three different kernels and compare how they will perform.

```
kernels=['linear', 'poly', 'rbf']
for kernel in kernels:
    svc = svm.SVC(kernel=kernel)
    C_s = np.logspace(-15, 0, 15)
    scores = list()
    scores_std = list()
```

```
for C in C_s:
    svc.C = C
    this_scores = cross_val_score(svc, X, y, n_jobs=1)
    scores.append(np.mean(this_scores))
    scores_std.append(np.std(this_scores))
```

You must plot your results.

```
Title="Kernel:>" + kernel
fig=plt.figure(1, figsize=(8, 6))
plt.clf()
fig.suptitle(Title, fontsize=20)
plt.semilogx(C_s, scores)
plt.semilogx(C_s, np.array(scores) + np.array(scores_std), 'b--')
plt.semilogx(C_s, np.array(scores) - np.array(scores_std), 'b--')
locs, labels = plt.yticks()
plt.yticks(locs, list(map(lambda x: "%g" % x, locs)))
plt.ylabel('Cross-Validation Score')
plt.xlabel('Parameter C')
plt.ylim(0, 1.1)
plt.show()
```

Well done. You can now perform cross-validation of your results.

Univariate Analysis

Univariate analysis is the simplest form of analyzing data. *Uni* means "one," so your data has only one variable. It doesn't deal with causes or relationships, and its main purpose is to describe. It takes data, summarizes that data, and finds patterns in the data.

The patterns found in univariate data include central tendency (mean, mode, and median) and dispersion, range, variance, maximum, minimum, quartiles (including the interquartile range), and standard deviation.

Example:

How many students are graduating with a data science degree? You have several options for describing data using a univariate approach. You can use frequency distribution tables, frequency polygons, histograms, bar charts, or pie charts.

Bivariate Analysis

Bivariate analysis is when two variables are analyzed together for any possible association or empirical relationship, such as, for example, the correlation between gender and graduation with a data science degree? Canonical correlation in the experimental context is to take two sets of variables and see what is common between the two sets.

Graphs that are appropriate for bivariate analysis depend on the type of variable. For two continuous variables, a scatterplot is a common graph. When one variable is categorical and the other continuous, a box plot is common, and when both are categorical, a mosaic plot is common.

Multivariate Analysis

Multivariate data analysis refers to any statistical technique used to analyze data that arises from more than one variable. This essentially models reality, in which each situation, product, or decision involves more than a single variable.

More than two variables are analyzed together for any possible association or interactions.

Example:

What is the correlation between gender, country of residence, and graduation with a data science degree? Any statistical modeling exercise, such as regression, decision tree, SVM, and clustering are multivariate in nature. The analysis is used when more than two variables determine the final outcome.

Linear Regression

Linear regression is a statistical modeling technique that endeavors to model the relationship between an explanatory variable and a dependent variable, by fitting the observed data points on a linear equation, for example, modeling the body mass index (BMI) of individuals by using their weight.

Warning   A regression analysis with one dependent variable and four independent variables is *not* a multivariate regression. It's a multiple regression. Multivariate analysis *always* refers to the dependent variable.

Simple Linear Regression

Linear regression is used if there is a relationship or significant association between the variables. This can be checked by scatterplots. If no linear association appears between the variables, fitting a linear regression model to the data will not provide a useful model.

A linear regression line has equations in the following form:

$Y = a + bX$,

Where, X = explanatory variable and

Y = dependent variable

b = slope of the line

a = intercept (the value of y when x = 0)

Example:

```
################################################################
# -*- coding: utf-8 -*-
################################################################
import sys
import os
import pandas as pd
import sqlite3 as sq
import matplotlib.pyplot as plt
import numpy as np
################################################################
if sys.platform == 'linux':
    Base=os.path.expanduser('~') + '/VKHCG'
else:
    Base='C:/VKHCG'
print('################################')
print('Working Base :',Base, ' using ', sys.platform)
print('################################')
################################################################
################################################################
Company='01-Vermeulen'
################################################################
sDataBaseDir=Base + '/' + Company + '/04-Transform/SQLite'
if not os.path.exists(sDataBaseDir):
```

```
    os.makedirs(sDataBaseDir)
################################################################
sDatabaseName=sDataBaseDir + '/Vermeulen.db'
conn1 = sq.connect(sDatabaseName)
################################################################
sDataVaultDir=Base + '/88-DV'
if not os.path.exists(sDataVaultDir):
    os.makedirs(sDataVaultDir)
################################################################
sDatabaseName=sDataVaultDir + '/datavault.db'
conn2 = sq.connect(sDatabaseName)
################################################################
sDataWarehouseDir=Base + '/99-DW'
if not os.path.exists(sDataWarehouseDir):
    os.makedirs(sDataWarehouseDir)
################################################################
sDatabaseName=sDataWarehouseDir + '/datawarehouse.db'
conn3 = sq.connect(sDatabaseName)
################################################################
t=0
tMax=((300-100)/10)*((300-30)/5)
for heightSelect in range(100,300,10):
    for weightSelect in range(30,300,5):
        height = round(heightSelect/100,3)
        weight = int(weightSelect)
        bmi = weight/(height*height)
        if bmi <= 18.5:
            BMI_Result=1
        elif bmi > 18.5 and bmi < 25:
            BMI_Result=2
        elif bmi > 25 and bmi < 30:
            BMI_Result=3
        elif bmi > 30:
            BMI_Result=4
        else:
            BMI_Result=0
```

```
        PersonLine=[('PersonID', [str(t)]),
                    ('Height', [height]),
                    ('Weight', [weight]),
                    ('bmi', [bmi]),
                    ('Indicator', [BMI_Result])]
        t+=1
        print('Row:',t,'of',tMax)
        if t==1:
            PersonFrame = pd.DataFrame.from_items(PersonLine)
        else:
            PersonRow = pd.DataFrame.from_items(PersonLine)
            PersonFrame = PersonFrame.append(PersonRow)
################################################################
DimPerson=PersonFrame
DimPersonIndex=DimPerson.set_index(['PersonID'],inplace=False)
################################################################
sTable = 'Transform-BMI'
print('\n################################')
print('Storing :',sDatabaseName,'\n Table:',sTable)
print('\n################################')
DimPersonIndex.to_sql(sTable, conn1, if_exists="replace")
################################################################
################################################################
sTable = 'Person-Satellite-BMI'
print('\n################################')
print('Storing :',sDatabaseName,'\n Table:',sTable)
print('\n################################')
DimPersonIndex.to_sql(sTable, conn2, if_exists="replace")
################################################################
################################################################
sTable = 'Dim-BMI'
print('\n################################')
print('Storing :',sDatabaseName,'\n Table:',sTable)
print('\n################################')
DimPersonIndex.to_sql(sTable, conn3, if_exists="replace")
################################################################
```

584

```
fig = plt.figure()

PlotPerson=DimPerson[DimPerson['Indicator']==1]
x=PlotPerson['Height']
y=PlotPerson['Weight']
plt.plot(x, y, ".")
PlotPerson=DimPerson[DimPerson['Indicator']==2]
x=PlotPerson['Height']
y=PlotPerson['Weight']
plt.plot(x, y, "o")
PlotPerson=DimPerson[DimPerson['Indicator']==3]
x=PlotPerson['Height']
y=PlotPerson['Weight']
plt.plot(x, y, "+")
PlotPerson=DimPerson[DimPerson['Indicator']==4]
x=PlotPerson['Height']
y=PlotPerson['Weight']
plt.plot(x, y, "^")
plt.axis('tight')
plt.title("BMI Curve")
plt.xlabel("Height(meters)")
plt.ylabel("Weight(kg)")
plt.plot()
```

Now that we have identified the persons at risk, we can study the linear regression of these diabetics.

Note   You will use the standard diabetes data sample set that is installed with the `sklearn` library, the reason being the protection of medical data. As this data is in the public domain, you are permitted to access it.

Warning   When you process people's personal information, you are accountable for any issues your processing causes. So, work with great care.

Note   In the next example, we will use a medical data set that is part of the standard `sklearn` library. This ensures that you are not working with unauthorized medical results.

You set up the ecosystem, as follows:

```
import matplotlib.pyplot as plt
import numpy as np
from sklearn import datasets, linear_model
from sklearn.metrics import mean_squared_error, r2_score
```

Load the data set.

```
# Load the diabetes dataset
diabetes = datasets.load_diabetes()
```

Perform feature development.

```
# Use only one feature
diabetes_X = diabetes.data[:, np.newaxis, 2]
```

Split the data into train and test data sets.

```
diabetes_X_train = diabetes_X[:-30]
diabetes_X_test = diabetes_X[-50:]
```

Split the target into train and test data sets.

```
diabetes_y_train = diabetes.target[:-30]
diabetes_y_test = diabetes.target[-50:]
```

Generate a linear regression model.

```
regr = linear_model.LinearRegression()
```

Train the model using the training sets.

```
regr.fit(diabetes_X_train, diabetes_y_train)
```

Create predictions, using the testing set.

```
diabetes_y_pred = regr.predict(diabetes_X_test)
```

Display the coefficients.

```
print('Coefficients: \n', regr.coef_)
```

Display the mean squared error.

```
print("Mean squared error: %.2f"
      % mean_squared_error(diabetes_y_test, diabetes_y_pred))
```

Display the variance score. (Tip: A score of 1 is perfect prediction.)

```
print('Variance score: %.2f' % r2_score(diabetes_y_test, diabetes_y_pred))
```

Plot outputs.

```
plt.scatter(diabetes_X_test, diabetes_y_test,  color='black')
plt.plot(diabetes_X_test, diabetes_y_pred, color='blue', linewidth=3)

plt.xticks(())
plt.yticks(())

plt.axis('tight')
plt.title("Diabetes")
plt.xlabel("BMI")
plt.ylabel("Age")
plt.show()
```

Well done. You have successfully calculated the BMI and determined the diabetes rate of our staff.

RANSAC Linear Regression

RANSAC (RANdom SAmple Consensus) is an iterative algorithm for the robust estimation of parameters from a subset of inliers from the complete data set. An advantage of RANSAC is its ability to do robust estimation of the model parameters, i.e., it can estimate the parameters with a high degree of accuracy, even when a significant number of outliers is present in the data set. The process will find a solution, because it is so robust.

Generally, this technique is used when dealing with image processing, owing to noise in the domain. See http://scikit-learn.org/stable/modules/generated/sklearn.linear_model.RANSACRegressor.html.

Example:

```
import numpy as np
from matplotlib import pyplot as plt

from sklearn import linear_model, datasets


n_samples = 1000
n_outliers = 50

X, y, coef = datasets.make_regression(n_samples=n_samples, n_features=1,
                                       n_informative=1, noise=10,
                                       coef=True, random_state=0)

# Add outlier data
np.random.seed(0)
X[:n_outliers] = 3 + 0.5 * np.random.normal(size=(n_outliers, 1))
y[:n_outliers] = -3 + 10 * np.random.normal(size=n_outliers)

# Fit line using all data
lr = linear_model.LinearRegression()
lr.fit(X, y)

# Robustly fit linear model with RANSAC algorithm
ransac = linear_model.RANSACRegressor()
ransac.fit(X, y)
inlier_mask = ransac.inlier_mask_
outlier_mask = np.logical_not(inlier_mask)

# Predict data of estimated models
line_X = np.arange(X.min(), X.max())[:, np.newaxis]
line_y = lr.predict(line_X)
line_y_ransac = ransac.predict(line_X)

# Compare estimated coefficients
print("Estimated coefficients (true, linear regression, RANSAC):")
print(coef, lr.coef_, ransac.estimator_.coef_)

lw = 2
plt.scatter(X[inlier_mask], y[inlier_mask], color='yellowgreen', marker='.',
```

```
                label='Inliers')
plt.scatter(X[outlier_mask], y[outlier_mask], color='gold', marker='.',
                label='Outliers')
plt.plot(line_X, line_y, color='navy', linewidth=lw, label='Linear regressor')
plt.plot(line_X, line_y_ransac, color='cornflowerblue', linewidth=lw,
            label='RANSAC regressor')
plt.legend(loc='lower right')
plt.xlabel("Input")
plt.ylabel("Response")
plt.show()
```

This regression technique is extremely useful when using robotics and robot vision in which the robot requires the regression of the changes between two data frames or data sets.

Hough Transform

The Hough transform is a feature extraction technique used in image analysis, computer vision, and digital image processing. The purpose of the technique is to find imperfect instances of objects within a certain class of shapes, by a voting procedure. This voting procedure is carried out in a parameter space, from which object candidates are obtained as local maxima in a so-called accumulator space that is explicitly constructed by the algorithm for computing the Hough transform.

See http://scikit-image.org/docs/dev/api/skimage.transform.html#skimage. transform.hough_line, http://scikit-image.org/docs/dev/api/skimage. transform.html#skimage.transform.hough_line_peaks, and http://scikit-image. org/docs/dev/api/skimage.transform.html#skimage.transform.probabilistic_ hough_line.

With the help of the Hough transformation, this regression improves the resolution of the RANSAC technique, which is extremely useful when using robotics and robot vision in which the robot requires the regression of the changes between two data frames or data sets to move through an environment.

Logistic Regression

Logistic regression is the technique to find relationships between a set of input variables and an output variable (just like any regression), but the output variable, in this case, is a binary outcome (think of 0/1 or yes/no).

Simple Logistic Regression

I will guide you through a simple logistic regression that only compares two values. A real-word business example would be the study of a traffic jam at a certain location in London, using a binary variable. The output is a categorical: yes or no. Hence, is there a traffic jam? Yes or no?

The probability of occurrence of traffic jams can be dependent on attributes such as weather condition, day of the week and month, time of day, number of vehicles, etc. Using logistic regression, you can find the best-fitting model that explains the relationship between independent attributes and traffic jam occurrence rates and predicts probability of jam occurrence.

This process is called binary logistic regression. The state of the traffic changes for No = Zero to Yes = One, by moving along a curve modeled by the following code, is illustrated in Figure 10-10.

```
for x in range(-10,10,1): print(math.sin(x/10))
```

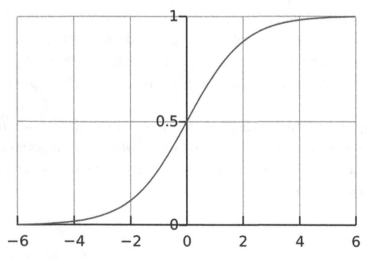

Figure 10-10. *Binary logistic regression*

I will now discuss the logistic regression, using a sample data set.

Example:

```
from sklearn import datasets, neighbors, linear_model
```

Load the data.

```
digits = datasets.load_digits()
X_digits = digits.data
y_digits = digits.target
n_samples = len(X_digits)
```

Select the train data set.

```
X_train = X_digits[:int(.9 * n_samples)]
y_train = y_digits[:int(.9 * n_samples)]
X_test = X_digits[int(.9 * n_samples):]
y_test = y_digits[int(.9 * n_samples):]
```

Select the K-Neighbor classifier.

```
knn = neighbors.KNeighborsClassifier()
```

Select the logistic regression model.

```
logistic = linear_model.LogisticRegression()
```

Train the model to perform logistic regression.

```
print('KNN score: %f' % knn.fit(X_train, y_train).score(X_test, y_test))
```

Apply the trained model against the test data set.

```
print('LogisticRegression score: %f'
      % logistic.fit(X_train, y_train).score(X_test, y_test))
```

Well done. You have just completed your next transform step, by successfully deploying a logistic regression model with a K-Neighbor classifier against the sample data set.

Tip   Using this simple process, I have discovered numerous thought-provoking correlations or busted myths about relationships between data values.

Multinomial Logistic Regression

Multinomial logistic regression (MLR) is a form of linear regression analysis conducted when the dependent variable is nominal with more than two levels. It is used to describe data and to explain the relationship between one dependent nominal variable and one or more continuous-level (interval or ratio scale) independent variables. You can consider the nominal variable as a variable that has no intrinsic ordering.

This type of data is most common in the business world, as it generally covers most data entries within the data sources and directly indicates what you could expect in the average data lake. The data has no intrinsic order or relationship.

I will now guide you through the example, to show you how to deal with these data sets. You need the following libraries for the ecosystem:

```
import time
import matplotlib.pyplot as plt
import numpy as np
from sklearn.datasets import fetch_mldata
from sklearn.linear_model import LogisticRegression
from sklearn.model_selection import train_test_split
from sklearn.preprocessing import StandardScaler
from sklearn.utils import check_random_state
```

Set up the results target.

```
sPicNameOut='../VKHCG/01-Vermeulen/04-Transform/01-EDS/02-Python/Letters.
png'
```

You must tune a few parameters.

```
t0 = time.time()
train_samples = 5000
```

Get the sample data from simulated data lake.

```
mnist = fetch_mldata('MNIST original')
```

Data engineer the data lake data.

```
X = mnist.data.astype('float64')
y = mnist.target
```

```
random_state = check_random_state(0)
permutation = random_state.permutation(X.shape[0])
X = X[permutation]
y = y[permutation]
X = X.reshape((X.shape[0], -1))
```

Train the data model.

```
X_train, X_test, y_train, y_test = train_test_split(
    X, y, train_size=train_samples, test_size=10000)
```

Apply a scaler to training to inhibit overfitting.

```
scaler = StandardScaler()
X_train = scaler.fit_transform(X_train)
X_test = scaler.transform(X_test)
```

Turn the tolerance (tol = 0.1) for faster convergence.

```
clf = LogisticRegression(C=50. / train_samples,
                         multi_class='multinomial',
                         penalty='l2', solver='sag', tol=0.1)
```

Apply the model to the data set.

```
clf.fit(X_train, y_train)
sparsity = np.mean(clf.coef_ == 0) * 100
```

Score the model.

```
score = clf.score(X_test, y_test)
print('Best C % .4f' % clf.C_)
print("Sparsity with L1 penalty: %.2f%%" % sparsity)
print("Test score with L1 penalty: %.4f" % score)

coef = clf.coef_.copy()
```

Display the results.

```
Fig=plt.figure(figsize=(15, 6))
scale = np.abs(coef).max()
for i in range(10):
```

```
    l1_plot = plt.subplot(2, 5, i + 1)
    l1_plot.imshow(coef[i].reshape(28, 28), interpolation='nearest',
                cmap=plt.cm.RdBu, vmin=-scale, vmax=scale)
    l1_plot.set_xticks(())
    l1_plot.set_yticks(())
    l1_plot.set_xlabel('Letter %i' % i)
plt.suptitle('Classification vector for...')

run_time = time.time() - t0
print('Process run in %.3f s' % run_time)
plt.show()
```

Save results to disk.

```
Fig.savefig(sPicNameOut,dpi=300)
```

You're performing well with the examples. Now, you can handle data that has no intrinsic order. If you understand this process, you have successfully achieved a major milestone in your understanding of the transformation of data lakes via data vaults, by using a Transform step.

Tip I normally store any results back into the data warehouse as a sun model, which then gets physicalized as facts and dimensions in the data warehouse.

Ordinal Logistic Regression

Ordinal logistic regression is a type of binomial logistics regression. Ordinal regression is used to predict the dependent variable with ordered multiple categories and independent variables. In other words, it is used to facilitate the interaction of dependent variables (having multiple ordered levels) with one or more independent variables.

This data type is an extremely good data set to process, as you already have a relationship between the data entries that is known. Deploying your Transform step's algorithms will give you insights into how strongly or weakly this relationship supports the data discovery process.

Business Problem

Rate a student on a scale from 1 to 5, to determine if he or she has the requisite qualifications ("prestige") to join the university. The rating introduces a preference or likelihood that these items can be separated by a scale. I will take you through a data science solution, using a set of generated potential students. The question is "What criteria drives the final decision?"

Here is the example:

Open your Python editor and do the following.

Set up the ecosystem.

```
import sys
import os
import pandas as pd
import statsmodels.api as sm
import pylab as pl
import numpy as np

if sys.platform == 'linux':
    Base=os.path.expanduser('~') + '/VKHCG'
else:
    Base='C:/VKHCG'
```

Retrieve StudentData.

```
sFileName=Base + '/01-Vermeulen/00-RawData/StudentData.csv'
StudentFrame = pd.read_csv(sFileName,header=0)
StudentFrame.columns = ["sname", "gre", "gpa", "prestige","admit","QR","VR"
,"gpatrue"]
StudentSelect=StudentFrame[["admit", "gre", "gpa", "prestige"]]
print('Record select:',StudentSelect.shape[0])
df=StudentSelect
I add the following two lines if you want to speed-up the processing, but
it does make the predictions less accurate. So use it if you want.
#df=StudentSelect.drop_duplicates(subset=None, keep='first', inplace=False)
#print('Records Unique:', df.shape[0])
```

Here are the columns for the data set:

```
print(df.columns)
```

Here is a description of the data profile:

```
print(df.describe())
```

Did you spot the interesting "skewness" in the admit profile? You can investigate using this command:

```
print(df.std())
```

Note   The admit standard deviation is at +/-0.2, out of a range of 0 to 1.

---

admit	0.201710
gre	1.030166
gpa	145.214383
prestige	1.030166

This is a very selective admissions model! Not a lot of students make the cut. Let's investigate the influence prestige has on being selected.

```
print(pd.crosstab(df['admit'], df['prestige'], rownames=['admit']))
```

prestige	1	2	3	4
admit				
0	16810	15810	15810	7905
1	0	1000	1000	500

There is clearly a criterion related to prestige. A student requires a minimum of Prestige = 2. Can you see this also? If you create a plot of all the columns, you will see it better. I suggest that we perform a binning reduction for your plot, by using a histogram function, as follows:

```
df.hist()
```

Here are your results:

```
pl.tight_layout()
pl.show()
```

You should now create a rank for prestige.

```
dummy_ranks = pd.get_dummies(df['prestige'], prefix='prestige')
print(dummy_ranks.head())
```

You now have a grid that you can populate with the correct values, as per your data.
You need to create an empty data frame for the regression analysis as follows:

```
cols_to_keep = ['admit', 'gre', 'gpa']
data = df[cols_to_keep].join(dummy_ranks.loc[:, 'prestige_1':])
print(data.head())
```

Can you see how the data now transforms? This is called feature engineering.
Next, you must add the intercept value.

```
data['intercept'] = 1.00
```

You have successfully transformed the data into a structure the algorithm
understands. Well done!
You can now train the model.

```
train_cols = data.columns[1:]
logit = sm.Logit(data['admit'], data[train_cols])
```

You now have a trained model.
You can fit the model for the data set. Note that the maxiter = 500 drives the
processing time. You could increase or reduce it, to control the processing effort of the
model. Experiment on what works for you.

```
result = logit.fit(maxiter=500)
```

You can now investigate the results.

```
print('Results')
print(result.summary())
```

Note   The interesting discovery is that GPA (grade point average) score determines admittance.

You will now investigate at the confidence interval of each coefficient.

```
print(result.conf_int())
```

The model identifies GPA. Well done, as that is the true driver: nRank = int(GPA), a.k.a. prestige.

You can now investigate the odds ratios.

```
print(np.exp(result.params))
```

Can you identify the trend?

	Odds	Odds(Real)
gre	1.02E-02	0.010
gpa	1.06E+00	1.063
prestige_1	1.13E-18	0.000
prestige_2	2.17E-01	0.217
prestige_3	2.14E+01	21.376
prestige_4	2.10E+03	2101.849

The trend is that if a student has a GPA of > 4, he or she has 2100 to 1 odds of being admitted.

Let's investigate these odds ratios and 95% CI.

```
params = result.params
conf = result.conf_int()
conf['OR'] = params
conf.columns = ['2.5%', '97.5%', 'OR']
print(np.exp(conf))
```

Can you spot the true nature of the process? GPA is the main driver with higher GPA securing admittance.

You can now generate all possible values of GRE (Graduate Record Examinations) and GPA, but I suggest you only use an evenly spaced range of ten values, from the minimum to the maximum. This a simple way of binning data to reduce complexity. Transforming the story results to a smaller result set, while keeping the core findings, is a skill a data scientist requires.

Here are your bins:

```
gres = np.linspace(data['gre'].min(), data['gre'].max(), 10)
print(gres)
gpas = np.linspace(data['gpa'].min(), data['gpa'].max(), 10)
print(gpas)
```

Now you must perform some more data engineering. You have to define a Cartesian function to assist with your transform process.

```
def cartesian(arrays, out=None):
    arrays = [np.asarray(x) for x in arrays]
    dtype = arrays[0].dtype

    n = np.prod([x.size for x in arrays])
    if out is None:
        out = np.zeros([n, len(arrays)], dtype=dtype)

    m = int(n / arrays[0].size)
    out[:,0] = np.repeat(arrays[0], m)
    if arrays[1:]:
        cartesian(arrays[1:], out=out[0:m,1:])
        for j in range(1, arrays[0].size):
            out[j*m:(j+1)*m,1:] = out[0:m,1:]
    return out
```

Now that you have your new function, I suggest you apply it to your data, as follows: You must enumerate all possibilities for the investigation.

```
combos = pd.DataFrame(cartesian([gres, gpas, [1, 2, 3, 4], [1,]]))
```

Now you re-create the dummy variables.

```
combos.columns = ['gre', 'gpa', 'prestige', 'intercept']
dummy_ranks = pd.get_dummies(combos['prestige'], prefix='prestige')
dummy_ranks.columns = ['prestige_1', 'prestige_2', 'prestige_3', 'prestige_4']
```

I now suggest that you keep only what you need for making the predictions.

```
cols_to_keep = ['gre', 'gpa', 'prestige', 'intercept']
combos = combos[cols_to_keep].join(dummy_ranks.loc[:, 'prestige_1':])
```

Now calculate the predictions on the enumerated data set.

```
combos['admit_pred'] = result.predict(combos[train_cols])
```

Here are your results:

```
print(combos.head())
```

You need a further function. This function will isolate and plot specific characteristics of the data set.

```
def isolate_and_plot(variable):
    # isolate gre and class rank
    grouped = pd.pivot_table(combos, values=['admit_pred'],
index=[variable, 'prestige'],
                  aggfunc=np.mean)
    # make a plot
    colors = 'rbgyrbgy'
    for col in combos.prestige.unique():
        plt_data = grouped.loc[grouped.index.get_level_values(1)==col]
        pl.plot(plt_data.index.get_level_values(0), \
        plt_data['admit_pred'], color=colors[int(col)])
    pl.xlabel(variable)
    pl.ylabel("P(admit=1)")
    pl.legend(['1', '2', '3', '4'], loc='upper left', title='Prestige')
    pl.title("Prob(admit=1) isolating " + variable + " and presitge")
    pl.show()
```

Now that you have your function, I suggest you investigate these two values:

```
isolate_and_plot('gre')
isolate_and_plot('gpa')
```

The story is simple. If your grades (Grade Point Average [GPA]) is low, your Graduate Record Examinations (GRE) (capability) can get you into the university. But a high GPA indicates a successful ambition to achieve success, as the capability is assumed by the university.

Congratulations! You have completed the ordinal logistic regression section.

Warning   Ordinal logistic regression requires massive training data sets, as they have the extra requirement of ordering the steps.

Clustering Techniques

Clustering (or segmentation) is a kind of unsupervised learning algorithm, in which a data set is grouped into unique, differentiated clusters. Let's say we have customer data spanning 1000 rows. Using clustering, we can group the customers into separate clusters or segments, based on the variables. In the case of customers' data, the variables can be demographic information or purchasing activities.

Clustering is an unsupervised learning algorithm, because the input is unknown to the data scientist as no training set is available to pre-train a model of the solution. You do not train the algorithm on any past input–output information, but you let the algorithm define the output for itself. Therefore, there is no right solution to a clustering algorithm, only a reasonably best-fit solution, based on business usability. Clustering is also known as unsupervised classification.

There are two basic types of clustering techniques.

Hierarchical Clustering

Hierarchical clustering is a method of cluster analysis whereby you build a hierarchy of clusters. This works well for data sets that are complex and have distinct characteristics for separated clusters of data.

The following would be an example. People on a budget are more attracted to your sale items and multi-buy combinations, while more prosperous shoppers are more brand-orientated. These are two clearly different clusters, with poles-apart driving forces to buy an item.

There are two design styles

Agglomerative

This is a bottom-up approach. Each observation starts in its own cluster, and pairs of clusters are merged as one moves up the hierarchy.

Divisive

This is a top-down approach. All observations start in one cluster, and splits are performed recursively as one moves down the hierarchy. I will take you through the transform process to generate a hierarchical cluster.

Note   Clustering is applied in the field of biostatistics to assess cluster-based models of DNA sequences. Mapping the human genome was one of humanity's greatest scientific breakthroughs, but data science is busy taking it to the next level. Investigate bioinformatics's applications for clusters.

Open your Python editor and set up a new ecosystem.

```
from matplotlib import pyplot as plt
from scipy.cluster.hierarchy import dendrogram, linkage
import numpy as np
```

You will now generate two clusters: (a), with 100 points, and (b), with 50.

```
np.random.seed(4711)
a = np.random.multivariate_normal([10, 0], [[3, 1], [1, 4]], size=[100,])
b = np.random.multivariate_normal([0, 20], [[3, 1], [1, 4]], size=[50,])
X = np.concatenate((a, b),)
```

This creates 150 samples with 2 dimensions.

```
print( X.shape)
```

Let's quickly display it.

```
plt.scatter(X[:,0], X[:,1])
plt.show()
```

You must generate the linkage matrix. The matrix contains the linkage criterion that determines the distance between sets of observations as a function of the pairwise distances between observations. For more information on this matrix, see https://docs.scipy.org/doc/scipy-0.19.1/reference/generated/scipy.cluster.hierarchy.linkage.html.

```
Z = linkage(X, 'ward')
```

Enhance the ecosystem, by adding extra libraries.

```
from scipy.cluster.hierarchy import cophenet
from scipy.spatial.distance import pdist
```

Trigger the cophenetic correlation coefficient. The cophenetic correlation coefficient is a measure of how faithfully a dendrogram preserves the pairwise distances between the original unmodeled data points. In simple terms, how accurate is the measure.

```
c, coph_dists = cophenet(Z, pdist(X))
print('Cophenetic Correlation Coefficient:',c)
```

You will now calculate a full dendrogram.

```
plt.figure(figsize=(25, 10))
plt.title('Hierarchical Clustering Dendrogram')
plt.xlabel('Sample Index')
plt.ylabel('Distance')
dendrogram(
    Z,
    leaf_rotation=90,
    leaf_font_size=8,
)
plt.show()
```

Now, you can truncate the cluster (show only the last p merged clusters).

```
plt.title('Hierarchical Clustering Dendrogram (truncated)')
plt.xlabel('sample index')
plt.ylabel('distance')
dendrogram(
    Z,
```

```
    truncate_mode='lastp',
    p=12,
    show_leaf_counts=False,
    leaf_rotation=90,
    leaf_font_size=12,
    show_contracted=True,
)
plt.show()
```

You now must define a new function, to improve the diagram's display.

```
def fancy_dendrogram(*args, **kwargs):
    max_d = kwargs.pop('max_d', None)
    if max_d and 'color_threshold' not in kwargs:
        kwargs['color_threshold'] = max_d
    annotate_above = kwargs.pop('annotate_above', 0)
    ddata = dendrogram(*args, **kwargs)
    if not kwargs.get('no_plot', False):
        plt.title('Hierarchical Clustering Dendrogram (truncated)')
        plt.xlabel('sample index or (cluster size)')
        plt.ylabel('distance')
        for i, d, c in zip(ddata['icoord'], ddata['dcoord'], ddata
        ['color_list']):
            x = 0.5 * sum(i[1:3])
            y = d[1]
            if y > annotate_above:
                plt.plot(x, y, 'o', c=c)
                plt.annotate("%.3g" % y, (x, y), xytext=(0, -5),
                             textcoords='offset points',
                             va='top', ha='center')
        if max_d:
            plt.axhline(y=max_d, c='k')
    return ddata
```

You can now use the new function against your clusters.

```
fancy_dendrogram(
    Z,
```

```
    truncate_mode='lastp',
    p=12,
    leaf_rotation=90.,
    leaf_font_size=12.,
    show_contracted=True,
    annotate_above=10,  # useful in small plots so annotations don't overlap
)
plt.show()
```

Let's set the cutoff to 50.

```
max_d = 50
```

Now, you just replot the new data.

```
fancy_dendrogram(
    Z,
    truncate_mode='lastp',
    p=12,
    leaf_rotation=90.,
    leaf_font_size=12.,
    show_contracted=True,
    annotate_above=10,
    max_d=max_d,
)
plt.show()
```

Change the cut to 16.

```
fancy_dendrogram(
    Z,
    truncate_mode='lastp',
    p=12,
    leaf_rotation=90.,
    leaf_font_size=12.,
    show_contracted=True,
    annotate_above=10,
    max_d=16,
)
plt.show()
```

Now, you add extra functions to investigate your transformation.

```
from scipy.cluster.hierarchy import inconsistent
You can investigate at a depth of five?
depth = 5
incons = inconsistent(Z, depth)
print(incons[-10:])
```

What are you seeing?

Move to depth of three.

```
depth = 3
incons = inconsistent(Z, depth)
print(incons[-10:])
```

What do you see? You will see it better with a graph.

```
last = Z[-10:, 2]
last_rev = last[::-1]
idxs = np.arange(1, len(last) + 1)
plt.plot(idxs, last_rev)
```

You should now look at the acceleration.

```
acceleration = np.diff(last, 2)
acceleration_rev = acceleration[::-1]
plt.plot(idxs[:-2] + 1, acceleration_rev)
plt.show()
k = acceleration_rev.argmax() + 2  # if idx 0 is the max of this we want 2
clusters
print ("Clusters:", k)
c = np.random.multivariate_normal([40, 40], [[20, 1], [1, 30]],
size=[200,])
d = np.random.multivariate_normal([80, 80], [[30, 1], [1, 30]], size=[200,])
e = np.random.multivariate_normal([0, 100], [[100, 1], [1, 100]], size=[200,])
X2 = np.concatenate((X, c, d, e),)
plt.scatter(X2[:,0], X2[:,1])
plt.show()
```

Can you see the clusters? Here is a proper cluster diagram:

```
Z2 = linkage(X2, 'ward')
plt.figure(figsize=(10,10))
fancy_dendrogram(
    Z2,
    truncate_mode='lastp',
    p=30,
    leaf_rotation=90.,
    leaf_font_size=12.,
    show_contracted=True,
    annotate_above=40,
    max_d=170,
)
plt.show()
Well done you can now see the clusters.
```

Let's look at the data in more detail.

```
last = Z2[-10:, 2]
last_rev = last[::-1]
idxs = np.arange(1, len(last) + 1)
plt.plot(idxs, last_rev)
```

You can now perform more analysis.

```
acceleration = np.diff(last, 2)  # 2nd derivative of the distances
acceleration_rev = acceleration[::-1]
plt.plot(idxs[:-2] + 1, acceleration_rev)
plt.show()
k = acceleration_rev.argmax() + 2  # if idx 0 is the max of this we want 2
clusters
print ("Clusters:", k)
print (inconsistent(Z2, 5)[-10:])
```

Let's look at an F-cluster.

```
from scipy.cluster.hierarchy import fcluster
max_d = 50
clusters = fcluster(Z, max_d, criterion='distance')
print(clusters)
```

Can you see the clusters?

```
k=2
fcluster(Z, k, criterion='maxclust')
```

And now can you spot the clusters? It is not that easy to spot the clusters, is it? Let's try a different angle.

```
from scipy.cluster.hierarchy import fcluster
fcluster(Z, 8, depth=10)

plt.figure(figsize=(10, 8))
plt.scatter(X[:,0], X[:,1], c=clusters, cmap='winter'
plt.show()
```

You can see them now!

Note   Visualize insights. People understand graphics much more easily. Columned data sets require explaining and additional contextual information. In Chapter 11, I will discuss how to report your insights in an easy but effective manner.

You have successfully completed hierarchical clustering. This should enable you to understand that there are numerous subsets of clusters that can be combined to form bigger cluster structures.

Partitional Clustering

A partitional clustering is simply a division of the set of data objects into non-overlapping subsets (clusters), such that each data object is in exactly one subset. Remember when you were at school? During breaks, when you played games, you could only belong to either the blue team or the red team. If you forgot which team was yours, the game normally ended in disaster!

Open your Python editor, and let's perform a transformation to demonstrate how you can create a proper partitional clustering solution. As always, you will require the ecosystem.

```python
import numpy as np
from sklearn.cluster import DBSCAN
from sklearn import metrics
from sklearn.datasets.samples_generator import make_blobs
from sklearn.preprocessing import StandardScaler
```

You can generate some sample data, as follows:

```python
centers = [[4, 4], [-4, -4], [4, -4],[6,0],[0,0]]
X, labels_true = make_blobs(n_samples=750, centers=centers, cluster_std=0.5,
                            random_state=0)
```

Let's apply a scaler transform.

```python
X = StandardScaler().fit_transform(X)
```

You can now apply the DBSCAN transformation.

```python
db = DBSCAN(eps=0.3, min_samples=10).fit(X)
core_samples_mask = np.zeros_like(db.labels_, dtype=bool)
core_samples_mask[db.core_sample_indices_] = True
labels = db.labels_
```

Select the number of clusters in labels. You can ignore noise, if present.

```python
n_clusters_ = len(set(labels)) - (1 if -1 in labels else 0)
Here are your findings:
print('Estimated number of clusters: %d' % n_clusters_)
print("Homogeneity: %0.3f" % metrics.homogeneity_score(labels_true, labels))
print("Completeness: %0.3f" % metrics.completeness_score(labels_true, labels))
print("V-measure: %0.3f" % metrics.v_measure_score(labels_true, labels))
print("Adjusted Rand Index: %0.3f"
      % metrics.adjusted_rand_score(labels_true, labels))
print("Adjusted Mutual Information: %0.3f"
      % metrics.adjusted_mutual_info_score(labels_true, labels))
print("Silhouette Coefficient: %0.3f"
      % metrics.silhouette_score(X, labels))
```

You can also plot it. Remember: Graphics explains better.

```
import matplotlib.pyplot as plt
unique_labels = set(labels)
colors = [plt.cm.Spectral(each)
          for each in np.linspace(0, 1, len(unique_labels))]
for k, col in zip(unique_labels, colors):
    if k == -1:
        col = [0, 0, 0, 1]

    class_member_mask = (labels == k)

    xy = X[class_member_mask & core_samples_mask]
    plt.plot(xy[:, 0], xy[:, 1], 'o', markerfacecolor=tuple(col),
             markeredgecolor='k', markersize=14)

    xy = X[class_member_mask & ~core_samples_mask]
    plt.plot(xy[:, 0], xy[:, 1], 'o', markerfacecolor=tuple(col),
             markeredgecolor='k', markersize=6)

plt.title('Estimated number of clusters: %d' % n_clusters_)
plt.show()
```

If you see this layout, well done. You can perform partitional clustering. This type of clustering is commonly used, as it places any data entry in distinct clusters without overlay (see Figure 10-11).

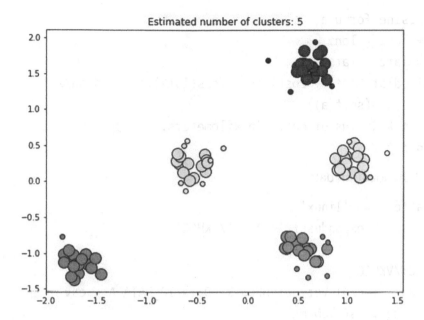

Figure 10-11. *Partitional clustering*

Open a new file in the Python editor and try the following location cluster. Set up the ecosystem.

```
import sys
import os
from math import radians, cos, sin, asin, sqrt

from scipy.spatial.distance import pdist, squareform
from sklearn.cluster import DBSCAN
import matplotlib.pyplot as plt
import pandas as pd
```

Remember this distance measures function? It calculates the great circle distance between two points on the earth, when specified in decimal degrees.

```
def haversine(lonlat1, lonlat2):
    # convert decimal degrees to radians
    lat1, lon1 = lonlat1
    lat2, lon2 = lonlat2
    lon1, lat1, lon2, lat2 = map(radians, [lon1, lat1, lon2, lat2])
```

```
    # haversine formula
    dlon = lon2 - lon1
    dlat = lat2 - lat1
    a = sin(dlat/2)**2 + cos(lat1) * cos(lat2) * sin(dlon/2)**2
    c = 2 * asin(sqrt(a))
    r = 6371 # Radius of earth in kilometers. Use 3956 for miles
    return c * r
```

Let's get the location data.

```
if sys.platform == 'linux':
    Base=os.path.expanduser('~') + '/VKHCG'
else:
    Base='C:/VKHCG'
sFileName=Base + '/01-Vermeulen/00-RawData/IP_DATA_ALL.csv'
print('Loading :',sFileName)
scolumns=("Latitude","Longitude")
RawData = pd.read_csv(sFileName,usecols=scolumns,header=0,low_memory=False)
print(RawData)
```

You now need a sample of 1000 data points.

```
X=RawData.sample(n=1000)
```

Now calculate the cluster distance parameter.

```
distance_matrix = squareform(pdist(X, (lambda u,v: haversine(u,v))))
print(distance_matrix)
```

Next, apply the clustering.

```
db = DBSCAN(eps=0.2, min_samples=2, metric='precomputed', algorithm='auto')
y_db = db.fit_predict(distance_matrix)
X['cluster'] = y_db
C = X.cluster.unique()
```

Let's visualize the data.

```
fig=plt.figure(1, figsize=(20, 20))
plt.title('Estimated number of clusters: %d' % len(C))
plt.scatter(X['Latitude'], X['Longitude'], c=X['cluster'],marker='D')
plt.show()
```

Now save your results.

```
sImageFile = Base + '/01-Vermeulen/04-Transform/01-EDS/02-Python/Location_
Cluster.jpg'
fig.savefig(sImageFile)
plt.close(fig)
```

You should see a layout similar to Figure 10-12.

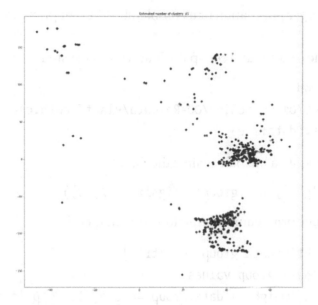

Figure 10-12. *Location cluster*

Note   The latent Dirichlet allocation (LDA), discussed earlier in this chapter, would be classed as a cluster algorithm and should be mentioned among other clustering algorithms.

You have completed the partitional clustering solution.

ANOVA

The one-way analysis of variance (ANOVA) test is used to determine whether the mean of more than two groups of data sets is significantly different from each data set.

Example:

A BOGOF (buy-one-get-one-free) campaign is executed on 5 groups of 100 customers each. Each group is different in terms of its demographic attributes. We would like to determine whether these five respond differently to the campaign. This would help us optimize the right campaign for the right demographic group, increase the response rate, and reduce the cost of the campaign.

The analysis of variance works by comparing the variance between the groups to that within the group. The core of this technique lies in assessing whether all the groups are in fact part of one larger population or a completely different population with different characteristics.

Open your Python editor and set up the following ecosystem:

```
import pandas as pd
datafile='../VKHCG/01-Vermeulen/00-RawData/PlantGrowth.csv'
data = pd.read_csv(datafile)
```

Now you must create a boxplot against the data.

```
data.boxplot('weight', by='group', figsize=(12, 8))
```

You must now perform feature extraction and engineering.

```
ctrl = data['weight'][data.group == 'ctrl']
grps = pd.unique(data.group.values)
d_data = {grp:data['weight'][data.group == grp] for grp in grps}
k = len(pd.unique(data.group))  # number of conditions
N = len(data.values)  # conditions times participants
n = data.groupby('group').size()[0] #Participants in each condition
You now need extra funtions from extra library:
from scipy import stats
```

Now activate the one-way ANOVA test for the null hypothesis that two or more groups have the same population mean transformation.

```
F, p = stats.f_oneway(d_data['ctrl'], d_data['trt1'], d_data['trt2'])
```

You need to set up a few parameters.

```
DFbetween = k - 1
DFwithin = N - k
DFtotal = N - 1
```

You can now further investigate the results from the transformation.

```
SSbetween = (sum(data.groupby('group').sum()['weight']**2)/n) \
    - (data['weight'].sum()**2)/N
sum_y_squared = sum([value**2 for value in data['weight'].values])
SSwithin = sum_y_squared - sum(data.groupby('group').sum()['weight']**2)/n
SStotal = sum_y_squared - (data['weight'].sum()**2)/N
MSbetween = SSbetween/DFbetween
MSwithin = SSwithin/DFwithin
F = MSbetween/MSwithin
eta_sqrd = SSbetween/SStotal
omega_sqrd = (SSbetween - (DFbetween * MSwithin))/(SStotal + MSwithin)
```

Here are the results of your investigation:

```
print(F,p,eta_sqrd,omega_sqrd)
```

Well done. You have performed a successful one-way ANOVA test.

Principal Component Analysis (PCA)

Dimension (variable) reduction techniques aim to reduce a data set with higher dimension to one of lower dimension, without the loss of features of information that are conveyed by the data set. The dimension here can be conceived as the number of variables that data sets contain.

Two commonly used variable reduction techniques follow.

Factor Analysis

The crux of PCA lies in measuring the data from the perspective of a principal component. A principal component of a data set is the direction with the largest variance. A PCA analysis involves rotating the axis of each variable to the highest Eigen vector/Eigen value pair and defining the principal components, i.e., the highest variance axis or, in other words, the direction that most defines the data. Principal components are uncorrelated and orthogonal.

PCA is fundamentally a dimensionality reduction algorithm, but it is just as useful as a tool for visualization, for noise filtering, for feature extraction, and engineering. Here is an example.

Open your Python editor and set up this ecosystem:

```
import matplotlib.pyplot as plt
from mpl_toolkits.mplot3d import Axes3D
from sklearn import datasets
from sklearn.decomposition import PCA
```

Import some data to apply your skills against.

```
iris = datasets.load_iris()
```

Take only the first two features.

```
X = iris.data[:, :2]
```

You need the following target:

```
y = iris.target
x_min, x_max = X[:, 0].min() - .5, X[:, 0].max() + .5
y_min, y_max = X[:, 1].min() - .5, X[:, 1].max() + .5
```

You can quickly visualize the data set.

```
plt.clf()
plt.figure(2, figsize=(8, 6))
```

Now plot the training points.

```
plt.scatter(X[:, 0], X[:, 1], c=y, cmap=plt.cm.Set1,
            edgecolor='k')
plt.xlabel('Sepal length')
plt.ylabel('Sepal width')
plt.xlim(x_min, x_max)
plt.ylim(y_min, y_max)
plt.xticks(())
plt.yticks(())
```

I suggest getting a better understanding of the interaction of the dimensions. Plot the first three PCA dimensions.

```
fig = plt.figure(1, figsize=(8, 6))
ax = Axes3D(fig, elev=-150, azim=110)
X_reduced = PCA(n_components=3).fit_transform(iris.data)
ax.scatter(X_reduced[:, 0], X_reduced[:, 1], X_reduced[:, 2], c=y,
           cmap=plt.cm.Set1, edgecolor='k', s=40)
ax.set_title("First three PCA directions")
ax.set_xlabel("1st eigenvector")
ax.w_xaxis.set_ticklabels([])
ax.set_ylabel("2nd eigenvector")
ax.w_yaxis.set_ticklabels([])
ax.set_zlabel("3rd eigenvector")
ax.w_zaxis.set_ticklabels([])
plt.show()
```

Can you see the advantage of reducing the features in a controlled manner? Let's deploy your new knowledge against a more complex data set.

Example:

Hillman Ltd has introduced a CCTV tracking process on the main gates, as it suspects unauthorized offloading of goods is happening through its harbor port's warehouse. You will load a set of digits from the cameras, to assist the company to read the numbers on the side of the containers that left the warehouse.

Open your Python editor and set up this ecosystem:

```
from sklearn.datasets import load_digits
from sklearn.decomposition import PCA
import matplotlib.pyplot as plt
import numpy as np
```

Load the digits from the preprocessed image software.

```
digits = load_digits()
```

You now apply the transformation to project from 64 to 2 dimensions.

```
pca = PCA(2)
projected = pca.fit_transform(digits.data)
print(digits.data.shape)
print(projected.shape)
```

Display your findings.

```
plt.figure(figsize=(10, 15))
plt.scatter(projected[:, 0], projected[:, 1],
            c=digits.target, edgecolor='none', alpha=0.5,
            cmap=plt.cm.get_cmap('nipy_spectral_r', 10))
plt.xlabel('Value 1')
plt.ylabel('Value 2')
plt.colorbar()
```

Apply the PCA transform.

```
pca = PCA().fit(digits.data)
plt.plot(np.cumsum(pca.explained_variance_ratio_))
plt.xlabel('Number of components')
plt.ylabel('Cumulative explained variance');
```

You will require the following function to plot the digits:

```
def plot_digits(data):
    fig, axes = plt.subplots(4, 10, figsize=(10, 4),
                             subplot_kw={'xticks':[], 'yticks':[]},
                             gridspec_kw=dict(hspace=0.1, wspace=0.1))
    for i, ax in enumerate(axes.flat):
        ax.imshow(data[i].reshape(8, 8),
                  cmap='binary', interpolation='nearest',
                  clim=(0, 16))
```

You can now plot results.

```
plot_digits(digits.data)
```

One of the cameras, however, has not generated perfect images.

```
np.random.seed(42)
noisy = np.random.normal(digits.data, 4)
plot_digits(noisy)
```

You can still use the data; it just needs more processing.

```
pca = PCA(0.50).fit(noisy)
pca.n_components_
```

You can determine the distortion by filtering the noise factors.

```
components = pca.transform(noisy)
filtered = pca.inverse_transform(components)
Visualize your results:
plot_digits(filtered)
```

Question   Can you see the digits from the cameras?

Do you think we have a workable solution?

You have now proven that you can take a complex 64-dimension data set, and using PCA, reduce it to 2 dimensions and still achieve good data science results. Good progress on your side.

Conjoint Analysis

Conjoint analysis is widely used in market research to identify customers' preference for various attributes that make up a product. The attributes can be various features, such as size, color, usability, price, etc.

Using conjoint (trade-off) analysis, brand managers can identify which features customers would trade off for a certain price point. Thus, such analysis is a highly used technique in new product design or pricing strategies.

Example:

The data is a ranking of three different features (TV size, TV type, TV color).

TV size options are 42", 47", or 60".

TV type options are LCD or Plasma.

TV color options are red, blue, or pink.

The data rates the different stimuli types for each customer. You are tasked with determining which TVs to display on Krennwallner's billboards.

Open your Python editor and set up the following ecosystem:

```
import numpy as np
import pandas as pd
```

Retrieve the customer buy choices.

```
sFile='C:/VKHCG/01-Vermeulen/00-RawData/BuyChoices.txt'
caInputeDF = pd.read_csv(sFile, sep = ";")
```

You can display the choices.

```
print(caInputeDF)
```

You need a new data structure, and you must create dummy variables for every stimulus. In total, you require 9 different stimuli and 18 different combinations.

```
ConjointDummyDF = pd.DataFrame(np.zeros((18,9)), columns=["Rank","A1",
"A2", "A3","B1","B2","C1", "C2","C3"])
You now need to feature engineer the data choices into the new structure:
ConjointDummyDF.Rank = caInputeDF.Rank
for index, row in caInputeDF.iterrows():
    stimuli1 = str(caInputeDF["Stimulus"].iloc[index][:2])
    stimuli2 = str(caInputeDF["Stimulus"].iloc[index][2:4])
    stimuli3 = str(caInputeDF["Stimulus"].iloc[index][4:6])
    stimuliLine=[stimuli1,stimuli2,stimuli3]
    Columns=ConjointDummyDF.columns
    for stimuli in stimuliLine:
        for i in range(len(Columns)):
            if stimuli == Columns[i]:
                ConjointDummyDF.iloc[index, i] = 1
```

Well done. Let's look at your new structure.

```
print(ConjointDummyDF.head())
```

You only have unknown numbers. I suggest you add suitable stimulus names. Be mindful of the Transform step you normally present to businesspeople. They understand plasma TV is not good with TVType3.

```
fullNames = {"Rank":"Rank", "A1": "42\" (106cm)","A2": "47\" (120cm)","A3":
"60\" (152cm)","B1": "Plasma","B2":"LCD","C1":"Blue","C2":"Red","C3":
"Pink",}
ConjointDummyDF.rename(columns=fullNames, inplace=True)
```

That's better. Now look at the data structure.

```
#ConjointDummyDF.head()
```

Next, you estimate the main effects with a linear regression.
You need extra libraries to achieve this action.

```
import statsmodels.api as sm
```

You can select any of these columns for your model.

```
print(ConjointDummyDF.columns)
```

I suggest you select these:

```
X = ConjointDummyDF[[u'42" (106cm)', u'47" (120cm)',\
 u'60" (152cm)', u'Plasma', u'LCD', u'Red', u'Blue', u'Pink']]
```

You now need to set up the model.

```
X = sm.add_constant(X)
Y = ConjointDummyDF.Rank
```

You now activate the model against your new data structure.

```
linearRegression = sm.OLS(Y, X).fit()
```

You can now look at the results.

```
print(linearRegression.summary())
```

There are numerous indicators that you can research. I am not going to cover any of these in detail, as several require complex mathematics to explain them fully. The indicators we are interested in are the part worth values and relative importance of the stimuli: Importance of Stimuli = Max(beta) - Min(beta)

Relative Importance of Stimuli = Importance of Stim/ Sum(Importance of All Stimuli)

You will now investigate these indicators. To do this, you need some data engineering. You now require several basic indicator variables to start the process.

```
importance = []
relative_importance = []
rangePerFeature = []
begin = "A"
tempRange = []
```

You now need to load the stimuli.

```
for stimuli in fullNames.keys():
    if stimuli[0] == begin:
        tempRange.append(linearRegression.params[fullNames[stimuli]])
    elif stimuli == "Rank":
        rangePerFeature.append(tempRange)
    else:
        rangePerFeature.append(tempRange)
        begin = stimuli[0]
        tempRange = [linearRegression.params[fullNames[stimuli]]]
```

Then, you need the feature ranges.

```
for item in rangePerFeature:
    importance.append( max(item) - min(item))
```

You then calculate the importance of a feature.

```
for item in importance:
    relative_importance.append(100* round(item/sum(importance),3))
```

622

Start a base data structure for the part worth values.

```
partworths = []
item_levels = [1,3,5,8]
```

You now calculate the part worth values.

```
for i in range(1,4):
    part_worth_range = linearRegression.params[item_levels[i-1]:item_
levels[i]]
    print (part_worth_range)
```

You need to determine the mean rank of the data set.

```
meanRank = []
for i in ConjointDummyDF.columns[1:]:
    newmeanRank = ConjointDummyDF["Rank"].loc[ConjointDummyDF[i] ==
1].mean()
    meanRank.append(newmeanRank)
```

I suggest you use total mean known as "basic utility" to be used as the "zero alternative."

```
totalMeanRank = sum(meanRank) / len(meanRank)
```

You will now rank the value of each part of the stimuli, i.e., what features of the TV are important?

```
partWorths = {}
for i in range(len(meanRank)):
    name = fullNames[sorted(fullNames.keys())[i]]
    partWorths[name] = meanRank[i] - totalMeanRank
```

You now have a result set.

```
print(partWorths)
```

Now, I will help you to develop the results into insights and deliver the basis for a summary and results report.

```
print ("Relative Importance of Feature:\n\nMonitor Size:",relative_
importance[0], "%","\nType of Monitor:", relative_importance[1], "%", "\
nColor of TV:", relative_importance[2], "%\n\n")
print ("--"*30)
print ("Importance of Feature:\n\nMonitor Size:",importance[0],"\nType of
Monitor:", importance[1],  "\nColor of TV:", importance[2])
```

So, what is the optimal product bundle? I suggest 60", LCD, Red. Do you agree? Let's test that assumed fact.

```
optBundle = [1,0,0,1,0,1,0,1,0]
print ("The best possible Combination of Stimuli would have the highest
rank:",linearRegression.predict(optBundle)[0])
optimalWorth = partWorths["60\" (152cm)"] + partWorths["LCD"] +
partWorths["Red"]
print ("Choosing the optimal Combination brings the Customer an extra ",
optimalWorth, "'units' of utility")
```

Does your science support my assumed fact?

Congratulations, you have just completed a conjoint analysis.

Tip I suggest you try the equivalent analysis with other data sets, as this is a common question I am asked. What makes people pick one product and not another? You should practice with data sets of all sizes, to understand how the formulas react to changes in the stimuli factors.

I have confidence that there is a future for you at Krennwallner as a data scientist.

Decision Trees

Decision trees, as the name suggests, are a tree-shaped visual representation of routes you can follow to reach a particular decision, by laying down all options and their probability of occurrence. Decision trees are exceptionally easy to understand and interpret. At each node of the tree, one can interpret what would be the consequence of selecting that node or option. The series of decisions leads you to the end result, as shown in Figure 10-13.

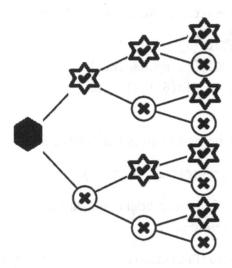

Figure 10-13. *Simple decision tree*

Before you start the example, I must discuss a common add-on algorithm to decision trees called AdaBoost. AdaBoost, short for "adaptive boosting," is a machine-learning meta-algorithm.

The classifier is a meta-estimator, because it begins by fitting a classifier on the original data set and then fits additional copies of the classifier on the same data set, but where the weights of incorrectly classified instances are adjusted, such that subsequent classifiers focus more on difficult cases. It boosts the learning impact of less clear differences in the specific variable, by adding a progressive weight to boost the impact. See http://scikit-learn.org/stable/modules/generated/sklearn.ensemble. AdaBoostClassifier.html.

This boosting enables the data scientist to force decisions down unclear decision routes for specific data entities, to enhance the outcome.

Example of a Decision Tree using AdaBoost:

Open your Python editor and set up this ecosystem:

```
import numpy as np
import matplotlib.pyplot as plt
from sklearn.tree import DecisionTreeRegressor
from sklearn.ensemble import AdaBoostRegressor
```

You need a data set. So, you can build a random data set.

```
rng = np.random.RandomState(1)
X = np.linspace(0, 6, 1000)[:, np.newaxis]
y = np.sin(X).ravel() + np.sin(6 * X).ravel() + rng.normal(0, 0.1,
X.shape[0])
```

You then fit the normal regression model to the data set.

```
regr_1 = DecisionTreeRegressor(max_depth=4)
```

You then also apply an AdaBoost model to the data set. The parameters can be changed, if you want to experiment.

```
regr_2 = AdaBoostRegressor(DecisionTreeRegressor(max_depth=4),
                        n_estimators=300, random_state=rng)
```

You then train the model.

```
regr_1.fit(X, y)
regr_2.fit(X, y)
```

You activate the predict.

```
y_1 = regr_1.predict(X)
y_2 = regr_2.predict(X)
```

You plot the results.

```
plt.figure(figsize=(15, 10))
plt.scatter(X, y, c="k", label="Training Samples")
plt.plot(X, y_1, c="g", label="n_Estimators=1", linewidth=2)
plt.plot(X, y_2, c="r", label="n_Estimators=300", linewidth=2)
plt.xlabel("Data")
plt.ylabel("Target")
plt.title("Boosted Decision Tree Regression")
plt.legend()
plt.show()
```

Congratulations! You just built a decision tree. You can now test your new skills against a more complex example.

Once more, open your Python editor and set up this ecosystem:

```python
import numpy as np
import matplotlib.pyplot as plt
from sklearn.ensemble import AdaBoostClassifier
from sklearn.tree import DecisionTreeClassifier
from sklearn.datasets import make_gaussian_quantiles
```

You are building a more complex data set.

```python
X1, y1 = make_gaussian_quantiles(cov=2.,
                                 n_samples=2000, n_features=2,
                                 n_classes=2, random_state=1)
X2, y2 = make_gaussian_quantiles(mean=(3, 3), cov=1.5,
                                 n_samples=3000, n_features=2,
                                 n_classes=2, random_state=1)
X = np.concatenate((X1, X2))
y = np.concatenate((y1, - y2 + 1))
```

You now have a data set ready. Create and fit an AdaBoosted decision tree to the data set.

```python
bdt = AdaBoostClassifier(DecisionTreeClassifier(max_depth=1),
                         algorithm="SAMME",
                         n_estimators=200)
```

You train the model.

```python
bdt.fit(X, y)
```

You now need to set up a few parameters that you will require.

```python
plot_colors = "br"
plot_step = 0.02
class_names = "AB"
```

You now create a visualization of the results.

```python
plt.figure(figsize=(10, 5))
```

Add the plot for the decision boundaries.

627

```
plt.subplot(121)
x_min, x_max = X[:, 0].min() - 1, X[:, 0].max() + 1
y_min, y_max = X[:, 1].min() - 1, X[:, 1].max() + 1
xx, yy = np.meshgrid(np.arange(x_min, x_max, plot_step),
                     np.arange(y_min, y_max, plot_step))

Z = bdt.predict(np.c_[xx.ravel(), yy.ravel()])
Z = Z.reshape(xx.shape)
cs = plt.contourf(xx, yy, Z, cmap=plt.cm.Paired)
plt.axis("tight")
```

Plot your training points.

```
for i, n, c in zip(range(2), class_names, plot_colors):
    idx = np.where(y == i)
    plt.scatter(X[idx, 0], X[idx, 1],
                c=c, cmap=plt.cm.Paired,
                s=20, edgecolor='k',
                label="Class %s" % n)
plt.xlim(x_min, x_max)
plt.ylim(y_min, y_max)
plt.legend(loc='upper right')
plt.xlabel('x')
plt.ylabel('y')
plt.title('Decision Boundary')
```

Plot the two-class decision scores.

```
twoclass_output = bdt.decision_function(X)
plot_range = (twoclass_output.min(), twoclass_output.max())
plt.subplot(122)
for i, n, c in zip(range(2), class_names, plot_colors):
    plt.hist(twoclass_output[y == i],
             bins=10,
             range=plot_range,
             facecolor=c,
             label='Class %s' % n,
             alpha=.5,
```

```
            edgecolor='k')
x1, x2, y1, y2 = plt.axis()
plt.axis((x1, x2, y1, y2 * 1.2))
plt.legend(loc='upper right')
plt.ylabel('Samples')
plt.xlabel('Score')
plt.title('Decision Scores')

plt.tight_layout()
plt.subplots_adjust(wspace=0.35)
plt.show()
```

Well done, you have just completed a complex decision tree.

Support Vector Machines, Networks, Clusters, and Grids

The support vector machine (SVM) constructs a hyperplane or set of hyperplanes in a high- or infinite-dimensional space, which can be used for classification and regression. The support vector network (SVN) daisy chains more than one SVM together to form a network. All the data flows through the same series of SVMs.

The support vector cluster (SVC) runs SVM on different clusters of the data in parallel. Hence, not all data flows through all the SVMs.

The support vector grid (SVG) is an SVC of an SVN or an SVN of an SVC. This solution is the most likely configuration you will develop at a customer site. It uses SVMs to handle smaller clusters of the data, to apply specific transform steps. As a beginner data scientist, you only need to note that they exist.

Support Vector Machines

A support vector machine is a discriminative classifier formally defined by a separating hyperplane. The method calculates an optimal hyperplane (Figure 10-14) with a maximum margin, to ensure it classifies the data set into separate clusters of data points.

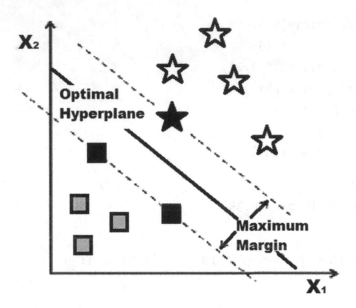

Figure 10-14. *Simple support vector machine's optimal hyperplane*

I will guide you through a sample SVM. Open your Python editor and create this ecosystem:

```
import numpy as np
import matplotlib.pyplot as plt
from sklearn import svm
```

Here is your data set and targets.

```
X = np.c_[(.4, -.7),
          (-1.5, -1),
          (-1.4, -.9),
          (-1.3, -1.2),
          (-1.1, -.2),
          (-1.2, -.4),
          (-.5, 1.2),
          (-1.5, 2.1),
          (1, 1),

          (1.3, .8),
          (1.2, .5),
          (.2, -2),
```

```
                  (.5, -2.4),
                  (.2, -2.3),
                  (0, -2.7),
                  (1.3, 2.1)].T
Y = [0] * 8 + [1] * 8
```

You display the results as Figure Number 1.

```
fignum = 1
```

You have several kernels you can use to fit the data model. I will take you through all of them.

```
for kernel in ('linear', 'poly', 'rbf'):
    clf = svm.SVC(kernel=kernel, gamma=2)
    clf.fit(X, Y)
```

You now plot the line, the points, and the nearest vectors to the plane.

```
    plt.figure(fignum, figsize=(8, 6))
    plt.clf()

    plt.scatter(clf.support_vectors_[:, 0], clf.support_vectors_[:, 1],
s=80,
                facecolors='none', zorder=10, edgecolors='k')
    plt.scatter(X[:, 0], X[:, 1], c=Y, zorder=10, cmap=plt.cm.Paired,
                edgecolors='k')

    plt.axis('tight')
    x_min = -3
    x_max = 3
    y_min = -3
    y_max = 3

    XX, YY = np.mgrid[x_min:x_max:200j, y_min:y_max:200j]
    Z = clf.decision_function(np.c_[XX.ravel(), YY.ravel()])
```

You now apply the result into a color plot.

```
    Z = Z.reshape(XX.shape)
    plt.figure(fignum, figsize=(8, 6))
    plt.pcolormesh(XX, YY, Z > 0, cmap=plt.cm.Paired)
    plt.contour(XX, YY, Z, colors=['k', 'k', 'k'], linestyles=['--', '-',
'--'],
                levels=[-.5, 0, .5])

    plt.xlim(x_min, x_max)
    plt.ylim(y_min, y_max)

    plt.xticks(())
    plt.yticks(())
    fignum = fignum + 1
```

You now show your plots.

```
plt.show()
```

Well done. You have just completed three different types of SVMs.

Support Vector Networks

The support vector network is the ensemble of a network of support vector machines that together classify the same data set, by using different parameters or even different kernels. This a common method of creating feature engineering, by creating a chain of SVMs.

Tip   You can change kernels on the same flow to expose new features. Practice against the data sets with the different kernels, to understand what they give you.

Support Vector Clustering

Support vector clustering is used were the data points are classified into clusters, with support vector machines performing the classification at the cluster level. This is commonly used in highly dimensional data sets, where the clustering creates a grouping that can then be exploited by the SVM to subdivide the data points, using different kernels and other parameters.

I have seen SVC, SVN, SVM, and SVG process in many of the deep-learning algorithms that I work with every day. The volume, variety, and velocity of the data require that the deep learning do multistage classifications, to enable the more detailed analysis of the data points to occur after a primary result is published.

Data Mining

Data mining is processing data to pinpoint patterns and establish relationships between data entities. Here are a small number of critical data mining theories you need to understand about data patterns, to be successful with data mining.

Association Patterns

This involves detecting patterns in which one occasion is associated to another. If, for example, a loading bay door is opened, it is fair to assume that a truck is loading goods. Pattern associations simply discover the correlation of occasions in the data. You will use some core statistical skills for this processing.

Warning   "Correlation does not imply causation."

Correlation is only a relationship or indication of behavior between two data sets. The relationship is not a cause-driven action.

Example:

If you discover a relationship between hot weather and ice cream sales, it does not mean high ice cream sales cause hot weather, or vice versa. It is only an observed relationship.

This is commonly used in retail basket analysis and recommender systems. I will discuss this causation factor later, in Chapter 11, when we start to prepare our insights and actionable knowledge extractions. I will guide you through an example now.

Please open your Python editor and create this ecosystem:

```
import pandas as pd
df1 = pd.DataFrame({'A': range(8), 'B': [2*i for i in range(8)]})
df2 = pd.DataFrame({'A': range(8), 'B': [-2*i for i in range(8)]})
```

Here is your data.

```
print('Positive Data Set')
print(df1)
print('Negative Data Set')
print(df2)
```

Here are your results.

```
print('Results')
print('Correlation Positive:', df1['A'].corr(df1['B']))
print('Correlation Negative:', df2['A'].corr(df2['B']))
```

You should see a correlation of either +1 or -1. If it is +1, this means there is a 100% correlation between the two values, hence they change at the same rate. If it is -1, this means there is a 100% negative correlation, indicating that the two values have a relationship if one increases while the other decreases.

Tip   In real-world data sets, such two extremes as +1 and -1 do not occur. The range is normally i-1 < C < +1.

You will now apply changes that will interfere with the correlation.

```
df1.loc[2, 'B'] = 10
df2.loc[2, 'B'] = -10
```

You can now see the impact.

```
print('Positive Data Set')
print(df1)
print('Negative Data Set')
print(df2)
print('Results')
print('Correlation Positive:', df1['A'].corr(df1['B']))
print('Correlation Negative:', df2['A'].corr(df2['B']))
```

Can you sense the impact a minor change has on the model?

So, let's add a bigger change.

```
df1.loc[3, 'B'] = 100
df2.loc[3, 'B'] = -100
```

You check the impact.

```
print('Positive Data Set')
print(df1)
print('Negative Data Set')
print(df2)
print('Results')
print('Correlation Positive:', df1['A'].corr(df1['B']))
print('Correlation Negative:', df2['A'].corr(df2['B']))
```

Well done, if you understood the changes that interfere with the relationship. If you see the relationship, you have achieved an understanding of association patterns.

These "What if?" analyses are common among data scientists' daily tasks. For example, one such analysis might relate to the following: What happens when I remove 300 of my 1000 staff? And at what point does the removal of staff have an impact? These small-step increases of simulated impact can be used to simulate progressive planned changes in the future.

Classification Patterns

This technique discovers new patterns in the data, to enhance the quality of the complete data set. Data classification is the process of consolidating data into categories, for its most effective and efficient use by the data processing. For example, if the data is related to the shipping department, you must then augment a label on the data that states that fact.

A carefully planned data-classification system creates vital data structures that are easy to find and retrieve. You do not want to scour your complete data lake to find data every time you want to analyze a new data pattern.

Clustering Patterns

Clustering is the discovery and labeling of groups of specifics not previously known. An example of clustering is if, when your customers buy bread and milk together on a Monday night, you group, or cluster, these customers as "start-of-the-week small-size shoppers," by simply looking at their typical basic shopping basket.

Any combination of variables that you can use to cluster data entries into a specific group can be viewed as some form of clustering. For data scientists, the following clustering types are beneficial to master.

Connectivity-Based Clustering

You can discover the interaction between data items by studying the connections between them. This process is sometimes also described as hierarchical clustering.

Centroid-Based Clustering (K-Means Clustering)

"Centroid-based" describes the cluster as a relationship between data entries and a virtual center point in the data set. K-means clustering is the most popular centroid-based clustering algorithm. I will guide you through an example.

Open your Python editor and set up this ecosystem:

```
import numpy as np
import matplotlib.pyplot as plt
from mpl_toolkits.mplot3d import Axes3D
from sklearn.cluster import KMeans
from sklearn import datasets
```

You now set up a data set to study.

```
np.random.seed(5)
iris = datasets.load_iris()
X = iris.data
y = iris.target
```

Configure your estimators for the clustering.

```
estimators = [('k_means_iris_8', KMeans(n_clusters=8)),
              ('k_means_iris_3', KMeans(n_clusters=3)),
              ('k_means_iris_bad_init', KMeans(n_clusters=3, n_init=1,
                                        init='random'))]
```

Get ready to virtualize your results.

```
fignum = 1
titles = ['8 clusters', '3 clusters', '3 clusters, bad initialization']
```

```
for name, est in estimators:
    fig = plt.figure(fignum, figsize=(4, 3))
    ax = Axes3D(fig, rect=[0, 0, .95, 1], elev=48, azim=134)
    est.fit(X)
    labels = est.labels_
    ax.scatter(X[:, 3], X[:, 0], X[:, 2],
                c=labels.astype(np.float), edgecolor='k')
    ax.w_xaxis.set_ticklabels([])
    ax.w_yaxis.set_ticklabels([])
    ax.w_zaxis.set_ticklabels([])
    ax.set_xlabel('Petal width')
    ax.set_ylabel('Sepal length')
    ax.set_zlabel('Petal length')
    ax.set_title(titles[fignum - 1])
    ax.dist = 12
    fignum = fignum + 1
```

Plot your results.

```
fig = plt.figure(fignum, figsize=(4, 3))
ax = Axes3D(fig, rect=[0, 0, .95, 1], elev=48, azim=134)
for name, label in [('Setosa', 0),
                    ('Versicolour', 1),
                    ('Virginica', 2)]:
    ax.text3D(X[y == label, 3].mean(),
            X[y == label, 0].mean(),
            X[y == label, 2].mean() + 2, name,
            horizontalalignment='center',
            bbox=dict(alpha=.2, edgecolor='w', facecolor='w'))
```

Reorder the labels to have colors matching the cluster results.

```
y = np.choose(y, [1, 2, 0]).astype(np.float)
ax.scatter(X[:, 3], X[:, 0], X[:, 2], c=y, edgecolor='k')
ax.w_xaxis.set_ticklabels([])
ax.w_yaxis.set_ticklabels([])
ax.w_zaxis.set_ticklabels([])
```

```
ax.set_xlabel('Petal width')
ax.set_ylabel('Sepal length')
ax.set_zlabel('Petal length')
ax.set_title('Ground Truth')
ax.dist = 12
fig.show()
```

Well done. You should see a set of results for your hard work. Can you see how the connectivity-based clustering uses the interaction between data points to solve the transformation steps?

Distribution-Based Clustering

This type of clustering model relates data sets with statistics onto distribution models. The most widespread density-based clustering technique is DBSCAN.

Density-Based Clustering

In density-based clustering, an area of higher density is separated from the remainder of the data set. Data entries in sparse areas are placed in separate clusters. These clusters are considered to be noise, outliers, and border data entries.

Grid-Based Method

Grid-based approaches are common for mining large multidimensional space clusters having denser regions than their surroundings. The grid-based clustering approach differs from the conventional clustering algorithms in that it does not use the data points but a value space that surrounds the data points.

Bayesian Classification

Naive Bayes (NB) classifiers are a group of probabilistic classifiers established by applying Bayes's theorem with strong independence assumptions between the features of the data set. There is one more specific Bayesian classification you must take note of, and it is called tree augmented naive Bayes (TAN).

Tree augmented naive Bayes is a semi-naive Bayesian learning method. It relaxes the naive Bayes attribute independence assumption by assigning a tree structure, in which each attribute only depends on the class and one other attribute. A maximum weighted spanning tree that maximizes the likelihood of the training data is used to perform classification. The naive Bayesian (NB) classifier and the tree augmented naive Bayes (TAN) classifier are well-known models that will be discussed next. Here is an example:

Open your Python editor and start with this ecosystem.

```
import numpy as np
import urllib
```

Load data via the web interface.

```
url = "http://archive.ics.uci.edu/ml/machine-learning-databases/pima-
indians-diabetes/pima-indians-diabetes.data"
```

Download the file.

```
raw_data = urllib.request.urlopen(url)
```

Load the CSV file into a numpy matrix.

```
dataset = np.loadtxt(raw_data, delimiter=",")
```

Separate the data from the target attributes in the data set.

```
X = dataset[:,0:8]
y = dataset[:,8]
```

Add extra processing capacity.

```
from sklearn import metrics
from sklearn.naive_bayes import GaussianNB
model = GaussianNB()
model.fit(X, y)
print(model)
```

Produce predictions.

```
expected = y
predicted = model.predict(X)
# summarize the fit of the model
print(metrics.classification_report(expected, predicted))
print(metrics.confusion_matrix(expected, predicted))
```

You have just built your naive Bayes classifiers. Good progress!

Sequence or Path Analysis

This identifies patterns in which one event leads to another, later resulting in insights into the business. Path analysis is a chain of consecutive events that a given business entity performs during a set period, to understand behavior, in order to gain actionable insights into the data.

I suggest you use a combination of tools to handle this type of analysis. I normally model the sequence or path with the help of a graph database, or, for smaller projects, I use a library called networkx in Python.

Example:

Your local telecommunications company is interested in understanding the reasons or flow of events that resulted in people churning their telephone plans to their competitor.

Open your Python editor and set up this ecosystem.

```
################################################################
# -*- coding: utf-8 -*-
################################################################
import sys
import os
import pandas as pd
import sqlite3 as sq
import networkx as nx
import datetime
pd.options.mode.chained_assignment = None
################################################################
if sys.platform == 'linux':
    Base=os.path.expanduser('~') + '/VKHCG'
else:
    Base='C:/VKHCG'
print('################################')
print('Working Base :',Base, ' using ', sys.platform)
print('################################')
################################################################
```

```
Company='01-Vermeulen'
###################################################################
sDataBaseDir=Base + '/' + Company + '/04-Transform/SQLite'
if not os.path.exists(sDataBaseDir):
    os.makedirs(sDataBaseDir)
###################################################################
sDatabaseName=sDataBaseDir + '/Vermeulen.db'
conn = sq.connect(sDatabaseName)
###################################################################
```

You must create a new graph to track the individual paths of each customer through their journey over the last nine months.

```
G=nx.Graph()
###################################################################
```

The following loop structure enables you to compare two sequential months and determine the changes:

```
for M in range(1,10):
    print('Month: ', M)
    MIn = str(M - 1)
    MOut = str(M)
    sFile0 = 'ConnectionsChurn' + MIn + '.csv'
    sFile1 = 'ConnectionsChurn' + MOut + '.csv'
    sTable0 = 'ConnectionsChurn' + MIn
    sTable1 = 'ConnectionsChurn' + MOut

    sFileName0=Base + '/' + Company + '/00-RawData/' + sFile0
    ChurnData0=pd.read_csv(sFileName0,header=0,low_memory=False,
    encoding="latin-1")

    sFileName1=Base + '/' + Company + '/00-RawData/' + sFile1
    ChurnData1=pd.read_csv(sFileName1,header=0,low_memory=False,
    encoding="latin-1")

    ###############################################################
```

Owing to an error during extraction, the files dates are not correct, so you perform a data-quality correction.

```
dt1 = datetime.datetime(year=2017, month=1, day=1)
dt2 = datetime.datetime(year=2017, month=2, day=1)
ChurnData0['Date'] = dt1.strftime('%Y/%m/%d')
ChurnData1['Date'] = dt2.strftime('%Y/%m/%d')
################################################################
```

You now compare all the relevant features of the customer, to see what actions were taken during the month under investigation.

```
TrackColumns=['SeniorCitizen', 'Partner',
        'Dependents', 'PhoneService', 'MultipleLines',
        'OnlineSecurity', 'OnlineBackup', 'DeviceProtection',
'TechSupport',
        'StreamingTV', 'PaperlessBilling', 'StreamingMovies',
'InternetService',
        'Contract', 'PaymentMethod', 'MonthlyCharges']
################################################################
for i in range(ChurnData0.shape[0]):
    for TColumn in TrackColumns:
        t=0
        if ChurnData0[TColumn][i] == 'No':
            t+=1
        if t > 4:
            ChurnData0['Churn'][i] = 'Yes'
        else:
            ChurnData0['Churn'][i] = 'No'
################################################################
for i in range(ChurnData1.shape[0]):
    for TColumn in TrackColumns:
        t=0
        if ChurnData1[TColumn][i] == 'No':
            t+=1
        if t > 4:
            ChurnData1['Churn'][i] = 'Yes'
```

```
        else:
            ChurnData1['Churn'][i] = 'No'
####################################################################
print('Store CSV Data')
ChurnData0.to_csv(sFileName0, index=False)
ChurnData1.to_csv(sFileName1, index=False)
####################################################################
print('Store SQLite Data')
ChurnData0.to_sql(sTable0, conn, if_exists='replace')
ChurnData1.to_sql(sTable1, conn, if_exists='replace')
####################################################################
for TColumn in TrackColumns:
    for i in range(ChurnData0.shape[0]):
```

You always start with a "Root" node and then attach the customers. Then you connect each change in status and perform a complete path analysis.

```
        Node0 = 'Root'
        Node1 = '(' + ChurnData0['customerID'][i] + '-Start)'
        G.add_edge(Node0, Node1)
        Node5 = '(' + ChurnData0['customerID'][i] + '-Stop)'
        G.add_node(Node5)

        if ChurnData0['Churn'][i] == 'Yes':
            NodeA = '(' + ChurnData0['customerID'][i] + '-Start)'
            NodeB = '(' + ChurnData0['customerID'][i] + '-Stop)'
            if nx.has_path(G, source=NodeA, target=NodeB) == False:
                NodeC = '(' + ChurnData0['customerID'][i] + '):
                (Churn)=>(' + ChurnData1['Churn'][i] + ')'
                G.add_edge(NodeA, NodeC)
                G.add_edge(NodeC, NodeB)
        else:
            if ChurnData0[TColumn][i] != ChurnData1[TColumn][i]:
                #print(M,ChurnData0['customerID'][i],ChurnData0['Date']
                [i],ChurnData1['Date'][i],TColumn, ChurnData0[TColumn]
                [i], ChurnData1[TColumn][i])
```

```
                         Node2 = '(' + ChurnData0['customerID'][i] + ')-
                         (' + ChurnData0['Date'][i] + ')'
                         G.add_edge(Node1, Node2)
                         Node3 = Node2 + '-(' + TColumn + ')'
                         G.add_edge(Node2, Node3)
                         Node4 = Node3 + ':(' + ChurnData0[TColumn][i] + ')=>
                         (' + ChurnData1[TColumn][i] + ')'
                         G.add_edge(Node3, Node4)

                         if M == 9:
                             Node6 = '(' + ChurnData0['customerID'][i] + '):
                             (Churn)=>(' + ChurnData1['Churn'][i] + ')'
                             G.add_edge(Node4, Node6)
                             G.add_edge(Node6, Node5)
                         else:
                             G.add_edge(Node4, Node5)
```

You can use these lines to investigate the nodes and the edges you created.

```
for n in G.nodes():
    print(n)

for e in G.edges():
    print(e)
```

You must now store your graph for future use.

```
sGraphOutput=Base + '/' + Company + \
    '/04-Transform/01-EDS/02-Python/Transform_ConnectionsChurn_Graph.gml.gz'
nx.write_gml(G, sGraphOutput)
```

You now investigate the paths taken by a customer over the nine months and produce an output of all the steps taken by the customer over the period.

```
sFile0 = 'ConnectionsChurn9.csv'
sFileName0=Base + '/' + Company + '/00-RawData/' + sFile0
ChurnData0=pd.read_csv(sFileName0,header=0,low_memory=False,
encoding="latin-1")
c=0
```

```
for i in range(ChurnData0.shape[0]):
    sCustomer = ChurnData0['customerID'][i]
    NodeX = '(' + ChurnData0['customerID'][i] + '-Start)'
    NodeY = '(' + ChurnData0['customerID'][i] + '-Stop)'
    if nx.has_path(G, source=NodeX, target=NodeY) == False:
        NodeZ = '(' + ChurnData0['customerID'][i] + '):(Churn)=>
        (' + ChurnData0['Churn'][i] + ')'
        G.add_edge(NodeX, NodeZ)
        G.add_edge(NodeZ, NodeX)

    if nx.has_path(G, source=NodeX, target=NodeY) == True:
```

This function enables you to expose all the paths between the two nodes you created for each customer.

```
            pset = nx.all_shortest_paths(G, source=NodeX, target=NodeY,
            weight=None)
            t=0
            for p in pset:
                t=0
                ps = 'Path: ' + str(p)
                for s in p:
                    c+=1
                    t+=1
                    ts = 'Step: ' + str(t)
                    #print(NodeX, NodeY, ps, ts, s)
                    if c == 1:
                        pl = [[sCustomer, ps, ts, s]]
                    else:
                        pl.append([sCustomer, ps, ts, s])
```

You now store the path analysis results into a CSV for later use.

```
sFileOutput=Base + '/' + Company + \
    '/04-Transform/01-EDS/02-Python/Transform_ConnectionsChurn.csv'
df = pd.DataFrame(pl, columns=['Customer', 'Path', 'Step', 'StepName'])
df.index.name = 'RecID'
```

```
sTable = 'ConnectionsChurnPaths'
df.to_sql(sTable, conn, if_exists='replace')
df.to_csv(sFileOutput)
################################################################
print('### Done!! ####################################')
################################################################
```

Well done. You have just completed your first path analysis over a period of nine months.

Insights:

Can you understand why the routes are different for each customer but still have the same outcome? Can you explain why one customer churns and another does not?

The common activity you will spot is that customers begin to remove services from your client as they start to churn, to enable their own new churned services, they nolonger will support you services 100%. If they have fewer than five services, customers normally churn, as they are now likely with the other telecommunications company.

If you were advising the client, I would suggest you highlight that the optimum configuration for the trigger that a customer is about to churn is if the customer has changed his or her configuration to include fewer services from your client. You can still prevent a churn, if you intervene before the customer hits the five-minimum level.

Well done. You can now perform Transform steps for path analysis, get a simple list of node and edges (relationships) between them, and model a graph and ask pertinent questions.

Forecasting

This technique is used to discover patterns in data that result in practical predictions about a future result, as indicated, by predictive analytics of future probabilities and trends. We have been performing forecasting at several points in the book.

Pattern Recognition

Pattern recognition identifies regularities and irregularities in data sets. The most common application of this is in text analysis, to find complex patterns in the data.

I will guide you through an example for text extraction. The example will extract text files from a common 20 newsgroups data set and then create categories of text that was found together in the same document. This will provide you with the most common word that is used in the newsgroups.

Open your Python editor and set up this ecosystem:

```
from pprint import pprint
from time import time
import logging
from sklearn.datasets import fetch_20newsgroups
from sklearn.feature_extraction.text import CountVectorizer
from sklearn.feature_extraction.text import TfidfTransformer
from sklearn.linear_model import SGDClassifier
from sklearn.model_selection import GridSearchCV
from sklearn.pipeline import Pipeline
```

Modify your logging, to display progress logs on stdout (standard output).

```
logging.basicConfig(level=logging.INFO,
                    format='%(asctime)s %(levelname)s %(message)s')
You can now load your categories from the training set
categories = [
    'sci.space',
    'comp.sys.ibm.pc.hardware',
    'sci.electronics',
    'sci.crypt',
    'rec.autos',
    'rec.motorcycles',
]
```

You could use this for all the categories, but I would perform it as a later experiment, as it slows down your processing, by using larger volumes of data from your data lake.

```
#categories = None
```

You can now investigate what you loaded.

```
print("Loading 20 newsgroups dataset for categories:")
print(categories)
```

You must now load the training data.

```
data = fetch_20newsgroups(subset='train', categories=categories)
print("%d documents" % len(data.filenames))
print("%d categories" % len(data.target_names))
```

You now have to define a pipeline, combining a text feature extractor with a simple classifier.

```
pipeline = Pipeline([
    ('vect', CountVectorizer()),
    ('tfidf', TfidfTransformer()),
    ('clf', SGDClassifier()),
])
```

Warning   Uncommenting more of the following parameters will provide improved exploring power but will increase processing time in a combinatorial way.

```
parameters = {
    'vect__max_df': (0.5, 0.75, 1.0),
    #'vect__max_features': (None, 5000, 10000, 50000),
    'vect__ngram_range': ((1, 1), (1, 2)),
    #'tfidf__use_idf': (True, False),
    #'tfidf__norm': ('l1', 'l2'),
    'clf__alpha': (0.00001, 0.000001),
    'clf__penalty': ('l2', 'elasticnet'),
    #'clf__n_iter': (10, 50, 80),
}
```

You can now build the main processing engine.

```
if __name__ == "__main__":
    grid_search = GridSearchCV(pipeline, parameters, n_jobs=-1, verbose=1)

    print("Performing grid search...")
    print("pipeline:", [name for name, _ in pipeline.steps])
    print("parameters:")
```

```
pprint(parameters)
t0 = time()
grid_search.fit(data.data, data.target)
print("done in %0.3fs" % (time() - t0))
print()

print("Best score: %0.3f" % grid_search.best_score_)
print("Best parameters set:")
best_parameters = grid_search.best_estimator_.get_params()
for param_name in sorted(parameters.keys()):
    print("\t%s: %r" % (param_name, best_parameters[param_name]))
```

When you execute the program, you will have successfully completed a text extraction. The data sources for this type of extract are typically documents, e-mail, Twitter, or note fields in databases. Any test source can receive a transform step.

Machine Learning

The business world is bursting with activities and philosophies about machine learning and its application to various business environments. Machine learning is the capability of systems to learn without explicit software development. It evolved from the study of pattern recognition and computational learning theory.

The impact is that with the appropriate processing and skills, you can amplify your own data capabilities, by training a processing environment to accomplish massive amounts of discovery of data into actionable knowledge, while you have a cup of coffee, for example. This skill is an essential part of achieving major gains in shortening the data-to-knowledge cycle.

I will cover a limited rudimentary theory, but machine learning encompasses a wide area of expertise that merits a book by itself. So, I will introduce you only to the core theories.

Supervised Learning

Supervised learning is the machine-learning task of inferring a function from labeled training data. The training data consists of a set of training examples. In supervised learning, each example is a pair consisting of an input object and a desired output value. You use this when you know the required outcome for a set of input features.

649

Example:

If you buy bread and jam, you can make a jam sandwich. Without either, you have no jam sandwich.

If you investigate this data set, you can easily spot from the indicators what is bread and what is jam. A data science model could perform the same task, using supervised learning.

BaseFoodName	FillingName	IsBread	IsJam	IsJamSandwich
Aish Merahrah	Apple Pie	1	1	1
Aish Merahrah	Apple & Flowering Quince	1	1	1
Ajdov Kruh	Apple Pie	1	1	1
Ajdov Kruh	Apple & Flowering Quince	1	1	1
Aish Merahrah	Almogrote	1	0	0
Aish Merahrah	Alouette cheese	1	0	0
Ajdov Kruh	Almogrote	1	0	0
Ajdov Kruh	Alouette cheese	1	0	0
Angel cake	Apple Pie	0	1	0
Angel cake	Apple & Flowering Quince	0	1	0
Angel food cake	Apple Pie	0	1	0
Angel food cake	Apple & Flowering Quince	0	1	0
Angel cake	Almogrote	0	0	0
Angel cake	Alouette cheese	0	0	0
Angel food cake	Almogrote	0	0	0
Angel food cake	Alouette cheese	0	0	0

If you investigate the bigger data set stored in directory .. \VKHCG\03-Hillman\00-RawData in file KitchenData.csv, you can see how the preceding knowledge can be used to infer what the correct selection is for items that will result in a jam sandwich.

Unsupervised Learning

Unsupervised learning is the machine-learning task of inferring a function to describe hidden structures from unlabeled data. This encompasses many other techniques that seek to summarize and explain key features of the data.

Example:

You can take a bag of marble with different colors, sizes, and materials and split them into three equal groups, by applying a set of features and a model. You do not know up front what the criteria are for splitting the marbles.

Reinforcement Learning

Reinforcement learning (RL) is an area of machine learning inspired by behavioral psychology that is concerned with how software agents should take actions in an environment, so as to maximize, more or less, a notion of cumulative reward. This is used in several different areas, such as game theory, swarm intelligence, control theory, operations research, simulation-based optimization, multi-agent systems, statistics, and genetic algorithms (Figure 10-15).

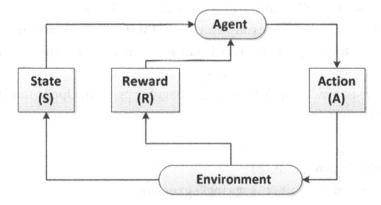

Figure 10-15. *Reinforced learning diagram*

The process is simple. Your agent extracts features from the environment that are either "state" or "reward." State features indicate that something has happened. Reward features indicate that something happened that has improved or worsened to the perceived gain in reward. The agent uses the state and reward to determine actions to change the environment.

This process of extracting state and reward, plus responding with action, will continue until a pre-agreed end reward is achieved. The real-world application for these types of reinforced learning is endless. You can apply reinforced learning to any environment which you can control with an agent.

I build many RL systems that monitor processes, such as a sorting system of purchases or assembly of products. It is also the core of most robot projects, as robots are physical agents that can interact with the environment. I also build many "soft-robots" that take decisions on such data processing as approval of loans, payments of money, and fixing of data errors.

Bagging Data

Bootstrap aggregating, also called bagging, is a machine-learning ensemble meta-algorithm that aims to advance the stability and accuracy of machine-learning algorithms used in statistical classification and regression. It decreases variance and supports systems to avoid overfitting.

I want to cover this concept, as I have seen many data science solutions over the last years that suffer from overfitting, because they were trained with a known data set that eventually became the only data set they could process. Thanks to inefficient processing and algorithms, we naturally had a lead way for variance in the data.

The new GPU (graphics processing unit)-based systems are so accurate that they overfit easily, if the training data is a consistent set, with little or no major changes in the patterns within the data set.

You will now see how to perform a simple bagging process. Open your Python editor and create this ecosystem:

```
import numpy as np
import matplotlib.pyplot as plt
from sklearn.ensemble import BaggingRegressor
from sklearn.tree import DecisionTreeRegressor
```

You need a few select settings.

```
n_repeat = 100         # Number of iterations for processing
n_train = 100          # Size of the Training Data set
n_test = 10000         # Size of the Test Data set
noise = 0.1            # Standard deviation of the noise introduced
np.random.seed(0)
```

You will select two estimators to compare.

```
estimators = [("Tree", DecisionTreeRegressor()),
              ("Bagging(Tree)", BaggingRegressor(DecisionTreeRegressor()))]
n_estimators = len(estimators)
```

You will need a set of data to perform the bagging against, so generate a random data set.

```
def f(x):
    x = x.ravel()
    return np.exp(-x ** 2) - 2 * np.exp(-(x - 2) ** 2)
```

You can experiment with other data configurations, if you want to see the process working.

You need to create a function to add the noise to the data.

```
def generate(n_samples, noise, n_repeat=1):
    X = np.random.rand(n_samples) * 10 - 5
    X = np.sort(X)
    if n_repeat == 1:
        y = f(X) + np.random.normal(0.0, noise, n_samples)
    else:
        y = np.zeros((n_samples, n_repeat))
        for i in range(n_repeat):
            y[:, i] = f(X) + np.random.normal(0.0, noise, n_samples)
    X = X.reshape((n_samples, 1))
    return X, y
```

You can now train the system using these Transform steps.

```
X_train = []
y_train = []
```

You train the system with the bagging data set, by taking a sample each cycle. This exposes the model to a more diverse data-training spread of the data.

```
for i in range(n_repeat):
    X, y = generate(n_samples=n_train, noise=noise)
    X_train.append(X)
    y_train.append(y)
```

You can now test your models.

```
X_test, y_test = generate(n_samples=n_test, noise=noise, n_repeat=n_repeat)
```

You can now loop over estimators to compare the results, by computing your predictions.

```
for n, (name, estimator) in enumerate(estimators):
    y_predict = np.zeros((n_test, n_repeat))
    for i in range(n_repeat):
        estimator.fit(X_train[i], y_train[i])
        y_predict[:, i] = estimator.predict(X_test)
    # Bias^2 + Variance + Noise decomposition of the mean squared error
    y_error = np.zeros(n_test)
    for i in range(n_repeat):
        for j in range(n_repeat):
            y_error += (y_test[:, j] - y_predict[:, i]) ** 2
    y_error /= (n_repeat * n_repeat)
    y_noise = np.var(y_test, axis=1)
    y_bias = (f(X_test) - np.mean(y_predict, axis=1)) ** 2
    y_var = np.var(y_predict, axis=1)
```

You can now display your results.

```
    print("{0}: {1:.4f} (error) = {2:.4f} (bias^2) "
          " + {3:.4f} (var) + {4:.4f} (noise)".format(name,
        np.mean(y_error),
        np.mean(y_bias),
        np.mean(y_var),
        np.mean(y_noise)))
```

You can now plot your results.

```
    plt.subplot(2, n_estimators, n + 1)
    plt.plot(X_test, f(X_test), "b", label="$f(x)$")
    plt.plot(X_train[0], y_train[0], ".b", label="LS ~ $y = f(x)+noise$")
    for i in range(n_repeat):
        if i == 0:
            plt.plot(X_test, y_predict[:, i], "r", label="$\^y(x)$")
```

```
    else:
        plt.plot(X_test, y_predict[:, i], "r", alpha=0.05)
    plt.plot(X_test, np.mean(y_predict, axis=1), "c",
            label="$\mathbb{E}_{LS} \^y(x)$")
    plt.xlim([-5, 5])
    plt.title(name)
    if n == 0:
        plt.legend(loc="upper left", prop={"size": 11})
    plt.subplot(2, n_estimators, n_estimators + n + 1)
    plt.plot(X_test, y_error, "r", label="$error(x)$")
    plt.plot(X_test, y_bias, "b", label="$bias^2(x)$"),
    plt.plot(X_test, y_var, "g", label="$variance(x)$"),
    plt.plot(X_test, y_noise, "c", label="$noise(x)$")
    plt.xlim([-5, 5])
    plt.ylim([0, 0.1])
    if n == 0:
        plt.legend(loc="upper left", prop={"size": 11})
```

Display your hard work!

```
plt.show()
```

Well done. You have completed you bagging example.

Remember: The bagging enables the training engine to train against different sets of the data you expect it to process. This eliminates the impact of outliers and extremes on the data model. So, remember that you took a model and trained it against several training sets sampled from the same population of data.

Random Forests

Random forests, or random decision forests, are an ensemble learning method for classification and regression that works by constructing a multitude of decision trees at training time and outputting the results the mode of the classes (classification) or mean prediction (regression) of the individual trees. Random decision forests correct for decision trees' habit of overfitting to their training set.

The result is an aggregation of all the trees' results, by performing a majority vote against the range of results. So, if five trees return three yeses and two nos, it passes a yes out of the Transform step.

Sometimes, this is also called tree bagging, as you take a bagging concept to the next level by not only training the model on a range of samples from the data population but by actually performing the complete process with the data bag and then aggregating the data results.

Warning   If you perform a ten-factor random forest, it means you are performing ten times more processing. So, investigate what the optimum spread is for your forests.

Tip   I have found that a 15+ spread on the average data set sizes I process improves the result's optimal. But please experiment, as I have also had solutions in which 5 worked perfectly and a painful process in which, owing to remarkably unpredictable outliers, a 100+ spread in conclusion achieved proper results.

Let me guide you through this process. Open your Python editor and prepare the following ecosystem:

```
from sklearn.ensemble import RandomForestClassifier
import pandas as pd
import numpy as np
import sys
import os
import datetime as dt
import calendar as cal
```

Set up the data location and load the data.

```
if sys.platform == 'linux':
    Base=os.path.expanduser('~') + 'VKHCG'
else:
    Base='C:/VKHCG'
print('###############################')
```

```
print('Working Base :',Base, ' using ', sys.platform)
print('##############################')
basedate = dt.datetime(2018,1,1,0,0,0)
Company='04-Clark'
InputFileName='Process_WIKI_UPS.csv'
InputFile=Base+'/'+Company+'/03-Process/01-EDS/02-Python/' + InputFileName
ShareRawData=pd.read_csv(InputFile,header=0,\
                        usecols=['Open','Close','UnitsOwn'], \
                        low_memory=False)
```

You must perform some preprocessing to reveal features in the data.

```
ShareRawData.index.names = ['ID']
ShareRawData['nRow'] = ShareRawData.index
ShareRawData['TradeDate']=ShareRawData.apply(lambda row:\
            (basedate - dt.timedelta(days=row['nRow'])),axis=1)
ShareRawData['WeekDayName']=ShareRawData.apply(lambda row:\
            (cal.day_name[row['TradeDate'].weekday()])\
             ,axis=1)
ShareRawData['WeekDayNum']=ShareRawData.apply(lambda row:\
            (row['TradeDate'].weekday())\
             ,axis=1)
ShareRawData['sTarget']=ShareRawData.apply(lambda row:\
            'true' if row['Open'] < row['Close'] else 'false'\
             ,axis=1)
```

Here is your data set:

```
print(ShareRawData.head())
```

Select a data frame with the two feature variables.

```
sColumns=['Open','Close','WeekDayNum']
df = pd.DataFrame(ShareRawData, columns=sColumns)
```

Let's look at the top-five rows.

```
print(df.head())
You need to select the target column.
```

```
df2 = pd.DataFrame(['sTarget'])
df2.columns =['WeekDayNum']
```

You must select a training data set.

```
df['is_train'] = np.random.uniform(0, 1, len(df)) <= .75
```

Now create two new data frames, one with the training rows and the other with the test rows.

```
train, test = df[df['is_train']==True], df[df['is_train']==False]
```

Here is the number of observations for the test and training data frames:

```
print('Number of observations in the training data:', len(train))
print('Number of observations in the test data:',len(test))
```

Start processing the data by creating a list of the feature column's names
```
features = df.columns[:3]
```

Display your features.

```
print(features)
```

You must factorize your target to use the model I selected.

```
y = pd.factorize(train['WeekDayNum'])[0]
```

You can now view the target values.

```
print(y)
You now train The Random Forest Classifier
```

Create a random forest classifier.

Tip   By convention, data scientists use "clf" for "Classifier." That way, they can simply change the clf by changing the type of classifier.

```
clf = RandomForestClassifier(n_jobs=2, random_state=0)
```

You now train the classifier to take the training features and learn how they relate to the training y (weekday number).

```
clf.fit(train[features], y)
```

Now apply the classifier to test data. This action is called "scoring."

```
clf.predict(test[features])
```

You can look at the predicted probabilities of the first ten observations.

```
print(clf.predict_proba(test[features])[0:10])
```

Evaluate the classifier. Is it any good?

```
preds = clf.predict(test[features])[3:4]
```

Look at the PREDICTED Week Day Number for the first ten observations.

```
print('PREDICTED Week Number:',preds[0:10])
```

Look at the ACTUAL WeekDayName for the first ten observations.

```
print(test['WeekDayNum'].head(10))
```

I suggest you create a confusion matrix.

```
c=pd.crosstab(df2['WeekDayNum'], preds, rownames=['Actual Week Day
Number'], colnames=['Predicted Week Day Number'])
print(c)
```

You can also look at a list of the features and their importance scores.

```
print(list(zip(train[features], clf.feature_importances_)))
```

You have completed the Transform steps for a random forest solution. At his point, I want to explain an additional aspect of random forests. This is simply the daisy-chaining of a series of random forests to create a solution. I have found these to become more popular over the last two years, as solutions become more demanding and data sets become larger. The same principles apply; you are simply repeating them several times in a chain.

Computer Vision (CV)

Computer vision is a complex feature extraction area, but once you have the features exposed, it simply becomes a matrix of values.

Note   In this book, I offer only a basic introduction to the computer vision aspect of data science. This is a massive field in its own right. I simply want you to note that will be a part of your future field of work.

Open your Python editor and enter this quick example:

```
import matplotlib.pyplot as plt
from PIL import Image
import numpy as np
sPicNameIn='C:/VKHCG/01-Vermeulen/00-RawData/AudiR8.png'
imageIn = Image.open(sPicNameIn)
fig1=plt.figure(figsize=(10, 10))
fig1.suptitle('Audi R8', fontsize=20)
imgplot = plt.imshow(imageIn)
plt.show()
```

You should see a car.

```
imagewidth, imageheight = imageIn.size
imageMatrix=np.asarray(imageIn)
pixelscnt = (imagewidth * imageheight)
print('Pixels:', pixelscnt)
print('Size:', imagewidth, ' x', imageheight,)
print(imageMatrix)
```

This is what your computer sees!

You have achieved computer vision. Remember how I showed you that movies convert several frames? Each frame becomes a matrix's entry, and now you can make your data science "see."

Natural Language Processing (NLP)

Natural language processing is the area in data science that investigates the process we as humans use to communicate with each other. This covers mainly written and spoken words that form bigger concepts. Your data science is aimed at intercepting or interacting with humans, to react to the natural language.

Note   This is a major field of study in data science. I am only providing a basic introduction to this interesting subject.

There are two clear requirements in natural language processing. First is the direct interaction with humans, such as when you speak to your smartphone, and it responds with an appropriate answer. For example, you request "phone home," and the phone calls the number set as "home."

The second type of interaction is taking the detailed content of the interaction and understanding its context and relationship with other text or recorded information. Examples of these are news reports that are examined, and common trends are found among different news reports. This a study of the natural language's meaning, not simply a response to a human interaction.

Text-Based

If you want to process text, you must set up an ecosystem to perform the basic text processing. I recommend that you use library nltk (conda install -c anaconda nltk).

Warning   This process generates a large text reference library on your local PC. Make sure that you have more than 3.3GB to download the complete data set.

Open your Python editor and set up your ecosystem. You then require the base data.

```
import nltk
nltk.download()
```

You will see a program that enables you to download several text libraries, which will assist the process to perform text analysis against any text you submit for analysis. The basic principle is that the library matches your text against the text stored in the data libraries and will return the correct matching text analysis.

Open your Python editor and create the following ecosystem, to enable you to investigate this library:

```
from nltk.tokenize import sent_tokenize, word_tokenize

Txt = "Good Day Mr. Vermeulen,\
 how are you doing today?\
 The weather is great, and Data Science is awesome.\
 You are doing well!"

print(Txt,'\n')

print('Identify sentences')
print(sent_tokenize(Txt),'\n')

print('Identify Word')
print(word_tokenize(Txt))
```

Speech-Based

There is a major demand for speech-to-text conversion, to extract features. I suggest looking at the SpeechRecognition library (https://pypi.python.org/pypi/SpeechRecognition/). You can install it by using conda install -c conda-forge speechrecognition.

This transform area is highly specialized, and I will not provide any further details on this subject.

Neural Networks

Neural networks (also known as artificial neural networks) are inspired by the human nervous system. They simulate how complex information is absorbed and processed by the human system. Just like humans, neural networks learn by example and are configured to a specific application.

Neural networks are used to find patterns in extremely complex data and, thus, deliver forecasts and classify data points. Neural networks are usually organized in layers. Layers are made up of a number of interconnected "nodes." Patterns (features) are presented to the network via the "input layer," which communicates to one or more "hidden layers," where the actual processing is done. The hidden layers then link to an "output layer," where the answer is output, as shown in Figure 10-16.

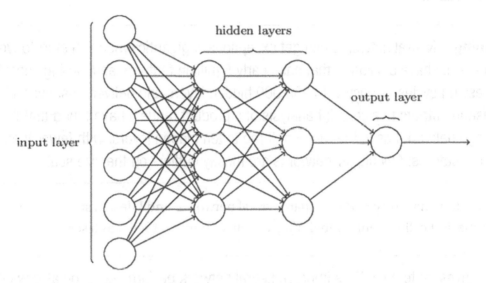

Figure 10-16 *General artificial neural network*

Tip   When I perform feature development for a neural network, I frame a question so as to receive a simple response (yes/no), but I create hundreds, or even thousands, of questions.

Let me offer an example of feature development for neural networks. Suppose you have to select three colors for your dog's new ball: blue, yellow, or pink. The features would be

- Is the ball blue? Yes/No

- Is the ball yellow? Yes/No

- Is the ball pink? Yes/No

When you feed these to the neural network, it will use them as simple 0 or 1 values, and this is what neural networks really excel at solving. Unfortunately, the most important feature when buying a ball for a dog is "Does the dog fit under the house?" It took me two hours to retrieve one dog and one black ball from a space I did not fit!

The lesson: You must change criteria as you develop the neural network. If you keep the question simple, you can just add more questions, or even remove questions, that result in features.

Warning   Neural networks can get exceptionally gigantic when you start to work with them. I have a solution that has nearly 5 million features and, during specific processing cycles, is busy with 250,000 hidden layers and delivers just over a thousand outputs to a conjoint analysis of a product range. I have found that the average network has in the range of three to ten hidden layers, with fewer than twenty features. Any bigger networks are mostly outliers on the size scale.

Note, too, that you can daisy-chain neural networks, and I design such systems on a regular basis. I call it "neural pipelining and clustering," as I sell it as a service.

Tip   Investigate how GPUs improve neural network performance. You also want to look at TensorFlow on GPUs (covered later in this chapter).

Before you start your example, you must understand two concepts.

Gradient Descent

Gradient descent is a first-order iterative optimization algorithm for finding the minimum of a function. To find a local minimum of a function using gradient descent, one takes steps proportional to the negative of the gradient of the function at the current point.

Example:

Open your Python editor and investigate this code:

```
cur_x = 3 # The algorithm starts at x=3
gamma = 0.01 # step size multiplier
precision = 0.00001
previous_step_size = cur_x
```

```
df = lambda x: 4 * x**3 - 9 * x**2

while previous_step_size > precision:
    prev_x = cur_x
    cur_x += -gamma * df(prev_x)
    previous_step_size = abs(cur_x - prev_x)
    print("Current X at %f" % cur_x, " with step at %f" % previous_step_
size )

print("The local minimum occurs at %f" % cur_x)
```

Can you see how the steps start to improve the process to find the local minimum? The gradient descent can take many iterations to compute a local minimum with a required accuracy. This process is useful when you want to find an unknown value that fits your model. In neural networks this supports the weighting of the hidden layers, to calculate to a required accuracy.

Regularization Strength

Regularization strength is the parameter that prevents overfitting of the neural network. The parameter enables the neural network to match the best set of weights for a general data set. The common name for this setting is the epsilon parameter, also known as the learning rate.

Simple Neural Network

Now that you understand these basic parameters, I will show you an example of a neural network. You can now open a new Python file and your Python editor.

Let's build a simple neural network.

```
Setup the eco-system.
import numpy as np
from sklearn import datasets, linear_model
import matplotlib.pyplot as plt
```

You need a visualization procedure.

```
def plot_decision_boundary(pred_func):
    # Set min and max values and give it some padding
    x_min, x_max = X[:, 0].min() - .5, X[:, 0].max() + .5
    y_min, y_max = X[:, 1].min() - .5, X[:, 1].max() + .5
    h = 0.01
    # Generate a grid of points with distance h between them
    xx, yy = np.meshgrid(np.arange(x_min, x_max, h), np.arange
    (y_min, y_max, h))
    # Predict the function value for the whole gid
    Z = pred_func(np.c_[xx.ravel(), yy.ravel()])
    Z = Z.reshape(xx.shape)
    # Plot the contour and training examples
    plt.contourf(xx, yy, Z, cmap=plt.cm.Spectral)
    plt.scatter(X[:, 0], X[:, 1], c=y, cmap=plt.cm.Spectral)
You will generate a data set for the example.
```

I suggest 200 points, but feel free to increase or decrease as you experiment with your neural network.

```
np.random.seed(0)
X, y = datasets.make_moons(200, noise=0.20)
```

You can plot the data to see what you generated.

```
plt.scatter(X[:,0], X[:,1], s=40, c=y, cmap=plt.cm.Spectral)
```

I suggest we train a logistic regression classifier to feed the features.

```
clf = linear_model.LogisticRegressionCV()
clf.fit(X, y)
 Now you can plot the decision boundary
plot_decision_boundary(lambda x: clf.predict(x))
plt.title("Logistic Regression")
```

You now configure the neural network. I kept it simple, with two inputs and two outputs.

```
num_examples = len(X) # training set size
nn_input_dim = 2 # input layer dimensionality
nn_output_dim = 2 # output layer dimensionality
```

You set the gradient descent parameters. This drives the speed at which you resolve the neural network.

```
Set learning rate for gradient descent and regularization strength,
experiment with this as it drives the transform speeds.

epsilon = 0.01
reg_lambda = 0.01 #
```

You must engineer a helper function, to evaluate the total loss on the data set.

```
def calculate_loss(model):
    W1, b1, W2, b2 = model['W1'], model['b1'], model['W2'], model['b2']
    # Forward propagation to calculate our predictions
    z1 = X.dot(W1) + b1
    a1 = np.tanh(z1)
    z2 = a1.dot(W2) + b2
    exp_scores = np.exp(z2)
    probs = exp_scores / np.sum(exp_scores, axis=1, keepdims=True)
    # Calculating the loss
    corect_logprobs = -np.log(probs[range(num_examples), y])
    data_loss = np.sum(corect_logprobs)
    # Add regulatization term to loss (optional)
    data_loss += reg_lambda/2 * (np.sum(np.square(W1)) + np.sum
(np.square(W2)))
    return 1./num_examples * data_loss
```

You also require a helper function, to predict an output (0 or 1).

```
def predict(model, x):
    W1, b1, W2, b2 = model['W1'], model['b1'], model['W2'], model['b2']
    # Forward propagation
```

```
z1 = x.dot(W1) + b1
a1 = np.tanh(z1)
z2 = a1.dot(W2) + b2
exp_scores = np.exp(z2)
probs = exp_scores / np.sum(exp_scores, axis=1, keepdims=True)
return np.argmax(probs, axis=1)
```

Your next function to engineer is central to the neural network. This function learns parameters for the neural network and returns the model.

- nn_hdim: Number of nodes in the hidden layer

- num_passes: Number of passes through the training data for gradient descent. I suggest 20000, but you can experiment with different sizes.

- print_loss: If True, print the loss every 1000 iterations.

```
def build_model(nn_hdim, num_passes=20000, print_loss=False):

    # Initialize the parameters to random values. We need to learn these.
    np.random.seed(0)
    W1 = np.random.randn(nn_input_dim, nn_hdim) / np.sqrt(nn_input_dim)
    b1 = np.zeros((1, nn_hdim))
    W2 = np.random.randn(nn_hdim, nn_output_dim) / np.sqrt(nn_hdim)
    b2 = np.zeros((1, nn_output_dim))

    # This is what we return at the end
    model = {}

    # Gradient descent. For each batch...
    for i in range(0, num_passes):

        # Forward propagation
        z1 = X.dot(W1) + b1
        a1 = np.tanh(z1)
        z2 = a1.dot(W2) + b2
        exp_scores = np.exp(z2)
        probs = exp_scores / np.sum(exp_scores, axis=1, keepdims=True)

        # Backpropagation
        delta3 = probs
```

```
delta3[range(num_examples), y] -= 1
dW2 = (a1.T).dot(delta3)
db2 = np.sum(delta3, axis=0, keepdims=True)
delta2 = delta3.dot(W2.T) * (1 - np.power(a1, 2))
dW1 = np.dot(X.T, delta2)
db1 = np.sum(delta2, axis=0)

# Add regularization terms (b1 and b2 don't have regularization
terms)
dW2 += reg_lambda * W2
dW1 += reg_lambda * W1

# Gradient descent parameter update
W1 += -epsilon * dW1
b1 += -epsilon * db1
W2 += -epsilon * dW2
b2 += -epsilon * db2

# Assign new parameters to the model
model = { 'W1': W1, 'b1': b1, 'W2': W2, 'b2': b2}

# Optionally print the loss.
# This is expensive because it uses the whole dataset, so we don't
want to do it too often.
if print_loss and i % 1000 == 0:
  print ("Loss after iteration %i: %f" %(i, calculate_loss(model)))

return model
```

You now define the model with a three-dimensional hidden layer.

```
model = build_model(3, print_loss=True)
```

You can now plot the decision boundary.

```
plot_decision_boundary(lambda x: predict(model, x))
plt.title("Decision Boundary for hidden layer size 3")
```

You can now visualize what you have achieved.

```
plt.figure(figsize=(16, 32))
hidden_layer_dimensions = [1, 2, 3, 4, 5, 20, 50]
for i, nn_hdim in enumerate(hidden_layer_dimensions):
    plt.subplot(5, 2, i+1)
    plt.title('Hidden Layer size %d' % nn_hdim)
    print('Model:',nn_hdim)
    model = build_model(nn_hdim, print_loss=True)
    plot_decision_boundary(lambda x: predict(model, x))
plt.show()
```

You can now build neural networks. Well done.

Advice   You should spend time on understanding how to construct neural networks, as I have experienced a massive increase in the amount of solutions that use neural networks to process data science. The rapid advances with GPU-driven processing is pushing this trend even faster. At a minimum, I must build 100 inputs and 10 hidden layers, to handle the smallest networks I receive requests for on a regular basis.   Also, look at daisy-chaining via other transform processes I discuss in this chapter.

The preceding is neural networking in its simplest form.

TensorFlow

TensorFlow is an open source software library for numerical computation using data-flow graphs. Nodes in the graph represent mathematical operations, while the graph edges represent the multidimensional data arrays (tensors) communicated between them. The flexible architecture allows you to deploy computation to one or more CPUs or GPUs.

TensorFlow was originally developed by researchers and engineers working on the Google Brain Team within Google's machine intelligence research organization, for the purposes of conducting machine learning and deep neural networks research, but the system is general enough to be applicable in a wide variety of other domains.

I will guide you through a few examples, to demonstrate the capability. I have several installations that use this ecosystem, and it is gaining popularity in the data science communities.

To use it, you will require a library named `tensorflow`. You install it using `conda install -c conda-forge tensorflow`. Details about the library are available at www.tensorflow.org.

The next big advantage is the Cloud Tensor Processing Unit (TPU) (https://cloud.google.com/tpu/) hardware product, which was specifically designed to calculate tensor processing at better performance levels than standard CPU or GPU hardware. The TPU supports the TensorFlow process with an extremely effective hardware ecosystem. The TensorFlow Research Cloud provides users with access to these second-generation cloud TPUs, each of which provides 180 teraflops of machine-learning acceleration.

Note   I use the open research platform at www.tensorflow.org/tfrc/ to test any new model I have, and it provides a great R&D ecosystem. I suggest you investigate it for your own ecosystem.

Basic TensorFlow

I will take you through a basic example by explaining, as a starting point, how to convert the following mathematical equation into a TensorFlow:

$$a=(b+c)*(c+2)a=(b+c)*(c+2)$$

I will calculate the following for you:

- B = 2.5
- C = 10

Open your Python editor and create the following ecosystem:

```
import tensorflow as tf
```

Create a TensorFlow constant.

```
const = tf.constant(2.0, name="const")
```

Create TensorFlow variables.

```
b = tf.Variable(2.5, name='b')
c = tf.Variable(10.0, name='c')
```

You must now create the operations.

```
d = tf.add(b, c, name='d')
e = tf.add(c, const, name='e')
a = tf.multiply(d, e, name='a')
```

Next, set up the variable initialization.

```
init_op = tf.global_variables_initializer()
```

You can now start the session.

```
with tf.Session() as sess:
    # initialise the variables
    sess.run(init_op)
    # compute the output of the graph
    a_out = sess.run(a)
    print("Variable a is {}".format(a_out))
```

Well done. You have just successfully deployed a TensorFlow solution. I will now guide you through a more advance example: how to feed a range of values into a TensorFlow.

$$a=(b+c)*(c+22)a=(b+c)*(c+22)$$

I will calculate for the following:

- B = range (-5,-4,-3,-2,-1,0,1,2,3,4,5)

- C = 3

Open your Python editor and create the following ecosystem:

```
import tensorflow as tf
import numpy as np
```

Create a TensorFlow constant.

```
const = tf.constant(22.0, name="const")
```

Now create the TensorFlow variables. Note the range format for variable b.

```
b = tf.placeholder(tf.float32, [None, 1], name='b')
c = tf.Variable(3.0, name='c')
```

You will create the required operations next.

```
d = tf.add(b, c, name='d')
e = tf.add(c, const, name='e')
a = tf.multiply(d, e, name='a')
```

Start the setup of the variable initialization.

```
init_op = tf.global_variables_initializer()
```

Start the session construction.

```
with tf.Session() as sess:
    # initialise the variables
    sess.run(init_op)
    # compute the output of the graph
    a_out = sess.run(a, feed_dict={b: np.arange(-5, 5)[:, np.newaxis]})
    print("Variable a is {}".format(a_out))
```

Did you notice how with minor changes TensorFlow handles larger volumes of data with ease? The advantage of TensorFlow is the simplicity of the basic building block you use to create it, and the natural graph nature of the data pipelines, which enable you to easily convert data flows from the real world into complex simulations within the TensorFlow ecosystem.

I will use a fun game called the One-Arm Bandits to offer a sample real-world application of this technology. Open your Python editor and create the following ecosystem:

```
import tensorflow as tf
import numpy as np
```

Let's construct a model for your bandits. There are four one-arm bandits, and currently, bandit 4 is set to provide a positive reward most often.

```
bandits = [0.2,0.0,-0.2,-2.0]
num_bandits = len(bandits)
```

You must model the bandit by creating a pull-bandit action.

```
def pullBandit(bandit):
    #Get a random number.
    result = np.random.randn(1)
    if result > bandit:
        #return a positive reward.
        return 1
    else:
        #return a negative reward.
        return -1
```

Now, you reset the ecosystem.

```
tf.reset_default_graph()
```

You need the following two lines to establish the feed-forward part of the network. You perform the actual selection using this formula.

```
weights = tf.Variable(tf.ones([num_bandits]))
chosen_action = tf.argmax(weights,0)
```

These next six lines establish the training procedure. You feed the reward and chosen action into the network by computing the loss and using it to update the network.

```
reward_holder = tf.placeholder(shape=[1],dtype=tf.float32)
action_holder = tf.placeholder(shape=[1],dtype=tf.int32)
responsible_weight = tf.slice(weights,action_holder,[1])
loss = -(tf.log(responsible_weight)*reward_holder)
optimizer = tf.train.GradientDescentOptimizer(learning_rate=0.001)
update = optimizer.minimize(loss)
```

Now, you must train the system to perform as a one-arm bandit.

```
total_episodes = 1000 #Set total number of episodes to train the bandit.
total_reward = np.zeros(num_bandits) #Set scoreboard for bandits to 0.
e = 0.1 #Set the chance of taking a random action.
```

Initialize the ecosystem now.

```
init = tf.initialize_all_variables()
```

Launch the TensorFlow graph processing.

```
with tf.Session() as sess:
    sess.run(init)
    i = 0
    while i < total_episodes:

        #Choose either a random action or one from our network.
        if np.random.rand(1) < e:
            action = np.random.randint(num_bandits)
        else:
            action = sess.run(chosen_action)

        reward = pullBandit(bandits[action])
```

Collect your reward from picking one of the bandits and update the network.

```
        _,resp,ww = sess.run([update,responsible_weight,weights], feed_
        dict={reward_holder:[reward],action_holder:[action]})
```

Update your running tally of the scores.

```
        total_reward[action] += reward
        if i % 50 == 0:
            print ("Running reward for the " + str(num_bandits) + "
            bandits: " + str(total_reward))
        i+=1
print ("The agent thinks bandit " + str(np.argmax(ww)+1) + " is the most
promising....")
if np.argmax(ww) == np.argmax(-np.array(bandits)):
    print ("...and it was right!")
else:
    print ("...and it was wrong!")
```

Congratulations! You have a fully functional TensorFlow solution. Can you think of three real-life examples you can model using this ecosystem?

Real-World Uses of TensorFlow

Over the last 12 months, I have used the TensorFlow process for the following three models:

- *Basket analysis*: What do people buy? What do they buy together?

- *Forex trading*: Providing recommendations on when to purchase forex for company requirements.

- *Commodities trading*: Buying and selling futures.

The One-Arm Bandits (Contextual Version)

To try your hand at a more complex bandit, I have prepared the following example. Let's play Bandits.

Open your Python editor and create the following ecosystem:

```
import tensorflow as tf
import tensorflow.contrib.slim as slim
import numpy as np
```

Create the contextual bandits.

```
class contextual_bandit():
    def __init__(self):
        self.state = 0
        #List out our bandits.
        self.bandits = np.array([\
                                [0.4,0,-0.0,-8],\
                                [0.9,-55,1,0.75],\
                                [-15,5,9,5],\
                                [-5,1,9,55],\
                                [15,51,9,5],\
                                [-15,51,9,55],\
                                [-5,51,99,55],\
                                [-65,51,9,55],\
                                [0.9,-55,1,0.75]\
                                ])
        self.num_bandits = self.bandits.shape[0]
        self.num_actions = self.bandits.shape[1]
```

Create the bandit.

```
def getBandit(self):
    self.state = np.random.randint(0,len(self.bandits)) #Returns a
    random state for each episode.
    return self.state
```

Create the pull-arm action.

```
def pullArm(self,action):
    #Get a random number.
    bandit = self.bandits[self.state,action]
    result = np.random.randn(1)
    if result > bandit:
        #return a positive reward.
        return 1
    else:
        #return a negative reward.
        return -1
```

Configure the rules for the bandits.

```
class agent():
    def __init__(self, lr, s_size,a_size):
        #These lines established the feed-forward part of the network.
        The agent takes a state and produces an action.
        self.state_in= tf.placeholder(shape=[1],dtype=tf.int32)
        state_in_OH = slim.one_hot_encoding(self.state_in,s_size)
        output = slim.fully_connected(state_in_OH,a_size,\
            biases_initializer=None,activation_fn=tf.nn.sigmoid,weights_
            initializer=tf.ones_initializer())
        self.output = tf.reshape(output,[-1])
        self.chosen_action = tf.argmax(self.output,0)
```

The next six lines establish the training procedure. Once again, feed the reward and chosen action into the network.

```
        #to compute the loss, and use it to update the network.
        self.reward_holder = tf.placeholder(shape=[1],dtype=tf.float32)
```

```
        self.action_holder = tf.placeholder(shape=[1],dtype=tf.int32)
        self.responsible_weight = tf.slice(self.output,self.action_
        holder,[1])
        self.loss = -(tf.log(self.responsible_weight)*self.reward_holder)
        optimizer = tf.train.GradientDescentOptimizer(learning_rate=lr)
        self.update = optimizer.minimize(self.loss)
```

Train your agent now.

```
tf.reset_default_graph() #Clear the Tensorflow graph.

cBandit = contextual_bandit() #Load the bandits.
myAgent = agent(lr=0.001,s_size=cBandit.num_bandits,a_size=cBandit.num_
actions) #Load the agent.
weights = tf.trainable_variables()[0] #The weights we will evaluate to look
into the network.

total_episodes = 10000 #Set total number of episodes to train agent on.
total_reward = np.zeros([cBandit.num_bandits,cBandit.num_actions]) #Set
scoreboard for bandits to 0.
e = 0.1 #Set the chance of taking a random action.

init = tf.initialize_all_variables()
```

Launch the TensorFlow graph.

```
with tf.Session() as sess:
    sess.run(init)
    i = 0
    while i < total_episodes:
        s = cBandit.getBandit() #Get a state from the environment.

        #Choose either a random action or one from our network.
        if np.random.rand(1) < e:
            action = np.random.randint(cBandit.num_actions)
        else:
            action = sess.run(myAgent.chosen_action,feed_dict={myAgent.
            state_in:[s]})
```

```
reward = cBandit.pullArm(action) #Get our reward for taking an
action given a bandit.

#Update the network.
feed_dict={myAgent.reward_holder:[reward],myAgent.action_
holder:[action],myAgent.state_in:[s]}
_,ww = sess.run([myAgent.update,weights], feed_dict=feed_dict)

#Update our running tally of scores.
total_reward[s,action] += reward
if i % 500 == 0:
    print ("Mean reward for each of the " + str(cBandit.num_
    bandits) + " bandits: " + str(np.mean(total_reward,axis=1)))
    i+=1
for a in range(cBandit.num_bandits):
    print ("The agent thinks action " + str(np.argmax(ww[a])+1) + " for
bandit " + str(a+1) + " is the most promising....")
    if np.argmax(ww[a]) == np.argmin(cBandit.bandits[a]):
        print ("...and it was right!")
    else:
        print ("...and it was wrong!")
```

Can you understand how the process assists you to model the real-life activities?

Processing the Digits on a Container

Here is a more practical application. You are required to identify the digits on the side of a container as it moves around your shipping yard at Hillman Ltd's Edinburgh warehouse. As it rains most of the time, the scanning equipment is not producing optimum images as the containers move past the camera recording the digits. Can you resolve the problem of identifying the digits?

Open your Python editor and create the following ecosystem:

```
import tensorflow as tf
from tensorflow.examples.tutorials.mnist import input_data
mnist = input_data.read_data_sets("MNIST_data/", one_hot=True)
```

You must have these optimization variables set. You can later experiment with different numbers to understand what impact they will have on your results.

```
learning_rate = 0.5
epochs = 10
batch_size = 100
```

You must declare the training data format. Set input x—for 28 × 28 pixels = 784.

```
x = tf.placeholder(tf.float32, [None, 784])
```

Now declare the output data placeholder—ten digits

```
y = tf.placeholder(tf.float32, [None, 10])
```

Declare the weights connecting the input to the hidden layer. Yes, you guessed right; this is a neural network.

```
W1 = tf.Variable(tf.random_normal([784, 300], stddev=0.03), name='W1')
b1 = tf.Variable(tf.random_normal([300]), name='b1')
```

Now add the weights connecting the hidden layer to the output layer.

```
W2 = tf.Variable(tf.random_normal([300, 10], stddev=0.03), name='W2')
b2 = tf.Variable(tf.random_normal([10]), name='b2')
```

Calculate the output of the hidden layer.

```
hidden_out = tf.add(tf.matmul(x, W1), b1)
hidden_out = tf.nn.relu(hidden_out)
```

Use a Softmax regression activated output layer.

```
y_ = tf.nn.softmax(tf.add(tf.matmul(hidden_out, W2), b2))

y_clipped = tf.clip_by_value(y_, 1e-10, 0.9999999)
cross_entropy = -tf.reduce_mean(tf.reduce_sum(y * tf.log(y_clipped)
                        + (1 - y) * tf.log(1 - y_clipped), axis=1))
```

You will need to add an optimizer.

```
optimizer = tf.train.GradientDescentOptimizer(learning_rate=learning_rate).
minimize(cross_entropy)
```

Next, set up the initialization operator.

```
init_op = tf.global_variables_initializer()
```

Set up the accuracy assessment operation.

```
correct_prediction = tf.equal(tf.argmax(y, 1), tf.argmax(y_, 1))
accuracy = tf.reduce_mean(tf.cast(correct_prediction, tf.float32))
```

All you need now is to start the session.

```
with tf.Session() as sess:
    # initialise the variables
    sess.run(init_op)
    total_batch = int(len(mnist.train.labels) / batch_size)
    for epoch in range(epochs):
        avg_cost = 0
        for i in range(total_batch):
            batch_x, batch_y = mnist.train.next_batch
            (batch_size=batch_size)
            _, c = sess.run([optimiser, cross_entropy],
                        feed_dict={x: batch_x, y: batch_y})
            avg_cost += c / total_batch
        print("Epoch:", (epoch + 1), "cost =", "{:.3f}".format(avg_cost))
    print(sess.run(accuracy, feed_dict={x: mnist.test.images, y: mnist.
    test.labels}))
```

Can you see how the TensorFlow assists you to build complex operations with ease?

Linear Regression Using TensorFlow

Do you feel secure enough to perform a linear regression in TensorFlow? Open your Python editor and create the following ecosystem:

```
import tensorflow as tf
import numpy
import matplotlib.pyplot as plt
rng = numpy.random
```

Set the parameters.

```
learning_rate = 0.01
training_epochs = 10000
display_step = 50
```

Get your training data.

```
train_X = numpy.
asarray([3.3,4.4,5.5,6.71,6.93,4.168,9.779,6.182,7.59,2.167,
                        7.042,10.791,5.313,7.997,5.654,9.27,3.1])
train_Y = numpy.
asarray([1.7,2.76,2.09,3.19,1.694,1.573,3.366,2.596,2.53,1.221,
                        2.827,3.465,1.65,2.904,2.42,2.94,1.3])
n_samples = train_X.shape[0]
```

Set up the TensorFlow graph input.

```
X = tf.placeholder("float")
Y = tf.placeholder("float")
```

Set the model weights.

```
W = tf.Variable(rng.randn(), name="weight")
b = tf.Variable(rng.randn(), name="bias")
```

Now construct a linear model.

```
pred = tf.add(tf.multiply(X, W), b)
```

Calculate a mean squared error.

```
cost = tf.reduce_sum(tf.pow(pred-Y, 2))/(2*n_samples)
```

Calculate a gradient descent.

```
optimizer = tf.train.GradientDescentOptimizer(learning_rate).minimize(cost)
```

Initialize the variables.

```
init = tf.global_variables_initializer()
```

Launch the graph.

```python
with tf.Session() as sess:
    sess.run(init)

    # Fit all training data
    for epoch in range(training_epochs):
        for (x, y) in zip(train_X, train_Y):
            sess.run(optimizer, feed_dict={X: x, Y: y})

        # Display logs per epoch step
        if (epoch+1) % display_step == 0:
            c = sess.run(cost, feed_dict={X: train_X, Y:train_Y})
            print("Epoch:", '%04d' % (epoch+1), "cost=", "{:.9f}".
            format(c), \
                "W=", sess.run(W), "b=", sess.run(b))

    print("Optimization Finished!")
    training_cost = sess.run(cost, feed_dict={X: train_X, Y: train_Y})
    print("Training cost=", training_cost, "W=", sess.run(W),
    "b=", sess.run(b), '\n')

    # Graphic display
    plt.plot(train_X, train_Y, 'ro', label='Original data')
    plt.plot(train_X, sess.run(W) * train_X + sess.run(b), label=
    'Fitted line')
    plt.legend()
    plt.show()

    # Testing example, as requested (Issue #2)
    test_X = numpy.asarray([6.83, 4.668, 8.9, 7.91, 5.7, 8.7, 3.1, 2.1])
    test_Y = numpy.asarray([1.84, 2.273, 3.2, 2.831, 2.92, 3.24, 1.35,
    1.03])

    print("Testing... (Mean square loss Comparison)")
    testing_cost = sess.run(
        tf.reduce_sum(tf.pow(pred - Y, 2)) / (2 * test_X.shape[0]),
        feed_dict={X: test_X, Y: test_Y})  # same function as cost above
    print("Testing cost=", testing_cost)
```

683

```
print("Absolute mean square loss difference:", abs(
    training_cost - testing_cost))

plt.plot(test_X, test_Y, 'bo', label='Testing data')
plt.plot(train_X, sess.run(W) * train_X + sess.run(b), label='Fitted
line')
plt.legend()
plt.show()
```

Congratulations! You just completed the linear model in TensorFlow with the Transform step. The use of TensorFlow enhances your capacity in a very positive manner. The TensorFlow ecosystem has opened up many new opportunities for processing large and complex data sets in the cloud, returning results to the customer's equipment.

Summary

This chapter was a marathon through complex and important Transform steps. So, let's review what you have learned.

- You learned that you can convert the data vault from the Process step into a data warehouse with dimensions and facts.

- You were introduced to sun models and how they can help convey your investigation results, by using the model as the basis of your discussions with customers and subject-matter experts.

- You have been introduced to several Transform steps and other techniques that you can apply against the data vault, or even the data warehouse, to achieve new insights into the meaning of the data to the business.

At this point, I want to congratulate you on your progress, as you now can build a stable pipeline from the data lake, via the Retrieve step, Assess step, and Process step, into the data vault. You now can transform data into the warehouse.

The next chapter will show you how to take your insights and organize them into specific groupings for our business entities. You will also see how to report the insights to the business community.

You can now celebrate with a well-deserved refreshment and then start Chapter 11.

CHAPTER 11

Organize and Report Supersteps

This chapter will cover the Organize superstep first, then proceed to the Report superstep. The two sections will enable you, as the data scientist, first to collect the relevant information from your prepared data warehouse, to match the requirement of a specific segment of the customer's decision makers. For example, you will use the same data warehouse to report to the chief financial officer (CFO) and the accountant in Wick, Scotland, but the CFO will receive an overall view, in addition to detailed views of all regions, while the accountant in Wick will see only details related to Wick. This is called organizing your data into a smaller data structure called a "data mart."

The second part of the chapter will introduce you to various reporting techniques that you can use to visualize the data for the various customers. I will discuss what each is used for and supply examples of the individual report techniques.

So, let's start our organization of data the warehouse's data into data marts.

Organize Superstep

The Organize superstep takes the complete data warehouse you built at the end of the Transform superstep and subsections it into business-specific data marts. A data mart is the access layer of the data warehouse environment built to expose data to the users. The data mart is a subset of the data warehouse and is generally oriented to a specific business group.

Any source code or other supplementary material referenced by me in this book is available to readers on GitHub, via this book's product page, located at www.apress.com/9781484230534. Please note that this source code assumes you have completed the source code setup outlined in Chapter 2.

© Andreas François Vermeulen 2018
A. F. Vermeulen, *Practical Data Science*, https://doi.org/10.1007/978-1-4842-3054-1_11

Horizontal Style

Performing horizontal-style slicing or subsetting of the data warehouse is achieved by applying a filter technique that forces the data warehouse to show only the data for a specific preselected set of filtered outcomes against the data population. The horizontal-style slicing selects the subset of rows from the population while preserving the columns. That is, the data science tool can see the complete record for the records in the subset of records.

Let's look at the technique, using the following example:

Start with the standard ecosystem.

```
################################################################
# -*- coding: utf-8 -*-
################################################################
import sys
import os
import pandas as pd
import sqlite3 as sq
################################################################

    If sys.platform == 'linux' or sys.platform == ' darwin':

      Base=os.path.expanduser('~') + '/VKHCG'
else:
      Base='C:/VKHCG'
print('###############################')
print('Working Base :',Base, ' using ', sys.platform)
print('###############################')
################################################################
################################################################
Company='01-Vermeulen'
################################################################
sDataWarehouseDir=Base + '/99-DW'
if not os.path.exists(sDataWarehouseDir):
    os.makedirs(sDataWarehouseDir)
```

```
###############################################################
sDatabaseName=sDataWarehouseDir + '/datawarehouse.db'
conn1 = sq.connect(sDatabaseName)
###############################################################
sDatabaseName=sDataWarehouseDir + '/datamart.db'
conn2 = sq.connect(sDatabaseName)
###############################################################
```

Load the complete BMI data set from the data warehouse.

Note   You are loading the data into memory for processing. The code assumes you have enough memory to perform the analysis in memory. To assist with removing unwanted or invalid data, you will reduce the memory requirements.

The next query loads all the data into memory, and that means you will have the complete data set ready in memory, but it also means you will have allocated memory to the data set that you may not need for your further analysis.

```
print('###############')
sTable = 'Dim-BMI'
print('Loading :',sDatabaseName,' Table:',sTable)
sSQL="SELECT * FROM [Dim-BMI];"
PersonFrame0=pd.read_sql_query(sSQL, conn1)
```

The following introduces a solution, by using a data-slicing technique. Let's look first at "horizontal" slicing, by loading only any person taller than 1.5 meters and having an indicator of 1. This will reduce the data set, to make it less demanding on memory.

Load the horizontal data slice for body mass index (BMI) from the data warehouse.

```
print('###############')
sTable = 'Dim-BMI'
print('Loading :',sDatabaseName,' Table:',sTable)
sSQL="SELECT PersonID,\
      Height,\
      Weight,\
      bmi,\
      Indicator\
```

```
FROM [Dim-BMI]\
WHERE \
Height > 1.5 \
and Indicator = 1\
ORDER BY \
     Height,\
     Weight;"
PersonFrame1=pd.read_sql_query(sSQL, conn1)
###############################################################
DimPerson=PersonFrame1
DimPersonIndex=DimPerson.set_index(['PersonID'],inplace=False)
###############################################################
```

Note   You will now replace the data in the database with the reduced data slice.
This will make all future queries result in a reduced memory requirement.

Store the horizontal data slice for BMI into the data warehouse.

```
sTable = 'Dim-BMI'
print('\n#################################')
print('Storing :',sDatabaseName,'\n Table:',sTable)
print('\n#################################')
DimPersonIndex.to_sql(sTable, conn2, if_exists="replace")
###############################################################
print('################')
sTable = 'Dim-BMI'
print('Loading :',sDatabaseName,' Table:',sTable)
sSQL="SELECT * FROM [Dim-BMI];"
PersonFrame2=pd.read_sql_query(sSQL, conn2)
```

You can show your results by printing the following code. You can see the
improvement you achieved.

```
print('Full Data Set (Rows):', PersonFrame0.shape[0])
print('Full Data Set (Columns):', PersonFrame0.shape[1])
print('Horizontal Data Set (Rows):', PersonFrame2.shape[0])
print('Horizontal Data Set (Columns):', PersonFrame2.shape[1])
```

You should return the following:

```
###############################
Full Data Set (Rows): 1080
Full Data Set (Columns): 5
###############################
Horizontal Data Set (Rows): 194
Horizontal Data Set (Columns): 5
###############################
```

This shows that you successfully reduced the records from 1080 to 194 relevant for your specific data slice, while keeping the 5 columns of data. Well done. You have successfully achieved horizontal-style slicing by modifying the amount of records in the data warehouse.

Warning   When performing horizontal-style slicing, make sure you do not overwrite the complete data set when you write your results back into the data warehouse.

Horizontal-style slicing is the most common organizational procedure that you will perform. This is because most dimensions are set up to normally contain all the columns the business requires.

Note   You can use the table-creation route or the view-creation route when generating the organize subsets. This ensures your personal preference on what works for your specific customer requirements.

In performing the example, you replaced the table permanently, but it is also possible to organize the creation of a view that then performs the horizontal slicing every time you call the view. This protects the original data set but requires extra database processing with every call into memory, while still reducing the memory requirements. I use both in equal measure.

Vertical Style

Performing vertical-style slicing or subsetting of the data warehouse is achieved by applying a filter technique that forces the data warehouse to show only the data for specific preselected filtered outcomes against the data population. The vertical-style slicing selects the subset of columns from the population, while preserving the rows. That is, the data science tool can see only the preselected columns from a record for all the records in the population.

Note   The use of vertical-style data slicing is common in systems in which specific data columns may not be shown to everybody, owing to security or privacy regulations. This is the most common way to impose different groups of security on the same data set.

Let's look at the technique, using the following example.

You will return only the weight, height, and indicator, to prevent the other columns from being seen, owing to privacy rules.

Start with the standard ecosystem.

```
###############################################################
# -*- coding: utf-8 -*-
###############################################################
import sys
import os
import pandas as pd
import sqlite3 as sq
###############################################################
if sys.platform == 'linux' or sys.platform == ' darwin':
    Base=os.path.expanduser('~') + '/VKHCG'
else:
    Base='C:/VKHCG'
print('###############################')
print('Working Base :',Base, ' using ', sys.platform)
print('###############################')
###############################################################
```

```
################################################################
Company='01-Vermeulen'
################################################################
sDataWarehouseDir=Base + '/99-DW'
if not os.path.exists(sDataWarehouseDir):
    os.makedirs(sDataWarehouseDir)
################################################################
sDatabaseName=sDataWarehouseDir + '/datawarehouse.db'
conn1 = sq.connect(sDatabaseName)
################################################################
sDatabaseName=sDataWarehouseDir + '/datamart.db'
conn2 = sq.connect(sDatabaseName)
################################################################
```

Load the complete BMI data set from the data warehouse.

```
print('################')
sTable = 'Dim-BMI'
print('Loading :',sDatabaseName,' Table:',sTable)
sSQL="SELECT * FROM [Dim-BMI];"
PersonFrame0=pd.read_sql_query(sSQL, conn1)
```

Load the vertical data slice for BMI from the data warehouse.

```
print('################')
sTable = 'Dim-BMI'
print('Loading :',sDatabaseName,' Table:',sTable)
sSQL="SELECT \
      Height,\
      Weight,\
      Indicator\
  FROM [Dim-BMI];"
PersonFrame1=pd.read_sql_query(sSQL, conn1)
################################################################
DimPerson=PersonFrame1
DimPersonIndex=DimPerson.set_index(['Indicator'],inplace=False)
################################################################
```

```
sTable = 'Dim-BMI'
print('\n###############################')
print('Storing :',sDatabaseName,'\n Table:',sTable)
print('\n###############################')
DimPersonIndex.to_sql(sTable, conn2, if_exists="replace")
################################################################
print('###############')
sTable = 'Dim-BMI'
print('Loading :',sDatabaseName,' Table:',sTable)
sSQL="SELECT * FROM [Dim-BMI];"
PersonFrame2=pd.read_sql_query(sSQL, conn2)
################################################################
```

You can show your results by printing the following:

```
print('Full Data Set (Rows):', PersonFrame0.shape[0])
print('Full Data Set (Columns):', PersonFrame0.shape[1])
print('Horizontal Data Set (Rows):', PersonFrame2.shape[0])
print('Horizontal Data Set (Columns):', PersonFrame2.shape[1])
```

You should return the following:

```
###############################
Full Data Set (Rows): 1080
Full Data Set (Columns): 5
###############################
Horizontal Data Set (Rows): 1080
Horizontal Data Set (Columns): 3
###############################
```

This shows that you successfully reduced the columns from 5 to 3 relevant for your specific data slice, while keeping the 1080 rows of data. Well done. You have successfully achieved vertical-style slicing by modifying the amount of columns in the data warehouse.

Note   Vertical slicing is common for data warehouses in which the dimensions are broad and contain dimensional attributes that cover several dissimilar business requirements.

Island Style

Performing island-style slicing or subsetting of the data warehouse is achieved by applying a combination of horizontal- and vertical-style slicing. This generates a subset of specific rows and specific columns reduced at the same time.

Note   These types of island slices are typical for snapshotting a reduced data set of data each month. Items may include unpaid accounts, overdue deliveries, and damaged billboards. You do not have to store the complete warehouse. You only want the specific subset of data.

The technique generates a set of data islands that are specific to particular requirements within the business.

Start with the standard ecosystem.

```
################################################################
# -*- coding: utf-8 -*-
################################################################
import sys
import os
import pandas as pd
import sqlite3 as sq
################################################################
if sys.platform == 'linux' or sys.platform == ' darwin':
    Base=os.path.expanduser('~') + '/VKHCG'
else:
    Base='C:/VKHCG'
print('################################')
print('Working Base :',Base, ' using ', sys.platform)
print('################################')
################################################################
################################################################
Company='01-Vermeulen'
################################################################
```

```
sDataWarehouseDir=Base + '/99-DW'
if not os.path.exists(sDataWarehouseDir):
    os.makedirs(sDataWarehouseDir)
################################################################
sDatabaseName=sDataWarehouseDir + '/datawarehouse.db'
conn1 = sq.connect(sDatabaseName)
################################################################
sDatabaseName=sDataWarehouseDir + '/datamart.db'
conn2 = sq.connect(sDatabaseName)
```

Load the complete BMI data set from the data warehouse.

```
print('###############')
sTable = 'Dim-BMI'
print('Loading :',sDatabaseName,' Table:',sTable)
sSQL="SELECT * FROM [Dim-BMI];"
PersonFrame0=pd.read_sql_query(sSQL, conn1)
```

Load the island of BMI from the data warehouse.

```
print('###############')
sTable = 'Dim-BMI'
print('Loading :',sDatabaseName,' Table:',sTable)
```

Note   Here is the island. You are taking only selective columns of the data warehouse that are bigger than an indicator of two and creating an island.

```
sSQL="SELECT \
       Height,\
       Weight,\
       Indicator\
  FROM [Dim-BMI]\
  WHERE Indicator > 2\
  ORDER BY  \
       Height,\
       Weight;"
PersonFrame1=pd.read_sql_query(sSQL, conn1)
```

694

```
################################################################
DimPerson=PersonFrame1
DimPersonIndex=DimPerson.set_index(['Indicator'],inplace=False)
################################################################
sTable = 'Dim-BMI'
print('\n###############################')
print('Storing :',sDatabaseName,'\n Table:',sTable)
print('\n###############################')
DimPersonIndex.to_sql(sTable, conn2, if_exists="replace")
################################################################
print('################')
sTable = 'Dim-BMI-Vertical'
print('Loading :',sDatabaseName,' Table:',sTable)
sSQL="SELECT * FROM [Dim-BMI-Vertical];"
PersonFrame2=pd.read_sql_query(sSQL, conn2)
```

You can show your results by printing the following:

```
print('Full Data Set (Rows):', PersonFrame0.shape[0])
print('Full Data Set (Columns):', PersonFrame0.shape[1])
print('Horizontal Data Set (Rows):', PersonFrame2.shape[0])
print('Horizontal Data Set (Columns):', PersonFrame2.shape[1])
```

You should return the following:

```
###############################
Full Data Set (Rows): 1080
Full Data Set (Columns): 5
###############################
Horizontal Data Set (Rows): 771
Horizontal Data Set (Columns): 3
###############################
```

Observe that you have successfully organized an island of data that fits your requirements. You reduced both the records and the columns.

Secure Vault Style

The secure vault is a version of one of the horizontal, vertical, or island slicing techniques, but the outcome is also attached to the person who performs the query. This is common in multi-security environments, where different users are allowed to see different data sets.

This process works well, if you use a role-based access control (RBAC) approach to restricting system access to authorized users. The security is applied against the "role," and a person can then, by the security system, simply be added or removed from the role, to enable or disable access.

The security in most data lakes I deal with is driven by an RBAC model that is an approach to restricting system access to authorized users by allocating them to a layer of roles that the data lake is organized into to support security access.

It is also possible to use a time-bound RBAC that has different access rights during office hours than after hours.

Warning   RBAC is a security process that is in effect at most of the customers I deal with daily. Make sure you understand this process, to ensure security compliance.

Start with the standard ecosystem.

```
###############################################################
# -*- coding: utf-8 -*-
###############################################################
import sys
import os
import pandas as pd
import sqlite3 as sq
###############################################################
if sys.platform == 'linux' or sys.platform == ' darwin':
    Base=os.path.expanduser('~') + '/VKHCG'
else:
    Base='C:/VKHCG'
print('###############################')
```

```
print('Working Base :',Base, ' using ', sys.platform)
print('##############################')
################################################################
################################################################
Company='01-Vermeulen'
################################################################
sDataWarehouseDir=Base + '/99-DW'
if not os.path.exists(sDataWarehouseDir):
    os.makedirs(sDataWarehouseDir)
################################################################
sDatabaseName=sDataWarehouseDir + '/datawarehouse.db'
conn1 = sq.connect(sDatabaseName)
################################################################
sDatabaseName=sDataWarehouseDir + '/datamart.db'
conn2 = sq.connect(sDatabaseName)
```

Load the complete BMI data set from the data warehouse.

```
print('################')
sTable = 'Dim-BMI'
print('Loading :',sDatabaseName,' Table:',sTable)
sSQL="SELECT * FROM [Dim-BMI];"
PersonFrame0=pd.read_sql_query(sSQL, conn1)
```

Load the security BMI data set from the data warehouse.

```
print('################')
sTable = 'Dim-BMI'
print('Loading :',sDatabaseName,' Table:',sTable)

sSQL="SELECT \
    Height,\
    Weight,\
    Indicator,\
    CASE Indicator\
    WHEN 1 THEN 'Pip'\
    WHEN 2 THEN 'Norman'\
    WHEN 3 THEN 'Grant'\
```

```
        ELSE 'Sam'\
        END AS Name\
  FROM [Dim-BMI]\
  WHERE Indicator > 2\
  ORDER BY  \
        Height,\
        Weight;"
PersonFrame1=pd.read_sql_query(sSQL, conn1)
##################################################################
DimPerson=PersonFrame1
DimPersonIndex=DimPerson.set_index(['Indicator'],inplace=False)
##################################################################
sTable = 'Dim-BMI-Secure'
print('\n#################################')
print('Storing :',sDatabaseName,'\n Table:',sTable)
print('\n#################################')
DimPersonIndex.to_sql(sTable, conn2, if_exists="replace")
```

Load Sam's view of the BMI data set from the data warehouse.

```
print('#################################')
sTable = 'Dim-BMI-Secure'
print('Loading :',sDatabaseName,' Table:',sTable)
print('#################################')
sSQL="SELECT * FROM [Dim-BMI-Secure] WHERE Name = 'Sam';"
PersonFrame2=pd.read_sql_query(sSQL, conn2)
##################################################################
print('#################################')
print('Full Data Set (Rows):', PersonFrame0.shape[0])
print('Full Data Set (Columns):', PersonFrame0.shape[1])
print('#################################')
print('Horizontal Data Set (Rows):', PersonFrame2.shape[0])
print('Horizontal Data Set (Columns):', PersonFrame2.shape[1])
print('Only Sam Data')
print(PersonFrame2.head())
print('#################################')
##################################################################
```

```
###############################
Full Data Set (Rows): 1080
Full Data Set (Columns): 5
###############################
Horizontal Data Set (Rows): 692
Horizontal Data Set (Columns): 4
```

Only Sam's data appears, as follows:

Indicator	Height	Weight	Name
4	1	35	Sam
4	1	40	Sam
4	1	45	Sam
4	1	50	Sam
4	1	55	Sam

Note   It is better to create roles than named references, as done in the example. The principle is the same, but the security is more flexible.

Association Rule Mining

Association rule learning is a rule-based machine-learning method for discovering interesting relations between variables in large databases, similar to the data you will find in a data lake. The technique enables you to investigate the interaction between data within the same population.

This example I will discuss is also called "market basket analysis." It will investigate the analysis of a customer's purchases during a period of time.

The new measure you need to understand is called "lift." Lift is simply estimated by the ratio of the joint probability of two items x and y, divided by the product of their individual probabilities:

$$Lift = \frac{P(x,y)}{P(x)P(y)}$$

If the two items are statistically independent, then P(x,y) = P(x)P(y), corresponding to Lift = 1, in that case. Note that anti-correlation yields lift values less than 1, which is also an interesting discovery, corresponding to mutually exclusive items that rarely co-occur.

You will require the following additional library: `conda install -c conda-forge mlxtend`.

The general algorithm used for this is the Apriori algorithm for frequent item set mining and association rule learning over the content of the data lake. It proceeds by identifying the frequent individual items in the data lake and extends them to larger and larger item sets, as long as those item sets appear satisfactorily frequently in the data lake. The frequent item sets determined by Apriori can be used to determine association rules that highlight common trends in the overall data lake. I will guide you through an example.

Start with the standard ecosystem.

```
################################################################
# -*- coding: utf-8 -*-
################################################################
import sys
import os
import pandas as pd
from mlxtend.frequent_patterns import apriori
from mlxtend.frequent_patterns import association_rules
################################################################
if sys.platform == 'linux' or sys.platform == ' darwin':
    Base=os.path.expanduser('~') + '/VKHCG'
else:
    Base='C:/VKHCG'
print('###############################')
print('Working Base :',Base, ' using ', sys.platform)
print('###############################')
################################################################
Company='01-Vermeulen'
InputFileName='Online-Retail-Billboard.xlsx'
EDSAssessDir='02-Assess/01-EDS'
InputAssessDir=EDSAssessDir + '/02-Python'
################################################################
```

```
sFileAssessDir=Base + '/' + Company + '/' + InputAssessDir
if not os.path.exists(sFileAssessDir):
    os.makedirs(sFileAssessDir)
################################################################
sFileName=Base+'/'+ Company + '/00-RawData/' + InputFileName
################################################################
```

Import the Excel worksheet into the ecosystem.

```
df = pd.read_excel(sFileName)
print(df.shape)
```

Perform some feature engineering to formulate the basket's simulation in your model.

```
df['Description'] = df['Description'].str.strip()
df.dropna(axis=0, subset=['InvoiceNo'], inplace=True)
df['InvoiceNo'] = df['InvoiceNo'].astype('str')
df = df[~df['InvoiceNo'].str.contains('C')]

basket = (df[df['Country'] =="France"]
          .groupby(['InvoiceNo', 'Description'])['Quantity']
          .sum().unstack().reset_index().fillna(0)
          .set_index('InvoiceNo'))
################################################################
def encode_units(x):
    if x <= 0:
        return 0
    if x >= 1:
        return 1
################################################################
basket_sets = basket.applymap(encode_units)
basket_sets.drop('POSTAGE', inplace=True, axis=1)
```

Apply the Apriori algorithm to the data model.

```
frequent_itemsets = apriori(basket_sets, min_support=0.07, use_
colnames=True)
```

```
rules = association_rules(frequent_itemsets, metric="lift", min_
threshold=1)
print(rules.head())

rules[ (rules['lift'] >= 6) &
        (rules['confidence'] >= 0.8) ]
```

You can now check what the results are for a "Green Clock."

```
sProduct1='ALARM CLOCK BAKELIKE GREEN'
print(sProduct1)
print(basket[sProduct1].sum())
```

And a "Red Clock" . . .

```
sProduct2='ALARM CLOCK BAKELIKE RED'
print(sProduct2)
print(basket[sProduct2].sum())
```

You can now check what the results for a basket for "Germany" are.

```
basket2 = (df[df['Country'] =="Germany"]
            .groupby(['InvoiceNo', 'Description'])['Quantity']
            .sum().unstack().reset_index().fillna(0)
            .set_index('InvoiceNo'))

basket_sets2 = basket2.applymap(encode_units)
basket_sets2.drop('POSTAGE', inplace=True, axis=1)
frequent_itemsets2 = apriori(basket_sets2, min_support=0.05, use_
colnames=True)
rules2 = association_rules(frequent_itemsets2, metric="lift", min_
threshold=1)

print(rules2[ (rules2['lift'] >= 4) &
        (rules2['confidence'] >= 0.5)])
```

You have successfully used a complex Apriori algorithm to organize the data science into a model that you can use to extract insights into the purchasing behavior of customers.

Note   There are several algorithms and techniques that a good data scientist can deploy to organize the data into logical data sets. I advise you to investigate algorithms that your customer's data fits best and practice against them, to become an expert.

Engineering a Practical Organize Superstep

Now that I have explained the various aspects of the Organize superstep, I will show you how to help our company with their processing.

Vermeulen PLC

I will show you how to apply the Organize superstep to our VKHCG companies. So, I will start with a networking requirement. Can you generate a network routing diagram for the VKHCG companies?

Create a Network Routing Diagram

I will guide you through a possible solution for the requirement, by constructing an island-style Organize superstep that uses a graph data model to reduce the records and the columns on the data set. As a bonus, it creates a visually improved graph that can be used to illustrate the routes.

Open your Python editor and start with a standard ecosystem.

```
################################################################
import sys
import os
import pandas as pd
import networkx as nx
import matplotlib.pyplot as plt
################################################################
pd.options.mode.chained_assignment = None
################################################################
if sys.platform == 'linux' or sys.platform == ' darwin':
    Base=os.path.expanduser('~') + 'VKHCG'
```

```
else:
    Base='C:/VKHCG'
################################################################
print('##############################')
print('Working Base :',Base, ' using ', sys.platform)
print('##############################')
################################################################
sInputFileName='02-Assess/01-EDS/02-Python/Assess-Network-Routing-Company.csv'
################################################################
sOutputFileName1='05-Organise/01-EDS/02-Python/Organise-Network-Routing-
Company.gml'
sOutputFileName2='05-Organise/01-EDS/02-Python/Organise-Network-Routing-
Company.png'
Company='01-Vermeulen'
################################################################
```

Now load the company information from a comma-delimited value file.

```
################################################################
### Import Country Data
################################################################
sFileName=Base + '/' + Company + '/' + sInputFileName
print('##############################')
print('Loading :',sFileName)
print('##############################')
CompanyData=pd.read_csv(sFileName,header=0,low_memory=False,
encoding="latin-1")
print('##############################')
```

Now, you can inspect the company information.

```
################################################################
print(CompanyData.head())
print(CompanyData.shape)
################################################################
```

```
G=nx.Graph()
for i in range(CompanyData.shape[0]):
    for j in range(CompanyData.shape[0]):
        Node0=CompanyData['Company_Country_Name'][i]
        Node1=CompanyData['Company_Country_Name'][j]
        if Node0 != Node1:
            G.add_edge(Node0,Node1)

for i in range(CompanyData.shape[0]):
    Node0=CompanyData['Company_Country_Name'][i]
    Node1=CompanyData['Company_Place_Name'][i] + '('+ CompanyData['Company_
    Country_Name'][i] + ')'
    if Node0 != Node1:
        G.add_edge(Node0,Node1)
```

Now you can inspect the company information within the graph.

```
print('Nodes:', G.number_of_nodes())
print('Edges:', G.number_of_edges())
```

I suggest you store your graph to disk now.

```
################################################################
sFileName=Base + '/' + Company + '/' + sOutputFileName1
print('###############################')
print('Storing :',sFileName)
print('###############################')
nx.write_gml(G, sFileName)
################################################################
```

You can now supply insights against your requirement. I suggest you create an image of the graph, to support your findings.

```
sFileName=Base + '/' + Company + '/' + sOutputFileName2
print('###############################')
print('Storing Graph Image:',sFileName)
print('###############################')
```

```
plt.figure(figsize=(15, 15))
pos=nx.spectral_layout(G,dim=2)
nx.draw_networkx_nodes(G,pos, node_color='k', node_size=10, alpha=0.8)
nx.draw_networkx_edges(G, pos,edge_color='r', arrows=False, style='dashed')
nx.draw_networkx_labels(G,pos,font_size=12,font_family='sans-serif',font_
color='b')
plt.axis('off')
plt.savefig(sFileName,dpi=600)
plt.show()
###############################################################
print('###############################')
print('### Done!! ####################')
print('###############################')
###############################################################
```

Congratulations! You can now answer the requirement. Can you generate a network routing diagram for the VKHCG companies? Yes, you just did. Your network is a core of three nodes with a hub structure like that of a wheel (Figure 11-1).

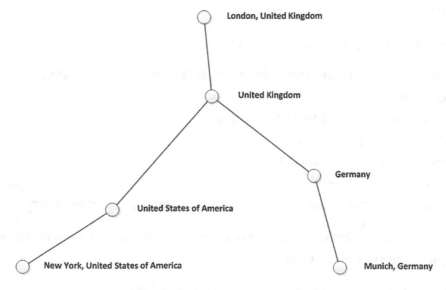

Figure 11-1. *Node diagram for United States, United Kingdom, and Germany*

Krennwallner AG

The Krennwallner group requires picking content for the billboards. Can you organize a model to support them with their inquiry?

Picking Content for Billboards

To enable the marketing salespeople to sell billboard content, they will require a diagram to show what billboards connect to which office content publisher. Each of Krennwallner's billboards has a proximity sensor that enables the content managers to record when a registered visitor points his/her smartphone at the billboard content or touches the near-field pad with a mobile phone.

So, I will assist you in building an organized graph of the billboards' locations data to help you to gain insights into the billboard locations and content picking process.

Open your Python editor and start with a standard ecosystem.

```
################################################################
import sys
import os
import pandas as pd
import networkx as nx
import matplotlib.pyplot as plt
import numpy as np
################################################################
pd.options.mode.chained_assignment = None
################################################################
if sys.platform == 'linux' or sys.platform == ' darwin':
    Base=os.path.expanduser('~') + 'VKHCG'
else:
    Base='C:/VKHCG'
################################################################
print('##############################')
print('Working Base :',Base, ' using ', sys.platform)
print('##############################')
################################################################
sInputFileName='02-Assess/01-EDS/02-Python/Assess-DE-Billboard-Visitor.csv'
################################################################
```

```
sOutputFileName1='05-Organise/01-EDS/02-Python/Organise-Billboards.gml'
sOutputFileName2='05-Organise/01-EDS/02-Python/Organise-Billboards.png'
Billboard='02-Krennwallner'
################################################################
```

Load the proximity sensor data for each billboard.

```
################################################################
### Import Proximity Sensor Data
################################################################
sFileName=Base + '/' + Billboard + '/' + sInputFileName
print('##############################')
print('Loading :',sFileName)
print('##############################')
BillboardDataRaw=pd.read_csv(sFileName,header=0,low_memory=False,
encoding="latin-1")
print('##############################')
```

Let's get insights into the data.

```
################################################################
print(BillboardDataRaw.head())
print(BillboardDataRaw.shape)
BillboardData=BillboardDataRaw
```

I suggest extracting a sample from the population, to create a training and test set of data.

```
sSample=list(np.random.choice(BillboardData.shape[0],20))
```

I suggest generating a graph data set from the sample.

```
G=nx.Graph()
for i in sSample:
    for j in sSample:
        Node0=BillboardData['BillboardPlaceName'][i] + '('+ BillboardData
        ['BillboardCountry'][i] + ')'
        Node1=BillboardData['BillboardPlaceName'][j] + '('+ BillboardData
        ['BillboardCountry'][i] + ')'
```

```
    if Node0 != Node1:
        G.add_edge(Node0,Node1)

for i in sSample:
    Node0=BillboardData['BillboardPlaceName'][i] + '('+ BillboardData
    ['VisitorPlaceName'][i] + ')'
    Node1=BillboardData['BillboardPlaceName'][i] + '('+ BillboardData
    ['VisitorCountry'][i] + ')'
    if Node0 != Node1:
        G.add_edge(Node0,Node1)
```

You can now get various insights from the graph.

```
print('Nodes:', G.number_of_nodes())
print('Edges:', G.number_of_edges())
```

Save your hard work.

```
################################################################
sFileName=Base + '/' + Billboard + '/' + sOutputFileName1
print('##############################')
print('Storing :',sFileName)
print('##############################')
nx.write_gml(G, sFileName)
```

I will now guide you through a visualization process.

```
################################################################
sFileName=Base + '/' + Billboard + '/' + sOutputFileName2
print('##############################')
print('Storing Graph Image:',sFileName)
print('##############################')
plt.figure(figsize=(15, 15))
pos=nx.circular_layout(G,dim=2)
nx.draw_networkx_nodes(G,pos, node_color='k', node_size=150, alpha=0.8)
nx.draw_networkx_edges(G, pos,edge_color='r', arrows=False, style='solid')
nx.draw_networkx_labels(G,pos,font_size=12,font_family='sans-serif',
font_color='b')
plt.axis('off')
```

```
plt.savefig(sFileName,dpi=600)
plt.show()
################################################################
print('##############################')
print('### Done!! ####################')
print('##############################')
################################################################
```

Success! You have a graph that can support any further inquiries and a picture you can distribute to the marketing departments (Figure 11-2).

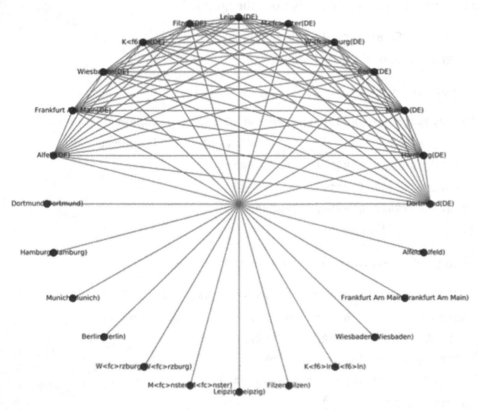

Figure 11-2. *Krennwallner's marketing departments*

Hillman Ltd

Hillman is now expanding its operations to cover a new warehouse and shops. Can you assist the company with a standard planning pack for activating the logistics for Warehouse 13?

Tip   The system records Warehouse 13 as WH-KA13. Have you seen this before? Yes, you are right; you processed it earlier in the book.

Create a Delivery Route

Hillman requires a new delivery route plan from HQ-KA13's delivery region. The managing director has to know the following:

- What his most expensive route is, if the cost is £1.50 per mile and two trips are planned per day

- What the average travel distance in miles is for the region per 30-day month

With your newfound knowledge in building the technology stack for turning data lakes into business assets, can you convert the graph stored in the Assess step called "Assess_Best_Logistics" into the shortest path between the two points? Open your Python editor and start with a standard ecosystem.

```
# -*- coding: utf-8 -*-
################################################################
import sys
import os
import pandas as pd
################################################################
if sys.platform == 'linux' or sys.platform == ' darwin':
    Base=os.path.expanduser('~') + 'VKHCG'
else:
    Base='C:/VKHCG'
################################################################
print('###############################')
print('Working Base :',Base, ' using ', sys.platform)
print('###############################')
################################################################
sInputFileName='02-Assess/01-EDS/02-Python/Assess_Shipping_Routes.txt'
```

```
###############################################################
sOutputFileName='05-Organise/01-EDS/02-Python/Organise-Routes.csv'
Company='03-Hillman'
###############################################################
```

You now import the possible routes that you processed before and stored to disk earlier in the book.

```
###############################################################
### Import Routes Data
###############################################################
sFileName=Base + '/' + Company + '/' + sInputFileName
print('###############################')
print('Loading :',sFileName)
print('###############################')
RouteDataRaw=pd.read_csv(sFileName,header=0,low_memory=False, sep='|',
encoding="latin-1")
print('###############################')
```

Isolate with a horizontal-style slice the specific WH-KA13-related data.

```
###############################################################
RouteStart=RouteDataRaw[RouteDataRaw['StartAt']=='WH-KA13']
###############################################################
```

With a secondary horizontal-style slice, isolate WH-KA13's distance in miles data.

```
RouteDistance=RouteStart[RouteStart['Cost']=='DistanceMiles']
```

Now organize it into a logical order.

```
RouteDistance=RouteDistance.sort_values(by=['Measure'], ascending=False)
```

Determine the maximum distance, and calculate the cost for performing two trips at £1.50 per mile.

```
###############################################################
RouteMax=RouteStart["Measure"].max()
RouteMaxCost=round(((((RouteMax/1000)*1.5*2)),2)
```

```
print('###############################')
print('Maximum (£) per day:')
print(RouteMaxCost)
print('###############################')
```

Determine the mean distance, and calculate the 30-day monthly distance in miles.

```
###############################################################
RouteMean=RouteStart["Measure"].mean()
RouteMeanMonth=round(((((RouteMean/1000)*2*30)),6)
print('###############################')
print('Mean per Month (Miles):')
print(RouteMeanMonth)
print('###############################')
```

You have just used existing data science outcomes and a series of extra organize processes to answer the logistics questions from the MD. Well done. Warehouse 13 will be a big success, if your numbers are correct.

Clark Ltd

Our financial services company has been tasked to investigate the options to convert 1 million pounds sterling into extra income. Mr. Clark Junior suggests using the simple variance in the daily rate between the British pound sterling and the US dollar, to generate extra income from trading. Your chief financial officer wants to know if this is feasible?

Simple Forex Trading Planner

Your challenge is to take 1 million US dollars or just over six hunderd thou sand pounds sterling and, by simply converting it between pounds sterling and US dollars, achieve a profit. Are you up to this challenge?

I will show you how to model this problem and achieve a positive outcome. The forex data has been collected on a daily basis by Clark's accounting department, from previous overseas transactions.

Open your Python editor and start with a standard ecosystem.

```python
# -*- coding: utf-8 -*-
################################################################
import sys
import os
import pandas as pd
import sqlite3 as sq
import re
################################################################
if sys.platform == 'linux' or sys.platform == ' darwin':
    Base=os.path.expanduser('~') + 'VKHCG'
else:
    Base='C:/VKHCG'
################################################################
print('################################')
print('Working Base :',Base, ' using ', sys.platform)
print('################################')
################################################################
sInputFileName='03-Process/01-EDS/02-Python/Process_ExchangeRates.csv'
################################################################
sOutputFileName='05-Organise/01-EDS/02-Python/Organise-Forex.csv'
Company='04-Clark'
################################################################
sDatabaseName=Base + '/' + Company + '/05-Organise/SQLite/clark.db'
conn = sq.connect(sDatabaseName)
#conn = sq.connect(':memory:')
################################################################
```

Import the forex data into the ecosystem.

```python
################################################################
### Import Forex Data
################################################################
sFileName=Base + '/' + Company + '/' + sInputFileName
print('################################')
print('Loading :',sFileName)
```

```
print('###############################')
ForexDataRaw=pd.read_csv(sFileName,header=0,low_memory=False,
encoding="latin-1")
print('###############################')
###################################################################
ForexDataRaw.index.names = ['RowID']
sTable='Forex_All'
print('Storing :',sDatabaseName,' Table:',sTable)
ForexDataRaw.to_sql(sTable, conn, if_exists="replace")
###################################################################
```

You now seed the process with £1,000,000.00.

```
sSQL="SELECT 1 as Bag\
      , CAST(min(Date) AS VARCHAR(10)) as Date \
      ,CAST(1000000.0000000 as NUMERIC(12,4)) as Money \
      ,'USD' as Currency \
      FROM Forex_All \
      ;"
sSQL=re.sub("\s\s+", " ", sSQL)
nMoney=pd.read_sql_query(sSQL, conn)

###################################################################
nMoney.index.names = ['RowID']
sTable='MoneyData'
print('Storing :',sDatabaseName,' Table:',sTable)
nMoney.to_sql(sTable, conn, if_exists="replace")
###################################################################
sTable='TransactionData'
print('Storing :',sDatabaseName,' Table:',sTable)
nMoney.to_sql(sTable, conn, if_exists="replace")
###################################################################
```

You now simulate the trading model Clark Jr. suggested.

> **Note**   Clark Jr.'s hypothesis, or proposed explanation prepared on the basis of limited evidence as a starting point for further investigation by you, is that if you simply keep on exchanging the same "bag" of money, you will produce a profit.

```
ForexDay=pd.read_sql_query("SELECT Date FROM Forex_All GROUP BY Date;", conn)
################################################################
t=0
for i in range(ForexDay.shape[0]):
    sDay=ForexDay['Date'][i]
    sSQL='\
    SELECT M.Bag as Bag, \
            F.Date as Date, \
            round(M.Money * F.Rate,6) AS Money, \
            F.CodeIn AS PCurrency, \
            F.CodeOut AS Currency \
    FROM MoneyData AS M \
    JOIN \
    ( \
        SELECT \
        CodeIn, CodeOut, Date, Rate \
        FROM \
        Forex_All \
        WHERE\
        CodeIn = "USD" AND CodeOut = "GBP" \
        UNION \
        SELECT \
        CodeOut AS CodeIn, CodeIn AS CodeOut,  Date, (1/Rate) AS Rate \
        FROM \
        Forex_All \
        WHERE\
        CodeIn = "USD" AND CodeOut = "GBP" \
    ) AS F \
    ON \
    M.Currency=F.CodeIn \
```

```
AND \
F.Date ="' + sDay + '";'
sSQL=re.sub("\s\s+", " ", sSQL)

ForexDayRate=pd.read_sql_query(sSQL, conn)
for j in range(ForexDayRate.shape[0]):
    sBag=str(ForexDayRate['Bag'][j])
    nMoney=str(round(ForexDayRate['Money'][j],2))
    sCodeIn=ForexDayRate['PCurrency'][j]
    sCodeOut=ForexDayRate['Currency'][j]

sSQL='UPDATE MoneyData SET Date= "' + sDay + '", '
sSQL= sSQL + ' Money = ' + nMoney + ', Currency="' + sCodeOut + '"'
sSQL= sSQL + ' WHERE Bag=' + sBag + ' AND Currency="' + sCodeIn + '";'

sSQL=re.sub("\s\s+", " ", sSQL)
cur = conn.cursor()
cur.execute(sSQL)
conn.commit()
t+=1
print('Trade :', t, sDay, sCodeOut, nMoney)

sSQL=' \
INSERT INTO TransactionData ( \
                            RowID, \
                            Bag, \
                            Date, \
                            Money, \
                            Currency \
                    ) \
SELECT ' + str(t) + ' AS RowID, \
   Bag, \
   Date, \
   Money, \
   Currency \
FROM MoneyData \
;'
```

```
    sSQL=re.sub("\s\s+", " ", sSQL)


    cur = conn.cursor()
    cur.execute(sSQL)
    conn.commit()
############################################################
```

Let's load the transaction log, to investigate the insights into doing this process of achieving extra income.

```
sSQL="SELECT RowID, Bag, Date, Money, Currency FROM TransactionData ORDER
BY RowID;"
sSQL=re.sub("\s\s+", " ", sSQL)
TransactionData=pd.read_sql_query(sSQL, conn)

OutputFile=Base + '/' + Company + '/' + sOutputFileName
TransactionData.to_csv(OutputFile, index = False)
############################################################
```

The challenge was to take $1,000,000 or £603,189.41 on January 4, 1999, and make a profit by converting the complete amount into either British pounds or US dollars. The end result: You made $119,828.98, or a 19.87% return on investment, by May 5, 2017. Well done.

Warning   This is a simple example. Trading foreign exchange is a high-risk enterprise. Please seek professional advice before attempting this for your customers.

Report Superstep

The Report superstep is the step in the ecosystem that enhances the data science findings with the art of storytelling and data visualization. You can perform the best data science, but if you cannot execute a respectable and trustworthy Report step by turning your data science into actionable business insights, you have achieved no advantage for your business. Let me guide you through an example, to show you what you have learned up to now in this book.

Summary of the Results

The most important step in any analysis is the summary of the results. Your data science techniques and algorithms can produce the most methodically, most advanced mathematical or most specific statistical results to the requirements, but if you cannot summarize those into a good story, you have not achieved your requirements.

Understand the Context

What differentiates good data scientists from the best data scientists are not the algorithms or data engineering; it is the ability of the data scientist to apply the context of his findings to the customer.

Example:

The bar has served last rounds at 23:45 and there is one person left. The person drinking has less than 5% of the beer still in his glass at 23:50.

From the context of the drinker, he will have to drink slower to stay till midnight or drink the rest immediately, if he has to catch the midnight bus. From the context of the bar staff, they want to close at 23:45, and they have to get this last patron out the door to lock up, as they, too, want to catch the midnight bus.

It is clear that if you can determine if both parties want to be on the midnight bus, you will have the context of the amount of beer in the drinker's glass. The sole data science measure of the 5% left in the glass is of no value.

I had a junior data scientist try to use image processing from a brewery's CCTV system to determine the levels of people's beers, to enable his data science to notify the bar staff what their rate of beer consumption was during the night. The intention was to generate 3% more profit on the late shift.

The issue was more sinister, as every night, several local regulars would stay after midnight, to catch the new subsidized "Drive-Safe" night bus from the bus terminal across the street home after 12:30, hence not having to pay the normal bus fares. So, they got beer for less than their fare home and traveled at no cost, because they smelled like beer.

The brewery ended paying extra hours of overtime at double rates to six staff members in their three bars, owing to it being after midnight, and staff are not allowed to be in the bar alone with customers, for security reasons.

The staff was always late returning home, as, not qualifying for the late charities' bus, they had to use taxis. It was only after the brewery supplied the data scientist with the external CCTV that we found the real issue. The solution was to supply a coupon for a normal bus trip if customers took it before 23:30 and had beer at the brewery. The total profit equaled +7.3% on the late shift. No complex data science needs only some context.

I have seen too many data scientists spend hours producing great results using the most complex algorithms but being unable to articulate their findings to the business in a manner it understood. Or even worse, not able to get their results into production, because at a bigger scale, they simply did not work as designed by the data scientist— which immediately made the whole exercise pointless in the first place.

Caution   Experience has taught me that this is the one area where data scientists achieve success or failure. Always determine the complete contextual setting of your results. Understand the story behind the results.

Appropriate Visualization

It is true that a picture tells a thousand words. But in data science, you only want your visualizations to tell one story: the findings of the data science you prepared. It is absolutely necessity to ensure that your audience will get your most important message clearly and without any other meanings.

Warning   Beware of a lot of colors and unclear graph format choices. You will lose the message! Keep it simple and to the point.

Practice with your visual tools and achieve a high level of proficiency. I have seen numerous data scientists lose the value of great data science results because they did not perform an appropriate visual presentation.

Eliminate Clutter

Have you ever attended a presentation where the person has painstakingly prepared 50 slides to feedback his data science results? The most painful image is the faces of the people suffering through such a presentation for over two hours.

Tip   On average, it takes five minutes to cover one slide. You should never present more than ten slides.

The biggest task of a data scientist is to eliminate clutter in the data sets. There are various algorithms, such as principal component analysis (PCA), multicollinearity using the variance inflation factor to eliminate dimensions and impute or eliminate missing values, decision trees to subdivide, and backward feature elimination, but the biggest contributor to eliminating clutter is good and solid feature engineering.

Tip   If you do not need it, lose it! Applying appropriate feature engineering wins every time.

Draw Attention Where You Want It

Remember: Your purpose as a data scientist is to deliver insights to your customer, so that they can implement solutions to resolve a problem they may not even know about. You must place the attention on the insight and not the process. However, you must ensure that your process is verified and can support an accredited algorithm or technique.

Warning   Do not get into a situation in which your findings are questioned solely because your methods and procedures are not clear and precise.

Tell only the story that is important right now. You can always come back to tell the next story later.

Telling a Story (Freytag's Pyramid)

Under Freytag's pyramid, the plot of a story consists of five parts: exposition, rising action, climax, falling action, and resolution. This is used by writers of books and screenplays as the basic framework of any story. In the same way, you must take your business through the data science process.

Exposition is the portion of a story that introduces important background information to the audience. In data science, you tell the background of the investigation you performed.

Rising action refers to a series of events that build toward the point of greatest interest. In data science, you point out the important findings or results. Keep it simple and to the point.

The climax is the turning point that determines a good or bad outcome for the story's characters. In data science, you show how your solution or findings will change the outcome of the work you performed.

During the falling action, the conflict between what occurred before and after the climax takes place. In data science, you prove that after your suggestion has been implemented in a pilot, the same techniques can be used to find the issues now proving that the issues can inevitably be resolved.

Resolution is the outcome of the story. In data science, you produce the solution and make the improvements permanent.

Decide what repeatable sound bite you can use to help your core message stick with your audience. People remember the core things by your repeating them. You must ensure one thing: that the customer remembers your core message.

Make sure you deliver that message clearly and without any confusion. Drive the core message home by repeating sound bites throughout the meeting.

The art of a good data scientist lies in the ability to tell a story about the data and get people to remember its baseline.

Tip   Practice your storytelling with friends and family. Get them to give you feedback. Make sure you are comfortable speaking in front of people.

Now that you can tell the story with ease, I will supply you with more examples of how you create the Report step for the data science.

Engineering a Practical Report Superstep

Any source code or other supplementary material referenced by me in this book is available to readers on GitHub, via this book's product page, located at www.apress.com/ 9781484230534. Please note that this source code assumes you have completed the source code setup outlined in Chapter 2.

Let's check your graphics capacity.

```
import matplotlib.pyplot as plt
import matplotlib.patheffects as path_effects

fig = plt.figure(figsize=(21, 3))
t = fig.text(0.2, 0.5, 'Practical Data Science', fontsize=75, weight=1000,
va='center')
t.set_path_effects([path_effects.PathPatchEffect(offset=(4, -4),
hatch='xxxx',
          facecolor='gray'),
       path_effects.PathPatchEffect(edgecolor='white', linewidth=1.1,
          facecolor='black')])
plt.plot()
plt.show()
```

Now that I have explained the various aspects of the Report superstep, I will show you how to help VKHC group companies with their processing. Try this code to find out your progress on the data science company scale:

```
import matplotlib.pyplot as plt
from numpy.random import randn
z = randn(8)
plt.figure(figsize=(15, 5))
red_dot, = plt.plot(z, "ro-", markersize=18)
white_cross, = plt.plot(z[:5], "w+", markeredgewidth=3, markersize=15)
black_cross, = plt.plot(z[:2], "y+", markeredgewidth=3, markersize=15)
plt.legend([red_dot, (red_dot, white_cross),(red_dot,black_cross)],\
          ["Krennwallner", "Clark","Hillman"])
```

Run it a few times, if you want better results.

Vermeulen PLC

Vermeulen requires a map of all their customers' data links. Can you provide a report to deliver this? I will guide you through an example that delivers this requirement.

Creating a Network Routing Diagram

Let's load the customers' data and create a graph database. You can use the graph database to then create an image to support your story about the customers.

Open your Python editor and create the standard ecosystem.

```python
################################################################
import sys
import os
import pandas as pd
import networkx as nx
import matplotlib.pyplot as plt
################################################################
pd.options.mode.chained_assignment = None
################################################################
if sys.platform == 'linux' or sys.platform == ' darwin':
    Base=os.path.expanduser('~') + 'VKHCG'
else:
    Base='C:/VKHCG'
################################################################
print('################################')
print('Working Base :',Base, ' using ', sys.platform)
print('################################')
################################################################
sInputFileName='02-Assess/01-EDS/02-Python/Assess-Network-Routing-Customer.csv'
################################################################
sOutputFileName1='06-Report/01-EDS/02-Python/Report-Network-Routing-
Customer.gml'
sOutputFileName2='06-Report/01-EDS/02-Python/Report-Network-Routing-
Customer.png'
Company='01-Vermeulen'
################################################################
```

Import the data from the Access step for the networking customers.

```
################################################################
### Import Country Data
################################################################
sFileName=Base + '/' + Company + '/' + sInputFileName
print('###############################')
print('Loading :',sFileName)
print('###############################')
CustomerDataRaw=pd.read_csv(sFileName,header=0,low_memory=False,
encoding="latin-1")
CustomerData=CustomerDataRaw.head(100)
print('Loaded Country:',CustomerData.columns.values)
print('###############################')
################################################################
print(CustomerData.head())
print(CustomerData.shape)
################################################################
G=nx.Graph()
for i in range(CustomerData.shape[0]):
    for j in range(CustomerData.shape[0]):
        Node0=CustomerData['Customer_Country_Name'][i]
        Node1=CustomerData['Customer_Country_Name'][j]
        if Node0 != Node1:
            G.add_edge(Node0,Node1)

for i in range(CustomerData.shape[0]):
    Node0=CustomerData['Customer_Country_Name'][i]
    Node1=CustomerData['Customer_Place_Name'][i] + '('+
    CustomerData['Customer_Country_Name'][i] + ')'
    Node2='('+ "{:.9f}".format(CustomerData['Customer_Latitude'][i]) + ')\
    ('+ "{:.9f}".format(CustomerData['Customer_Longitude'][i]) + ')'
    if Node0 != Node1:
        G.add_edge(Node0,Node1)
    if Node1 != Node2:
        G.add_edge(Node1,Node2)
```

Here are your insights into the networking customers:

```
print('Nodes:', G.number_of_nodes())
print('Edges:', G.number_of_edges())
#############################################################
```

Export the data for the networking customers.

```
sFileName=Base + '/' + Company + '/' + sOutputFileName1
print('###############################')
print('Storing :',sFileName)
print('###############################')
nx.write_gml(G, sFileName)
#############################################################
sFileName=Base + '/' + Company + '/' + sOutputFileName2
print('###############################')
print('Storing Graph Image:',sFileName)
print('###############################')
```

You can now prepare an image for the networking customers, to explain the insights you achieved.

```
plt.figure(figsize=(25, 25))
pos=nx.spectral_layout(G,dim=2)
nx.draw_networkx_nodes(G,pos, node_color='k', node_size=10, alpha=0.8)
nx.draw_networkx_edges(G, pos,edge_color='r', arrows=False, style='dashed')
nx.draw_networkx_labels(G,pos,font_size=12,font_family='sans-serif',font_
color='b')
plt.axis('off')
plt.savefig(sFileName,dpi=600)
plt.show()
#############################################################
print('###############################')
print('### Done!! ####################')
print('###############################')
#############################################################
```

Well done. You have a report for the management of Vermeulen on their network requirements for their customers.

Krennwallner AG

The Krennwallner marketing department wants to deploy the locations of the billboards onto the company web server. Can you prepare three versions of the locations' web pages?

- Locations clustered into bubbles when you zoom out

- Locations as pins

- Locations as heat map

To achieve this, I must introduce you to a new library: folium. This library assists with the creation of web pages by using locations. Install it using conda install -c conda-forge folium.

Picking Content for Billboards

Open your Python editor and set up the following ecosystem:

```
################################################################
# -*- coding: utf-8 -*-
################################################################
import sys
import os
import pandas as pd
from folium.plugins import FastMarkerCluster, HeatMap
from folium import Marker, Map
import webbrowser
################################################################
if sys.platform == 'linux' or sys.platform == ' darwin':
    Base=os.path.expanduser('~') + '/VKHCG'
else:
    Base='C:/VKHCG'
print('###############################')
print('Working Base :',Base, ' using ', sys.platform)
print('###############################')
################################################################
```

Import the required data for the companies' billboards in Germany.

```
sFileName=Base+'/02-Krennwallner/01-Retrieve/01-EDS/02-Python/Retrieve_DE_
Billboard_Locations.csv'
df = pd.read_csv(sFileName,header=0,low_memory=False, encoding="latin-1")
df.fillna(value=0, inplace=True)
print(df.shape)
```

The data requires some missing data treatment. I suggest simply removing the "bad"-quality records by performing a try-except process.

```
################################################################
t=0
for i in range(df.shape[0]):
    try:
        sLongitude=df["Longitude"][i]
        sLongitude=float(sLongitude)
    except Exception:
        sLongitude=float(0.0)

    try:
        sLatitude=df["Latitude"][i]
        sLatitude=float(sLatitude)
    except Exception:
        sLatitude=float(0.0)

    try:
        sDescription=df["Place_Name"][i] + ' (' + df["Country"][i]+')'
    except Exception:
        sDescription='VKHCG'

    if sLongitude != 0.0 and sLatitude != 0.0:
        DataClusterList=list([sLatitude, sLongitude])
        DataPointList=list([sLatitude, sLongitude, sDescription])
        t+=1
        if t==1:
            DataCluster=[DataClusterList]
            DataPoint=[DataPointList]
```

```
        else:
            DataCluster.append(DataClusterList)
            DataPoint.append(DataPointList)
data=DataCluster
```

You now have a data set that can be used to create the web content. You can start with the cluster web page first.

```
pins=pd.DataFrame(DataPoint)
pins.columns = [ 'Latitude','Longitude','Description']
################################################################
billbords_map1 = Map(location=[48.1459806, 11.4985484], zoom_start=5)
marker_cluster = FastMarkerCluster(data).add_to(billbords_map1)
sFileNameHtml=Base+'/01-Vermeulen/06-Report/01-EDS/02-Python/Billboard1.
html'
billbords_map1.save(sFileNameHtml)
```

Now that you saved the web page, let's open it and see your achievement.

```
webbrowser.open('file://' + os.path.realpath(sFileNameHtml))
```

Next, you can produce the pins or marker-only web page.

```
################################################################
billbords_map2 = Map(location=[48.1459806, 11.4985484], zoom_start=5)
for name, row in pins.iloc[:100].iterrows():
    Marker([row["Latitude"],row["Longitude"]], popup=row["Description"]).
add_to(billbords_map2)
sFileNameHtml=Base+'/01-Vermeulen/06-Report/01-EDS/02-Python/Billboard2.
html'
billbords_map2.save(sFileNameHtml)
webbrowser.open('file://' + os.path.realpath(sFileNameHtml))
```

We're making good progress with the example. You can also prepare the heat map version of the data into a web page.

```
################################################################
billbords_heatmap = Map(location=[48.1459806, 11.4985484], zoom_start=5)
billbords_heatmap.add_child(HeatMap([[row["Latitude"], row["Longitude"]]
for name, row in pins.iloc[:100].iterrows()]))
```

729

```
sFileNameHtml=Base+'/01-Vermeulen/06-Report/01-EDS/02-Python/Billboard_
heatmap.html'
billbords_heatmap.save(sFileNameHtml)
webbrowser.open('file://' + os.path.realpath(sFileNameHtml))
################################################################
print('### Done!! ######################################')
################################################################
```

Hillman Ltd

Dr. Hillman Sr. has just installed a camera system that enables the company to
capture video and, therefore, indirectly, images of all containers that enter or leave the
warehouse. Can you convert the number on the side of the containers into digits?

Reading the Containers

I will assist you with a quick experiment to prove that you can perform the task required.
Open your Python editor and set up this ecosystem:

```
from time import time

import numpy as np
import matplotlib.pyplot as plt
from matplotlib import offsetbox
from sklearn import (manifold, datasets, decomposition, ensemble,
                     discriminant_analysis, random_projection)

digits = datasets.load_digits(n_class=6)
X = digits.data
y = digits.target
n_samples, n_features = X.shape
n_neighbors = 30
```

Create a function for scaling and visualizing the embedding vectors.

```
def plot_embedding(X, title=None):
    x_min, x_max = np.min(X, 0), np.max(X, 0)
    X = (X - x_min) / (x_max - x_min)
```

```
plt.figure(figsize=(10, 10))
ax = plt.subplot(111)
for i in range(X.shape[0]):
    plt.text(X[i, 0], X[i, 1], str(digits.target[i]),
            color=plt.cm.Set1(y[i] / 10.),
            fontdict={'weight': 'bold', 'size': 9})

if hasattr(offsetbox, 'AnnotationBbox'):
    # only print thumbnails with matplotlib > 1.0
    shown_images = np.array([[1., 1.]])  # just something big
    for i in range(digits.data.shape[0]):
        dist = np.sum((X[i] - shown_images) ** 2, 1)
        if np.min(dist) < 4e-3:
            # don't show points that are too close
            continue
        shown_images = np.r_[shown_images, [X[i]]]
        imagebox = offsetbox.AnnotationBbox(
            offsetbox.OffsetImage(digits.images[i], cmap=plt.cm.gray_r),
            X[i])
        ax.add_artist(imagebox)
plt.xticks([]), plt.yticks([])
if title is not None:
    plt.title(title)
```

You can now plot images of the digits you identified.

```
n_img_per_row = 20
img = np.zeros((10 * n_img_per_row, 10 * n_img_per_row))
for i in range(n_img_per_row):
    ix = 10 * i + 1
    for j in range(n_img_per_row):
        iy = 10 * j + 1
        img[ix:ix + 8, iy:iy + 8] = X[i * n_img_per_row + j].reshape((8, 8))

plt.figure(figsize=(10, 10))
plt.imshow(img, cmap=plt.cm.binary)
plt.xticks([])
```

```
plt.yticks([])
plt.title('A selection from the 64-dimensional digits dataset')
```

You can process with a random 2D projection using a random unitary matrix.

```
print("Computing random projection")
rp = random_projection.SparseRandomProjection(n_components=2, random_
state=42)
X_projected = rp.fit_transform(X)
plot_embedding(X_projected, "Random Projection of the digits")
```

Now add a projection onto the first two principal components.

```
print("Computing PCA projection")
t0 = time()
X_pca = decomposition.TruncatedSVD(n_components=2).fit_transform(X)
plot_embedding(X_pca,
               "Principal Components projection of the digits (time %.2fs)" %
               (time() - t0))
```

Add a projection on to the first two linear discriminant components.

```
print("Computing Linear Discriminant Analysis projection")
X2 = X.copy()
X2.flat[::X.shape[1] + 1] += 0.01  # Make X invertible
t0 = time()
X_lda = discriminant_analysis.LinearDiscriminantAnalysis(n_components=2).
fit_transform(X2, y)
plot_embedding(X_lda,
               "Linear Discriminant projection of the digits (time %.2fs)" %
               (time() - t0))
```

Add an Isomap projection of the digits data set.

```
print("Computing Isomap embedding")
t0 = time()
X_iso = manifold.Isomap(n_neighbors, n_components=2).fit_transform(X)
print("Done.")
plot_embedding(X_iso,
```

```
                      "Isomap projection of the digits (time %.2fs)" %
                      (time() - t0))
```

Add a locally linear embedding of the digits data set.

```
print("Computing LLE embedding")
clf = manifold.LocallyLinearEmbedding(n_neighbors, n_components=2,
                                      method='standard')
t0 = time()
X_lle = clf.fit_transform(X)
print("Done. Reconstruction error: %g" % clf.reconstruction_error_)
plot_embedding(X_lle,
                      "Locally Linear Embedding of the digits (time %.2fs)" %
                      (time() - t0))
```

Add a modified locally linear embedding of the digits data set.

```
print("Computing modified LLE embedding")
clf = manifold.LocallyLinearEmbedding(n_neighbors, n_components=2,
                                      method='modified')
t0 = time()
X_mlle = clf.fit_transform(X)
print("Done. Reconstruction error: %g" % clf.reconstruction_error_)
plot_embedding(X_mlle,
                      "Modified Locally Linear Embedding of the digits (time %.2fs)" %
                      (time() - t0))
```

Add a Hessian LLE embedding of the digits data set.

```
print("Computing Hessian LLE embedding")
clf = manifold.LocallyLinearEmbedding(n_neighbors, n_components=2,
                                      method='hessian')
t0 = time()
X_hlle = clf.fit_transform(X)
print("Done. Reconstruction error: %g" % clf.reconstruction_error_)
plot_embedding(X_hlle,
                      "Hessian Locally Linear Embedding of the digits (time %.2fs)" %
                      (time() - t0))
```

Add an LTSA embedding of the digits data set.

```
print("Computing LTSA embedding")
clf = manifold.LocallyLinearEmbedding(n_neighbors, n_components=2,
                                        method='ltsa')
t0 = time()
X_ltsa = clf.fit_transform(X)
print("Done. Reconstruction error: %g" % clf.reconstruction_error_)
plot_embedding(X_ltsa,
                "Local Tangent Space Alignment of the digits (time %.2fs)" %
                (time() - t0))
```

Add an MDS embedding of the digits data set.

```
print("Computing MDS embedding")
clf = manifold.MDS(n_components=2, n_init=1, max_iter=100)
t0 = time()
X_mds = clf.fit_transform(X)
print("Done. Stress: %f" % clf.stress_)
plot_embedding(X_mds,
                "MDS embedding of the digits (time %.2fs)" %
                (time() - t0))
```

Add a random trees embedding of the digits data set.

```
print("Computing Totally Random Trees embedding")
hasher = ensemble.RandomTreesEmbedding(n_estimators=200, random_state=0,
                                        max_depth=5)
t0 = time()
X_transformed = hasher.fit_transform(X)
pca = decomposition.TruncatedSVD(n_components=2)
X_reduced = pca.fit_transform(X_transformed)

plot_embedding(X_reduced,
                "Random forest embedding of the digits (time %.2fs)" %
                (time() - t0))
```

Add a spectral embedding of the digits data set.

```
print("Computing Spectral embedding")
embedder = manifold.SpectralEmbedding(n_components=2, random_state=0,
                                      eigen_solver="arpack")
t0 = time()
X_se = embedder.fit_transform(X)

plot_embedding(X_se,
               "Spectral embedding of the digits (time %.2fs)" %
               (time() - t0))
```

Add a t-SNE embedding of the digits data set.

```
print("Computing t-SNE embedding")
tsne = manifold.TSNE(n_components=2, init='pca', random_state=0)
t0 = time()
X_tsne = tsne.fit_transform(X)

plot_embedding(X_tsne,
               "t-SNE embedding of the digits (time %.2fs)" %
               (time() - t0))
```

```
plt.show()
```

Great job! You have successfully completed the container experiment. Which display format do you think is the best?

The right answer is your choice, as it has to be the one that matches your own insight into the data, and there is not really a wrong answer.

Clark Ltd

The financial company in VKHCG is the Clark accounting firm that VKHCG owns with a 60% stake. The accountants are the financial advisers to the group and handle everything to do with the complex work of international accounting.

Financials

The VKHCG companies did well last year, and the teams at Clark must prepare a balance sheet for each company in the group. The companies require a balance sheet for each company, to be produced using the template (`Balance-Sheet-Template.xlsx`) that can be found in the example directory (`..\VKHCG\04-Clark\00-RawData`). Figure 11-3 shows the basic layout.

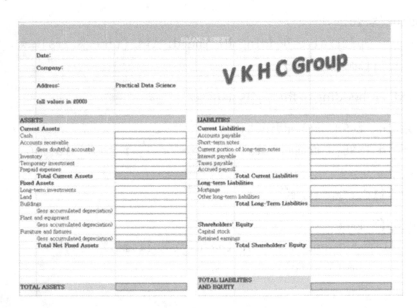

Figure 11-3. *Basic internal balance sheet for VKHCG companies*

I will guide you through a process that will enable you to merge the data science with preformatted Microsoft Excel template, to produce a balance sheet for each of the VKHCG companies.

Note A new library named openpyxl supports you in editing Microsoft Excel workbooks.

Install the new library by using `"conda install -c anaconda openpyxl"`. Open your Python ecosystem and build the following code into it.

Set up a standard ecosystem.

```
# -*- coding: utf-8 -*-
################################################################
import sys
import os
import pandas as pd
import sqlite3 as sq
import re
from openpyxl import load_workbook
################################################################
if sys.platform == 'linux' or sys.platform == ' darwin':
    Base=os.path.expanduser('~') + 'VKHCG'
else:
    Base='C:/VKHCG'
################################################################
print('################################')
print('Working Base :',Base, ' using ', sys.platform)
print('################################')
################################################################
```

Set the input data for the example.

```
sInputTemplateName='00-RawData/Balance-Sheet-Template.xlsx'
```

Set up the preamble for the output of the example.

```
sOutputFileName='05-Organise/01-EDS/02-Python/Report-Balance-Sheet'
Company='04-Clark'
################################################################
sDatabaseName=Base + '/' + Company + '/06-Report/SQLite/clark.db'
```

You have a choice here. You can use a disk-bound SQLite database that will store the data to the hard disk for use at later stages but is not as fast as a second option.

```
conn = sq.connect(sDatabaseName)
```

The second option is keeping the data in a memory-bound database, which is faster to access but not permanent.

```
#conn = sq.connect(':memory:')
################################################################
### Import Balance Sheet Data
################################################################
for y in range(1,13):
    sInputFileName='00-RawData/BalanceSheets' + str(y).zfill(2) + '.csv'
    sFileName=Base + '/' + Company + '/' + sInputFileName
    print('###############################')
    print('Loading :',sFileName)
    print('###############################')
    ForexDataRaw=pd.read_csv(sFileName,header=0,low_memory=False,
    encoding="latin-1")
    print('###############################')
    ################################################################
    ForexDataRaw.index.names = ['RowID']
    sTable='BalanceSheets'
    print('Storing :',sDatabaseName,' Table:',sTable)
    if y == 1:
        print('Load Data')
        ForexDataRaw.to_sql(sTable, conn, if_exists="replace")
    else:
        print('Append Data')
        ForexDataRaw.to_sql(sTable, conn, if_exists="append")
################################################################
sSQL="SELECT \
        Year, \
        Quarter, \
        Country, \
        Company, \
        CAST(Year AS INT) || 'Q' || CAST(Quarter AS INT) AS sDate, \
        Company || ' (' || Country || ')' AS sCompanyName , \
        CAST(Year AS INT) || 'Q' || CAST(Quarter AS INT) || '-' ||\
        Company || '-' || Country AS sCompanyFile \
```

```
        FROM BalanceSheets \
        GROUP BY \
            Year, \
            Quarter, \
            Country, \
            Company \
        HAVING Year is not null \
        ;"
sSQL=re.sub("\s\s+", " ", sSQL)
sDatesRaw=pd.read_sql_query(sSQL, conn)
print(sDatesRaw.shape)
sDates=sDatesRaw.head(5)
###################################################################
## Loop Dates
###################################################################
for i in range(sDates.shape[0]):
    sFileName=Base + '/' + Company + '/' + sInputTemplateName
    wb = load_workbook(sFileName)
    ws=wb.get_sheet_by_name("Balance-Sheet")
    sYear=sDates['sDate'][i]
    sCompany=sDates['sCompanyName'][i]
    sCompanyFile=sDates['sCompanyFile'][i]
    sCompanyFile=re.sub("\s+", "", sCompanyFile)

    ws['D3'] = sYear
    ws['D5'] = sCompany

    sFields = pd.DataFrame(
            [
        ['Cash','D16', 1],
        ['Accounts_Receivable','D17', 1],
        ['Doubtful_Accounts','D18', 1],
        ['Inventory','D19', 1],
        ['Temporary_Investment','D20', 1],
        ['Prepaid:Expenses','D21', 1],
        ['Long_Term_Investments','D24', 1],
        ['Land','D25', 1],
```

```
        ['Buildings','D26', 1],
        ['Depreciation_Buildings','D27', -1],
        ['Plant_Equipment','D28', 1],
        ['Depreciation_Plant_Equipment','D29', -1],
        ['Furniture_Fixtures','D30', 1],
        ['Depreciation_Furniture_Fixtures','D31', -1],
        ['Accounts_Payable','H16', 1],
        ['Short_Term_Notes','H17', 1],
        ['Current_Long_Term_Notes','H18', 1],
        ['Interest_Payable','H19', 1],
        ['Taxes_Payable','H20', 1],
        ['Accrued_Payroll','H21', 1],
        ['Mortgage','H24', 1],
        ['Other_Long_Term_Liabilities','H25', 1],
        ['Capital_Stock','H30', 1]
        ]
         )

nYear=str(int(sDates['Year'][i]))
nQuarter=str(int(sDates['Quarter'][i]))
sCountry=str(sDates['Country'][i])
sCompany=str(sDates['Company'][i])

sFileName=Base + '/' + Company + '/' + sOutputFileName + \
'-' + sCompanyFile + '.xlsx'

print(sFileName)

for j in range(sFields.shape[0]):

    sSumField=sFields[0][j]
    sCellField=sFields[1][j]
    nSumSign=sFields[2][j]

    sSQL="SELECT   \
        Year, \
        Quarter, \
```

```
            Country, \
            Company, \
            SUM(" + sSumField + ") AS nSumTotal \
        FROM BalanceSheets \
        GROUP BY \
            Year, \
            Quarter, \
            Country, \
            Company \
        HAVING \
            Year=" + nYear + " \
        AND \
            Quarter=" + nQuarter + " \
        AND \
            Country='" + sCountry + "' \
        AND \
            Company='" + sCompany + "' \
        ;"
    sSQL=re.sub("\s\s+", " ", sSQL)
    sSumRaw=pd.read_sql_query(sSQL, conn)

    ws[sCellField] = sSumRaw["nSumTotal"][0] * nSumSign

    print('Set cell',sCellField,' to ', sSumField,'Total')

 wb.save(sFileName)
```

Well done. You now have all the reports you need.

Graphics

Until now, you have seen little graphical visualization in the examples. I will now guide you through a number of visualizations that I find particularly useful when I present my data to my customers.

Plot Options

First, let's look at several plotting options within the ecosystem. Open your Python editor and set up the following ecosystem:

```python
################################################################
# -*- coding: utf-8 -*-
################################################################
import sys
import os
import pandas as pd
import matplotlib as ml
from matplotlib import pyplot as plt
################################################################
if sys.platform == 'linux' or sys.platform == ' darwin':
    Base=os.path.expanduser('~') + '/VKHCG'
else:
    Base='C:/VKHCG'
print('#############################')
print('Working Base :',Base, ' using ', sys.platform)
print('#############################')
################################################################
GBase = Base+'/01-Vermeulen/06-Report/01-EDS/02-Python/'
ml.style.use('ggplot')
```

Here is a special data set to demonstrate the visualization.

```python
data=[
['London',     29.2, 17.4],
['Glasgow',     18.8, 11.3],
['Cape Town',     15.3, 9.0],
['Houston',     22.0, 7.8],
['Perth',     18.0, 23.7],
['San Francisco',     11.4, 33.3]
]
```

```
os_new=pd.DataFrame(data)
pd.Index(['Item', 'Value', 'Value Percent', 'Conversions', 'Conversion
Percent',
        'URL', 'Stats URL'],
      dtype='object')

os_new.rename(columns = {0 : "Warehouse Location"}, inplace=True)
os_new.rename(columns = {1 : "Profit 2016"}, inplace=True)
os_new.rename(columns = {2 : "Profit 2017"}, inplace=True)
```

Pie Graph

Let's start with a simple pie graph.

```
explode = (0, 0, 0, 0, 0, 0.1)
labels=os_new['Warehouse Location']
colors_mine = ['yellowgreen', 'gold', 'lightskyblue', 'lightcoral',
'lightcyan','lightblue']
os_new.plot(figsize=(10, 10),kind="pie", y="Profit 2017",autopct='%.2f%%', \
            shadow=True, explode=explode, legend = False, colors = colors_mine,\
            labels=labels, fontsize=20)
sPicNameOut1=GBase+'pie_explode.png'
plt.savefig(sPicNameOut1,dpi=600)
```

Figure 11-4 shows the resulting pie graph.

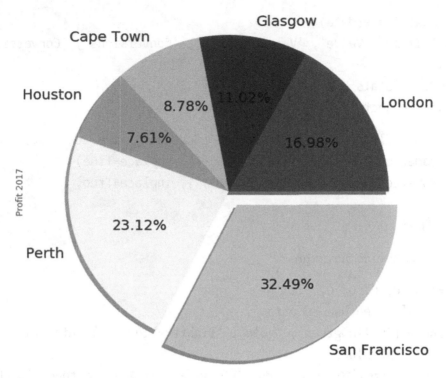

Figure 11-4.   *Pie graph*

Double Pie

With the following, you can put two pie graphs side by side:

```
explode = (0, 0, 0, 0, 0, 0)
colors_mine = ['yellowgreen', 'gold', 'lightskyblue', 'lightcoral',
'lightcyan','lightblue']
os_new.plot(figsize=(10, 5),kind="pie", y=['Profit 2016','Profit
2017'],autopct='%.2f%%', \
          shadow=True, explode=explode, legend = False, colors =
          colors_mine,\
          subplots=True, labels=labels, fontsize=10)
sPicNameOut2=GBase+'pie.png'
plt.savefig(sPicNameOut2,dpi=600)
```

Figure 11-5 shows the resulting double pie graph.

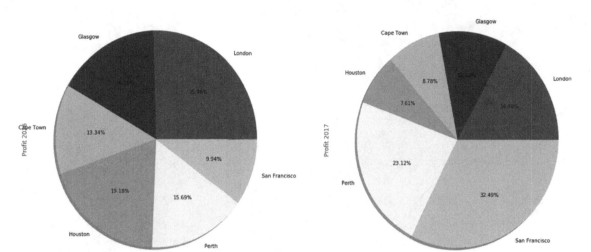

Figure 11-5. *Double pie graph*

Line Graph

We can also visualize the data in the form of a line graph.

```
os_new.iloc[:5].plot(figsize=(10, 10),kind='Line',x='Warehouse Location',\
           y=['Profit 2016','Profit 2017']);
sPicNameOut3=GBase+'line.png'
plt.savefig(sPicNameOut3,dpi=600)
```

Figure 11-6 shows the resulting line graph.

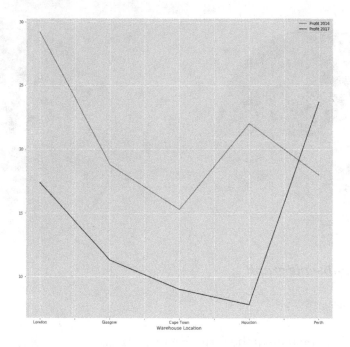

Figure 11-6. *Line graph*

Bar Graph

The following results in a bar graph:

```
os_new.iloc[:5].plot(figsize=(10, 10),kind='bar',x='Warehouse Location',\
        y=['Profit 2016','Profit 2017']);
sPicNameOut4=GBase+'bar.png'
plt.savefig(sPicNameOut4,dpi=600)
```

Figure 11-7 shows the resulting bar graph.

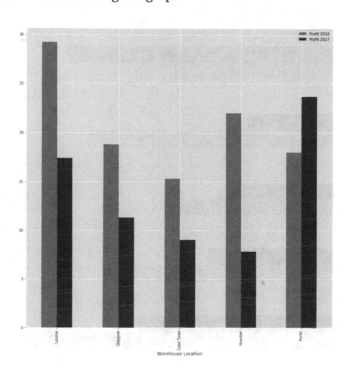

Figure 11-7. *Bar graph*

Horizontal Bar Graph

You can also create a horizontal bar graph.

```
os_new.iloc[:5].plot(figsize=(10, 10),kind='barh',x='Warehouse Location',\
          y=['Profit 2016','Profit 2017']);
sPicNameOut5=GBase+'hbar.png'
plt.savefig(sPicNameOut5,dpi=600)
```

Figure 11-8 shows the resulting horizontal bar graph.

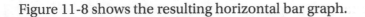

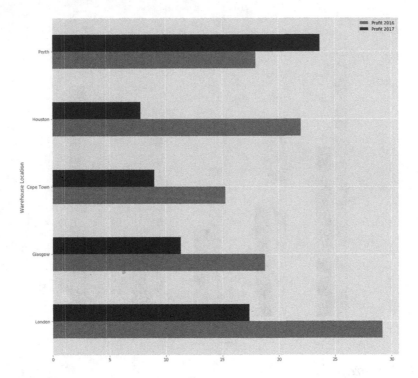

Figure 11-8. *Horizontal bar graph*

Area Graph

An area graph format is also possible.

```
os_new.iloc[:5].plot(figsize=(10, 10),kind='area',x='Warehouse Location',\
        y=['Profit 2016','Profit 2017'],stacked=False);
sPicNameOut6=GBase+'area.png'
plt.savefig(sPicNameOut6,dpi=600)
```

Figure 11-9 shows the resulting area graph.

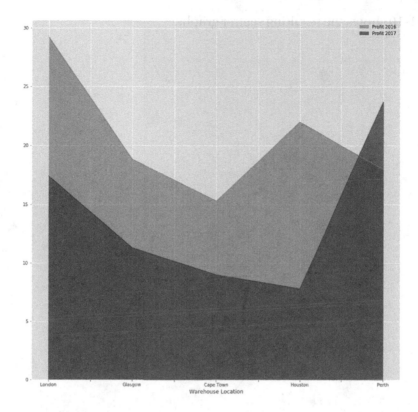

Figure 11-9. *Area graph*

Scatter Graph

We could also visualize using scatter graphs, if you want.

```
os_new.iloc[:5].plot(figsize=(10, 10),kind='scatter',x='Profit 2016',\
        y='Profit 2017',color='DarkBlue',marker='D');
sPicNameOut7=GBase+'scatter.png'
plt.savefig(sPicNameOut7,dpi=600)
```

Figure 11-10 shows the resulting scatter graph.

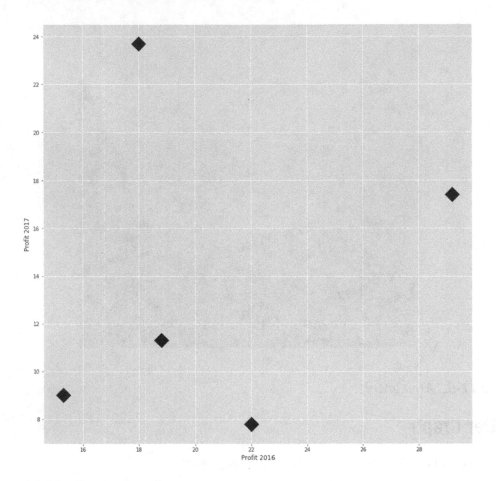

Figure 11-10. *Scatter graph*

Hex Bin Graph

Here is something interesting I discovered—a hex bin graph. The hex graph is included because, if you can perform the data science, you can

- Easily judge distances, owing to the built-in scale

- Easily judge the density of points of interest, also owing to the built-in scale

- Easily scale the map up or down and put things roughly where they need to be for geospatial reporting formats

Let's look at a simple example for a hex graph. We have a farm of about 25 × 25 miles. If we measure the profit generated by each square mile of farming, we can discover the areas of the farm that are profitable or not.

```
os_new.iloc[:5].plot(figsize=(13, 10),kind='hexbin',x='Profit 2016',\
         y='Profit 2017', gridsize=25);
sPicNameOut8=GBase+'hexbin.png'
plt.savefig(sPicNameOut8,dpi=600)
```

Figure 11-11 shows the resulting hex bin graph, which indicates that there are five fields that generated profits both in 2016 and 2017.

Figure 11-11. *Hex bin graph for farm*

More Complex Graphs

I will next show you some more complex graph techniques that I have used over my time working with data from various data sources. Often, a normal pie, line, or bar graph simply does not convey the full visualization of the results, or the data is so complex that it cannot be visualized without some extra data science applied to resolve the complexities.

I will now guide you through a few of the more common techniques I use regularly.

Kernel Density Estimation (KDE) Graph

Kernel density estimation is an essential data-smoothing technique in which interpretations about the population are prepared, based on a finite data sample.

```python
###############################################################
# -*- coding: utf-8 -*-
###############################################################
import sys
import os
import pandas as pd
import matplotlib as ml
import numpy as np
from matplotlib import pyplot as plt
###############################################################
if sys.platform == 'linux' or sys.platform == ' darwin':
    Base=os.path.expanduser('~') + '/VKHCG'
else:
    Base='C:/VKHCG'
print('###############################')
print('Working Base :',Base, ' using ', sys.platform)
print('###############################')
###############################################################

ml.style.use('ggplot')

fig1=plt.figure(figsize=(10, 10))
ser = pd.Series(np.random.randn(1000))
ser.plot(figsize=(10, 10),kind='kde')
```

```
sPicNameOut1=Base+'/01-Vermeulen/06-Report/01-EDS/02-Python/kde.png'
plt.savefig(sPicNameOut1,dpi=600)
plt.tight_layout()
plt.show()
```

Figure 11-12 shows the resulting KDE graph, which illustrates the estimate of the probability density function of a random variable. It reduces the random thousand points into a curve that indicates how the values change across the population, without displaying each value.

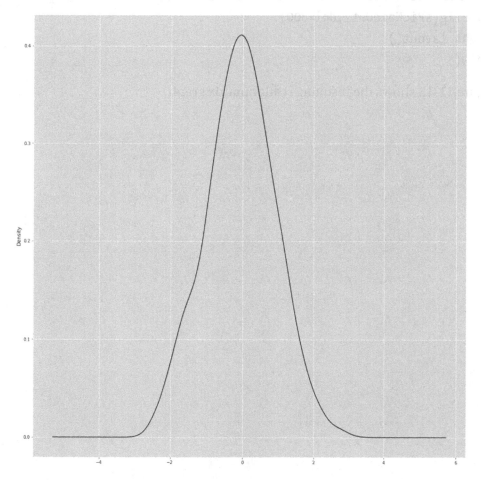

Figure 11-12. *Kernel density estimation graph*

Scatter Matrix Graph

Scatter matrix is a statistic tool that is used to estimate the covariance matrix.

```
fig2=plt.figure(figsize=(10, 10))
from pandas.plotting import scatter_matrix
df = pd.DataFrame(np.random.randn(1000, 5), columns=['Y2014','Y2015',
'Y2016', 'Y2017', 'Y2018'])
scatter_matrix(df, alpha=0.2, figsize=(10, 10), diagonal='kde')
sPicNameOut2=Base+'/01-Vermeulen/06-Report/01-EDS/02-Python/scatter_matrix.png'
plt.savefig(sPicNameOut2,dpi=600)
plt.tight_layout()
plt.show()
```

Figure 11-13 shows the resulting scatter matrix graph.

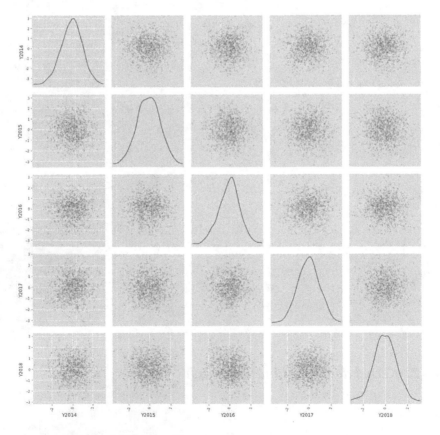

Figure 11-13. *Scatter matrix graph*

Andrews' Curves

Visualization with Andrews' curve is a way to visualize structure in high-dimensional data. Andrews' curves are examples of the space transformed visualization (STV) techniques for visualizing multivariate data, which represent k-dimensional data points by a profile line curve in two- or three-dimensional space, using orthogonal basis functions. Andrews' curves are based on a Fourier series in which the coefficients are the observation's values. In simple terms, they allow you, in a structured way, to reduce dimensional results of more than three dimensions into a visualized format.

Open your Python editor and set up this basic ecosystem:

```python
import sys
import os
import pandas as pd
from matplotlib import pyplot as plt
################################################################
if sys.platform == 'linux' or sys.platform == ' darwin':
    Base=os.path.expanduser('~') + '/VKHCG'
else:
    Base='C:/VKHCG'
print('###############################')
print('Working Base :',Base, ' using ', sys.platform)
print('###############################')
################################################################
sDataFile=Base+'/01-Vermeulen/00-RawData/irisdata.csv'

data = pd.read_csv(sDataFile)

from pandas.plotting import andrews_curves
plt.figure(figsize=(10, 10))
andrews_curves(data, 'Name')
sPicNameOut1=Base+'/01-Vermeulen/06-Report/01-EDS/02-Python/andrews_curves.png'
plt.savefig(sPicNameOut1,dpi=600)
plt.tight_layout()
plt.show()
```

Figure 11-14 shows the result: a four- to two-dimensional reduction graph.

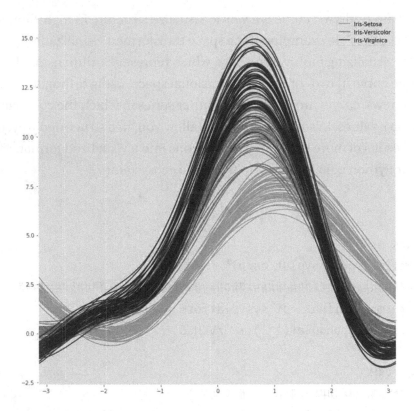

Figure 11-14. *Graph produced using Andrews' curves*

Parallel Coordinates

Parallel coordinates are a regularly used for visualizing high-dimensional geometry and analyzing multivariate data.

```
from pandas.plotting import parallel_coordinates
plt.figure(figsize=(10, 10))
parallel_coordinates(data, 'Name')
sPicNameOut2=Base+'/01-Vermeulen/06-Report/01-EDS/02-Python/parallel_
coordinates.png'
plt.savefig(sPicNameOut2,dpi=600)
plt.tight_layout()
plt.show()
```

Figure 11-15 shows the results of this graph.

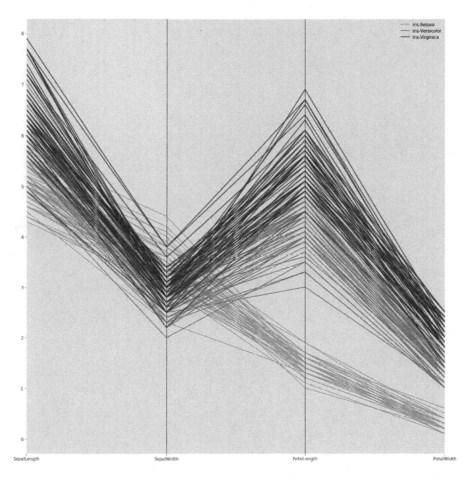

Figure 11-15. *Graph produced using parallel coordinates*

RADVIZ Method

The RADVIZ method maps a set of multidimensional points onto two-dimensional space.

```
from pandas.plotting import radviz
plt.figure(figsize=(10, 10))
radviz(data, 'Name')
sPicNameOut3=Base+'/01-Vermeulen/06-Report/01-EDS/02-Python/radviz.png'
plt.savefig(sPicNameOut3,dpi=600)
plt.tight_layout()
plt.show()
```

Figure 11-16 shows the results.

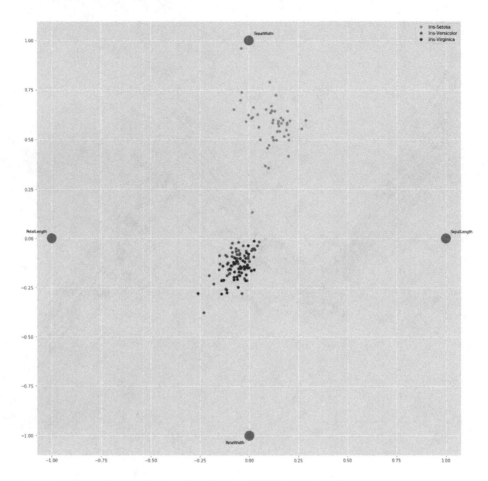

Figure 11-16. *Graph produced by the RADVIZ method*

Lag Plot

A lag plot allows the data scientist to check whether a data set is random.

Creating a lag plot enables you to check for randomness. Random data will spread fairly evenly, both horizontally and vertically. If you cannot see a pattern in the graph, your data is most probably random.

This example will show you how it works in practical data science:

```
################################################################
# -*- coding: utf-8 -*-
################################################################
import sys
import os
import pandas as pd
from matplotlib import style
from matplotlib import pyplot as plt
import numpy as np
################################################################
if sys.platform == 'linux' or sys.platform == ' darwin':
    Base=os.path.expanduser('~') + '/VKHCG'
else:
    Base='C:/VKHCG'
print('###############################')
print('Working Base :',Base, ' using ', sys.platform)
print('###############################')
################################################################
style.use('ggplot')

from pandas.plotting import lag_plot
plt.figure(figsize=(10, 10))
data = pd.Series(0.1 * np.random.rand(1000) + \
                0.9 * np.sin(np.linspace(-99 * np.pi, 99 * np.pi, num=1000)))
lag_plot(data)

sPicNameOut1=Base+'/01-Vermeulen/06-Report/01-EDS/02-Python/lag_plot.png'
plt.savefig(sPicNameOut1,dpi=600)
plt.tight_layout()
plt.show()
```

Figure 11-17 shows the results.

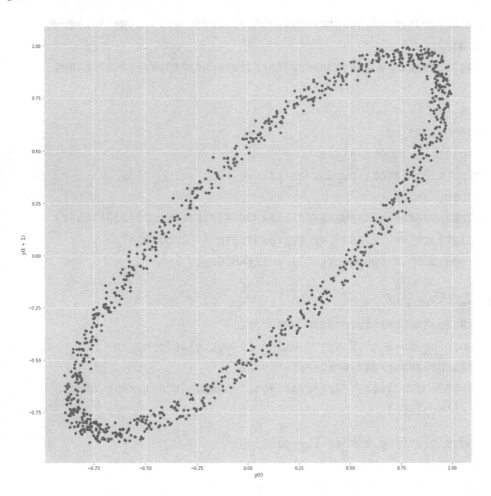

Figure 11-17. *Lag plot graph*

Autocorrelation Plot

An autocorrelation plot is a design of the sample autocorrelations vs. the time lags.

```
from pandas.plotting import autocorrelation_plot
plt.figure(figsize=(10, 10))
data = pd.Series(0.7 * np.random.rand(1000) + \
                 0.3 * np.sin(np.linspace(-9 * np.pi, 9 * np.pi, num=1000)))
autocorrelation_plot(data)

sPicNameOut2=Base+'/01-Vermeulen/06-Report/01-EDS/02-Python/
autocorrelation_plot.png'
```

```
plt.savefig(sPicNameOut2,dpi=600)
plt.tight_layout()
plt.show()
```

Figure 11-18 shows the results.

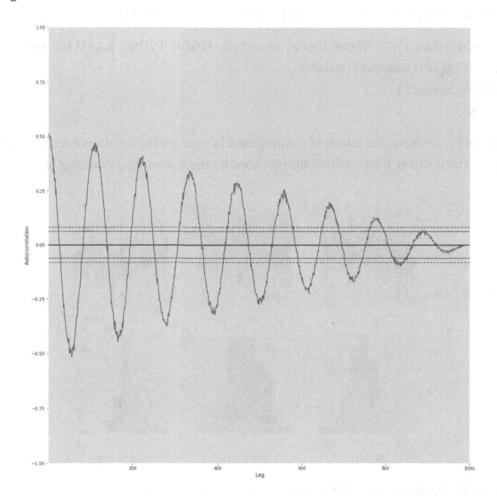

Figure 11-18. *Autocorrelation plot graph*

Bootstrap Plot

To generate a bootstrap uncertainty estimate for a given statistic from a set of data, a subsample of a size less than or equal to the size of the data set is generated from the data, and the statistic is calculated. This subsample is generated with replacement, so that any data point can be sampled multiple times or not sampled at all. This process is repeated for many subsamples.

761

Here is an example for 500 samples of 50 elements out of a population of 1000 points.

```
from pandas.plotting import bootstrap_plot
data = pd.Series(np.random.rand(1000))
plt.figure(figsize=(10, 10))
bootstrap_plot(data, size=50, samples=500, color='grey')
```

```
sPicNameOut3=Base+'/01-Vermeulen/06-Report/01-EDS/02-Python/bootstrap_plot.png'
plt.savefig(sPicNameOut3,dpi=600)
plt.tight_layout()
plt.show()
```

Figure 11-19 shows the results of the proposed bootstrap. This enables the data scientist to show what the distribution of the samples used for the bootstrap processing of the data is.

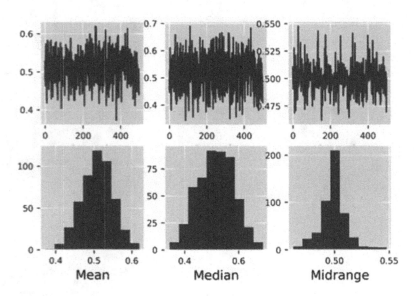

Figure 11-19. *Bootstrap plot graph*

Contour Graphs

A contour plot is a graphical technique for demonstrating a three-dimensional surface by plotting constant z slices, called contours, on a two-dimensional format. The latest geospatial insights for new technology such as 5G and robotics requires data scientists to start taking the terrain around the customers into consideration when performing their data analysis. Data science has also shown that previous beliefs about concentric circles with a common center of influence were incorrect. The latest findings show that results similar to the contours of the earth as the patterns of the influence.

Following is an example to show you how to visualize these contour maps. Open your Python editor and set up this ecosystem:

```
################################################################
# -*- coding: utf-8 -*-
################################################################
import sys
import os
import matplotlib
import numpy as np
import matplotlib.cm as cm
import matplotlib.mlab as mlab
import matplotlib.pyplot as plt
################################################################
if sys.platform == 'linux' or sys.platform == ' darwin':
    Base=os.path.expanduser('~') + '/VKHCG'
else:
    Base='C:/VKHCG'
print('##############################')
print('Working Base :',Base, ' using ', sys.platform)
print('##############################')
################################################################

matplotlib.rcParams['xtick.direction'] = 'out'
matplotlib.rcParams['ytick.direction'] = 'out'
```

Here is your data set.

```
delta = 0.025
x = np.arange(-3.0, 3.0, delta)
y = np.arange(-2.0, 2.0, delta)
X, Y = np.meshgrid(x, y)
Z1 = mlab.bivariate_normal(X, Y, 1.0, 1.0, 0.0, 0.0)
Z2 = mlab.bivariate_normal(X, Y, 1.5, 0.5, 1, 1)
# difference of Gaussians
Z = 10.0 * (Z2 - Z1)
```

You can now create a simple contour graph with labels.

```
plt.figure(figsize=(10, 10))
CS = plt.contour(X, Y, Z)
plt.clabel(CS, inline=1, fontsize=10)
plt.title('Simply default with labels')

sPicNameOut0=Base+'/01-Vermeulen/06-Report/01-EDS/02-Python/contour0.png'
plt.savefig(sPicNameOut0,dpi=600)
plt.tight_layout()
plt.show()
```

Here is the output (Figure 11-20). A clear contour was achieved.

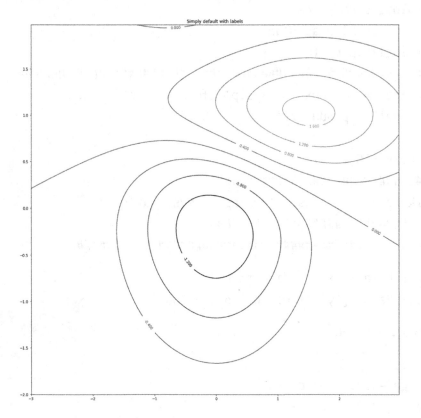

Figure 11-20. *Contour plot graph*

Can you see how `matplotlib` assists you with creating a complex visualization? This type of contour can be used for the normal mapping of the contours of the land. I also use them to show the area of impact of a specific event or action, for example, Hillman opening a new warehouse or Krennwallner putting up a new billboard. The other use is when you can tie the data set to move across a period of time. In that case, I simply generate the same code with a loop that generates, say, 12 images. Then I simply perform a command to produce a movie in MP4 format.

Look at `ffmpeg -f image2 -r 1/5 -i yourimage%03d.png -vcodec mpeg4 -y yourmovie.mp4`. It simply takes images with three-digit padding (`yourimage001.png`) into an PEG4 QuickTime movie.

Tip   People can follow a quick movie clip quickly, to visualize the effects or the context of the results you are presenting.

Now we can look at changing our contour format. I found the forcing of manual locations for the labels a useful feature, as it supports the better control of the presentation format when you make images for your movie series.

```
plt.figure(figsize=(10, 10))
CS = plt.contour(X, Y, Z)
manual_locations = [(-1, -1.4), (-0.62, -0.7), (-2, 0.5),\
                    (1.7, 1.2), (2.0, 1.4), (2.4, 1.7)]
plt.clabel(CS, inline=1, fontsize=10, manual=manual_locations)
plt.title('Labels at selected locations')

sPicNameOut1=Base+'/01-Vermeulen/06-Report/01-EDS/02-Python/contour1.png'
plt.savefig(sPicNameOut1,dpi=600)
plt.tight_layout()
plt.show()
```

In Figure 11-21, can you see how the contour labels are now pinned to a specific location?

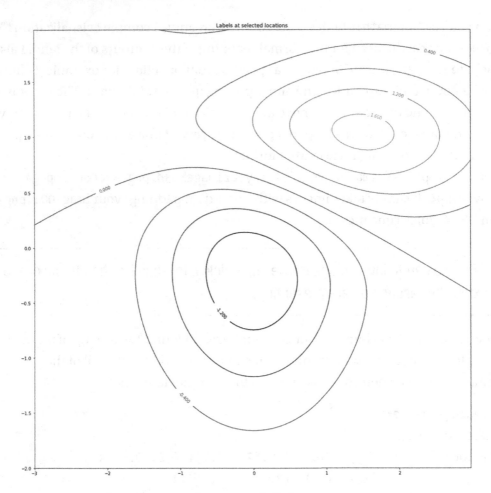

Figure 11-21. *Graph with pinned contour labels*

Tip   Only move or change items that you want to communicate. A person's eye picks up on the most minor changes, and they can distract the person's interest.

You can also change the line types, to communicate new aspects of the process.

```
plt.figure(figsize=(10, 10))
CS = plt.contour(X, Y, Z, 6,
                 colors='k',  # negative contours will be dashed by default
                 )
plt.clabel(CS, fontsize=9, inline=1)
plt.title('Single color - negative contours dashed')
```

766

```
sPicNameOut2=Base+'/01-Vermeulen/06-Report/01-EDS/02-Python/contour2.png'
plt.savefig(sPicNameOut2,dpi=600)
plt.tight_layout()
plt.show()
```

The results are shown in Figure 11-22.

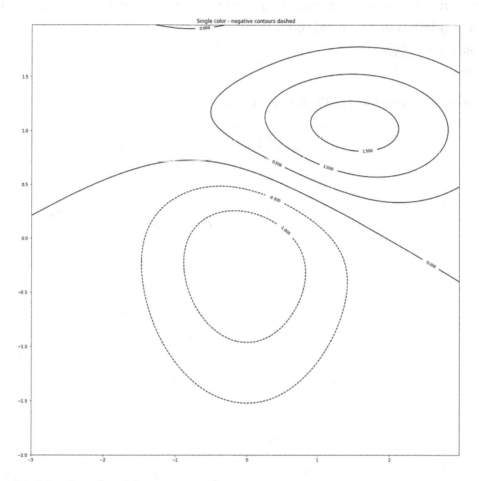

Figure 11-22. *Graph with contrasted contours*

As you can see, you can tune many aspects of the graph.

```
plt.figure(figsize=(10, 10))
matplotlib.rcParams['contour.negative_linestyle'] = 'solid'
plt.figure(figsize=(10, 10))
```

```
CS = plt.contour(X, Y, Z, 6,
                 colors='k',
                 )
plt.clabel(CS, fontsize=9, inline=1)
plt.title('Single color - negative contours solid')

sPicNameOut3=Base+'/01-Vermeulen/06-Report/01-EDS/02-Python/contour3.png'
plt.savefig(sPicNameOut3,dpi=600)
plt.tight_layout()
plt.show()
```

More results are shown in Figure 11-23.

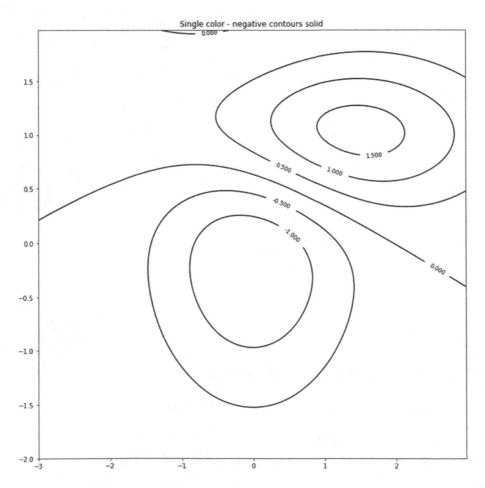

Figure 11-23. *Graph with solid contours*

So, let's construct a contour to show you what you could achieve.

```
plt.figure(figsize=(10, 10))
CS = plt.contour(X, Y, Z, 6,
                 linewidths=np.arange(.5, 4, .5),
                 colors=('r', 'green', 'blue', \
                         (1, 1, 0), '#afeeee', '0.5')
                 )
plt.clabel(CS, fontsize=9, inline=1)
plt.title('Crazy lines')

sPicNameOut4=Base+'/01-Vermeulen/06-Report/01-EDS/02-Python/contour4.png'
plt.savefig(sPicNameOut4,dpi=600)
plt.tight_layout()
plt.show()
```

What do you think? Can you see the benefits of ensuring a change in line color or line type in Figure 11-24? I find it a good communication tool.

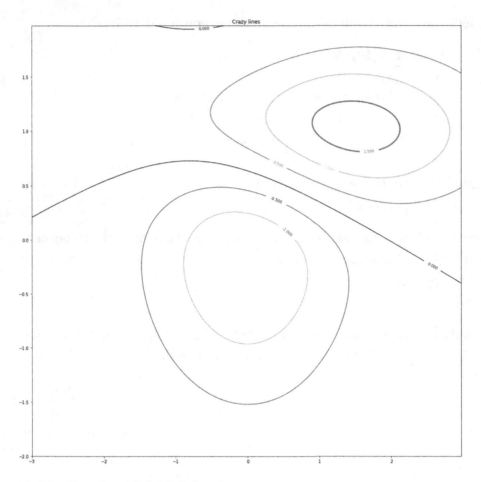

Figure 11-24. *Graph with highlighted contour*

Tip   Gray out but never remove unimportant information, as it gives the audience a point of comparison about what you think is important or not, without losing the data set.

You can do some amazing graphs.

```
plt.figure(figsize=(12, 10))
im = plt.imshow(Z, interpolation='bilinear', origin='lower',
            cmap=cm.gray, extent=(-3, 3, -2, 2))
levels = np.arange(-1.2, 1.6, 0.2)
```

```
CS = plt.contour(Z, levels,
                 origin='lower',
                 linewidths=2,
                 extent=(-3, 3, -2, 2))
```

Let's now thicken the zero contours.

```
zc = CS.collections[6]
plt.setp(zc, linewidth=4)

plt.clabel(CS, levels[1::2],  # label every second level
           inline=1,
           fmt='%1.1f',
           fontsize=14)
```

Now add a color bar for the contour lines.

```
CB = plt.colorbar(CS, shrink=0.8, extend='both')


plt.title('Lines with colorbar')
#plt.hot()  # Now change the colormap for the contour lines and colorbar
plt.flag()
```

You can still add a color bar for the image also.

```
CBI = plt.colorbar(im, orientation='horizontal', shrink=0.8)
```

That makes the original color bar look a bit out of place, so let's recover its location.

```
l, b, w, h = plt.gca().get_position().bounds
ll, bb, ww, hh = CB.ax.get_position().bounds
CB.ax.set_position([ll, b + 0.1*h, ww, h*0.8])

sPicNameOut5=Base+'/01-Vermeulen/06-Report/01-EDS/02-Python/contour5.png'
plt.savefig(sPicNameOut5,dpi=600)
plt.tight_layout()
plt.show()
```

Wow, you have created a super contour graph (Figure 11-25).

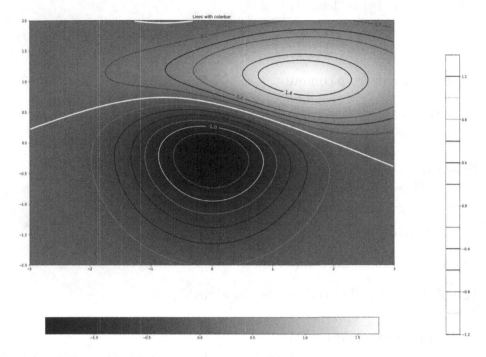

Figure 11-25. *Multidimensional complex contour graph*

I have a question. Did the extra work to create this super contour graph add any more value to the graph's message? I would not create graphs of this complexity, unless I want to demonstrate my ability to generate complex visualization.

Warning   Always make sure that your message is passed on!

I am, however, also a scientist at heart, so when you have the spare time, make different changes, try other colors, and explore what each parameter controls. In short, have fun with the process.

3D Graphs

Let's go 3D with our graphs. Many data sets have more than two influencing factors, and a 3D graph will assist in communicating those.

Open your Python editor and build this ecosystem:

```
################################################################
# -*- coding: utf-8 -*-
################################################################
import sys
import os
import numpy as np
import matplotlib.pyplot as plt
from mpl_toolkits.mplot3d import Axes3D
from sklearn import decomposition
from sklearn import datasets
if sys.platform == 'linux' or sys.platform == ' darwin':
    Base=os.path.expanduser('~') + '/VKHCG'
else:
    Base='C:/VKHCG'
print('################################')
print('Working Base :',Base, ' using ', sys.platform)
print('################################')
```

Now generate a data set.

```
np.random.seed(5)

centers = [[1, 1], [-1, -1], [1, -1]]
iris = datasets.load_iris()
X = iris.data
y = iris.target
```

Now let's go 3D.

```
fig = plt.figure(1, figsize=(16, 12))
plt.clf()
ax = Axes3D(fig, rect=[0, 0, .95, 1], elev=48, azim=134)

plt.cla()
pca = decomposition.PCA(n_components=3)
pca.fit(X)
X = pca.transform(X)
```

```
for name, label in [('Setosa', 0), ('Versicolour', 1), ('Virginica', 2)]:
    ax.text3D(X[y == label, 0].mean(),
              X[y == label, 1].mean() + 1.5,
              X[y == label, 2].mean(), name,
              horizontalalignment='center',
              bbox=dict(alpha=.5, edgecolor='w', facecolor='w'))
```

I suggest you reorder the labels, to have colors matching the cluster results.

```
y = np.choose(y, [1, 2, 0]).astype(np.float)
ax.scatter(X[:, 0], X[:, 1], X[:, 2], c=y, cmap=plt.cm.spectral,
           edgecolor='k',marker='p',s=300)

ax.w_xaxis.set_ticklabels([])
ax.w_yaxis.set_ticklabels([])
ax.w_zaxis.set_ticklabels([])

sPicNameOut0=Base+'/01-Vermeulen/06-Report/01-EDS/02-Python/3DPlot.png'
plt.savefig(sPicNameOut0,dpi=600)
plt.show()
```

Well done. You can now create 3D graphs. See Figure 11-26 for the result.

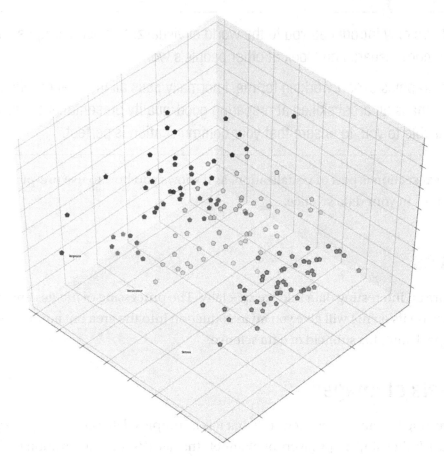

Figure 11-26. Three-dimensional graph

Summary

You have successfully progressed thought the graph section of this chapter. Let's summarize what you have learned.

- You can now create various 2D and 3D graph types.

- You can convert your graphs into movies, by using the techniques I discussed earlier in the book. You could also simply add a series of pictures together on your favorite presentation software.

- You can enhance parts of your graphs, by changing colors, line types, or other criteria.

> **Tip**   I have only introduced you to the world of visualization using graphs. Please perform more research and look at other people's work.
>
> I have a graphics expert working for me. I normally pass all my final graphs to her to fix, as she is an artist skilled at preparing good-quality presentations. Use every skill available to you, to ensure that your communication is perfect.

The next section is part of visualization, as it shows you how to use pre-prepared visualizations in your data science.

Pictures

Pictures are an interesting data science specialty. The processing of movies and pictures is a science on its own. I will give you an introduction into the area but admit that it is merely a peek into this subfield of data science.

Channels of Images

The interesting fact about any picture is that it is a complex data set in every image. Pictures are built using many layers or channels that assists the visualization tools to render the required image.

Open your Python editor, and let's investigate the inner workings of an image.

```python
import sys
import os
import matplotlib.pyplot as plt
import matplotlib.image as mpimg
################################################################
if sys.platform == 'linux' or sys.platform == ' darwin':
    Base=os.path.expanduser('~') + '/VKHCG'
else:
    Base='C:/VKHCG'
print('###############################')
```

```
print('Working Base :',Base, ' using ', sys.platform)
print('###############################')
################################################################
```

Here is a typical image in Portable Network Graphics (PNG) format:

```
sPicName=Base+'/01-Vermeulen/00-RawData/AudiR8.png'
```

You can load it into a matrix of value.

```
t=0
img=mpimg.imread(sPicName)
print('Size:', img.shape)
plt.figure(figsize=(10, 10))
t+=1
sTitle= '(' + str(t) + ') Original'
plt.title(sTitle)
plt.imshow(img)
plt.show()
for c in range(img.shape[2]):
    t+=1

    plt.figure(figsize=(10, 10))
    sTitle= '(' + str(t) + ') Channel: ' + str(c)
    plt.title(sTitle)
    lum_img = img[:,:,c]
    plt.imshow(lum_img)
    plt.show()
```

You can clearly see that the image is of the following size: 1073, 1950, 4. This means you have 1073 × 1950 pixels per channel. That is a total of 2,092,350 pixels per layer. As you have four layers, that equals a total of 8,369,400 values describing image size. You can get this size by executing img.size.

Let's look into transforming the images.

Cutting the Edge

One of the most common techniques that most data science projects require is the determination of the edge of an item's image. This is useful in areas such as robotics object selection and face recognition, to cite two special cases.

I will show you how to determine the edge of the image you saw before. Open your Python editor and create this ecosystem:

```
################################################################
# -*- coding: utf-8 -*-
################################################################
import sys
import os
import matplotlib.pyplot as plt
from PIL import Image
################################################################
if sys.platform == 'linux' or sys.platform == ' darwin':
    Base=os.path.expanduser('~') + '/VKHCG'
else:
    Base='C:/VKHCG'
print('##############################')
print('Working Base :',Base, ' using ', sys.platform)
print('##############################')
################################################################
sPicNameIn=Base+'/01-Vermeulen/00-RawData/AudiR8.png'
sPicNameOut=Base+'/01-Vermeulen/06-Report/01-EDS/02-Python/AudiR8Edge1.png'
```

You can now open the image.

```
imageIn = Image.open(sPicNameIn)
fig1=plt.figure(figsize=(10, 10))
fig1.suptitle('Audi R8', fontsize=20)
imgplot = plt.imshow(imageIn)
```

You now convert the image to black and white.

```
mask=imageIn.convert("L")
```

The next value must be adjusted for an image with which to experiment with outcomes. You can select a range between 0 and 255, to inspect the outcome of your investigation.

```
th=49
imageOut = mask.point(lambda i: i < th and 255)
```

You can save your experiment or try another.

```
imageOut.save(sPicNameOut)

imageTest = Image.open(sPicNameOut)
fig2=plt.figure(figsize=(10, 10))
fig2.suptitle('Audi R8 Edge', fontsize=20)
imgplot = plt.imshow(imageTest)
```

Figure 11-27 shows the result.

Figure 11-27. *Image with edges extracted*

I will now show you a slightly different route. This method is common in robotics, because it is a quick way to identify the existence of an object in the line of sight.

Open your Python editor and set up this ecosystem:

```
################################################################
# -*- coding: utf-8 -*-
################################################################
import sys
import os
import numpy
```

```
import scipy
from scipy import ndimage
#################################################################
if sys.platform == 'linux' or sys.platform == ' darwin':
    Base=os.path.expanduser('~') + '/VKHCG'
else:
    Base='C:/VKHCG'
print('###############################')
print('Working Base :',Base, ' using ', sys.platform)
print('###############################')
#################################################################
sPicNameIn=Base+'/01-Vermeulen/00-RawData/AudiR8.png'
sPicNameOut=Base+'/01-Vermeulen/06-Report/01-EDS/02-Python/AudiR8Edge2.png'
#################################################################
```

You now load the image.

```
im = scipy.misc.imread(sPicNameIn)
```

You can now apply limited mathematics calculations, and you have an edge image.

```
im = im.astype('int32')
dx = ndimage.sobel(im, 0) # horizontal derivative
dy = ndimage.sobel(im, 1) # vertical derivative
mag = numpy.hypot(dx, dy) # magnitude
mag *= 255.0 / numpy.max(mag) # normalize (Q&D)
```

You can now save your image.

```
scipy.misc.imsave(sPicNameOut, mag)
```

You can now compare your two images.

```
import matplotlib.pyplot as plt
from PIL import Image

imageIn = Image.open(sPicNameIn)
plt.figure(figsize=(10, 10))
imgplot = plt.imshow(imageIn)
```

```
imageTest = Image.open(sPicNameOut)
plt.figure(figsize=(10, 10))
imgplot = plt.imshow(imageTest)
```

The result is shown in Figure 11-28.

Figure 11-28. *Edged image*

Well done. You can now do edges and compare results. These are two important skills for processing images.

One Size Does Not Fit All

The images we get to process are mostly of different sizes and quality. You will have to size images to specific sizes for most of your data science.

Warning   If you change an image, make sure you save your result under a new name. I have seen too much image processing that simply overwrites the originals.

The following examples will demonstrate what happens to an image if you reduce the pixel quality. Open your Python editor and set up this ecosystem:

```
################################################################
# -*- coding: utf-8 -*-
################################################################
import sys
import os
import matplotlib.pyplot as plt
from PIL import Image
################################################################
if sys.platform == 'linux' or sys.platform == ' darwin':
    Base=os.path.expanduser('~') + '/VKHCG'
else:
    Base='C:/VKHCG'
print('#############################')
print('Working Base :',Base, ' using ', sys.platform)
print('#############################')
################################################################
sPicName=Base+'/01-Vermeulen/00-RawData/AudiR8.png'
nSize=4
################################################################
```

Let's load the original image.

```
img = Image.open(sPicName)
plt.figure(figsize=(nSize, nSize))
sTitle='Unchanges'
plt.title(sTitle)
imgplot = plt.imshow(img)
```

You now apply a thumbnail function that creates a 64 × 64 pixel thumbnail image.

```
img.thumbnail((64, 64), Image.ANTIALIAS) # resizes image in-place

plt.figure(figsize=(nSize, nSize))
sTitle='Resized'
plt.title(sTitle)
imgplot = plt.imshow(img)
```

```
plt.figure(figsize=(nSize, nSize))
sTitle='Resized with Bi-Cubic'
plt.title(sTitle)
imgplot = plt.imshow(img, interpolation="bicubic")
################################################################
print('### Done!! ####################################')
################################################################
```

Can you see the impact it has on the quality of the image?

Showing the Difference

I want to show you this technique, to enable you to produce two overlaying results but still show that both exist. In a data science presentation showing that two sets of nodes on a graph are the same, you must make each set marginally different, in an orderly manner, to facilitate their visualization. Without this slight shift, the two sets will simply overlay each other.

```
import numpy as np

from matplotlib import pyplot as plt
from matplotlib.collections import LineCollection

from sklearn import manifold
from sklearn.metrics import euclidean_distances
from sklearn.decomposition import PCA

n_samples = 25
seed = np.random.RandomState(seed=3)
X_true = seed.randint(0, 20, 2 * n_samples).astype(np.float)
X_true = X_true.reshape((n_samples, 2))
# Center the data
X_true -= X_true.mean()

similarities = euclidean_distances(X_true)
```

You simply add noise to the similarities.

```
noise = np.random.rand(n_samples, n_samples)
noise = noise + noise.T
noise[np.arange(noise.shape[0]), np.arange(noise.shape[0])] = 0
similarities += noise

mds = manifold.MDS(n_components=2, max_iter=3000, eps=1e-9, random_state=seed,
                   dissimilarity="precomputed", n_jobs=1)
pos = mds.fit(similarities).embedding_

nmds = manifold.MDS(n_components=2, metric=False, max_iter=3000, eps=1e-12,
                    dissimilarity="precomputed", random_state=seed, n_jobs=1,
                    n_init=1)
npos = nmds.fit_transform(similarities, init=pos)
```

You then rescale the data.

```
pos *= np.sqrt((X_true ** 2).sum()) / np.sqrt((pos ** 2).sum())
npos *= np.sqrt((X_true ** 2).sum()) / np.sqrt((npos ** 2).sum())
```

Next, you rotate the data by a small margin.

```
clf = PCA(n_components=2)
X_true = clf.fit_transform(X_true)

pos = clf.fit_transform(pos)

npos = clf.fit_transform(npos)

fig = plt.figure(1)
ax = plt.axes([0., 0., 1., 1.])

s = 100
plt.scatter(X_true[:, 0], X_true[:, 1], color='navy', s=s, lw=0,
            label='True Position')
plt.scatter(pos[:, 0], pos[:, 1], color='turquoise', s=s, lw=0, label='MDS')
plt.scatter(npos[:, 0], npos[:, 1], color='darkorange', s=s, lw=0,
label='NMDS')
plt.legend(scatterpoints=1, loc='best', shadow=False)
```

```
similarities = similarities.max() / similarities * 100
similarities[np.isinf(similarities)] = 0

# Plot the edges
start_idx, end_idx = np.where(pos)
# a sequence of (*line0*, *line1*, *line2*), where::
#              linen = (x0, y0), (x1, y1), ... (xm, ym)
segments = [[X_true[i, :], X_true[j, :]]
            for i in range(len(pos)) for j in range(len(pos))]
values = np.abs(similarities)
lc = LineCollection(segments,
                    zorder=0, cmap=plt.cm.Blues,
                    norm=plt.Normalize(0, values.max()))
lc.set_array(similarities.flatten())
lc.set_linewidths(0.5 * np.ones(len(segments)))
ax.add_collection(lc)

plt.show()
```

You should see the same result as that shown in Figure 11-29.

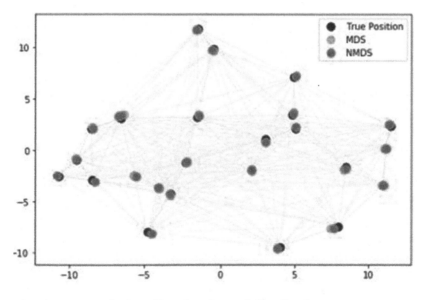

Figure 11-29. *Improved visualization by a shift of values*

Success, you have improved the visualization.

Summary

Congratulations! You have reached the end of the process I use for my own data science projects. You should now be able to construct a pipeline for any data in the data lake, via the complete Retrieve ➤ Assess ➤ Process ➤ Transform ➤ Organize ➤ Report steps to building the technology stack for turning data lakes into business assets.

Closing Words

You have successfully reached the end of this book. Well done on all your efforts and achievements.

Thank you for following me along the entire pipeline, from the data lake to the insights for the business. I hope that your practical data science provides you with everything you wish for in the future.

The biggest achievement you could accomplish is . . .

Have lots of fun along the way!

Index

A

Absolute location, 500
Adders utilities, 105
Akka, 9, 21
Altitude, 499
Amazon Redshift, 271
Amazon Simple Storage Service
      (Amazon S3)
  Boto package, 271
  Python's s3 module, 271
Amazon Web Services, 272–273
Analysis of variance (ANOVA), 613–615
Analytical models
  columns, 154
  data field name verification, 155–156
  data pattern, 168–171
  data type of data column, 157
  histograms of column, 157–160
  maximum value, 162–163
  mean, 163
  median, 164
  minimum value, 161–162
  missing/unknown values, 168
  mode, 164–165
  quartiles, 165–166
  range, 165
  sample data set, 153
  skewness, 166–168
  standard deviation, 166
  unique identifier, data entry, 156–157

Andrews' curves, 755–756
Animals
  class, 477
  families, 477
  genera, 477
  kingdoms, 476
  orders, 477
  phyla, 477
  species, 477–490
ANOVA, *see* Analysis of variance (ANOVA)
Apache Cassandra, 21, 269
Apache Hadoop
  Luigi, 271
  Pydoop, 270
Apache Hive, 270
Apache Mesos, 21
Apache Spark, 20, 269
Apriori algorithm, 700–702
Area graph, 748–749
Assess_Best_Logistics, 711
Assess superstep
  Clark Ltd (*see* Clark Ltd)
  data analysis, 277–279
  data profiling, 276
  data quality, 275
  erroneous data, 275
  errors
    accept, 276
    correct, 277
    default value, 277

© Andreas François Vermeulen 2018
A. F. Vermeulen, *Practical Data Science*, https://doi.org/10.1007/978-1-4842-3054-1

Assess superstep (*cont.*)
    reject, 276
    Hillman Ltd
        (*see* Hillman Ltd)
    Krennwallner AG
        (*see* Krennwallner AG)
    pandas, missing values
        Assess-Good-Bad-01.py, 280–281
        Assess-Good-Bad-02.py, 284–288
        Assess-Good-Bad-03.py, 288–290
        Assess-Good-Bad-05.py, 292–296
        basic concepts, 279
        Good-or-Bad-01.csv, 282–283
        Good-or-Bad-02.csv, 286–288
        Good-or-Bad-03.csv, 290–292
        Good-or-Bad-05.csv, 296
    Vermeulen PLC (*see* Vermeulen PLC)
Assess utilities, *see* Feature
        engineering
Association rule learning
    Apriori algorithm, 700–702
    lift, 699
    market basket analysis, 699
Audit
    basic logging, 127
    built-in logging
        data science tools, 126
        debug watcher, 126
        error watcher, 127
        fatal watcher, 127
        information watcher, 126
        warning watcher, 126
    data lineage, 130
    data provenance, 130
    definition, 125
    process tracking, 130
    statistics, 125
Autocorrelation plot, 760–761

B

Bagging data, 652–655
Balance layer, 131
Bar graph, 746–747
Billboards, 209–210
    Assess-DE-Billboard.py, 339–344
    Assess-DE-Billboard-Visitor.csv, 344
    DE_Billboard_Locations,
        loading, 187–188
    IP addresses, 339
    online visitor data, 344
    Organize superstep, 707–709
    Report superstep, 727–729
    sqlite3 and pandas
        packages, 339
Binary logistic regression, 590
Binning, 108–109
Bivariate analysis, 581
Body mass index (BMI)
    horizontal-style, 687–689
    island-style, 693–695
    secure vault style, 696–699
    vertical-style, 690–692
Bootstrap plot, 761
Business dynamics,
        data lake
    Clark Ltd
        Euro_EchangeRates,
            loading, 195–198
        Profit_And_Loss, loading, 198–200
    Hillman Ltd
        GB_Postcode_Full,
            loading, 189–190
        GB_Postcode_Shops,
            loading, 192–194
        GB_Postcode_Warehouse,
            loading, 190–192

Krennwallner AG
  DE_Billboard_Locations,
    loading, 187–188
Python Retrieve solution, 200–202
R Retrieve solution, 173
Vermeulen PLC
  Country_Codes, loading, 184–186
  IP_DATA_ALL, loading, 174–177
  IP_DATA_CORE, loading, 181–183
  IP_DATA_C_VKHCG, loading,
    178–181
Business layer
  ambiguity, 80
  description, 53
  functional requirements
    data mapping matrix, 55
    general, 54
    MoSCoW options, 54
    sun models (*see* Sun models)
    "Will of the Business", 54
  implicit collections, 79
  nonfunctional requirements
    accessibility, 63
    audit and control, 63
    availability, 64–65
    backup, 65
    capacity, 66
    concurrency, 66–67
    configuration management, 69
    deployment, 70
    disaster recovery (DR), 70
    disk storage, 68
    documentation, 70
    effectiveness, 70
    efficiency, 70
    extensibility, 71
    failure management, 71
    fault tolerance, 71
    interoperability, 72
    latency, 72
    maintainability, 72
    memory storage, 67
    modifiability, 72
    network topology, 73
    privacy, 73
    quality, 74
    recovery/recoverability, 74
    reliability, 74
    resilience, 74
    resource constraints, 74
    reusability, 75
    scalability, 75–76
    security, 76–77
    storage (GPU), 69
    testability, 77–78
    throughput capacity, 67
    year-on-year growth, 69
  optionality, 80
  practical
    primary structures, 81
    requirements, 81
    requirements registry, 81–82
    traceability matrix, 82–83
  subjectivity, 80
  unbounded lists, 79
  under-reference, 80
  under-specification, 80
  vagueness, 80
  weak words, 78

C

Cassandra, 9
Causal loop diagram (CLD), 515–517
Cause-and-effect analysis system, 140
CentOS/RHEL, 23–25, 27

Centroid-based clustering, 636–638

Chief financial officer (CFO), 685

Chi-square test, 565–566

Clark Ltd, 18

    Euro_EchangeRates, loading, 195–198

    financial base data, 259

    financial calendar, 412–414

    financials, 410–412

    forex, 259

    forex base data, 259, 406

    Organize superstep, 713

    people, 414–420

    person base data, 259, 261

    Profit_And_Loss, loading, 198–200

    Python Retrieve solution, 200–202

    Report superstep, 735–741

Cleanup utilities, 113

Cloud Tensor Processing Unit (TPU), 671

Clustering

    centroid-based, 636–638

    connectivity, 636

    definition, 601

    density, 638

    distribution, 638

    grid-based, 638

    hierarchical (*see* Hierarchical
        clustering)

    partitional, 608–613

    variables, 636

Computer vision (CV), 660

Configuration management (CM), 69

Consumer, 133

Contour graphs

    construct, 769

    contour labels, 765–766

    contrasted, 766–767

    definition, 762

    ecosystem, 763–764

    highlighted, 769–770

    multidimensional complex, 771–772

    solid, 767–768

Control layer, 131

Coordinated Universal
    Time (UTC), 447–452

Cross-Industry Standard Process for Data
    Mining (CRISP-DM)

    business understanding, 41

    data preparation, 42

    data understanding, 42

    deployment, 43

    description, 40

    evaluation, 42

    flow chart, 41

    modeling, 42

Cross-validation test, 579–580

Customer data sets, 35

CV, *see* Computer vision (CV)

D

Data analysis

    accuracy, 278

    completeness, 277

    consistency, 279

    timeliness, 278

    uniqueness, 278

    validity, 278

Data lakes

    actionable business knowledge, 202

    business customer, 148

    business dynamics (*see* Business
        dynamics, data lake)

    characterize, 148

    definition, 4

Data lineage, 130, 147

Data mapping matrix, 55

Data mart, 685

Data mining
    association patterns, 633–635
    Bayesian classification, 638–639
    clustering patterns (*see* Clustering)
    data classification, 635
    definition, 633
    forecasting, 646
    sequence/path analysis, 640

Data processing utilities
    data science (*see* Data science
        utilities)
    data vault, 106–107
    description, 116
    feature engineering, 104–106
    organize, 112
    report, 112
    retrieve utilities (*see* Homogeneous
        Ontology for Recursive Uniform
        Schema (HORUS))
    transform, 107

Data provenance, 130

Data science, 719, 722, 762
    basic steps, 524–525
    CLD, 515–517
    correlation analysis, 519
    customers and subject matter
        experts, 142
    data traceability matrix, 547
    feature extraction (*see* Feature
        extraction techniques)
    fishbone diagrams, 511–512
    forecasting, 519–524
    guess, potential pattern, 141
    missing value treatments
        feature engineering, 556
        inconsistency, data lake, 547
        methods, 548

outlier detection (*see* Outlier
    detection techniques|)
    reasons, 548
    Monte Carlo simulation, 515
    observations and hypothesis, 142
    Pareto charts, 517–518
    real-world evidence, 142
    start with what-if question, 141
    5 Whys, 510, 513–514

Data science framework
    definition, 40
    top layers, 45

Data science processing tools
    Akka, 9
    Cassandra, 9–10
    elastic search, 11
    Kafka, 10
    Mesos, 9
    MQTT, 13
    Python, 12–13
    R, 11–12
    Scala, 12
    Spark (*see* Spark)

Data Science Technology Stack
    data lake, 4
    data vault, 4
    data warehouse bus matrix, 6
    description, 13
    hubs, 5
    links, 5
    processing tools (*see* Data science
        processing tools)
    Rapid Information Factory
        ecosystem, 1
    satellites, 5
    storage tools
        schema-on-read ecosystems, 3
        schema-on-write ecosystems, 2–3

Data science utilities
  averaging of data, 110
  binning/bucketing, 108–109
  outlier detection, 110
Data sources
  Amazon Redshift, 271
  Amazon S3 storage, 271
  Amazon Web Services, 272–273
  Apache Cassandra, 269
  Apache Hadoop
    Luigi, 271
    Pydoop, 270
  Apache Hive, 270
  Apache Spark, 269
  databases, 266
  date and time, 264–266
  Microsoft Excel, 268–269
  Microsoft SQL server, 267
  MySQL, 267
  Oracle, 268
  PostgreSQL, 267
  SQLite, 261–264
Data swamps
  audit and version management, 172
  concrete business questions, 149–150
  data governance
    analytical model usage, 153–171
    business glossary, 151–152
    data source catalog, 151
    role of, 150
  data quality, 172
  trainer model, 172
Data.Table package, 28
Data vault
  basic, 106
  configuration, 106
  definition, 4, 106
  hubs, 5, 107, 422

links, 107, 422
modeling, 4, 422
satellites, 107, 422
Data warehouse, 1, 4, 6, 542–547
Data warehouse bus matrix, 6
Day, 442–443
Day of the year, 458–463
Debug watcher, 126
Decision Model and
        Notation (DMN), 41
Decision trees, 624–629
Delivery route plan, 711–713
Density-based clustering, 638
3D graphs, 772–775
Dijkstra's algorithm, 489–490
Directed acyclic graph (DAG)
  definition, 309
  job scheduling (see Job
        scheduling, DAG)
  not a, 309
  one path, 308, 309
  Spark, Pig, and Tez, 309
Disaster recovery (DR), 70
Documentation, 70
Double pie, 744
Dutch culture, naming, 471–476

E

Edged image, 778–781
Elastic search, 11
Enterprise Bus Matrix, 6
Error watcher, 127
Event, process superstep
  data entities, 507
  explicit, 507
  implicit, 507–510
Exchange rate data set, 37

F

Failure management, 71
Fatal watcher, 127
Fault tolerance, 71
Feature engineering
    adders utilities, 105
    fixers utilities, 104–105
    process utilities, 106
Feature extraction techniques
    averaging, 558–560
    binning, 557–558
    LDA, 560–561
Financial calendar, 412–414
Fishbone diagrams, 511–512
Fixers utilities, 104–105
Forecasting, 646
Forex trading planner, 406–410, 713–718
Fraction of a second, 444–445
Freytag's pyramid, 721–722
Functional layer
    data models, 143
    data schemas and data
        formats, 143
    infrastructure provision, 143
    processing algorithms
        assess, 50
        organize, 50
        process, 50
        report, 50
        retrieve, 50
        transform, 50
    processing data lake
        assess, 143
        organize, 144
        process, 144
        report, 144
        retrieve, 143
        transform, 144

G

Genera, 477
General Data Protection Regulation
    (GDPR), 85–87, 278
Geospatial, 498
Ggplot2 package, 29
Golden nominal record, 464–467
Gradient descent, 664–665
Graphics
    Andrews' curves, 755–756
    autocorrelation plot, 760–761
    bootstrap, 761–762
    contour (see Contour
        graphs)
    3D, 772–775
    KDE, 752–753
    lag plot, 758–760
    parallel coordinates, 756–757
    plot options
        area graph, 748–749
        bar graph, 746–747
        double pie, 744–745
        ecosystem, 742–743
        hex bin graph, 750–751
        horizontal bar graph, 747–748
        line graph, 745–746
        pie graph, 743–744
        scatter graph, 749–750
    RADVIZ method, 757–758
    scatter matrix, 754
Graphics processing
    unit (GPU), 652
Graph theory
    DAG (see Directed acyclic graph
        (DAG))
    description, 307
    edges, 307–308
    nodes, 307

GraphX, 8
Gregorian calendar, 439
Grid-based clustering, 638

H

Health Insurance Portability and
    Accountability Act of 1996
    (HIPAA), 87
Hex bin graph, 750–751
Hierarchical clustering
    agglomerative, 602
    definition, 601
    divisive, 602
Hillman Ltd, 18
    delivery routes
        Assess_Shipping_Routes.gml, 406
        Assess-Shipping-Routes.py, 387
        create, 254–257
        graph theory, 393
        HQ nodes to graph, 393
        HQ routes with Vincenty's
            distance, 395
        HQ-to-warehouse routes, 397
        intra-shop routes, 400–403
        intra-warehouse routes, 398
        nx.shortest_path(G,x,y), 403–406
        shops' nodes, 394
        warehouse nodes, 393
        warehouse-to-shop routes, 392, 399
        warehouse-to-warehouse
            routes, 391
    GB_Postcode_Full, loading, 189–190
    GB_Postcode_Shops, loading, 192–194
    GB_Postcode_Warehouse,
        loading, 190–192
    global post codes, 257, 259
    Incoterm 2010

        Carriage and Insurance (CIP),
            231–233
        Carriage Paid To (CPT), 229–230
        Delivered at Place (DAP), 235–237
        Delivered At Terminal (DAT),
            233–235
        Delivered Duty Paid (DDP), 238–240
        Ex Works (EXW), 225–226
        Free Carrier (FCA), 227–228
    Organize superstep, 710–713
    Report superstep, 730–735
    shipping chains
        call an expert, 240–241
        possible shipping
            routes, 242–247
    shipping containers (see Shipping
        containers)
    shipping information, 223–224
    shipping routes (see Shipping routes,
        best-fit international logistics)
    warehouses (see Warehouses)
Homogeneous Ontology for Recursive
    Uniform Schema (HORUS),
        39, 43–44
    audio, 101–103
    database, 95–96
    data format, 89
    data lake, 89
    data streams, 103
    JSON, 93–94
    picture, 96–97
    text-delimited, 90–91
    video
        frames to horus, 99–100
        movie to frames, 98–99
    XML, 91–93
Horizontal-style
    slicing, 686–689

HORUS, *see* Homogeneous Ontology for Recursive Uniform Schema (HORUS)

Hough transform, 589

Hour, 443

Hub-and-spoke data conversions, 43–44

Hubs, 5, 107

Hypothesis testing
 chi-square test, 565–566
 definition, 562
 t-test, 562–565

I

Images, channels of, 776–777

Incoterm 2010
 Carriage and Insurance (CIP), 231–233
 Carriage Paid To (CPT), 229–231
 Delivered at Place (DAP), 235–237
 Delivered At Terminal (DAT), 233–235
 Delivered Duty Paid (DDP), 238–240
 Ex Works (EXW), 225–226
 Free Carrier (FCA), 227–228

Information watcher, 126

Interoperability, 72

IP Addresses data sets, 32

Island-style, 693–695

ISO 8601-2004 standards, 439, 457

ISO 8601(European), 458

ISO 8601(US), 458

J

Job scheduling, DAG
 Assess-DAG-Company-Country-Place.png, 314
 Assess-DAG-Company-Country.png, 314

Assess-DAG-Company-GPS.png, 318

Assess-DAG-GPS.py, 314–317

Assess-DAG-Location.py, 310–313

Assess-DAG-Schedule.gml, 324

Assess-DAG-Schedule.py, 319–324
 basic validation, 319
 G.add_node() features, 321
 nodes_iter(G), 322
 Retrieve_IP_DATA.csv, 325
 SQLite database (*see* SQLite database)
 transport agents, 325
 valid routing network, 318–319

JSONLite package, 29

Julian calendar, 440

K

Kafka, 22
 Kafka Connect, 10
 Kafka Core, 10
 Kafka Streams, 10

Kernel density estimation (KDE) graph, 752–753

Kiribati, Republic of, 454

K-means clustering, 636–638

Krennwallner AG, 17–18
 achievements, 350, 356
 billboards (*see* Billboards)
 online visitor data, 210–221, 344
 Organize superstep, 707–708, 710
 processing offloading, top-ten customers, 350
 Report superstep, 727–729

L

Lag plot, 758–760

Latency, 72

Latent Dirichlet allocation (LDA), 560–561

Latitude, 499

Layered framework

    CRISP-DM (*see* Cross-Industry Standard Process for Data Mining (CRISP-DM))

    data science, definition of, 40

    high-level data science and engineering, 50

    HORUS, 43–44

    Linux, 51

    top layers

        audit, 48

        balance, 48

        business layer, 45–46

        Center of Excellence, 44

        control, 49

        data science framework, 45

        data science skills, 45

        functional layer, 49

        operational management layer, 47

        utility layer, 46–47

    warning, 39

    Windows, 51

LDA, *see* Latent Dirichlet allocation (LDA)

Leap years, 442–443

Linear regression

    definition, 581

    Hough transform, 589

    RANSAC, 587–589

    simple, 582

    TensorFlow, 681–684

Line graph, 745–746

Links, 5, 107

Linux, 15, 51

Local outlier factor (LOF), 553

Local time, 446–447

Location cluster, 613

Location, process superstep characteristics

    absolute location, 500

    human, 503

    natural, 502–503

    relative location, 501

    GLONASS, 500

    GPS, 500

    human-environment interaction, 503–506

    latitude, longitude, and altitude, 499

LOF, *see* Local outlier factor (LOF)

Logistic regression

    definition, 590

    MLR, 592–594

    ordinal (*see* Ordinal logistic regression)

    simple, 590–591

Logistics data sets

    exchange rate, 37

    post codes, 35–36

    profit-and-loss statement, 38

    shop, 36

    warehouse, 36

Longitude, 499

Luigi, 271

M

Machine learning

    definition, 649

    impact, 649

    RL, 651–652

    supervised, 649–650

    unsupervised, 651

Maintenance utilities

    backup and restores, 112

    checks data integrity, 112

    description, 115

history cleanup, 113

notify operator, 113

rebuild data structure, 113

reorganize indexing, 113

shrink/move data structure, 114

solution statistics, 114

Market basket analysis, 699

Matplotlib

CentOS/RHEL, 25

PIP, 25

Ubuntu, 25

Mean time between failures (MTBF), 74

Mean time to recovery (MTTR), 74

Mcsos, 9

Message Queue Telemetry
Transport (MQTT), 22

Metadata management, 147

Microsoft Excel, 268–269

Microsoft SQL server, 267

Minute, 443–444

MLR, *see* Multinomial logistic
regression (MLR)

Monte Carlo simulation, 515

Month, 440–441

MoSCoW, 54, 82

MQ Telemetry Transport (MQTT), 1, 13

Multinomial logistic regression (MLR),
592–594

Multivariate analysis, 581

MySQL, 267

text-based, 661–662

Network routing diagram

Assess-Network-Routing-
Company.csv, 301

Assess-Network-Routing-Company.py,
297–301

Assess-Network-Routing-
Customer.csv, 305

Assess-Network-Routing-Customer.py,
301–304

Assess-Network-Routing-Node.csv,
306–307

Assess-Network-Routing-Node.py,
305–306

company, 203–206

data issues, 297

graph theory (*see* Graph theory)

Organize superstep, 703–706

Report superstep, 724–726

Network topology, 73

Networkx, 310

Neural networks

daisy-chain, 664

definition, 662

features, 663–664

gradient descent, 664–665

layers, 663

regularization strength, 665

simple, 665

NumPy, 26

N

Naive Bayes (NB), 638–639

Natural language processing (NLP)

definition, 661

requirements, 661

speech-to-text converstion, 662

O

Object, process superstep

animals

class, 477

kingdoms, 476

orders, 477

Object, process superstep (*cont.*)
    phyla, 476
    species, 477–490
  chemical compounds, 497–499
  families, 477
  genera, 477
  vehicles, 491–497
Operational management layer
  alerting, 124
  communication, 124
  description, 119
  monitoring, 123–124
  parameters, 120–121
  processing-stream definition and
    management, 119
  scheduling, 121–123
Oracle, 268
Ordinal logistic regression
  business problem, 595–600
  data type, 594
  definition, 594
Organize superstep
  association rule learning, 699
  billboard content, 707–710
  Clark Ltd, 713
  data mart, 685
  delivery route plan, 711–713
  forex trading planner, 713–718
  Hillman Ltd, 710–713
  horizontal-style, 686–689
  island-style, 693–695
  Krennwallner AG, 707–710
  network routing diagram, 703–706
  secure vault style, 696–699
  Vermeulen PLC, 703–706
  vertical-style, 690–692
Organize utilities, 112
Outlier detection techniques

elliptic envelope, 549
  isolation forest, 552
  LOF, 553
  novelty detection, 552
Outliers, 110–111

P, Q

Pandas
  Centos/RHEL, 24
  PIP, 24
  Ubuntu, 24
Parallel coordinates, 756–757
Pareto charts, 517, 519
Partitional clustering, 608–613
Pattern recognition, 646–649
Person, process superstep
  golden nominal record, 464–467
  name
    Dutch culture, 471
    inheritance of, 470
    matronymic and
     patronymic, 468–469
    order of, 469–470
Phyla, 477
Pictures
  channels of images, 776–777
  definition, 776
  different sizes and quality, 781–783
  edged image, 778–781
  improved visualization, 784–785
Pie graph, 743–744
Plot options
  area graph, 748–749
  bar graph, 746–747
  double pie, 744
  ecosystem, 742
  hex bin graph, 750–751

horizontal bar graph, 747–748
line graph, 745–746
pie graph, 743
scatter graphs, 749–750
Portable Network Graphics (PNG), 777
Post codes, 35–36
PostgreSQL, 267
Precision-recall
    curve, 570
    F1-measure, 571
    ROC analysis curves, 577–579
    sensitivity and specificity, 571
Principal component analysis (PCA), 721
    conjoint analysis, 619–624
    factor analysis, 615–619
Processing ecosystem
    Akka, 21
    Apache Cassandra, 21
    Apache Mesos, 21
    Apache Spark, 20
    Kafka, 22
    MQTT, 22
    Python (see Python)
    R (see R)
    Scala, 20
    Vermeulen, 19
Processing utilities, 106
    description, 116
    monitoring, 115
    scheduling
        backlog, 114
        doing, 115
        done, 115
        to-do, 115
Process superstep
    categories of data, 421
    data science (see Data science)
    data vault, 422

event, 436–439, 507–510
location, 433–436, 499–507
objects, 430–433, 476–499
person, 427–430, 463–476
time, 424–427, 439–463
Producer, 132–133
Profit-and-loss statement data set, 38
Pydoop, 270
Python
    CentOS/RHEL, 23
    definition, 12, 23
    Ubuntu, 23
    Windows, 23
Python3
    Matplotlib, 25
    NumPy, 26
    Pandas, 24–25
    PyPI, 24
    Scikit-Learn, 26
    SymPy, 26
Python Package Index (PyPI), 24

R

R
    capabilities, 11
    dplyr, 11
    forecast, 11
    gbm, 12
    ggplot2, 12
    installation
        CentOS/RHEL, 27
        Ubuntu, 26
        Windows, 27
    lubridate, 11
    packages
        Data.Table, 28
        Ggplot2, 29

R *(cont.)*

    JSONLite, 29

    ReadR, 28

    Spark, 30

  randomForest, 12

  reshape2, 12

  RODBC, RSQLite, and RCassandra, 11

  RStudio, 27

  sqldf, 11

  stringr, 11

RADVIZ method, 757–758

Random forests, 655–659

RANdom SAmple Consensus (RANSAC)
    linear regression, 587–589

Rapid Information Factory ecosystem, 1

ReadR package, 28

Receiver operating characteristic (ROC)
    analysis curves, 577–579

Reference satellites, 423

Reinforcement learning (RL), 651–652

Relative location, 501

Report superstep

  appropriate visualization, 720

  billboards content, 728–729

  Clark Ltd, 735–741

  containers, 730–735

  data science, 718

  data science company scale, 723

  eliminate clutter, 721

  financials, 736–741

  Freytag's pyramid, 721–722

  graphics capacity, 723

  Hillman Ltd, 730–735

  Krennwallner AG, 727–729

  network routing diagram, 724–726

  understanding context, 719–720

  Vermeulen PLC, 723, 725–726

Report utilities, 112

Resilience, 74

Resource constraints, 74

Retrieve superstep, 147

  Clark Ltd, 259–261

  Hillman Ltd, 223–259

  Krennwallner AG, 209–221

  Vermeulen PLC, 203–208

Reusability, 75

Rising action, 722

RL, *see* Reinforcement learning (RL)

ROC analysis curves, *see* Receiver
    operating characteristic (ROC)
    analysis curves

Role-based access control (RBAC), 63, 696

RStudio

  CentOS/RHEL, 27

  Ubuntu, 27

  Windows, 27

S

Satellites, 5

  defined, 422

  reference, 423

  utilities, 107

Scala, 12, 20

Scalability, 75–76

Scatter graphs, 749–750

Scatter matrix, 754

Scikit-Learn, 26

Second, 444

Secure vault style, 696–699

Security

  access, 77

  physical, 76

  privacy, 76

Shipping containers

  adoption, 247–254

air freight volumetric (chargeable weight), 378

Assess-Shipping-Containers.py, 374–375

fit box in pallet, 381

fit product in box, 378

length value, 377

loading pallets, 384–385

1000 packing boxes, 376

packing material thickness, 379, 381

pallet-in-container option, 385–386

product in box and box on pallet options, 383

products, 375–376

road freight volumetric (chargeable weight), 379

size of pallet, 382

strength of boxes ratings, 380

Shipping information

    buyer, 224

    carrier, 223

    named place, 224

    port, 223

    seller, 223

    ship, 223

    shipping terms, 223

    terminal, 224

Shipping routes, best-fit international logistics

    Assess-Best-Fit-Logistics.py, 362

    Assess_RoutePointsCountry, 364

    Assess_RoutePointsPlaceName, 365

    Assess_RoutePointsPostCode, 364

    average latitude and longitude, 364–365

    country-to-country routes, 365

    country to post code routes, 367

    libraries, 362

    load data, 363

    map new locations

        countries, 369

        database mainteanance, 372–373

        hidden features, 372

        place-names, 371

        postcodes, 370

        shortest path between locations, 372

        stored in graph, 372

    parameters, 362

    post code to place-name routes, 368–369

    Vincenty's formulae, 366

Shop data set, 36

Simple linear regression, 582–587

Simple logistic regression, 590–591

Simple Network Time Protocol (SNTP), 447

Skewness, 166–168

Slowly Changing Dimensions (SCDs)

    SCD Type 1, only update, 57

    SCD Type 2

        Dr Jacob Roggeveen, location, 58–59

        effective date, 58

        flagging, 58

        versioning, 57

    SCD Type 3, transitions, 59–60

    SCD Type 4, fast-growing, 60

Space transformed visualization (STV), 755

Spark

    description, 6

    GraphX, 8

    IBM, 6

    MLlib machine learning library, 7

    SAP, Tableau, and Talend, 6

Spark (*cont.*)

    Spark Core, 7

    Spark MLlib, 7

    Spark SQL, 7

    Spark Streaming, 7

SQLite database, 261–264

    add_edge() link, 334

    Assess-DAG-Schedule-All.gml, 337–339

    Assess_IP_Country, 329, 334

    Assess_IP_GPS, 331–334

    Assess_IP_PlaceName, 329–330

    Assess_IP_PostCode, 330

    CompanyData, 328

    df.tosql() command, 327

    G=nx.Graph(), 327

    nodes, 334–337

    pandas data structure, 327

    remove nodes and edges, 338

    Retrieve-IP_2_SQLite_2_DAG.py, 326

    SQL DISTINCT command, 328

Standard deviation, 166

Subject matter experts, 142

Sun models

    basic knowledge, 56

    description, 534

    development, 55

    Dr Jacob Roggeveen record, 56

    facts, 60

    intra-sun model consolidation matrix

        dimensions and facts, 62

        Event, Date, and Time, 61

        Object, Person, Date, and Time, 62

        Person, Date, Time, and Location, 61

    PersonBornAtTime fact, 538

    person-to-event, 538

    person-to-location, 537

    person-to-object, 537

    person-to-time, 534–535

    SCDs (*see* Slowly Changing Dimensions (SCDs))

    template, 536

    to transform step, 538–542

    typical, 56

    version 2.0, 55

Supervised learning, 649–650

Support vector clustering (SVC), 629, 632–633

Support vector grid (SVG), 629

Support vector machine (SVM), 629–632

Support vector network (SVN), 629, 632

SymPy, 26

T

TensorFlow

    basic, 671–676

    basket analysis, 676

    commodities trading, 676

    definition, 670

    development, 670

    digits on container, 679–681

    forex trading, 676

    linear regression, 681–684

    one-arm bandits, 676–679

Testability

    controllability, 77

    definition, 77

    isolate ability, 77

    understandability, 78

Thaddeus Vincenty's formulae, 345

Time-Person-Object-Location-Event (T-P-O-L-E)

    design, 424

    event

hub, 436
links, 437
location hub, 438
object hub, 438
person hub, 438
satellites, 439
time hub, 437
location
event hub, 435
hub, 433
links, 433–434
object hub, 435
person hub, 435
satellites, 436
time hub, 434
object
event hub, 432
hub, 430
links, 430
location hub, 432
person hub, 432
satellites, 432–433
time hub, 431
person
event hub, 429
hub, 427–428
links, 428
location hub, 429
object hub, 429
satellites, 429
time hub, 428–429
time
event hubs, 426
hub, 424–425
links, 425
location hubs, 426
object hub, 426
person hubs, 425

satellites, 426
Time, process superstep
combining date and, 452
day, 442–443
defined, 439
fraction of a second, 444–445
hour, 443
Kiribati, Republic of, 454–455
local time, 446–447
minute, 443
month, 440–441
second, 444
special date formats
day of the week, 457
day of the year, 458–463
week dates, 457–458
start and end dates, 455–456
UTC, 447–452
year, 439–440
zones, 452–454
T-P-O-L-E, *see* Time-Person-Object-
Location-Event (T-P-O-L-E)
Traceability matrix, 82–83
Transform superstep
ANOVA, 613, 615
bagging data, 652–655
bivariate analysis, 581
clustering (*see* Clustering)
cross-validation test, 579–580
CV, 660
data categorization, 527
data mining (*see* Data mining)
data science (*see* Data science)
data warehouse, 542
decision trees, 624–629
dimension consolidation
data vault, 528–529
data warehouse, 530

Transform superstep (*cont.*)

    person, 532–534

    Python ecosystem, 528

    set up company, 529

    time, 530–532

    T-P-O-L-E high-level design, 528

    Transform-Gunnarsson_is_Born.py, 528

    working directory, 529

  hypothesis testing (*see* Hypothesis testing)

  linear regression, 581–589

  logistic regression, 590–601

  machine learning (*see* Machine learning)

  multivariate analysis, 581

  neural networks, 662–670

  NLP, 661–662

  overfitting and underfitting

    data-fitting issue, 568–570

    definition, 566–567

    polynomial features, 567–568

  pattern recognition, 646–649

  PCA (*see* Principal component analysis (PCA))

  precision-recall, 570–577

  random forests, 655–659

  source code, 527

  sun models (*see* Sun models)

  TensorFlow (*see* TensorFlow)

  univariate analysis, 580

Transform utilities

  dimensions, 107

  facts, 108

Tree augmented naive Bayes (TAN), 638

T-test, 562–563, 565

U

Ubuntu, 23–27

Univariate analysis, 580

Unsupervised learning, 651

Utilities

  algorithms, 85

  basic design, 88

  data processing (*see* Data processing utilities)

  description, 85

  GDPR, 85–87

  HIPAA, 87

  maintenance, 112

  processing, 114–115

  three-stage process, 87

  types, 88

V

Vagueness, 80

Vehicle identification number (VIN), 492

Vermeulen-Krennwallner-Hillman-Clark Group (VKHCG)

  Clark Ltd, 18

  customer data sets, 35

  Hillman Ltd, 18

  IP Addresses data sets, 32

  Krennwallner AG, 17–18

  Linux, 15

  logistics data sets (*see* Logistics data sets)

  processing ecosystem (*see* Processing ecosystem)

  sample data, 30–31

  Vermeulen PLC, 16–17

  Windows, 15

Vermeulen PLC, 173, 200–203
  Country_Codes, loading, 184–186
  diagram for job scheduling, 206–208
  graph theory, 307–309
  IP_DATA_ALL, loading, 174–177
  IP_DATA_CORE, loading, 181–184
  IP_DATA_C_VKHCG, loading, 178–181
  job scheduling (*see* Job scheduling, DAG)
  network routing (*see* Network routing
    diagram)
  Organize superstep, 703–706
  Report superstep, 723–726
Vincenty's formula, 502
VKHCG, *see* Vermeulen-Krennwallner-
    Hillman-Clark Group (VKHCG)

W, X

Warehouses
  achievements, 359
  Assess_GB_Warehouse_Address.csv,
    359
  Assess-Warehouse-Address.py,
    357–359
  data set, 36, 359
  global, 359–362
  GPS locations, 356
  locations, 358
  online services
    provider, 356
  parameter, address locator
    services, 358
Warning watcher, 126
Weak words, 78
Week dates, 457–458
Windows, 15, 23, 27, 51

Y, Z

Year, 439–440
Yoke solution
  consumer, 133
  DAG scheduling, 134
  Kafka, 132
  Master-Yoke.py, 136–138
  producer, 132–133
  Run-Yoke.py, 134–136
  Slave-Yoke.py, 138–140

Get the eBook for only $5!

Why limit yourself?

With most of our titles available in both PDF and ePUB format, you can access your content wherever and however you wish—on your PC, phone, tablet, or reader.

Since you've purchased this print book, we are happy to offer you the eBook for just $5.

To learn more, go to http://www.apress.com/companion or contact support@apress.com.

Apress®

Printed in the United States
By Bookmasters